Biology Today

VOLUME 2

SANDRA S. GOTTFRIED

Assistant Professor of Biology and Education
Departments of Biology and Educational Studies
University of Missouri—St. Louis

St. Louis Baltimore Boston Chicago London Philadelphia Sydney Toronto

Dedicated to Publishing Excellence

Editor-in-Chief: **James M. Smith**
Editor: **Robert J. Callanan**
Developmental Editor: **Kathleen Scogna**
Editorial Assistant: **Jennifer Collins**
Photo Researchers: **Holly Leicht,**
         Jennifer Collins
Project Manager: **Carol Sullivan Wiseman**
Senior Production Editor: **Linda McKinley**
Book Designer: **Susan Lane**
Cover Photographs: Landscape © Animals Animals/Richard Kolar
         Plongee diving © Animals Animals/Henry Ausloos
         Grizzlies © Animals Animals/Johnny Johnson
         Peacock © Animals Animals/Miriam Austerman

Art Coordinator: **E. Rhone Rudder**
Computer Art: **Pagecrafters, Inc.**
Artists: **Molly Babich, Scott Bodell,**
      Raychel Ciemma, Barbara Cousins,
      Carlyn Iverson, Christy Krames,
      Bill Ober, Lili Robins,
      Kevin Somerville, Kate Sweeney,
      Nadine B. Sokol

K.C.F.E. Library	
Class No.	574
Acc. No.	00051011
Date Rec	16-5-94
Order No.	F2594

Copyright © 1993 by Mosby–Year Book, Inc.

PRINTED ON RECYCLED PAPER

All rights reserved. No part of this publication may be reproduced, stored in a retrieval system, or transmitted, in any form or by any means, electronic, mechanical, photocopying, recording, or otherwise, without written permission of the publisher.

Permission to photocopy or reproduce solely for internal or personal use is permitted for libraries or other users registered with the Copyright Clearance Center, provided that the base fee of $4.00 per chapter plus $.10 per page is paid directly to the Copyright Clearance Center, 27 Congress Street, Salem, MA 01970. This consent does not extend to other kinds of copying, such as copying for general distribution, for advertising or promotional purposes, for creating new collected works, or for resale.

Printed in the United States of America

Mosby–Year Book, Inc.
11830 Westline Industrial Drive
St. Louis, MO 63146

International Standard Book Number: 0-8151-3844-X

93 94 95 96 97 GW/CD/VHP 9 8 7 6 5 4 3 2

Preface

Is human papilloma virus #16 a threat to your health? The answer is on p. 486. Why are anabolic steroids a health risk? Find out on p. 315. How is the AIDS epidemic affecting college students today? If you want to know, see pp. 344-345. If these questions intrigue you, then reading *Biology Today* may be more than a "textbook" experience. These topics and others, such as genetic engineering, the global population crisis, rain forest destruction, and the ozone hole, link the concepts of biology today to the students of today in a context of becoming an informed citizen of the world. Providing students with information that is vital to their health, well-being, and general education, *Biology Today* explores traditional biological content within societal, technological, and personal contexts.

I wrote *Biology Today* for the introductory biology student at a community college or university. Of my 20-plus year career in biology education, I spent 5 years teaching at a community college and have spent the last 4 years teaching introductory students at a university. I feel I am well-aware of my audience and their needs. My training in scientific research, biology, and education and experience as a science writer and teacher enabled me to write this book in a way that truly embodies the philosophy of educating introductory students to become better-informed citizens at personal, local, national, and international levels. To accomplish this task, I first assumed that my audience of students had no prior knowledge of biology. Therefore I used a writing style that is easy and direct, developing concepts within a context of rich description to provide a scaffold for learning and to make concepts more concrete. Today's introductory students need explanation, not simply a list of vocabulary terms.

In addition, *Biology Today* places traditional biology in a context both relevant and important to the student. This is done in two ways. First, everyday situations or concerns that students have encountered in the media are used as examples wherever possible. Second, the text has a strong human focus, with the human body serving as the focal point for the physiology section (Part Four, Human Biology: The Structure and Function of the Body). Students are interested in how their bodies function. They also need accurate information about their own physiology to make informed choices about activities that will affect their bodies, such as making decisions regarding birth control and choosing what foods to eat.

In summary, *Biology Today* portrays biology as a process rather than a product, helping students better understand how scientists do their work and develop theories about the natural world. Introductory students need a broad understanding and an appreciation of how the biological world affects their lives. *Biology Today* attempts to fill this need.

Organization

Biology Today is organized into eight parts. Chapters were designed to stand alone so that the professor may use the text in the order in which it was written or in an alternative order. The chapter opening pedagogy provides "Highlights" for each chapter that serve as a "menu" to the chapter content. These Highlights will help professors determine the chapter order that best suits the organization of their course. Page references to concepts previously defined and explained are sprinkled throughout the chapters.

In general, the book is organized in a traditional fashion. Basic biological concepts are presented in the first chapter with chemistry, cell biology, metabolism, physiology, and genetics composing the first five parts of the text. The last three parts present evolution, diversity, and ecology.

Part One
An introduction to biology and chemistry

This part begins with Chapter 1 (The Themes of Biology Today), which provides an overview of the scientific method and an introduction to biological themes such as unity and diversity that are emphasized throughout the book. In addition, other introductory concepts, such as the kingdoms of life, are introduced and explained. Chapter 2 (The Chemistry of Life) presents basic chemistry, with an emphasis on biologically relevant chemical principles.

Part Two
Cells: The basic unit of life

Part Two contains information about all aspects of cells: their basic structure and function (Chapter 3, Cell Structure and Function), the cell membrane and its function (Chapter 4, Cell Membranes), and the processes of mitosis and meiosis (Chapter 5, Cell Division).

Part Three
How living things transform energy

Cellular metabolism is the focus of this unit. A chapter on chemical reactions introduces this unit (Chapter 6, The Flow of Energy within Organisms) and prepares the students to study cellular respiration (Chapter 7, Cellular Respiration: How Cells Release Stored Energy from Food Molecules) and photosynthesis (Chapter 8, Photosynthesis: How Plants Capture and Store Energy from the Sun).

Part Four
Human biology: The structure and function of the body

This unit covers aspects of the structure and function of the human body. The chapters focus on homeostasis and how structure is related to function. The numerous boxed essays in this unit were designed to interest students. Boxes include such relevant issues "How to Avoid Heart Disease" in Chapter 12 (Circulation), "AIDS on the College Campus" in Chapter 13 (Defense Against Disease), and "Solving the Mystery of Alzheimer's Disease" in Chapter 16 (The Nervous System).

Part Five
How humans reproduce and pass on biological information

This unit covers human reproduction and genetics, beginning with Chapter 20 (Sex and Reproduction) and moving on to Chapter 21 (Human Development Before Birth). Mendelian genetics (Chapter 22, Patterns of Inheritance), human genetics (Chapter 23, Human Genetics), and molecular genetics (Chapter 24, The Molecular Basis of Inheritance) follow. Again, emphasis is placed on student-oriented information. In this unit, students learn about how they inherited their particular blood type, how color-blindness is passed along in the genes, and how genetic engineering is helping those with diabetes.

Part Six
Evolution: How living things change over time

Current information about evolution is presented in this unit. Chapter 25 (The Scientific Evidence for Evolution) presents the evidence contained in fossils, the geological record, and the molecular record that support the theory of evolution. Chapter 26 (The Evolution of the Five Kingdoms of Life) follows scientists' thinking as they attempt to construct scenarios of how life began. A separate, complete chapter on human evolution, including vertebrate evolution (Chapter 27), is a hallmark of this section.

Part Seven
The diversity and unity of living things

Structure, function, and evolutionary position of the five kingdoms of life are covered in this unit. The first two chapters describe the Kingdoms Monera, Protista, and Fungi (Chapter 28, viruses, bacteria, and genetic engineering and Chapter 29, protists and fungi). The reproductive patterns of plants are described and followed by their patterns of structure and function (Chapter 30, Plants: Reproductive Patterns and Diversity and Chapter 31, Plants: Patterns of Structure and Function). Chapters on invertebrates (Chapter 32, Invertebrate Animals: Patterns of Structure, Function and Reproduction) and vertebrates (Chapter 33, Vertebrate Animals: Patterns of Structure, Function, and Reproduction) describe patterns of body symmetry, body cavity structure, and embryological development, linking these concepts to the evolutionary history of these animal groups. Characteristics of animal phyla and vertebrate classes are discussed with a comparative approach.

Part Eight
How living things interact with each other and with their environment

These seven chapters discuss the interaction of organisms with the environment. The unit begins with two chapters on animal behavior: Chapter 34 (Innate Behavior and Learning in Animals) and Chapter 35 (Social Behavior in Animals). Chapter 36 (Population Ecology), Chapter 37 (Interactions within Communities of Organisms), and Chapter 38 (Ecosystems) present an overview of ecology. Chapter 39 (Biomes and Life Zones of the World) describes the large climatic areas of the world along with their distinctive plant and animal populations. This chapter includes a discussion of fresh water and saltwater environments. The unit ends with the ecological problems facing the world today in Chapter 40 (The Biosphere: Today and Tomorrow).

Features

Biology Today boasts several pedagogical features that make the book easy to use. Students should find these features a great help when reading their assignments for class and reviewing for exams and quizzes.

Each chapter opens with a short "vignette" that is designed to spark student interest. Sometimes these vignettes explain something that may already be familiar to the student in a context that relates to the chapter material. For example, Chapter 19 (Hormones) opens with a vignette about the dangers of anabolic steriods. Other vignettes focus on unusual phenomena that nevertheless have a biological explanation. Chapter 14 (Excretion) opens with a discussion about why some species of turtles "cry" and how this crying is related to salt and water balance.

In addition to the vignettes, the chapter openers contain a list of chapter "Highlights" that detail in short statements the chapter content. These highlights will help students condense the chapter material and provide a quick review. An "outline" of the chapter accompanies the highlights and also assists students with organizing and segmenting the chapter concepts.

Within each chapter are "Concept summaries" that provide spot summaries at key points in the chapter. They are short (usually no more than three sentences long) and effectively review the preceding material. Students will find them invaluable when reviewing for examinations.

"Boxed essays" are found throughout each chapter. The four themes of the boxed essays were specially designed with our students in mind. The "Biology and You" boxes link biology to students' everyday lives, highlighting the personal relevance of many biological concepts. "Biology, Technology, and Society" boxes highlight the links among biology, the fast-paced technology of today's world, and societal issues. "Biology in Focus" describes the work of the scientist in detail, often answering the question, "How do we know that?" And lastly, "Biology and Evolution" boxes emphasize the change of organisms over time and their evolutionary relationships to one another.

Each chapter closes with a "Summary" that lists all the key concepts in the chapter, a list of "Key terms" with the page number on which the term appears, a list of "Review questions" that tests student comprehension of the chapter content, and a few "Thought questions" that are designed to initiate class discussion and debate. Each chapter also has a short list of "Further Readings," chosen with the introductory student in mind. In addition to the usual *Scientific American* and *Science* articles, there are articles from *Smithsonian*, *National Wildlife*, and *National Geographic*.

Two appendixes provide additional support for the student. Appendix A provides answers to the review questions. Appendix B is a table of classification of the five kingdoms of life. Because of the human focus of this book, animals are classified through class, and the vertebrates are classified through order.

The glossary provides the pronunciation, definition, and derivation for each key term in the book and also gives the page reference for each term. This glossary is complete and easy to use and will help students immeasurably as they make their way through the text.

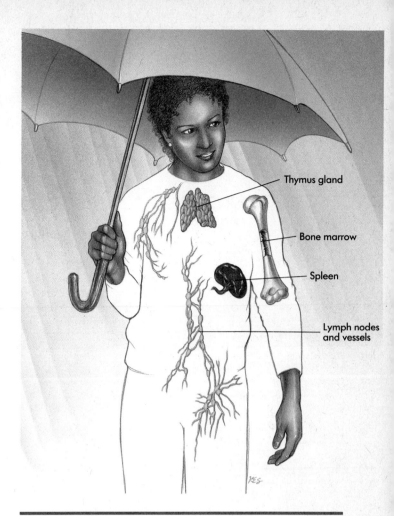

The art program

No biology text today can be a pedagogical success without a top-notch art program. Given the visual orientation of today's students—the computer graphics that they are bombarded with on television and the special effects that enliven today's feature films—it makes sense that a biology textbook use visual impact to teach concepts. *Biology Today* accomplishes this task by applying some basic principles to the art program:

- The illustrations and photographs should make abstract concepts more concrete to enhance student understanding.
- The content of text and art should match—there should not be concepts or structures presented in the art that are not discussed in the narrative.
- There should be an emphasis on simple, clear illustrations that teach process using layout and sometimes, for very complex processes, numbered steps.
- Legends should be a helpful guide to the illustration and should not introduce material that is not in the text.
- Art should be visually appealing so that students are prompted to "take a look."

We have followed these principles to the letter in *Biology Today*, with the result that the art program represents another important focus in the pedagogical program and is an element in itself.

Ancillaries

Biology Today has a full range of ancillary materials that meets the needs of both students and instructors. Carefully designed and executed, the ancillaries supplement and support *Biology Today* and enhance its student-oriented approach.

Instructor's resource manual

For the instructor, *Teaching Biology Today* is an invaluable teaching resource. Written by Dr. Ann Lumsden of Florida State University, each chapter contains a complete synopsis, suggested topics for class discussion, lecture outlines, teaching ideas, and a listing of applicable audiovisual resources. A text bank of over 2000 questions is also included in the manual. In addition, transparency masters of key illustrations in *Biology Today* are included in *Teaching Biology Today* to enhance visual learning.

Computerized test bank

The test questions included in *Teaching Biology Today* are also available as *Computest,* a computerized test generation system. *Computest* is available in MacIntosh and IBM formats. It has many features that make it easy for the instructor to design tests and quizzes. The instructor can browse and select questions for inclusion on an examination using several different criteria, including question type and level of difficulty.

Transparency acetates

Also available for the instructor are 150 full-color transparency acetates that reproduce the important illustrations in *Biology Today*. Labeling is clear, bold, and large enough for even students seated in the back of a large room to see.

Videodisc

Mechanisms of Life: Stability and Change Videodisc is a state-of-the art instructional medium that combines versatility, compactness, and ease of use. This videodisc contains thousands of still photographs, selected artwork, and film clips on biological processes to provide visual reinforcement in the classroom. Other features include the ability to search and display images or animated sequences and extensive use of full motion along with still images.

Laboratory manual

Exploring Biology Today is a unique laboratory manual written by Ann Wilke, Director of Undergraduate Laboratories at the University of Missouri—St. Louis. This manual responds to the needs of introductory students by guiding both formal and concrete thinkers through a laboratory experience that teaches them biology and how to become more critical thinkers. Based on the learning cycle, each of the 30 2-hour laboratory exercises in *Exploring Biology Today* leads students through their investigation of the biological world using three steps: exploration, in which students develop new understandings by means of hands-on experiences; concept introduction, in which the concepts students explored are articulated and labeled; and concept application, in which the students apply their knowledge and understanding to new situations. *Exploring Biology Today* is accompanied by an instructor's manual that offers support for both the laboratory set-up and the learning cycle approach.

Study guide

The student study guide, *Learning Biology Today,* by Dr. David Cotter of Georgia College, enhances and explores the concepts in the text by building from fact-based knowledge to the use of higher-order thinking skills. In each chapter, students are first tested for their mastery of basic concepts. Subsequent sections then test for deeper understanding of the chapter concepts, with the last section stressing application of these concepts to new situations. This "building block" approach facilitates the critical thinking necessary to solve problems using knowledge acquired from the text, without short-changing the straightforward learning of new concepts and ideas.

Acknowledgments

The highly professional and skilled staff at Mosby–Year Book, Inc. have helped make *Biology Today* an outstanding learning tool. Linda McKinley, production editor, made an extremely valuable contribution with her flawless copy editing and attention to detail. I also thank her for her talented page make-up. Susan Lane, book designer, provided a lovely design that is both simple and dynamic. I extend many thanks to Jennifer Collins and Holly Leicht for their persistent efforts and excellent work in acquiring the outstanding photographs used throughout the book. Very special thanks goes to the developmental editor, Kathleen Scogna, whose excellent editing and management skills played vital roles in *Biology Today*. I extend a sincere personal thanks to Kathleen for her exemplary work. My deepest gratitude goes to Bob Callanan, acquisitions editor, for his constant focus on excellence and unceasing support of innovation to make the entire *Biology Today* program a pedagogically sound and appropriate learning tool for today's students.

The reviewers of *Biology Today* provided countless suggestions that enhanced the quality of this textbook and ensured its accuracy. Their names are listed and to each one I extend a personal "Thank you." I would like to give special thanks to my colleagues at the University of Missouri—St. Louis who were always available to review manuscript pages and provide instant feedback: Ed Joern, Randy Nolan, Carl Thurman (now at the University of Northern Iowa), and Ann Wilke.

- William Barnes, *Clarion University*
- Kathy Burt-Utley, *University of New Orleans*
- James Conkey, *Truckee Meadows Community College*
- Neil Crenshaw, *Indian River Community College*
- Don Emmeluth, *Fulton-Montgomery Community College*
- Robert Grammar, *Belmont College*
- Ronald Hoham, *Colgate University*
- Ed Joern, *University of Missouri—St. Louis*
- Alan Karpoff, *University of Louisville*
- Miriam Kitrell, *Kingsborough Community College*
- John Knesel, *Northeast Louisiana University*
- Kenneth Mace, *Stephen F. Austin State University*
- Virginia Michelich, *Dekalb College*
- Randy Nolan, *University of Missouri—St. Louis*
- Michael Postula, *Parkland College*
- June Ramsey, *Pensacola Junior College*
- David Rayle, *San Diego State University*
- Michael Smiles, *State University of New York—Farmingdale*
- Carl Thurman, *University of Northern Iowa*
- Ann Wilke, *University of Missouri—St. Louis*
- Nancy Yurko, *Prince Georges Community College*

Personal thanks go to George Johnson of Washington University in St. Louis for contributing boxed essays to *Biology Today*. Also, a very special personal thank you goes to Arthur Tarrow of Melbourne, Florida, for publishing my first book, which was a step to further publications and, of course, to *Biology Today*.

This book is dedicated to my parents, Evo and Florence Sebastianelli of Farmington, Connecticut, and to my son, Marc Gottfried of St. Louis, Missouri.

SANDRA S. GOTTFRIED

Contents in Brief

VOLUME 1 PART ONE AN INTRODUCTION TO BIOLOGY AND CHEMISTRY

1 **The themes of biology today** 2

2 **The chemistry of life** 16

VOLUME 1 PART TWO CELLS THE BASIC UNIT OF LIFE

3 **Cell structure and function** 44

4 **Cell membranes** 67

5 **Cell division** 80

VOLUME 1 PART THREE HOW LIVING THINGS TRANSFORM ENERGY

6 **The flow of energy within organisms** 100

7 **Cellular respiration** 114
How cells release stored energy from food molecules

8 **Photosynthesis** 130
How plants capture and store energy from the sun

VOLUME 1 PART FOUR HUMAN BIOLOGY
THE STRUCTURE AND FUNCTION OF THE BODY

9 **Levels of organization in the human body** 146

10 **Digestion** 165

11 **Respiration** 184

12 **Circulation** 198

13 **Defense against disease** 216

14 **Excretion** 232

15 **Nerve cells and how they transmit information** 249

16 **The nervous system** 262

17 **The senses** 280

18 **Protection, support, and movement** 294

19 **Hormones** 314

VOLUME 1 **PART FIVE** HOW HUMANS REPRODUCE AND PASS ON BIOLOGICAL INFORMATION

20 **Sex and reproduction** 334
21 **Human development before birth** 354
22 **Patterns of inheritance** 372
23 **Human genetics** 388
24 **The molecular basis of inheritance** 404

VOLUME 2 **PART SIX** EVOLUTION HOW LIVING THINGS CHANGE OVER TIME

25 **The scientific evidence for evolution** 426
26 **The evolution of the five kingdoms of life** 444
27 **Human evolution** 463

VOLUME 2 **PART SEVEN** THE DIVERSITY AND UNITY OF LIVING THINGS

28 **Viruses, bacteria, and genetic engineering** 486
29 **Protists and fungi** 502
30 **Plants** 526
Reproductive patterns and diversity
31 **Plants** 544
Patterns of structure and function
32 **Invertebrate animals** 560
Patterns of structure, function, and reproduction
33 **Vertebrate animals** 580
Patterns of structure, function, and reproduction

VOLUME 2 **PART EIGHT** HOW LIVING THINGS INTERACT WITH EACH OTHER AND WITH THEIR ENVIRONMENT

34 **Innate behavior and learning in animals** 598
35 **Social behavior in animals** 610
36 **Population ecology** 624
37 **Interactions within communities of organisms** 638
38 **Ecosystems** 656
39 **Biomes and life zones of the world** 671
40 **The biosphere** 694
Today and tomorrow

Contents

PART ONE

AN INTRODUCTION TO BIOLOGY AND CHEMISTRY

 1 The themes of biology today 2

Scientific process: The unifying theme among all sciences 3
    The pathway of thinking in the scientific method 3
How biologists do their work 6
    Summarizing the steps in the scientific method 7
    Theory building 7
The unifying themes of biology today 7
    Living things display both diversity and unity 7
    Living things are composed of cells and are hierarchically organized 7
    Living things interact with each other and with their environments 8
    Living things transform energy and maintain a steady internal environment 10
    Living things exhibit forms that fit their functions 11
    Living things reproduce and pass on biological information to their offspring 11
    Living things change over time, or evolve 11
Classification: A reflection of evolutionary history 12
Biology today and you 14
Summary 14

 2 The chemistry of life 16

Atoms 17
Molecules 18
    Nature of the chemical bond 19
    Ions and ionic bonds 19
    Covalent bonds 20
The cradle of life: Water 22
    Water is a powerful solvent 24
    Water organizes nonpolar molecules 24
    Water ionizes 25
Inorganic versus organic chemistry 26
The chemical building blocks of life 28
    Carbohydrates 30
    Fats and lipids 32
    Proteins 34
    Nucleic acids 39
The use of isotopes in medicine 40
Summary 40

VOLUME 1 CHAPTERS 1-24

PART TWO

CELLS
THE BASIC UNIT OF LIFE

 3 Cell structure and function 44

The cell theory 45
Why aren't cells larger? 46
Two kinds of cells 47
Eukaryotic cells: An overview 49
   The plasma membrane 49
   The cytoplasm and cytoskeleton 50
   Membranous organelles 53
How cells crawl 54
   Bacteria-like organelles 58
   Cilia and flagella 60
   Cell walls 61
Prokaryotic cells 61
   Cell walls 62
   Simple interior organization 63
A comparison of prokaryotic and eukaryotic cells 65
Summary 65

 4 Cell membranes 67

The foundation of membranes 68
Membrane channels and cystic fibrosis 70
Membrane proteins 70
How substances move across cell membranes 72
   Movement that does not require energy 72
   Movement that does require energy 75
Summary 78

 5 Cell division 80

Chromosomes 81
Cell division, growth, and reproduction 83
Cancer and mitosis 84
Mitosis 86
   Prophase 86
   Metaphase 89
   Anaphase 90
   Telophase 90
Cytokinesis 91
   Cytokinesis in animal cells 91
   Cytokinesis in plant cells 91
Meiosis 92
   The stages of meiosis 93
The importance of meiotic recombination 97
Summary 97

PART THREE

HOW LIVING THINGS TRANSFORM ENERGY

6 The flow of energy within organisms 100

Starting chemical reactions 101
Energy flow and change in living systems 103
Regulating chemical reactions 105
   Enzymes: Biological catalysts 105
Nonfattening sweets 107
Storing and transferring energy 110
   ATP: The energy currency of living things 110
   Carrying and transferring energy using ATP 111
Summary 112

7 Cellular respiration 114
How cells release stored energy from food molecules

Using chemical energy to drive metabolism 115
How cells make ATP: Variations on a theme 116
How oxygen-using organisms release ATP from food molecules 116
Carbohydrates and winter depression 118
   Oxidation-reduction 119
   Electron carriers 119
   A more detailed look at cellular respiration 120
How organisms release ATP from food molecules without using oxygen 127
Summary 128

8 Photosynthesis 130
How plants capture and store energy from the sun

Producers and consumers of food 131
The energy in sunlight 132
Capturing light energy in chemical bonds 132
An overview of photosynthesis 134
C_4 and CAM photosynthesis 136
Light-dependent reactions: Making ATP and NADPH 136
   How photosystems capture light 136
Light-independent reactions: Making glucose 141
Relationships: Photosynthesis and cellular respiration 143
Summary 143

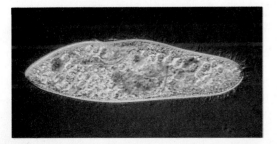

xiii

VOLUME 1 CHAPTERS 1-24

PART FOUR

HUMAN BIOLOGY
THE STRUCTURE AND FUNCTION OF THE BODY

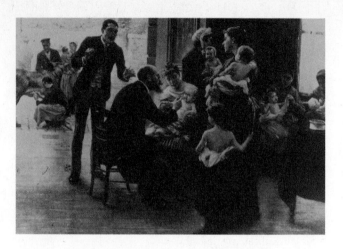

 Levels of organization in the human body 146

How the human body is organized 147
Tissues 149
    Epithelial tissue 149
    Connective tissue 152
    Muscle tissue 158
    Nervous tissue 159
Organs 160
Organ systems 160
The organism: Coordinating it all 162
Summary 163

 Digestion 165

The nutrition-digestion connection 166
Where it all begins: The mouth 170
The journey of food to the stomach 172
Preliminary digestion: The stomach 174
Terminal digestion and absorption: The small intestine 175
    Accessory organs that help digestion 175
Ulcers: What they are and how to prevent them 176
    Digestion 176
    Absorption 178
Concentration of solids: The large intestine 179
Diet and nutrition 180
Summary 182

 Respiration 184

Respiration 185
The pathway of air into and out of the body 186
    Nasal cavities 186
    The pharynx 186
    The larynx 186
    The trachea 186
    The bronchi and its branches 188
    The alveoli: Where gas exchange takes place 188
How to "catch" cancer 189
The mechanics of breathing 190
    Inspiration 190
    Expiration 191
    Deep breathing 191
    Lung volumes 192
Gas transport and exchange 192
    External respiration 192
    Internal respiration 194
Choking: A common respiratory emergency 195
Chronic obstructive pulmonary disease 196
Summary 196

 Circulation 198

Functions of the circulatory system 199
    Nutrient and waste transport 199
    Oxygen and carbon dioxide transport 199
    Temperature maintenance 199
    Hormone circulation 201
The heart and blood vessels 201
    Arteries and arterioles 201
    Capillaries 202
    Veins and venules 203
    The heart: A double pump 204
How to avoid heart disease 206
The blood 208
    Blood plasma 208
    Types of blood cells 208
The lymphatic system 211
Diseases of the heart and blood vessels 213
Summary 214

 Defense against disease 216

Nonspecific versus specific resistance to infection 217
Nonspecific resistance 217
Specific resistance 218
Discovery of specific resistance: The immune response 218
The cells of the immune system 221
The immune response: How it works 221
    Sounding the alarm 221
    The two branches of the immune response 222
Helping the immune system fight cancer 225
How do immune receptors recognize antigens? 226
Immunization: Protection against infection 227
Defeat of the immune system: AIDS 227
Allergy 229
Summary 230

VOLUME 1 CHAPTERS 1-24

14 Excretion 232
What substances does the body excrete? 233
The organs of excretion 234
An overview of how the kidney works 234
A closer look at how the kidney works 235
The anatomy of the kidney 235
    The workhorse of the kidney: The nephron 235
Testing urine for drugs 242
The urinary system 244
The kidney and homeostasis 244
Problems with kidney function 245
    Kidney stones 245
    Renal failure 246
    Treatments for renal failure 247
Summary 247

15 Nerve cells and how they transmit information 249
The communication systems of your body 250
The nerve cell, or neuron 250
The nerve impulse 252
    The neuron at rest: The resting potential 253
    Conducting an impulse: The action potential 254
    Transmission of the nerve impulse: A propagation of the action potential 255
    Speedy neurons: Saltatory conduction 256
Transmitting information between cells 257
    Neuron-to-muscle cell connections 259
    Neuron-to-neuron connections 259
Summary 260

16 The nervous system 262
The organization of the nervous system 263
The central nervous system 265
    The brain 265
Solving the mystery of Alzheimer's disease 268
    The spinal cord 271
Memories 273
The peripheral nervous system 273
    Sensory pathways 273
    Integration 275
    Motor pathways 275
Summary 278

17 The senses 280
The nature of sensory communication 281
Sensing the body's internal environment 282
Sensing the body's position in space 282
Sensing the external environment 283
    The general senses 283
    The special senses 284
Summary 292

18 Protection, support, and movement 294
How skin, bones, and muscles work together 295
Skin 295
Bones 296
    The skeletal system 298
Muscles 305
    The muscular system 305
Can special shoes make you run faster? 308
Summary 312

19 Hormones 314
Endocrine glands and their hormones 315
    The pituitary gland 318
Living with diabetes 321
    The thyroid gland 322
    The parathyroid glands 325
    The adrenal glands 325
    The pancreas 327
    The pineal gland 328
    The thymus gland 328
    The ovaries and testes 328
Nonendocrine hormones 328
Summary 330

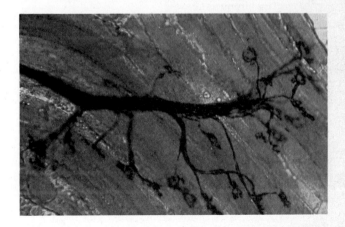

VOLUME 1 CHAPTERS 1-24

PART FIVE

HOW HUMANS REPRODUCE AND PASS ON BIOLOGICAL INFORMATION

 Sex and reproduction 334

Reproduction 335
The male reproductive system 335
  The production of sperm: The testes 335
  Maturation and storage: The epididymis and vas deferens 338
  Nourishment of the sperm: The accessory glands 338
  The penis 338
The female reproductive system 339
  The production of eggs: The ovaries 339
  Passage to the uterus: The uterine tubes 340
  The site of prenatal development: The uterus 341
  The reproductive cycle 341
  The vagina 343
  The external genitals 343
AIDS on the college campus 344-345
  The mammary glands 346
The sexual response 346
Contraception and birth control 347
  Abstinence 347
  Sperm blockage 350
  Sperm destruction 350
  Prevention of egg maturation 350
  Surgical intervention 351
Summary 352

 Human development before birth 354

Development 355
Fertilization 355
The first and second weeks of development: The pre-embryo 356
Early development of the extraembryonic membranes 360
Development from the third to eighth week: The embryo 361
  The third week 361
Fetal alcohol syndrome 362
  The fourth week 365
  The fifth through eighth week 366
Development from the ninth to thirty-eighth week: The fetus 366
  The third month 366
  The fourth through sixth month 367
  The seventh through ninth month 368
Birth 368
Physiological adjustments of the newborn 369
Summary 370

 Patterns of inheritance 372

Inheritance and variation within species 373
  Historical views of inheritance 373
  The birth of the study of inheritance 374
Gregor Mendel's experiments to determine inheritance patterns 374
  Conclusions Mendel drew from his experiments 377
  Analyzing Mendel's experiments 378
  How Mendel tested his conclusions 379
  Further questions Mendel asked 380
  Analyzing the results of Mendel's dihybrid crosses 381
The connection between Mendel's factors and chromosomes 383
Sex linkage 383
Solving the mysteries of inheritance 385
Summary 385

 Human genetics 388

Studying inheritance patterns in humans using karyotypes 389
  The inheritance of abnormal numbers of autosomes 389
  The inheritance of abnormal numbers of sex chromosomes 391
  Changes in chromosome structure 392
Curing genetic disorders by gene transfer 393
  Gene mutations 394
Studying inheritance patterns in humans using pedigrees 395
Dominant and recessive genetic disorders 398
Incomplete dominance and codominance 399
Multiple alleles 400
Genetic counseling 400
Using blood typing to test for paternity 401
Summary 402

VOLUME 2 CHAPTERS 25-40

PART SIX

EVOLUTION
HOW LIVING THINGS CHANGE OVER TIME

25 **The scientific evidence for evolution** 426

The development of Darwin's theory of evolution 427
   Darwin's observations 427
   Factors that influenced Darwin's thinking 429
   <u>Natural selection: A mechanism of evolution 432</u>
     An example of natural selection at work:
       Darwin's finches 432
The publication of Darwin's theory 433
Testing the theory 434
   The fossil record 434
   The age of the Earth: Rock and fossil dating 436
   Comparative anatomy 438
   Comparative embryology 440
   Molecular biology 440
Summary 442

26 **The evolution of the five kingdoms of life** 444

Theories about the origin of organic molecules 445
Theories about the origin of cells 447
The history of life 448
   The Archean Era: Oxygen-producing cells appear 448
   The Proterozoic Era: The first eukaryotes appear 452
   The Paleozoic Era: The occupation of the land 452
   Mass extinctions 455
   The Mesozoic Era: The age of reptiles 455
Was Protoavis the first bird? 457
   Changes in the Earth: The impact on evolution 458
   The Cenozoic Era: The age of mammals 458
Summary 461

24 **The molecular basis of inheritance** 404

The location of the hereditary material 405
The chemical nature of the hereditary material:
   Nucleic acids 406
   The nucleic acids: DNA and RNA 406
   A major scientific breakthrough: The structure of DNA 408
How DNA replicates 410
The human genome project 411
Genes: The units of hereditary information 412
Gene expression: How DNA directs the synthesis of polypeptides 413
   An overview 413
   Transcribing the DNA message to RNA 413
The Hershey-Chase experiment 414
   Translating the transcribed DNA message into a polypeptide 416
Differences in prokaryotic and eukaryotic genes 419
Differences in prokaryotic and eukaryotic gene expression 420
   Prokaryotic gene expression 420
   Eukaryotic gene expression 422
Gene therapy: Changing genes to cure disease 422
Summary 423

27 **Human evolution** 463

Evolution of the vertebrates 464
   Evolution of the fishes 465
   Evolution of the amphibians, reptiles, and birds 466
   Evolution of the mammals 468
All about Eve, or how scientists are attempting to construct the human family tree 469
Evolution of the primates 471
   The prosimians 473
   The anthropoids 474
Evolution of the anthropoids 476
Evolution of the hominids 477
   The first hominids: *Australopithecus* 477
   The first humans: *Homo habilis* 480
   Human evolution continues: *Homo erectus* 480
   Modern humans: *Homo sapiens* 480
Summary 482

VOLUME 2 CHAPTERS 25-40

PART SEVEN
THE DIVERSITY AND UNITY OF LIVING THINGS

28 Viruses, bacteria, and genetic engineering 486

Viruses 487
  The discovery of viruses 487
  Viral replication 489
  Viruses and cancer 490
  The classification of viruses 490
Bacteria 491
  Bacterial reproduction 491
  Bacterial diversity and classification 492
  Bacteria as disease producers 493
Genetic engineering 494
  Natural gene transfer among bacteria 494
The new agriculture 496
  Human-engineered gene transfer among bacteria 496
  Applications of recombinant DNA technology 498
  Medically important products 498
  Gene therapy 499
  Agriculture 499
Summary 500

29 Protists and fungi 502

Protists 503
Animal-like protists: protozoans 504
Malaria today 508
  Plant-like protists: Algae 509
  Fungus-like protists 514
Fungi 516
  Zygote-forming fungi 518
  Sac fungi 519
  Club fungi 520
  Imperfect fungi 522
  Lichens 523
Summary 524

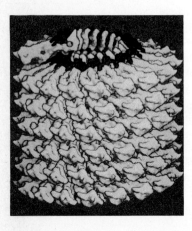

30 Plants 526
Reproductive patterns and diversity

Characteristics of plants 527
The general pattern of reproduction in plants 527
Patterns of reproduction in nonvascular plants 529
  The bryophytes 529
Patterns of reproduction in vascular plants 531
  Seedless vascular plants 531
  Vascular plants with naked seeds 533
  Vascular plants with protected seeds 534
  Mechanisms of pollination 535
  Fruits and their significance in sexual reproduction 536
Finding new food plants 537
  Seed formation and germination 538
  Types of vegetative propagation in plants 540
Regulating plant growth: Plant hormones 541
Summary 542

xviii   CONTENTS

VOLUME 2 CHAPTERS 25-40

PART EIGHT

HOW LIVING THINGS INTERACT WITH EACH OTHER AND WITH THEIR ENVIRONMENT

31 Plants 544
Patterns of structure and function

The organization of vascular plants 545
Tissues of vascular plants 545
   Fluid movement: Vascular tissue 546
   Food storage: Ground tissue 547
   Protection: Dermal tissue 547
   Growth: Meristematic Tissue 548
Organs of vascular plants 548
   Roots 548
   Shoots 551
Movement of water and dissolved substances in vascular plants 555
The organization of nonvascular plants 557
Summary 558

32 Invertebrate animals 560
Patterns of structure, function, and reproduction

Characteristics of animals 561
   Patterns in symmetry 562
   Patterns in body cavity structure 562
   Patterns in embryological development 564
Invertebrates: Diversity in symmetry and coelom 565
   Asymmetry: Sponges 565
   Radial symmetry: Hydra and jellyfish 566
   Bilateral symmetry: Variation on a theme in many phyla 568
Parasitic flatworms 572
Summary 578

33 Vertebrate animals 580
Patterns of structure, function, and reproduction

Chordates and vertebrates: Unity in symmetry and coelom 581
   Tunicates 581
   Lancelets 583
   Vertebrates 583
Summary 595

34 Innate behavior and learning in animals 598

Ethology: The biology of behavior 599
The link between genetics and behavior 600
Innate (instinctive) behaviors 600
   Kineses 600
   Taxes 600
   Reflexes 602
   Fixed action patterns 602
Learning 603
   Imprinting 603
   Habituation 605
   Trial-and-error learning (operant conditioning) 605
   Classical conditioning 606
   Insight 606
Yawn! 608
Summary 608

35 Social behavior in animals 610

Types of social behavior 611
   Competitive behaviors 611
Elephant talk 614
   Reproductive behaviors 615
   Parenting behaviors 618
   Group behaviors 618
Human behavior 621
Summary 622

36 Population ecology 624

An introduction to ecology 625
Population growth 625
   Exponential growth 625
   Carrying capacity 627
Population size 627
Population density and dispersion 628
Regulation of population size 628
   Density-independent limiting factors 628
   Density-dependent limiting factors 628
The urban explosion 629
Mortality and survivorship 630
Demography 631
The human population explosion 632
Summary 636

xix

VOLUME 2 CHAPTERS 25-40

37 Interactions within communities of organisms 638

Ecosystems 639
Communities 639
Types of interactions within communities 640
    Competition 640
    Predation 642
Competition and killer bees 646
    Symbiosis 649
Changes in communities over time: Succession 651
Summary 654

38 Ecosystems 656

Populations, communities, and ecosystems 657
The flow of energy through ecosystems 658
    Food chains and webs 659
    Food pyramids 661
Hubbard Brook and the cycling of nutrients in ecosystems 663
The cycling of chemicals within ecosystems 664
    The water cycle 664
    The carbon cycle 665
    The nitrogen cycle 666
    The phosphorus cycle 667
Summary 669

39 Biomes and life zones of the world 671

Biomes and climate 672
    The sun and its effects on climate 672
    Atmospheric circulation and its effects on climate 672
Life on land: The biomes of the world 675
    Tropical rain forests 676
The El Niño southern oscillation 678
    Savannas 678
    Deserts 680
    Temperate grasslands 681
    Temperate deciduous forests 682
    Taiga 682
    Tundra 684
Life in fresh water 684
Estuaries: Life between rivers and oceans 685
Life in the oceans 687
    The intertidal zone 687
    The neritic zone 688
    The open-sea zone 690
Summary 692

40 The biosphere 694
Today and tomorrow

The biosphere 695
The land 696
    Diminishing natural resources 696
    Species extinction 700
Ten things you can do to change the world 702
    Solid waste 702
The water 703
    Surface water pollution 703
    Ground water pollution 705
    Acid rain 705
The atmosphere 707
    Air pollution 707
    Ozone depletion 708
Overpopulation and environmental problems 708
Summary 709

APPENDIXES

A   Answers to review questions A-1

B   Classification of organisms B-1

Glossary G-1

Credits C-1

PART SIX

EVOLUTION
HOW LIVING THINGS CHANGE OVER TIME

25 ▶ The scientific evidence for evolution

Brought back to life with computer chips and fleshlike polymers, the eighteenth century English naturalist Charles Darwin speaks to visitors from his library at the St. Louis Zoo in St. Louis, Missouri. Darwin's library is located in The Hall of Animals, one of the indoor exhibit areas at The Living World, the Zoo's educational complex. Darwin explains:

> A finch, not very different from this one, led my mind to an idea of which you may have heard—that the great diversity of creatures on Earth, the wealth of form and color, is the result of evolution by natural selection.

Darwin then continues to explain the experiences that affected his thinking and led to the development of his theory:

> As a young man, I sailed around the world on Her Majesty's ship *Beagle,* and on the little Galapagos Islands—a thousand miles at sea—I saw finches. Some had big bills with which they cracked seeds like sparrows. Others darted around catching insects as a warbler might. There was even one that used a twig to dig for grubs, trying to be a woodpecker. And I noticed a peculiar thing . . . the birds that I saw on the Galapagos were not at all like our English Finches. Instead, they looked like finches I had seen before, earlier in my voyage in South America. This was the clue. An ancestor of these Galapagos finches must have come from South America to the islands, and these finches are its descendants. But how could Galapagos finches be so different from one another if they all have the same ancestor? Somehow, they must have changed. And that is the heart of it . . . descent with modification . . . evolution.

HIGHLIGHTS

▼
One of the central theories of biology is the theory of evolution, which states that living things change over time.

▼
Charles Darwin, a nineteenth century English naturalist, developed this theory based on observations of plants and animals of oceanic islands, geological evidence that the Earth has changed over time, the results of artificial breeding, and Malthus' ideas on population growth and decline.

▼
Darwin proposed that evolution occurs as a result of natural selection.

▼
Although proposed over 100 years ago, the theory of evolution is still upheld and is further supported by the fossil record, the molecular record, progressive changes in homologous structures, and the presence of vestigial structures.

OUTLINE

The development of Darwin's theory of evolution
  Darwin's observations
  Factors that influenced Darwin's thinking
    The influence of geology
    The results of artificial breeding
    The study of populations
  Natural selection: A mechanism of evolution
  An example of natural selection at work: Darwin's finches

The publication of Darwin's theory

Testing the theory
  The fossil record
  The age of the Earth: Rock and fossil dating
  Comparative anatomy
  Comparative embryology
  Molecular biology

Visitors at the St. Louis Zoo have a chance to better understand Darwin's statements. Likewise, this chapter describes the history of evolutionary thought and explains its scientific evidence. Then you, too, can better understand Darwin's statements and your own evolutionary past.

The development of Darwin's theory of evolution

The story of Darwin and his theory begins in 1831, when he was 22 years old. On the recommendation of one of his professors at Cambridge University, he was selected to serve as naturalist for a 5-year voyage (from 1831 to 1836) around the coasts of South America on the H.M.S. *Beagle* (Figure 25-1). Darwin had the chance to study plants and animals on continents, islands, and seas distant from his native England. He was able to experience firsthand the remarkable diversity of living things on the Galapagos Islands off the west coast of South America. Such an opportunity clearly played an important role in the development of his thought about the nature of life on Earth.

When the *Beagle* set sail, Darwin was fully convinced that species were unchanging. Indeed, he wrote that it was not until 2 or 3 years after his return that he began to consider seriously the possibility that species could change. And it was not until years after that time that Darwin began to formulate a theory integrating his observations of the trip and his understanding of geology, population biology, and the fossil record. Beginning in 1842, Darwin began to write his explanation of the diversity of life on earth and the ways in which living things are related to one another.

Darwin's observations

During his 5 years on the ship, Darwin observed many phenomena that were of central importance to the development of his theory of evolution. While in southern South America, for example, Darwin observed fossils (the preserved remains) of extinct armadillos that were similar to armadillos still living in that area (Figure 25-2). He found it interesting that such similar yet distinct living and fossil organisms were found in this same small geographical area. This observation suggested to Darwin that the fossilized armadillos were related to the present-day armadillos; that they were "distant" relatives.

Another observation made by Darwin was that geographical areas having similar climates, such as Australia, South Africa, California, and Chile, are each populated by different species of plants and animals. A **species** is a group of related organisms that share common characteristics and are able to interbreed and produce viable offspring. Organ-

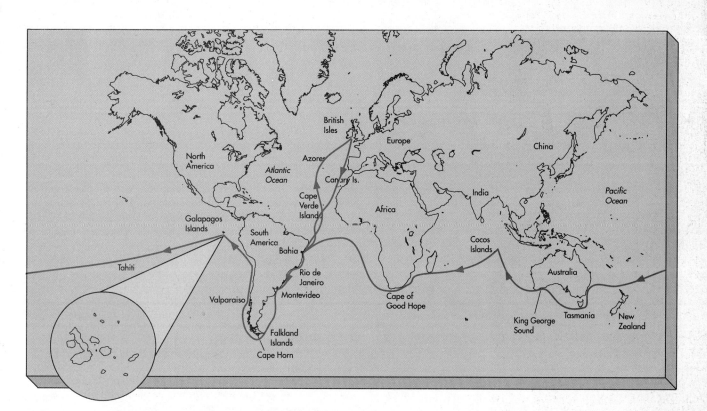

FIGURE 25-1 Voyage of the H.M.S. *Beagle*.
Most of Darwin's time was spent exploring the coasts and coastal islands of South America, such as the Galapagos Islands. Darwin's studies of the animals of the Galapagos Islands played a key role in his development of the theory of evolution by natural selection.

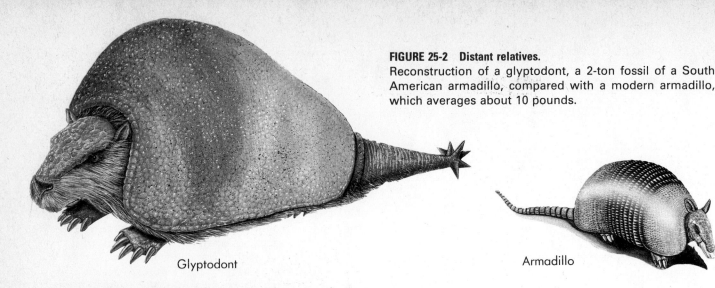

FIGURE 25-2 Distant relatives.
Reconstruction of a glyptodont, a 2-ton fossil of a South American armadillo, compared with a modern armadillo, which averages about 10 pounds.

FIGURE 25-3 Galapagos tortoises.
The tortoises of the Galapagos Islands exhibit variation. Tortoises with large, domed shells **(A)**, are found in relatively moist habitats, whereas those with lower, saddleback-type shells with the front bent up **(B)** are found in drier habitats. This pattern of physical variation suggested to Darwin that the island tortoises were related but had changed slightly after being isolated in different habitats.

isms found in different geographical areas having similar climates are often similar to one another, "shaped" by environmental similarities, but are, in fact, different and often unrelated species. These differences suggested to Darwin that factors other than or in addition to climate must play a role in plant and animal diversity. Otherwise, all lands having the same climate would have the same species of animals and plants.

On the Galapagos Islands, Darwin encountered giant land tortoises. The tortoises on the various islands were similar yet different from one another. In fact, local residents and sailors who captured the tortoises for food could tell which island a particular animal had come from just by looking at its shell (Figure 25-3). This pattern of physical variation suggested that all of the tortoises were related, but that they had changed slightly in appearance after becoming isolated on the different islands.

Darwin was also struck by the fact that the relatively young Galapagos Islands (formed by undersea volcanoes) were home to a profusion of living organisms resembling plants and animals that lived on the nearby coast of South America. Notice, for example, the similarity of the two birds in Figure 25-4. The bird on the left is a medium ground finch and is found on the Galapagos. The blue-black grassquit, shown on the right, is found in grasslands along the Pacific Coast from Mexico to Chile. These observations suggested to Darwin that the Galapagos organisms were related to ancestors who, long ago, flew, swam, or "hitchhiked" (were transported by other organisms) to the islands from the mainland.

> Darwin made observations during his 5-year voyage that suggested to him:
> 1. Organisms of the past and present are related to one another.
> 2. Factors other than or in addition to climate play a role in the development of plant and animal diversity.
> 3. Members of the same species often change slightly in appearance after becoming geographically isolated from one another.
> 4. Organisms living on oceanic islands often resembled organisms found living on a close mainland.

FIGURE 25-4 Evidence from the finches.
A One of Darwin's finches, the medium ground finch.
B The blue-black grassquit, which is found in grasslands along the Pacific coast from Mexico to Chile. This bird may have a common ancestor with Darwin's finches.

FIGURE 25-5 Artificial selection: another clue in the natural selection puzzle.
The differences that have been obtained by artificial selection of the wild European rock pigeon (**A**) and domestic races such as the red fantail (**B**) and the fairy swallow (**C**) are so great that birds probably would, if wild, be classified in entirely different major groups. In a way similar to that in which these races were derived, widely different species have originated in nature by means of natural selection.

Factors that influenced Darwin's thinking

As Darwin studied the data he collected during his voyage, he reflected on its significance in the context of what was known about geology, the breeding of domesticated animals, and population biology.

The influence of geology

In the late eighteenth and early nineteenth centuries, scientists studying the rock layers of the Earth noticed two things. First, scientists saw evidence that the Earth had changed over time, acted on by natural forces such as the winds, rain, heat, cold, and volcanic eruptions. Geologists began to hypothesize that the Earth was much older than the 6000 years originally suggested by a literal translation of the Bible. Second, they noticed that the fossils found within the Earth's rock layers were similar to but different in many ways from living organisms—an observation Darwin himself had made on his voyage. Not only had the Earth changed, thought scientists, but evidence existed that the organisms living on its surface had changed also.

> Geological evidence suggests that the Earth is much older than the 6000 years originally thought and that the Earth and organisms living on it have changed over time.

The results of artificial breeding

As he pondered these ideas that the earth and its organisms may have changed over time, Darwin reflected on the results of a process called **artificial selection.** In artificial selection a breeder selects for desired characteristics, such as those of the pigeons shown in Figure 25-5. At one time these pigeons came from the same stock, but through artificial breeding over successive generations, their offspring have changed dramatically.

Artificial selection is based on the *natural variation* all organisms exhibit. For example, looking back at p. 372, you can see that although individuals within this population of flamingos are similar, they possess characteristics that vary from individual to individual. Farmers and animal breeders, both today and in Darwin's time, take advantage of the natural variation within a population to select for characteristics they find valuable or useful. By choosing organisms

THE SCIENTIFIC EVIDENCE FOR EVOLUTION

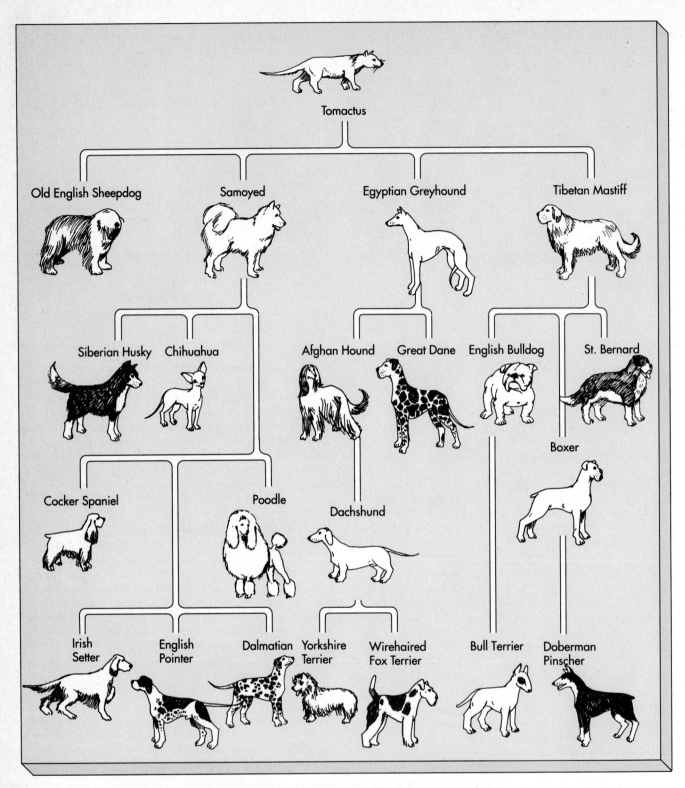

FIGURE 25-6 Artificial selection leads to different breeds of the same species.
All dogs belong to the species *Canis familiaris,* but through artificial selection, breeders have been able to choose which traits each breed should retain. The result is a myriad of different breeds within one species.

that naturally exhibit a particular trait and then breeding that organism with another exhibiting the same trait, breeders are able (over successive breedings) to produce animals or plants having a desired, inherited trait. This trait will breed true in successive generations when these organisms are bred with one another. For example, dogs have been artificially bred for centuries. Although your collie may look much different from your neighbor's terrier, both animals belong to the same species but have been artificially bred to retain traits that are characteristic of their breeds (Figure 25-6). Even the turkey you eat on Thanksgiving has been artificially bred for large cavities for stuffing.

> Breeders of plants and animals are able to alter the characteristics of organisms by selecting those with desired, inheritable traits and breeding them. After successive breedings, these inherited traits will consistently appear in offspring. Darwin hypothesized that a similar type of selection might take place in nature and result in changes within populations of organisms over time.

The study of populations

Pondering his observations, Darwin began to study Thomas Malthus' *Essay on the Principles of Population*. Malthus, an economist who lived from 1766 to 1834, pointed out that populations of plants and animals (including human beings) tend to increase geometrically. In a geometric progression, a population (for example) increases as its number is multiplied by a constant factor. In the geometric progression 2, 6, 18, 54, and so forth, each number is three times the preceding one. Figure 25-7 shows how the numbers in a geometric progression increase quickly! Malthus suggested that although populations grow geometrically, food supplies increase only arithmetically. An arithmetic progression, in contrast, is one in which the elements increase by a constant difference, as the progression 2, 6, 10, 14, and so forth. In this progression, each number is 4 greater than the preceding one. Figure 25-7 shows graphically how each type of progression increases.

Although Malthus suggested that populations grow at a geometric rate, he realized that factors existed to limit this astounding growth. If populations grew unchecked, organisms would cover the entire surface of the Earth within a surprisingly short time. But the world is not covered in ants, spiders, or poison ivy. Instead, populations of organisms vary in number within a certain limited range. Space and food are limiting factors of population growth; death limits infinite population growth. Malthus noted that in human populations, death was caused by famine, disease, and war. Sparked by Malthus' ideas, Darwin saw that in nature, although every organism has the potential to produce many offspring thereby contributing to a geometric growth rate of its population, only a limited number of organisms actually survive to reproductive age. Darwin realized that factors similar to those limiting human populations must also act to limit plant and animal populations in nature.

> A key contribution to Darwin's thinking was Malthus' concept of geometric population growth. Real populations do not expand at this rate, and this implies that nature acts to limit population numbers.

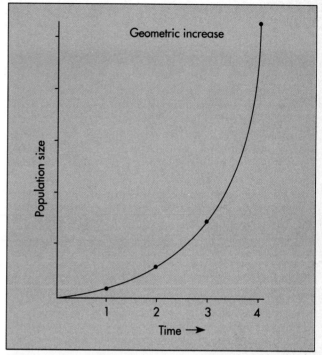

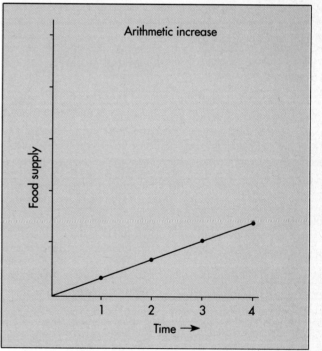

FIGURE 25-7 Types of mathematical progressions.
A In a geometric progression, a population increases as its number is multiplied by a constant factor. The numbers in a geometric progression increase rapidly.
B In an arithmetic progression, the numbers increase by a constant difference.

Natural selection: A mechanism of evolution

Darwin realized that environmental factors could influence which organisms in a population lived and which ones died. Berry-eating birds, for example, with variations in the structure of their beaks that allowed them to crush seeds, would survive longer than other birds of their species who did not have seed-crushing beaks, if the bushes bore few berries during a particular season. The non-seed-eating birds would die out rather quickly and would probably not live to reproductive age. The seed-eating birds, on the other hand, would survive the bad berry season and would likely reproduce. Many of their progeny would have seed-crushing beaks. If another season of few berries followed, these birds would have a survival advantage over other birds that were unable to live on an alternate food source.

Darwin made associations between the process of artificial breeding and reproduction within natural populations. His ideas were expressed in his autobiography:

> I soon perceived that selection was the keystone of man's success in making useful races of animals and plants. But how selection could be applied to organisms living in a state of nature remained for some time a mystery to me.
>
> In October 1838, that is, fifteen months after I had begun my systematic enquiry, I happened to read for amusement 'Malthus on Population' and being well prepared to appreciate the struggle for existence which everywhere goes on from long-continued observation of the habits of animals and plants, it at once struck me that under these circumstances favourable variations would tend to be preserved, and unfavourable ones to be destroyed. The result of this would be the formation of new species. Here then I had at last got a theory by which to work....

Darwin was saying that those individuals that possess physical, behavioral, or other attributes well-suited to their environment are more likely to survive than those that possess physical, behavioral, or other attributes less suited to their environment. The survivors have the opportunity to pass on their favorable characteristics to their offspring. These characteristics are naturally occurring inheritable traits found within populations and are called **adaptations.** (Populations are individuals of a particular species inhabiting a locale or region.)

Notice that the term *adaptation* is used differently than in its everyday sense. Here, it refers to naturally occurring inheritable traits present in a population of organisms rather than noninheritable traits in individuals. Adaptive traits are inherited characteristics that confer a reproductive advantage to the portion of the population possessing them. In its everyday sense, an *adaptation* refers to something a single individual does to change how it responds to the environment. For example, you may adapt to getting up early for an 8:00 class. But this is not an inherited trait in the entire population that confers a reproductive advantage!

As adaptive, or reproductively advantageous, traits are passed on from surviving individuals to their offspring, the individuals carrying these traits will increase in numbers within the population, and *the nature of the population as a whole will gradually change.* Darwin called this process, in which organisms having adaptive traits survive in greater numbers than those without such traits, **natural selection.** Change in populations of organisms therefore occurs over time because of natural selection: the environment imposes conditions that determine the results of the selection and thus the direction of change. The driving force of change—natural selection—is often referred to as *survival of the fittest.* Again, the term *fittest* does not have the everyday meaning of the healthiest, strongest, or most intelligent. You may be fit if you work out at the local health club regularly. But fitness in the context of natural selection refers to reproductive fitness—the ability of an organism to survive to reproductive age in a particular environment and produce viable offspring.

Natural selection provides a simple and direct explanation of biological diversity—why animals are different in different places. Environments differ, so organisms are "favored" by natural selection differently in different places. The nature of a population gradually changes as more individuals are born that possess the "selected" traits. **Evolution** by means of natural selection is this process of change over time by which existing populations of organisms develop from ancestral forms through modification of their characteristics.

An example of natural selection at work: Darwin's finches

Interestingly, the results of evolution by natural selection can actually be seen if the process takes place relatively quickly (over a period of years to several thousand years), resulting in the existence of groups of closely related species from an original ancestral species. The results that can be seen are clusters of these closely related species found living nearby one another. Such clusters of species are often found on a group of islands, in a series of lakes, or in other environments that are close to but separated from one another. Organisms living in such sharply discontinuous habitats are said to be **geographically isolated** from one another.

The Galapagos Islands are a particularly striking example of sharply discontinuous habitats, providing a "natural laboratory" to view the results of natural selection. The islands are all relatively young in geological terms (several million years) and have never been connected with the adjacent mainland of South America or with any other area. Made up of 13 major islands (and some very tiny islands), the Galapagos are separated from one another by distances of up to 100 miles and are 600 miles from the South American mainland (see Figure 25-1). As a group, they exhibit diverse habitats. For example, the lowlands of the Galapagos are covered with thorn scrub. At higher elevations, attained only on the larger islands, there are moist, dense forests.

Formed by undersea volcanoes, the Galapagos Islands were uninhabited when they appeared above the surface of the water. The ancestors of all the organisms found on the Galapagos today reached these islands by crossing the sea by water or wind or on the bodies of other organisms. Only eight species of land birds reached the islands. One of these species was the finch, which fascinated Darwin.

Presumably, the ancestor of Darwin's finches reached these islands earlier than any of the other birds. If so, all the types of habitats where birds occur on the mainland were unoccupied on the Galapagos—and the ancestral finches were able to take advantage of them all! As the finches moved into these vacant habitats, the ones best suited to each particular habitat were selected for by nature. In other words, those birds possessing naturally occurring variations in their characteristics that were beneficial to survival lived to reproduce. Their offspring also possessed these inheritable traits. Over time, the population of finches occupying each habitat changed, and the ancestral finches split into a series of diverse populations. This phenomenon, by which a population of a species changes as it is dispersed within a series of different habitats within a region, is referred to as **adaptive radiation**. Some of these populations became so changed from the others that interbreeding was no longer possible: new species of finches were formed. This process, by which new species are formed during the process of evolution, is termed **speciation**.

The evolution of Darwin's finches on the Galapagos Islands provides one of the classic examples of speciation. The descendants of the original finches that reached the Galapagos Islands now occupy many different kinds of habitats on the islands (Figure 25-8) and are found nowhere else in the world. Among the 13 species of Darwin's finches that inhabit the Galapagos, there are three main groups: ground finches, tree finches, and warbler finches. The ground finches feed on seeds of different sizes. The size of their bills is related to the size of the seeds on which the birds feed. The tree finches, as their name suggests, eat insects, buds, or fruit found in the trees. Again, the size and shape of their bills is related to their food. The most unusual member of this group is the woodpecker finch. This bird carries around a twig or a cactus spine, which it uses to probe for insects in deep crevices. It is an extraordinary example of a bird that uses a tool. And lastly, the warbler finches, named for their beautiful singing, search continually with their slender beaks over leaves and branches for insects.

> The evolution of Darwin's finches illustrates the same kinds of processes by which species are originating continuously in all groups of organisms. Isolated populations subjected to unique combinations of selective pressures (the conditions imposed by nature) diverge from one another and may ultimately become so different that they are distinct species.

The publication of Darwin's theory

Darwin drafted the overall argument for evolution by natural selection in 1842 and continued to refine it for many years. The stimulus that finally brought it into print was an essay that he received in 1858. A young English naturalist named Alfred Russel Wallace (1823-1913) sent the essay to Darwin from Malaysia; it concisely set forth the theory of evolution by means of natural selection (Figure 25-9). Like Darwin, Wallace had been influenced greatly in his development of this theory by reading Malthus' 1798 essay. After

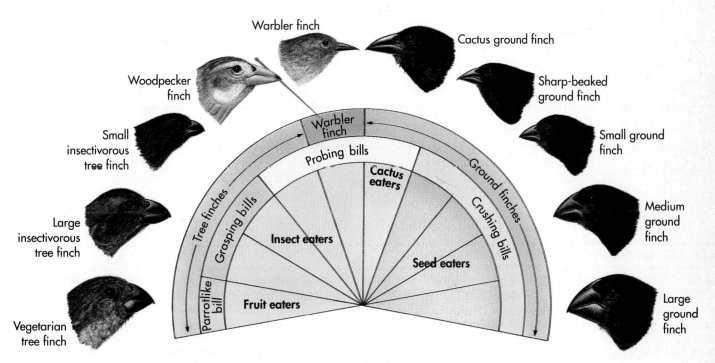

FIGURE 25-8 Darwin's finches.
Ten species of Darwin's finches from Indefatigable Island, one of the Galapagos Islands, showing differences in bills and feeding habits. The bills of several of these species resemble those of different, distinct families of birds on the mainland. All of these birds are thought to have been derived from a single common ancestor.

FIGURE 25-9 Alfred Russel Wallace.
Wallace arrived at the theory of natural selection independently of Darwin, but the ideas of the two were presented together at a seminar in 1859.

receiving Wallace's essay, Darwin arranged for a joint presentation of their ideas at a seminar in London. Darwin then proceeded to complete his own book, which he had been working on for some time, and submitted it for publication in what he considered an abbreviated version.

Darwin's book, *On the Origin of Species,* appeared in November 1859 and caused an immediate sensation. Some called the book "glorious" and were in complete agreement with Darwin's theories. Others criticized the book, admiring some parts and asserting that other parts were "totally false." Still others attacked Darwin on religious grounds because Darwin's ideas were different from the exact word of the Bible. Many clergy, however, openly agreed with Darwin, saying that the theory of evolution did not deny the existence of God.

> Darwin published his argument for the theory of evolution in 1859, in a book entitled *On the Origin of Species.* He also presented his ideas at a scholarly meeting with another scientist, Alfred Russel Wallace, who had independently developed a theory of evolution. These ideas were hotly debated at that time: the mechanism of evolution—natural selection—was not well accepted nor understood.

At the end of June 1860, a debate was held at a meeting of the British Association for the Advancement of Science. The debate lasted many days and attracted huge crowds of people. The debate was heated, but the outcome pleased Darwin: people read his book and gave his arguments serious consideration. Within 20 years or so after the publication of *On the Origin of Species,* the concept that species have changed over time was well accepted. The mechanism of this change—natural selection—was not. At that time, no one had any concept of genes or of how heredity works, and so it was impossible for Darwin to explain completely how evolution occurs. Gregor Mendel (see Chapter 22) had not yet begun his ground-breaking work in the study of inheritance. In fact, the science of genetics was not established until the beginning of the twentieth century, 40 years after the publication of Darwin's book. An understanding of the laws of inheritance and the mechanism by which inheritable traits are passed on from one generation to the next helped scientists understand the process of natural selection.

Testing the theory

More than a century has elapsed since Charles Darwin's death in 1882. During this period the evidence supporting his theory has grown progressively stronger. In fact, evolution is no longer considered a theory but is accepted as a **scientific law,** a theory that has been upheld countless times as it is tested and retested. By convention, the phrase *the theory of evolution* is still used, but its use does not suggest that evolution by means of natural selection is a highly tentative concept. Evolution is, rather, a scientific observation that is as well accepted in the scientific community as is the theory that the Earth revolves around the sun. Scientists are continuously learning more, however, about the intricacies of the mechanism of natural selection and the history of the evolution of life on Earth.

> The concept of evolution by means of natural selection is established as valid within the scientific community today and scientists are continually developing their understanding of natural selection.

What is the scientific evidence that upholds the theory of evolution? Scientists find evidence in the fossil record, using widely accepted techniques to assess the age of the rocks in which fossils are often found, while gathering a picture of the history of the earth and its organisms. In addition, the tools of comparative anatomy help researchers understand relationships among organisms alive today. Significant scientific advances, such as those in genetics and molecular biology, have given scientists tools Darwin did not have, so today scientists understand more fully than Darwin ever could how organisms change over time.

The fossil record

A **fossil** is any record of a dead organism. Fossils may be nearly complete impressions of organisms or merely burrows, tracks, molecules, or other traces of their existence. Unfortunately, only a minute fraction of the organisms living at any one time are preserved as fossils.

Most fossils are preserved in **sedimentary** rocks. Sedimentary rocks are made up of particles of other rocks, cast off as they weather and disintegrate. Running water, such as a river or stream, picks up these pieces of rock and carries them to lakes or oceans where they are deposited as **sediment,** better known as mud, sand, or gravel. Over time, the

FIGURE 25-10 The Grand Canyon.
In this photo, the layers of sedimentary rock can be clearly seen.

sediment hardens into rock. But while some sediment is hardening, other sediment is still being deposited, creating layers of rock formed one on top of the other. Therefore most sedimentary rock has a stratified appearance, like that seen in the Grand Canyon (Figure 25-10).

During the formation of sedimentary rock, dead organisms are sometimes washed along with the mud or sand and eventually reach the bottom of a pond or lake. Dead marine organisms fall to the bottom of the ocean. As the sediments harden into rock, they harden around the bodies of these dead organisms (Figure 25-11). The hard parts of these organisms, such as their skeletons, may become preserved or may be broken down and replaced with other minerals. Fossils of organisms having hard parts are the type most often found (Figure 25-12) rather than fossils formed from soft body parts, which usually decay quickly and leave no

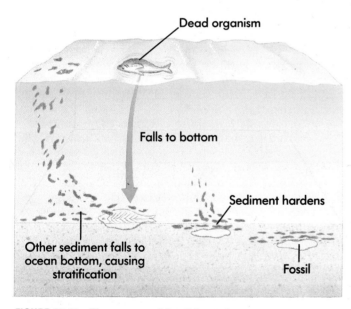

FIGURE 25-11 The process of fossil formation.
Fossils are formed when dead organisms are trapped in hardening sediment.

FIGURE 25-12 *Archaeopteryx,* **a prehistoric ancestor of modern birds.**
A well-preserved fossil of this bird, about 150 million years old, was discovered within 2 years of the publication of *On the Origin of Species.* This particular specimen is an excellent example of a fossil formed from the hard parts of the organism. The skeleton can be clearly seen.

THE SCIENTIFIC EVIDENCE FOR EVOLUTION

FIGURE 25-13 A fossil "mold."
Shown here is *Mawsonites spriggi*, a fossil jellyfish from South Australia. This fossil illustrates the remarkable preservation of soft-bodied fossils in fine-grained sedimentary rocks.

trace of their existence. Sometimes, however, soft-bodied animals are preserved in exceptionally fine-grained muds, in conditions where the supply of oxygen was poor while the muds were being deposited, thus slowing the decomposition of the organism. Eventually, the soft parts of an organism decay completely, leaving behind a **mold**, or impression, of its body. Molds may become filled with minerals, such as lime or silica found in underground water, forming **casts**, which resemble the original organism or body part (Figure 25-13). In general, however, fossils of soft-bodied organisms, such as worms, are rare. Even though soft-bodied animals undoubtedly evolved before their hard-bodied counterparts, there is comparatively little evidence of their history in the fossil record.

Fossils provide an actual record of organisms that once lived, an accurate understanding of where and when they lived, and some appreciation of the environment in which they lived. Limestone that contains corals, for example, would have been deposited when the location was an ocean. Oak leaves found in sandstone suggest that a location was once a continent. In this way, fossils and the rock in which they are embedded provide information about the history of an area, which gives scientists clues to the location of the continents, ponds, lakes, and oceans and how their positions have changed over time.

Scientists can determine the age of fossils and use this information to establish the broad patterns of the progression of life on Earth (Figure 25-14). One of the ways of determining the age of particular fossils is to compare the sequences in which they appear in different layers, or strata, of sedimentary rock. Sedimentation deposits new layers mostly on older ones, so the fossils in upper layers mostly represent younger species than the fossils in lower layers. Fossils found in the same strata are assumed to be of the same age. Such correlations, in fact, were known at the time of Darwin and were used by him to formulate the theory of evolution.

▶ **Fossils, impressions of organisms that once lived, provide a record of the past.** ◀

The age of the Earth: Rock and fossil dating

Direct methods of dating rocks and fossils first became available in the late 1940s. This process depends on naturally occurring isotopes of certain elements that are found in rock. Isotopes are atoms of an element that have the same number of protons but different numbers of neutrons in their nuclei (see Chapter 2). They therefore differ from one another in their atomic numbers. **Radioactive isotopes** are unstable; their nuclei decay, or break apart, at a steady rate, producing other isotopes and emitting energy. After decay some radioactive isotopes may give rise to elements. Many different isotopes are used in radioactive dating. Some methods give scientists information about the age of rocks; others measure the length of time since the death of an organism.

One of the most widely used methods of dating, the carbon-14 (^{14}C) method, estimates the relative amount of the different isotopes of carbon present in a fossil (or other organic material). Most carbon atoms have an atomic weight of 12 (6 protons and 6 neutrons); the symbol of this particular isotope of carbon is ^{12}C. A fixed proportion of the atoms in a given sample of carbon, however, consists of carbon with an atomic weight of 14 (^{14}C), an isotope that has two more neutrons than ^{12}C. Interestingly, ^{14}C is produced from ^{14}N (nitrogen-14) as these atoms are bombarded by cosmic rays—high-energy particles from space. The cosmic rays (usually protons) bump a proton from the nucleus of a nitrogen atom, leaving it with 6 protons (the atomic number of carbon) and 7 neutrons. As a result of the collision, the atom also captures a neutron and becomes an atom of ^{14}C. This newly created ^{14}C reacts with oxygen, becoming carbon dioxide, a gas commonly found in the air. Plants use this carbon during photosynthesis, incorporating it into the sugars and starches they make. By eating plants, animals incorporate ^{14}C into their bodies as well.

The carbon that is incorporated into the bodies of living organisms consists of the same fixed proportion of ^{14}C and ^{12}C that occurs in the atmosphere. After an organism dies, however, and is no longer incorporating carbon, the ^{14}C in it gradually decays back to nitrogen by emitting a beta particle. (A beta particle is an electron discharged from the nucleus when a neutron splits into a proton and an electron.) It takes 5730 years for half of the ^{14}C present in a sample to be converted by this process; this length of time is called the **half-life** of the ^{14}C isotope. By measuring the amount of ^{14}C in a fossil, scientists can estimate the proportion of ^{14}C to all other carbon that is still present and compare that to the ratio of these isotopes as they occur in the atmosphere. In this way, scientists can then estimate the length of time over which the ^{14}C has been decaying, which is the same as the length of time since the organism died.

For fossils older than 50,000 years, the amount of ^{14}C remaining is so small that it is not possible to measure it precisely enough to provide accurate estimates of age. These fossils may be dated using the isotope thorium-230, which has a half-life of 75,000 years, and decays from uranium-238. These techniques have been most useful to scientists who study deep sea sediments too old to be dated with ^{14}C.

With the use of radioactive dating methods, knowledge of the ages of various rocks has become more precise. The oldest rocks on Earth that have been dated include rocks

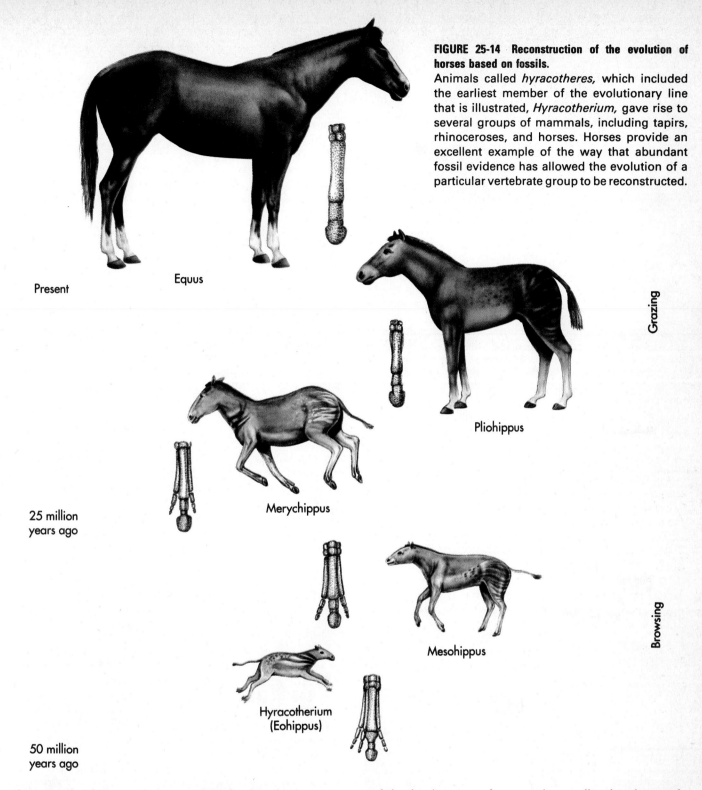

FIGURE 25-14 Reconstruction of the evolution of horses based on fossils.
Animals called *hyracotheres*, which included the earliest member of the evolutionary line that is illustrated, *Hyracotherium*, gave rise to several groups of mammals, including tapirs, rhinoceroses, and horses. Horses provide an excellent example of the way that abundant fossil evidence has allowed the evolution of a particular vertebrate group to be reconstructed.

from South Africa, southwestern Greenland, and Minnesota that are approximately 3.9 billion years old. Meteorites have been dated at about 4.6 billion years. Recently, rocks brought back to Earth from the moon have been dated from 3.3 to 4.6 billion years old. These pieces of evidence suggest that the Earth and the moon, most likely formed from the same processes at the same time, are about 4.6 billion years old.

What significance does the age of the Earth hold for the theory of evolution? The "accumulation" of adaptations and the development of new species usually takes thousands and probably millions of years. Until the time of Darwin, most held the belief that the Earth was approximately 6000 years old as described in the Judeo-Christian Bible. This time frame would not allow enough time for the process of evolution to take place. In fact, Sir Isaac Newton (1642-1727), an English physicist, calculated that it would take 50,000 years just for the Earth, after its formation (see Chapter 26), to cool to a temperature that would sustain life. Although scientists such as William Thomson, Lord Kelvin

THE SCIENTIFIC EVIDENCE FOR EVOLUTION

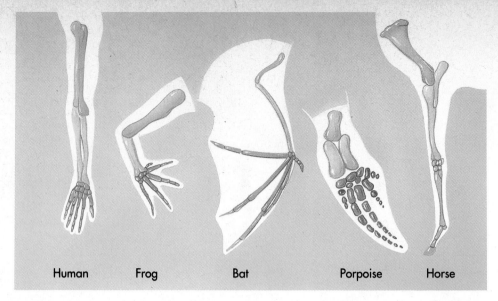

FIGURE 25-15 Homology among vertebrate limbs.
Homologies among the forelimbs of four mammals and a frog show the ways that the proportions of the bones have changed in relation to the particular way of life of the organism and that the forelimb of each animal has the same basic bone structure.

Human   Frog   Bat   Porpoise   Horse

(1824-1907) used the techniques available to him to date the Earth at about 100 million years, this time span still seemed too short to allow for evolution. Not until the present techniques of radioactive dating were developed could scientists begin to solve the "time problem" of evolution and accurately measure the geological age of the Earth, its rocks, and its fossils.

> By studying the comparative amounts of certain radioactive isotopes found within rocks and fossils, scientists can determine the age of fossils and use this information to establish patterns of life's progression. Studies using radioactive dating methods provide evidence that the Earth is 4.6 billion years old. This time is sufficient for organisms to have developed and evolved from "original" forms.

Comparative anatomy

Comparative studies of animal anatomy provide strong evidence for evolution. After Darwin proposed his theory, scientists began looking for evolutionary relationships in the anatomical structures of organisms. If derived from the same ancestor, organisms should possess similar structures with modifications reflecting adaptations to their environments. Such relationships have been shown most clearly in vertebrate animals.

Within the subphylum Vertebrata, the classes of organisms, such as birds, mammals, and amphibians, have the same basic anatomical plan of groups of bones (as well as nerves, muscles, and other organs and systems), but these bones are put to different uses among the classes. For example, the forelimbs seen in Figure 25-15 are all constructed from the same basic array of bones, modified in one way in the wing of a bat, in another way in the fin of a porpoise, and in yet another way in the leg of a horse. The bones are said to be **homologous** in the different vertebrates—that is, of the same evolutionary origin, now differing in structure and function. Although these vertebrates deviated from one another in their evolution, they all use the same bones in the same relative positions to do different jobs.

In some cases, homologous structures exist among related organisms but are no longer useful. These structures have diminished in size over time. Figure 25-16 shows tiny leg bones of the python that no longer serve a purpose. Such organs or structures that are present in an organism in a diminished size but are no longer useful are called **vestigial organs.** Humans have a tiny pouch called the appendix that hangs like a little worm at the junction of the small and large intestines. It serves no useful purpose in humans, but helps in digestion of plant material in organisms that are evolutionarily related to humans. And if you can wiggle your

FIGURE 25-16 A vestigial organ in a python.
Pythons possess tiny leg bones that serve no purpose in locomotion. Humans also possess a vestigial organ—the appendix—that is not needed for digestion or any other purpose.

FIGURE 25-17   Examples of convergent evolution.
The North American wolf, a placental animal **(A)** and the Tasmanian wolf, a marsupial animal **(B)** exhibit convergent evolution. The Tasmanian wolf has been extinct for approximately 110 years. Euphorbiaceae **(C)** found in Africa and the North American cactus **(D)** bear a striking resemblance to each other but belong to quite different families.

ears, you are using external ear muscles, which are also no longer useful except to amuse your friends, inherited from a distant ancestor.

> Comparative studies of animal anatomy show that many organisms have groups of bones, nerves, muscles, and organs with the same anatomical plan but with different functions. These homologous structures provide evidence of evolutionary relatedness.

In contrast to homologous structures, similar structures often evolve within organisms that have developed from different ancestors. Such body parts or organs, called **analogous** structures, have a similar form and function but have different evolutionary origins. The eyes of vertebrates and octopuses, which evolved independently but are similar in design, are analogous structures, as well as the wings of birds and insects.

Plants also show analogous structures. For example, three different families of flowering plants—the cacti, euphorbs, and milkweeds—have all developed thick, barrel-like fleshy stems that store water as an adaptation to a desert environment (Figure 25-17). In fact, these plants look so much like one another the casual observer might think they were all cacti. However, they evolved independently of one another in different parts of the world (southwest North America, Africa, and the Mediterranean, respectively).

To show that structures are indeed analogous, comparative anatomists do detailed dissections and study the embryological development of organisms. Analogous structures arise developmentally from different tissues. The presence of analogous structures shows how, in similar habitats, natural selection can lead to similar but not identical anatomical structures. Changes over time among different species of organisms having different ancestors that result in similar structures and adaptations is called **convergent evolution** (Figure 25-17).

THE SCIENTIFIC EVIDENCE FOR EVOLUTION

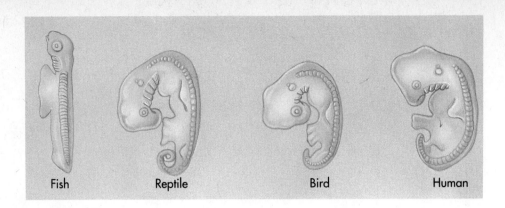

FIGURE 25-18 Embryos show our evolutionary history. The embryos of various groups of vertebrate animals show the primitive features that all vertebrate animals share early in development, such as gill arches and a tail.

Comparative embryology

Embryologists, scientists who study the development of organisms from conception to birth, noticed as early as the nineteenth century (around the time of Darwin) that various groups of organisms, although different as adults, possessed early developmental stages that were quite similar. For example, the embryological development of vertebrate animals is similar in that all vertebrate embryos have a similar number of gill arches, seen as pouches below the head. Only fish, however, actually develop gills. Likewise, the embryological development of the backbone of vertebrates is similar, but some organisms develop a tail, and others such as humans do not (Figure 25-18). "Tailbones"—the fused coccyx bones at the end of the spine—are actually vestigial structures.

These similar developmental forms tell scientists that similar genes are at work during the early developmental stages of related organisms. The genes active during development have been passed on to distantly related organisms from a common ancestor. Over time, new instructions are added to the old, but both are expressed at different times resulting in similar embryos that develop into organisms quite different from one another.

> Comparative embryological studies show that many organisms have early developmental stages that are quite similar. These similar developmental forms provide evidence of evolutionary relatedness.

Molecular biology

Today, biochemical tools provide additional evidence for evolution and give scientists new insights into the evolutionary relationships among organisms. Molecular biologists study the progressive evolution of organisms by looking at their hereditary material, DNA. According to evolutionary theory, every evolutionary change involves the formation of new alleles from the old by mutation; favorable new alleles persist because of natural selection. In this way, a series of evolutionary changes in a species involves a progressive accumulation of genetic change in its DNA. Organisms that are more distantly related will therefore have accumulated a greater number of changes in their DNA than organisms that more recently evolved from a common ancestor.

Within the last decade, molecular biologists have learned to study DNA by "reading" genes, much as you read this page. They have learned to recognize the order of the nitrogenous bases—the "letters"—of the long DNA molecules. By comparing the sequences of bases in the DNA of different groups of animals or plants, scientists can show the degree of relatedness among groups of organisms and develop detailed "family trees," called **phylogenetic trees** (Figure 25-19).

Interestingly, phylogenetic trees constructed from analyses of molecular differences are the same as those built using anatomical studies. For example, whales, dolphins, and porpoises cluster together using both approaches, as do the primates and the hoofed animals. By studying and interpreting the evidence from the fossil record, comparative anatomy, and genetic studies, it is often possible for scientists to estimate the rates at which evolution is occurring in different groups of organisms.

> The pattern of progressive change seen in the molecular record supports other scientific evidence for evolution and provides strong, direct evidence for change over time.

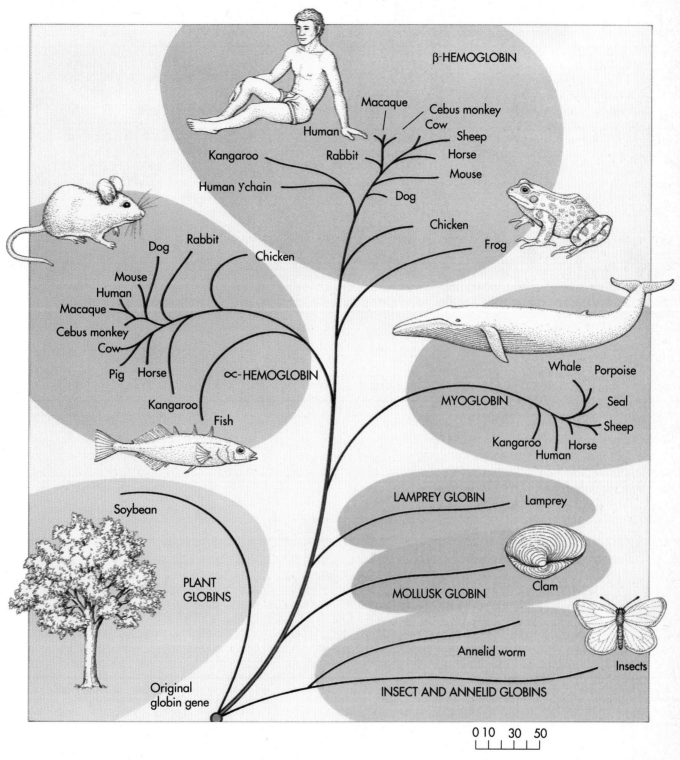

FIGURE 25-19 Evolution of the globin gene.
The length of the various lines is proportional to the number of nucleotide substitutions in the gene and measures the relatedness of the species.

THE SCIENTIFIC EVIDENCE FOR EVOLUTION

Summary

1. One of the central theories of biology is Darwin's theory of evolution, which states that living things change over time by means of natural selection. Proposed over a hundred years ago, this theory has stood up to a century of testing and questioning.

2. While studying the animals and plants of oceanic islands, Darwin accumulated a wealth of evidence that organisms have changed over time. Other factors that influenced Darwin's thinking were geological evidence that the earth had changed over time, the changes in organisms that farmers were able to attain by using artificial breeding methods, and Malthus' ideas on population dynamics.

3. Darwin proposed that evolution occurs as a result of natural selection: some individuals have traits that make them better-suited to a particular environment, allowing more of them to survive to reproductive age and produce more offspring than other individuals lacking these traits. These characteristics, or adaptations, are naturally occurring inherited traits found within populations.

4. Fitness is a measure of the tendency of some organisms to leave more offspring than competing members of the same population. Therefore genetic traits possessed by the fitter individuals will appear in greater proportions among members of succeeding generations. As a result of this process, the traits allowing greater reproduction will increase in frequency over time. The environment imposes conditions that determine the direction of selection and thus the direction of change.

5. Darwin published his theory in 1859 in a book titled *On the Origin of Species*. A wealth of evidence since Darwin's time has supported his proposals that evolution occurs and that its mechanism is natural selection. By the 1860s, natural selection was widely accepted as the correct explanation for the process of evolution, but the mechanism of natural selection was not understood. The field of evolution did not progress much further until the 1920s because of the lack of a suitable explanation of how hereditary traits are transmitted.

6. Two direct lines of evidence uphold the theory of evolution: (1) the fossil record, which exhibits a record of progressive change correlated with age, and (2) the molecular record, which exhibits a record of accumulated changes, the amount of change correlated with age as determined in the fossil record.

7. Several indirect lines of evidence uphold the theory of evolution, including progressive changes in homologous structures, the existence of vestigial structures, and changes in DNA sequences.

REVIEW QUESTIONS

1. Summarize four of Darwin's conclusions that were inspired by his observations during his voyage.
2. What three factors, in addition to his voyage on the H.M.S. *Beagle*, influenced Darwin's thinking on evolution?
3. A dachshund and a Siberian Husky are both dogs, but they look very different from each other. By what process did the two breeds come to look so different? What observable genetic principle is this process based on?
4. Distinguish between a geometric progression and an arithmetic progression. Which increases more quickly? Relate these concepts to the development of Darwin's theory of evolution.
5. What are adaptations? Explain their significance.
6. Explain the phrase *survival of the fittest*. What does "fit" mean in this context?
7. Most species of bears are black, brown, or gray. Why are polar bears white?
8. Distinguish between adaptive radiation and speciation.
9. Fill in the blanks: A _____ is a record of a dead organism, often preserved in _____ rocks. It can be dated using naturally occurring _____ _____ of certain elements found in rocks.
10. Explain how scientists can use the ^{14}C method to date a fossil.
11. How old is the Earth according to radioactive dating methods? Why is this significant?
12. What are vestigial organs? Give an example from the human body.
13. Distinguish between homologous and analogous structures.
14. Explain how studies of comparative anatomy and comparative embryology support the theory of evolution.
15. What is a phylogenetic tree? How does it support the concept of evolution?

KEY TERMS

adaptation 432
adaptive radiation 433
analogous 439
artificial selection 429
cast 436
convergent evolution 439
fossil 434
geographical isolation 432
half-life 436
homologous 438
mold 436
natural selection 432
phylogenetic tree 440
radioactive isotopes 436
scientific law 434
sediment 434
sedimentary 434
speciation 433
species 427
vestigial organs 438

THOUGHT QUESTIONS

1. Imagine you were attempting to date an old skull by the ^{14}C method. Assume the atmospheric ratio of ^{14}C to ^{12}C is one molecule of ^{14}C to every 999 molecules of ^{12}C (that is, the proportion of ^{14}C is 1 in a 1000, or 0.001), and the half-life of ^{14}C is 6000 years. How old is the skull if its proportion of ^{14}C is 0.00025?

FOR FURTHER READING

Carson, H. L. (1987). The process whereby species originate. *BioScience, 37,* 715-720.
One of the modern masters of evolution discusses a fascinating aspect of the origin of species.

Gould S. J. (1987, January). Darwinism defined: The difference between fact and theory. *Discover,* pp. 64-70.
This is a clear account of what biologists do and do not mean when they refer to the theory of evolution by natural selection.

Sibley, C., & Alquist, J. (1986, February). Reconstructing bird phylogenies by comparing DNAs. *Scientific American,* pp. 82-92.
This is a good introduction to the way in which biologists are beginning to use the tools of molecular biology to answer questions about evolution.

26 The evolution of the five kingdoms of life

HIGHLIGHTS

▼

Approximately 15 billion years ago, the universe condensed and then exploded in the "big bang." Earth and the other planets of the solar system were formed approximately 4.6 billion years ago.

▼

The long-held "primordial soup" theory states that life arose in the sea as elements and simple compounds reacted to form organic molecules about 3.5 billion years ago.

▼

The evolution of oxygen-producing bacteria was a key event, making possible an "explosion of life" during the Cambrian Period.

▼

The evolution of the five kingdoms of life has been influenced by the physical conditions of the Earth that have changed over time.

OUTLINE

Theories about the origin of organic molecules

Theories about the origin of cells

The history of life
    The Archean Era: Oxygen-producing cells appear
    The Proterozoic Era: The first eukaryotes appear
    The Paleozoic Era: The occupation of the land
    Mass extinctions
    The Mesozoic Era: The age of reptiles
    Changes in the Earth: The impact on evolution
    The Cenozoic Era: The age of mammals

Vast areas of gas and dust particles swirl in the Orion nebula M42. Orion is a specific constellation, or group of stars, that can be seen in the night sky of the northern hemisphere. Surrounding some of its stars are concentrations of gases and "stardust" called *nebulae,* a name derived from a Latin word meaning "cloud." In fact, nebulae are found in many sections of the Milky Way galaxy and are part of a cosmic life cycle—the birth, life, and death of stars.

Nebulae are "born" from the dust and gases hurled from unstable stars as they explode. Over time, this material condenses, giving birth to second- and later-generation stars. Then, in turn, these stars become unstable and explode, spewing dust and gases into space. And so the nebula life cycle continues.

The American astronomer, Edwin P. Hubble (1889-1953) observed that the galaxies in the universe are moving away from one another. Put simply, the universe is expanding. This movement suggests that at one time all the matter of the universe was in one spot, an assumption that leads scientists to hypothesize that the Earth (and the solar system) originated in the same way as nebulae do. Approximately 15 billion years ago, scientists theorize, *all* the material of the universe was condensed into a single small space. It then exploded in an event called the big bang. The universe—as it is known today—did not exist before that time. The gases and dust from the big bang produced an early generation of stars. Then, over billions of years, these stars exploded and their "space debris" formed other stars and planets. The solar system formed in this way some 4.6 billion years ago. For almost a billion years or so the molten Earth cooled, eventually forming a hardened, outer crust. Approximately 3.8 billion years ago, the oldest rocks on Earth formed. After that time, some 3.5 billion years ago, life began.

Theories about the origin of organic molecules

Primitive Earth was an incubator for life, but scientists know very little about what that incubator was like. Various hypotheses have been proposed that describe what the atmospheric and environmental conditions of "incubator Earth" might have been. But because scientists have no record of these conditions, their hypotheses are speculative and therefore highly controversial.

In trying to understand what the Earth was like 3.8 billion years ago, scientists often use information from events that are observable today. They use these pieces of information as starting points to build plausible hypotheses. For example, scientists have determined that the dust and gases ejected from unstable stars contain mostly hydrogen, as well as helium and varying amounts of other elements such as nitrogen, sodium, sulfur, and carbon. Some of these elements combine to form compounds such as hydrogen sulfide, methane, water, and ammonia. As the Earth coalesced, or came together into one mass, it may have contained some of these gases and vapors produced by the planet as it cooled. As the water vapor in this mix condensed, it could have caused millions of years of torrential rains extensive enough to form oceans. Other scientists suggest that water and gas deep within the earth were vented to the surface by volcanoes.

Scientists agree that the environment of primitive Earth was harsh and violent, bathed in ultraviolet radiation from the sun and subjected to violent electrical storms and constant volcanic eruptions. Under these conditions, many scientists hypothesize that the elements and simple compounds of the primitive atmosphere reacted with one another. New and more complex molecules were formed, capturing the surrounding energy within their bonds. In 1953, Stanley L. Miller and Harold C. Urey, working at the University of Chicago, designed an experiment that modeled an environment that included methane (CH_4), ammonia (NH_3), water vapor (H_2O), and hydrogen gas (H_2). They wondered if organic molecules—the molecules that make up the structure of all living things—could have spontaneously formed under such conditions.

Using an apparatus similar to that in Figure 26-1, Miller and Urey filled a glass chamber with the four gases. Electrodes within the chamber shot sparks of electricity through the mixture, while condensers cooled it. Miller and Urey wondered if complex molecules formed in their "atmosphere," dissolved in the water vapor, and fell into their "ocean." Within a week, they had their answer. A total of 15% of the carbon that was originally present as methane gas had been converted into other more complex compounds of carbon!

Miller and Urey, joined later by many other scientists in laboratories across the country, performed experiment after experiment using various combinations of simple compounds that might have made up the primitive atmosphere. In addition, scientists experimented with a variety of energy sources, such as ultraviolet light, heat, and radioactivity. The outcome was always the same: more complex organic molecules formed from simpler ones. These newly formed molecules included amino acids—the building blocks of protein. Sugars, including glucose, ribose, and deoxyribose, were also shown to form under the conditions of the primitive atmosphere. As scientists experimented further, they discovered that many of the molecules important to life—and important to the structure and function of cells—could be synthesized under certain primordial conditions. Not only were ribose and deoxyribose, components of nucleic acids, synthesized in their methane environment but so were purine and pyrimidine bases. Fatty acids, used in membranes and storage tissues of living organisms, were also made. From this array of experiments, scientists developed the primordial soup theory, which states that life arose in the primitive seas or in smaller lakes as complex organic molecules formed from the simple molecules in the ancient atmosphere.

During the past decade, some scientists have begun to doubt the primordial soup theory as described by the experiments of Miller, Urey, and others. Using computerized reconstructions of the atmosphere, James C. G. Walker of the University of Michigan at Ann Arbor suggests that the major components of the primitive atmosphere were carbon dioxide and nitrogen gases, spewed forth from volcanoes. But these gases could not have formed complex organic molecules—the precursors to life—in the seas or tidal pools of the Earth's surface.

Recently, some scientists have shifted from adherence to the primordial soup theory to thinking that life might

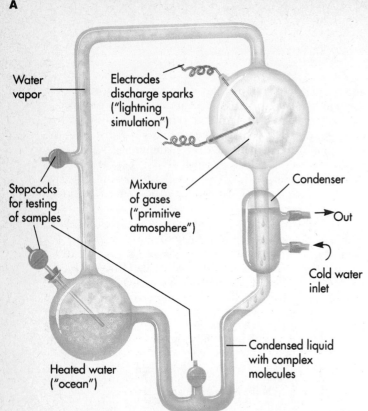
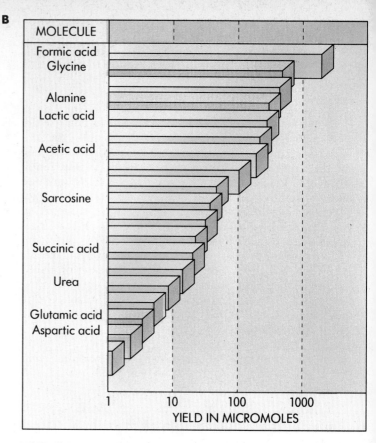
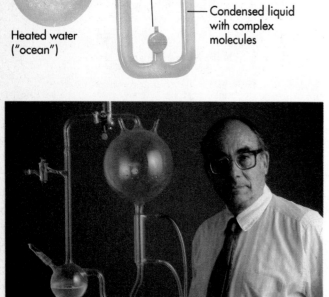

FIGURE 26-1 The Miller-Urey experiment.
A Miller and Urey's apparatus consisted of a closed tube connecting two chambers. The upper chamber contained a mixture of gases thought to resemble the Earth's atmosphere. Any complex molecules formed in the atmosphere chamber would be carried and dissolved in these droplets to the lower "ocean" chamber, from which samples were withdrawn for analysis.
B Some of the 20 most common complex molecules detected in the original Miller-Urey experiments are indicated. Among these 20 molecules are four amino acids (shown by green bars).
C Dr. Stanley Miller standing in front of the apparatus in 1991.

have arisen in hydrothermal vents, spots deep in the oceans where hot gases and sulfur compounds shoot from cracks in the Earth's crust (Figure 26-2). Some scientists hypothesize that the vents could have supplied the energy and nutrients needed for life to arise and be sustained. In fact, the hydrothermal vents in existence today support extensive communities of living organisms, such as tube worms, clams, and bacteria.

> The long-held primordial soup theory states that life arose in the sea as elements and simple compounds of the primitive atmosphere reacted with one another to form simple organic molecules such as amino acids and sugars. Recently, scientists have begun to reexamine the premises of this theory and have proposed alternate hypotheses.

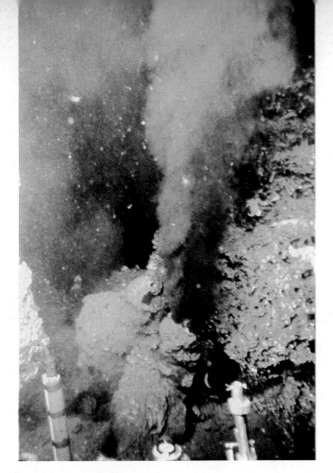

FIGURE 26-2 Hydrothermal vents.
Located under deep oceans, hydrothermal vents are cracks in the Earth's crust that release hot gases and sulfur compounds. Some researchers think that these vents could have provided enough energy to start and sustain life.

Theories about the origin of cells

Although scientists can replicate various plausible conditions of the primitive Earth, they cannot duplicate the millions of years that passed. Although scientists can observe the "birth" of simple organic molecules, they cannot generate life from nonlife. But at some point in the history of the Earth, organisms appeared. Scientists are still hard at work trying to unravel this mystery of life's beginnings. The work of many scientists over many years has led to various lines of reasoning regarding the evolution of the precursors to life, and life itself. However, most of these hypotheses have been based on the existence of a primordial soup.

The first step toward the evolution of life must have been the synthesis of even more complex organic molecules than the ones previously described. However, scientists know that complex molecules such as **polymers** (for example, polysaccharides and proteins) do not spontaneously develop from a mixture of their simpler building blocks, or **monomers** (in this case, sugars and amino acids). Most synthesis reactions are dehydrations and depend on the removal of water and the input of energy to chemically link one molecule with another. This process, called **dehydration synthesis** (see Chapter 2), could have taken place if condensing agents were present with the monomers in a primordial soup. Condensing agents are molecules that combine with water and release energy. If condensing agents were not present, heat and evaporation could also promote dehydration synthesis and the synthesis of polymers.

In the 1950s, the American scientist Sidney Fox and his co-workers developed a technique in which they used heat to produce polymers, which Fox called **proteinoids**, from dry mixtures of amino acids. Perhaps such complex molecules formed over 3.5 billion years ago as pools and puddles of a primordial soup evaporated under the heat of the sun or were heated near areas of volcanic activity. Interestingly, the proteinoid molecules synthesized by Fox act like enzymes because they are able to increase the rate of various organic reactions. Metal ions and clays (composed chiefly of minerals) could also have served as "early enzymes." When combined into sequences, enzymatic reactions can be thought of as the beginning of metabolic systems. To be considered "life" as scientists define it, metabolic systems must be organized within a cellular structure, be able to "carry" information about themselves, and be able to pass this information on by the process of replication.

> The first step toward the evolution of life is the synthesis of highly complex organic molecules. Such complex molecules might have formed as pools of a primordial soup evaporated under the heat of the sun, promoting the linking of small molecules. Another theory suggests that metal ions and clays might have served as early enzymes promoting such reactions.

Recently, scientists have begun to suspect that self-replicating systems of RNA molecules may have started the process of evolution. RNA molecules have the ability, as do all polynucleotides, of acting as a **template,** or guide, for the synthesis of a second polynucleotide based on the complementarity of its bases. DNA and RNA synthesis occurs in this fashion—DNA serves as a template for its own synthesis and for the synthesis of RNA (see Chapter 24). As mentioned previously, the building blocks of nucleotides were probably present in a primordial soup: deoxyribose and ribose sugars, purine and pyrimidine bases, and molecules bearing phosphate groups (PO_4^{2-}). These compounds could have joined to form nucleotides, which then could have linked to form polynucleotides. These polynucleotides could have served as templates for the synthesis of others—a process that many scientists think played a crucial role in the origin of life. In fact, theoretical work by Eigen and his co-workers suggests that short strands of self-replicating RNA eventually "worked" together to assemble enzymes from the amino acids present in the primordial soup.

For replicating RNA to be considered life, however, it must have become enclosed by some sort of a membrane forming the first cells. Chemical reactions could then take place in a closed environment—not the open environment of a primordial soup. Within the boundaries of a membrane, enzymes could be organized to carry on life functions. Chemicals could be selectively allowed into or kept out of the cell's interior. Wastes could be eliminated from the cell, and hereditary material could be passed on from cell to cell.

THE EVOLUTION OF THE FIVE KINGDOMS OF LIFE

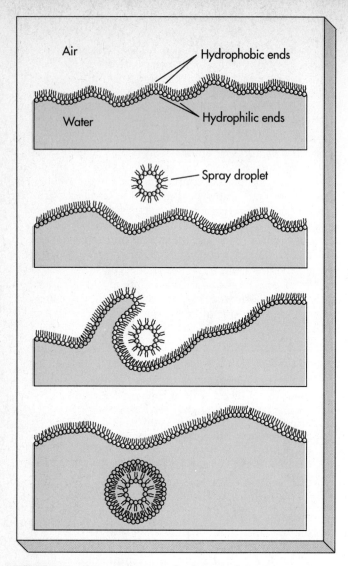

FIGURE 26-3 How membranous droplets can form.
The top diagram shows a single layer of lipid molecules floating on water. The sequence shows how wave action could cause these molecules, which have hydrophilic (water-loving) and hydrophobic (water-hating) ends, to form bilayered vesicles.

No one knows how much time passed before biochemical evolution became biological evolution. And no one knows how this evolution took place. Scientists do know, however, that single-celled life existed in the shallow seas of the primitive earth. How did these cells form? In the recent past, scientists thought that early precursors to cells—called **protocells**—were membranous droplets that had formed spontaneously. Scientists know (and can demonstrate) that many different kinds of molecules are attracted to one another when they are in solution because of forces such as electrical charge and surface tension. Sometimes these aggregates form microscopic, membranous droplets. Figure 26-3 shows how membranous droplets can form from molecules that have hydrophilic (water-loving) and hydrophobic (water-hating) portions, much like droplets of oil form in water when the two are mixed together. This droplet level of organization was thought to have been an important step in the origin of life. But although scientists have been able to produce two different types of spontaneously forming droplets in the laboratory, they recognize today that such droplets do not carry any genetic information and lack the internal organization of cells. Because of these facts, most scientists have abandoned the hypothesis that these droplets are the precursors to life.

> After the synthesis of complex organic molecules had taken place, self-replicating systems of RNA molecules may have begun the process of evolution. To be considered life, however, this "early" genetic material must have been organized within a cellular structure. And although scientists have shown that membrane-bounded droplets can form spontaneously in solution, they recognize that such droplets are devoid of cellular components and therefore doubt their role as the "first cells."

The history of life

When discussing the history of life on Earth, scientists divide the time from the formation of the Earth until the present day into five major time periods or **eras:**
- The **Archean Era,** which extends from the formation of the Earth 4.6 billion to 2.5 billion (2500 million) years ago
- The **Proterozoic Era,** which extends from 2500 to 590 million years ago
- The **Paleozoic Era,** which extends from 590 to 250 million years ago
- The **Mesozoic Era,** which extends from 250 to 65 million years ago
- The **Cenozoic Era,** which extends from 65 million years ago to the present

The eras are subdivided into shorter time units called **periods,** as shown in the table of geologic time (Table 26-1). Additionally, the periods of the Cenozoic era are further subdivided into **epochs.** However, each time unit (such as an era or a period) does not stand for a consistent length of time because early geologists defined and named these time units as they discovered them, and because the distinctive events that mark the beginning and end of a time unit occurred over varying amounts of time. A 1-year geological "calendar" is shown on the left of Table 26-1 on p. 450, which relates the geological time scale to a single year, making it easier to understand how long each geological time unit is in relation to the others.

The Archean Era: Oxygen-producing cells appear

Before the Archean Era (meaning "ancient life"), the Earth was coalescing from an original cloud of dust formed as the Earth was "born" 4.6 billion years ago. Approximately 800 million (0.8 billion) years passed before the first rocks were formed. Approximately 3.8 billion years ago, chemical evolution began, resulting, somehow, in the formation of the first cells 3.5 billion years ago. (Interestingly, some scientists have recently suggested that this 300 million year time span is *too short* a time for cells to have evolved from the chemicals of the primordial Earth!)

TABLE 26-1 Table of geological time

Era	Period	Epoch	(MYA)	Major biological and geological events
Cenozoic	Quaternary	Pleistocene	1-2	
	Tertiary	Pliocene	7	First humans.
		Miocene	26	Origin of first human-like forms.
		Oligocene	38	Monkey-like primates appear.
		Eocene	54	Origin of *Eohippus*.
		Paleocene	65	Small mammals undergo adaptive radiation.
Mesozoic	Cretaceous			Major extinction of the dinosaurs and many marine organisms. Flowering plants appear, insects become more diverse.
	Jurassic		130	Large dinosaurs dominate Earth. First birds.
	Triassic		210	Small dinosaurs appear. First mammals.
Paleozoic	Permian		250	Major extinction occurs. Most species disapppear. Conifers appear.
	Carboniferous		285	First reptiles and arthropods, coal deposits formed, horsetails, ferns, and seed-bearing plants abundant. Fungi appears.
	Devonian		370	"Age of the fishes." Fishes with bones and jaws appear. Amphibians also appear. Major extinction of marine invertebrates and fishes.
	Silurian		410	Notochord becomes flexible as single rod is replaced with separate pieces, as seen in the Ostracoderms (armored fish without bones, jaws, or teeth). Plants invade the land.
	Ordovician		430	First vertebrates appear. Major extinction of marine species.
	Cambrian		500	Major extinction of the trilobites. Origin of the main invertebrate phyla.
Proterozoic			590	Multicellular eukeryotic animals appear. First eukaryotic cells appear. Oxygen-producing bacteria present; atmosphere and oceans oxygenated.
Archean			2500	Earth is born. Chemical evolution resulting in formation of first cells. Stromatolites formed. First rock formed.

THE EVOLUTION OF THE FIVE KINGDOMS OF LIFE

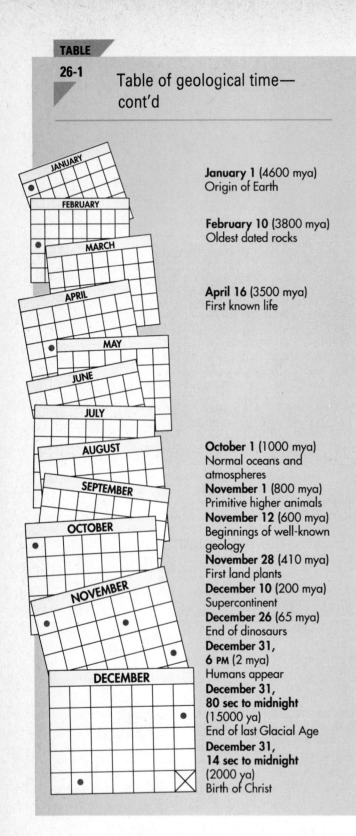

TABLE 26-1 Table of geological time—cont'd

January 1 (4600 mya)
Origin of Earth

February 10 (3800 mya)
Oldest dated rocks

April 16 (3500 mya)
First known life

October 1 (1000 mya)
Normal oceans and atmospheres
November 1 (800 mya)
Primitive higher animals
November 12 (600 mya)
Beginnings of well-known geology
November 28 (410 mya)
First land plants
December 10 (200 mya)
Supercontinent
December 26 (65 mya)
End of dinosaurs
December 31, 6 PM (2 mya)
Humans appear
December 31, 80 sec to midnight (15000 ya)
End of last Glacial Age
December 31, 14 sec to midnight (2000 ya)
Birth of Christ

Scientists speculate that the first cells to evolve fed on organic materials in the environment. Because there would have been a very limited amount of suitable organic food available, a type of nutrition must have soon evolved in which cells were able to capture energy from inorganic chemicals. The evolution of a pigment system that could capture the energy from sunlight and store it in chemical bonds most likely evolved after that. Then, oxygen-producing bacteria evolved. Probably similar to present-day **cyanobacteria** (formerly called blue-green algae), these single-celled organisms became key figures in the evolution of life as it is known today. Their photosynthesis gradually oxygenated the atmosphere and the oceans around 2 billion years ago as evidenced by Archean and Proterozoic sediments. Scientists have also discovered fossils of the cyanobacteria.

Although fossils of single-celled organisms are usually difficult to find, the cyanobacteria sometimes grew in "piles," creating fossilized columns of organisms and the sediments that collected around them. These ancient columns of cyanobacteria are called **stromatolites** and date back 3.5 billion years. In Figures 26-4, *A* and *C*, the tops of stromatolites are visible above the surface of the water in which they form. Like icebergs, much of the stromatolite is hidden under water, as shown in Figure 26-4, *B*. Actively developing stromatolite formations can be observed even today in the warm, shallow waters of places such as the Gulf of California, western Australia, and San Salvador, Bahamas. Interestingly, the shapes and sizes of the bacteria within present-day stromatolites look very much like the bacteria found within fossil stromatolites (Figure 26-5). In the absence of competing organisms, cyanobacteria produced stromatolites abundantly in all fresh water and marine communities until about 1.6 billion years ago.

> Scientists speculate that the first cells to evolve fed on organic materials in the environment. Because of limited organic material, the evolution of a pigment system that could capture the energy from sunlight and store it in chemical bonds most likely evolved relatively quickly. Then, oxygen-producing bacteria evolved, gradually oxygenating the atmosphere.

Until recently, all prokaryotes were considered to be genetically and evolutionarily similar. However, a small group of bacteria have been found to have biochemical differences from all other bacteria—differences that are as significant as the differences between prokaryotes and eukaryotes. This group of bacteria are fundamentally different from most bacteria in their ribosomal structure, RNA polymerase structure, cell membrane and cell wall composition, and other characteristics. Some of the "biochemically different" bacteria can multiply at higher temperatures than any other forms of life and require sulfur-containing compounds to survive. These unusual metabolic requirements remind many biologists of some of the habitats on Earth over 3 billion years ago. Such differences suggest an ancient evolutionary separation of these two "types" of bacteria. Speculation about the possible ancient origins of these cells led to naming them the **archaebacteria,** or "ancient bacteria." The larger group of more common bacteria are called

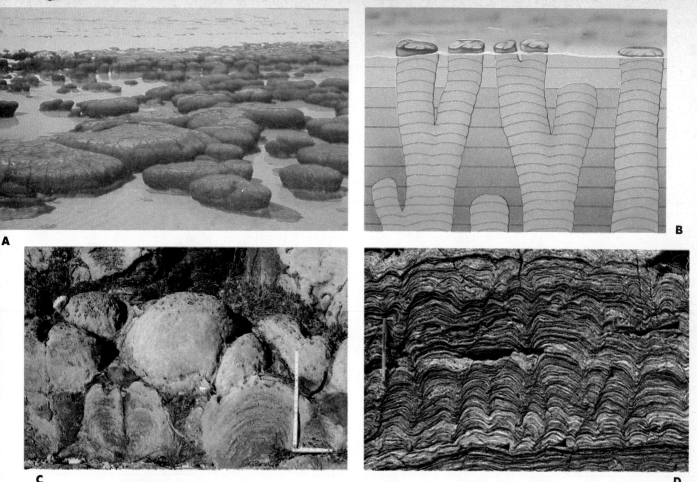

FIGURE 26-4 Stromatolites.
A These stromatolites are located in Shark Bay, Western Australia. The largest structures are about 1.5 meters (approximately 4½ feet) across. These stromatolites formed about 4000 years ago.
B This diagram shows that much of the stromatolite formation is underwater.
C Fossilized stromatolites (about 1.9 billion years old) in Great Slave Lake, Canada, showing some of the internal lamination typical of stromatolites.
D The internal, laminated structure of a fossilized stromatolite in Great Slave Lake, Canada.

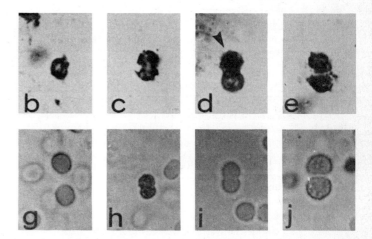

FIGURE 26-5 Fossilized bacteria.
The fossilized bacteria in **B** through **E** were found in stromatolites located in South Africa. These stromatolites are about 3.4 billion years old. The bacteria in **G** through **J** show living, similar bacteria. The fossilized bacteria bear a striking resemblance to the living bacteria.

THE EVOLUTION OF THE FIVE KINGDOMS OF LIFE

eubacteria, or "true bacteria." Little is known about the evolutionary history of the archaebacteria, but scientists think that archaebacteria and eubacteria most likely had a common ancestor.

Most scientists still classify the archaebacteria and the eubacteria in a single kingdom, usually called *Monera*. However, others place the archaebacteria in a kingdom separate from the eubacteria, creating two prokaryotic kingdoms, or six kingdoms in all. Still other scientists propose a three kingdom system of classification: archaebacteria, **eubacteria,** and eukaryotes. They then further subdivide the kingdom of eukaryotic organisms into familiar subkingdoms: the plants, animals, protists, and fungi.

The Proterozoic Era: The first eukaryotes appear

The beginning of Proterozoic time, which extends from 2.5 billion to 590 million years ago, is marked by the formation of a stable oxygen-containing atmosphere. The term *Proterozoic* is derived from two Greek words meaning "prior" (proteros) "life" (zoe) and is descriptive of the fact that, until a few decades ago, scientists did not find evidence of life during this time (or during Archean time) because they were not looking for microfossils.

During the Proterozoic Era, the first eukaryotic organisms appeared on Earth. A fossil stromatolite found in California dating back 1.4 to 1.2 billion years ago contains larger and more complex microfossils than the fossil cyanobacteria previously found. Scientists think that these microfossils may have been eukaryotes, organisms separate from the cyanobacteria that may have lived in harmony with stromatolites or have fossilized with them. Other fossils in rocks of the same time add additional evidence to the theory that eukaryotes appeared around 1.4 billion years ago.

How did the eukaryotes arise? The most widely accepted hypothesis at this time is called the **endosymbiotic theory.** The term *symbiosis* refers to a relationship between two different organisms in which both organisms benefit. According to this theory, bacteria became attached to or were engulfed by "pre-eukaryotic" (host prokaryotic) cells. The bacteria carried out chemical reactions necessary for living in an atmosphere that was increasing in its amounts of oxygen, which benefited their host cells. The bacteria, or endosymbionts, benefited by living within the protective environments of their hosts. The cells that engulfed the bacteria had a reproductive advantage over those cells not associated with bacteria and, therefore, flourished.

The cell membranes of the pre-eukaryotes are thought to have pouched inward during these symbiotic relationships, loosely surrounding the bacteria. These membranous invaginations are thought to have been precursors to nuclear envelopes and endoplasmic reticula. The bacteria are thought to be "early" mitochondria. Symbiotic events similar to those postulated for the origin of mitochondria also seem to have been involved in the origin of chloroplasts, which are thought to be derived from symbiotic photosynthetic bacteria (Figure 26-6).

Multicellular eukaryotic animals date back 630 million years. Scientists have found fossils of a diverse array of un-

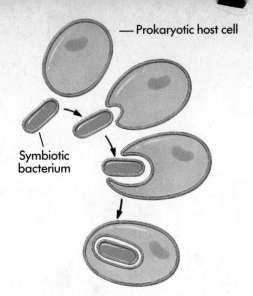

FIGURE 26-6 Endosymbiosis.
How pre-eukaryotes are thought to have engulfed bacteria, which may have become mitochondria (heterotrophic bacteria) and chloroplasts (photosynthetic bacteria) as eukaryotic cells evolved.

usual multicellular eukaryotes dating from that time. They regard these soft-bodied organisms as ancestors of the jellyfish, hydra, soft coral, and worms. However, other scientists suggest that they are not related to any currently existing species.

> Fossil evidence suggests that the first eukaryotes, larger and more complex than the cyanobacteria, appeared around 1.4 billion years ago. The chloroplasts and mitochondria of eukaryotic cells are thought to be derived from bacteria (referred to as *endosymbionts*) that came to live within pre-eukaryotic cells. Multicellular eukaryotes appeared approximately 630 million years ago.

The Paleozoic Era: The occupation of the land

The Paleozoic Era (590 to 250 million years ago) is marked by an abundance of easily visible, multicellular fossils. In fact, the earliest such fossils are found in rocks of southern Australia and are approximately 630 million years old. For many years such fossils were the oldest known, which led scientists to name the era in which these fossils became plentiful as the *Paleozoic*, meaning "old" (*paleos*) "life" (*zoos*). However, of the roughly 250,000 different kinds of fossils that have been identified, described, and named, only a few dozen are more than 630 million years old.

The Paleozoic Era is divided into six shorter time spans called *periods*. (These periods are shown in Table 26-1.) The **Cambrian Period,** which ended roughly 500 million years ago, is the oldest period within the Paleozoic Era. It represents an important point in the evolution of life; all of the main phyla and divisions of organisms that exist today (except for the chordates and land plants) evolved by the end of the Cambrian Period. Because so many new kinds of

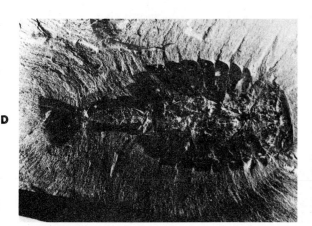

FIGURE 26-7 Cambrian Period fossils from the Burgess Shale, British Columbia, Canada, about 530 million years old.
A and **B** *Sidneyia inexpectans* and a model of the organism. This creature was an arthropod.
C *Hallucigenia sparsa.* This specimen is 12.5 millimeters (approximately ½ inch) long. This animal seems to have been supported on seven pairs of spines; its trunk bore seven long tentacles and an additional group of tentacles near the rear end. *H. sparsa* may have been a scavenger on the bottom of the sea.
D *Wiwaxia corrugata.* The body of this animal was covered with scales and also bore spines. This specimen is 30.5 millimeters (approximately 1¼ inches) across, excluding the spines. It may have been a distant relative of the mollusks.
E *Burgessochaeta setigera*, a segmented worm. The photograph shows the front end of the worm, which had a pair of tentacles. Foot-like structures are attached to the worm in pairs along the sides of the body. The specimen is 16.5 millimeters (approximately ⅝ inch) long.

organisms appeared in such a relatively short time span, paleontologists (scientists who study fossil life) speak of a Cambrian "explosion" of living forms and often refer to geological strata older than Cambrian time as **Precambrian**.

The evolution of Cambrian organisms took place in the sea. Figure 26-7 shows fossils of the organisms that lived on the seafloor during the Cambrian Period—quite unusual by today's standards! These fossils, together with well over 100 other species, have been found in the Burgess Shale, a geological strata formed from a fine-grained mud (Figure 26-8). During past movements of the Earth's crust, the Burgess Shale was uplifted to high elevations in the Rocky Mountains of British Columbia, Canada.

FIGURE 26-8 The Burgess Shale, located in the Rocky Mountains of British Columbia.
The stratification of this formation can be clearly seen.

THE EVOLUTION OF THE FIVE KINGDOMS OF LIFE

Many kinds of multicellular organisms that thrived in the early Paleozoic Era have no relatives living today. As multicellular organisms arose that were capable of moving from place to place, the forces of natural selection resulted in an array of organisms capable of filling previously unoccupied ecological niches, or roles, within their communities (see Chapter 37). These possibilities led to diversification along new evolutionary pathways; some resulted in "dead ends," and others ultimately led to the contemporary phyla of organisms. The trilobites, for example (Figure 26-9, *A*), appear to be derived from the same evolutionary line that gave rise to one living group of arthropods, the horseshoe crabs (Figure 26-9, *B*).

> During the Cambrian Period of the early Paleozoic Era, an array of species evolved in the sea. Some of these species still exist today.

As the Paleozoic Era continued into the **Ordovician Period** (500 to 430 million years ago), worm-like aquatic animals similar to lancelets and lampreys began to evolve. Characteristic of these organisms was the stiff, internal rod that ran down the back, parallel to the central nerve. This stiffening rod is called a **notochord** (meaning "back cord"); these organisms are therefore called **chordates.** As the **Silurian Period** began 430 million years ago, this rod was replaced by a sequence of pieces that provided more flexibility for the animal and protected the nerve cord. Eventually these organisms developed a bony, protective armor and cartilaginous internal skeletons. They are considered the first vertebrates—the **ostracoderms.** During this time, some groups of fishes developed jaws, bony internal skeletons, and scales, much like those of the fishes of today. During the **Devonian Period** (410 to 370 million years ago), these fishes became abundant and diverse, giving the Devonian the nickname "the age of fishes."

At the interface of the Silurian and Devonian Periods, the first terrestrial plants evolved. The earliest known fossil plants are about 410 million years old. The ancestors to these plants are believed to be the green algae, since their chloroplasts are biochemically similar to those of the plants. The green algae are an extremely varied group of more than 7000 protist species, found growing in the water as well as on tree trunks or in the soil. Many of these algae are microscopic and unicellular. However, the plants that appeared during the Silurian were multicellular and had developed mechanisms for transporting water within their bodies as well as for conserving it. Within the next 100 million years, plants became abundant and diverse on the land and eventually formed extensive forests.

Amphibians, the descendants of fishes, appear in the fossil record by the end of this period and subsequently became abundant in the great swamps that formed at that time. Leading "double lives," they spent part of their time on land and part of their time in the water. Then, about 300 million years ago in the late **Carboniferous Period,** the amphibians gave rise to the first reptiles. The reptiles, being better suited for a terrestrial existence than their am-

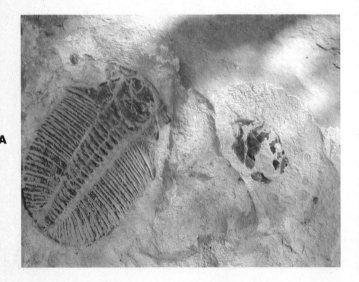

FIGURE 26-9 Trilobites and horseshoe crabs.
The trilobites flourished in the seas of the Cambrian Period, 500 to 550 million years ago. The example in **A** was found in the Burgess Shale. Trilobites appear to be directly related to modern-day horseshoe crabs (**B**).

phibian ancestors, became the dominant organisms over the next 50 million years.

The Carboniferous Period (370 to 285 million years ago) is named for the great coal deposits formed during this time. Much of the land was low and swampy due to a worldwide moist, warm climate—conditions that contributed to the fossil preservation of the still-prevalent forests. These coal deposits provide a relatively complete record of the horsetail plants, ferns, and primitive seed-bearing plants of the Carboniferous Period.

The abundant plant life of that time provided the opportunity for invasion of the land by insects, a type of arthropod. Arthropods are invertebrate animals having segmented bodies, jointed appendages, and a tough, outer "skin" called a *cuticle*. Arthropods evolved as a phylum during the Cambrian from segmented "worm-like" animals. During the subsequent history of this group, only a few kinds of insects have moved back into fresh water habitats, and a few have returned to marine habitats, living on the surface of the ocean.

The fungi, a kingdom of organisms that live on dead and decaying matter, also invaded the land in the late Carboniferous Period. One component of the success of the fungi on land has been related to the structure of their chitin-rich cell walls. Chitin, a stiff, hard substance that also characterizes the outer skeletons of the arthropods, helped both insects and fungi to remain "drought resistant." In addition, fungi are successful land organisms because of their roles (along with the bacteria) as decomposers.

> As the Paleozoic Era progressed, chordate and then vertebrate animals evolved in the sea. The first vertebrates appeared approximately 430 million years ago. Plants and insects colonized the land about 410 million years ago; amphibians did so about 50 million years later. Fungi may also have colonized the land at about the same time as plants.

The **Permian Period** (285 to 250 million years ago) was a period of drought and extensive glaciation, unlike the swamps of the preceding Carboniferous Period. The conifers—a group of seed-bearing plants that is represented today by pines, spruces, firs, and similar trees and shrubs—originated then. A major extinction event occurred at the end of the Permian, wiping out many of the species living at that time.

Mass extinctions

One of the most prominent features of the history of life on earth has been periodic **mass extinctions,** the global death of whole groups of organisms. There have been five such events. Four of these events occurred during the Paleozoic Era, the first occurring near the end of the Cambrian Period (about 500 million years ago). At that time, most of the existing families of trilobites became extinct. Additional mass extinctions occurred about 440 and about 360 million years ago. The fourth and most drastic mass extinction happened at the close of the Permian Period. It is estimated that approximately 96% of all species of marine animals living at that time died out! All of the trilobites, and many other groups of organisms as well, disappeared forever. The fifth major extinction event occurred at the close of the Mesozoic Era, 65 million years ago. This was the famous event when dinosaurs became extinct.

Major episodes of extinction produce conditions that can lead to the rapid evolution of those relatively few plants, animals, and microorganisms that survive. Habitats, or places to live, formerly occupied suddenly open up, providing the "raw material" for **adaptive radiation,** a phenomenon by which populations of a species change as they disperse within a series of different habitats within a region (see Chapter 25). In fact, many paleontologists suggest that extinction episodes have been key events in the "pattern" of evolution.

The Mesozoic Era: The age of reptiles

The Mesozoic Era, or "middle life" (250 to 65 million years ago), was a time of adaptive radiation of the terrestrial plants and animals that had been established during the mid-Paleozoic Era but that had not been "wiped out" by the mass extinction of the Permian Period. The radiation of these organisms led to the establishment of the major groups of organisms living today (Figure 26-10). This era was dom-

FIGURE 26-10 Basic body plans were established long ago. The fossil dragonfly, which is about 170 million years old, closely resembles its modern counterpart. It dramatically illustrates the establishment of modern groups of organisms in the Mesozoic Era; many of these have persisted to the present day.

inated by the adaptive radiation of the reptiles. During the course of the Mesozoic, one group of reptiles evolved into mammal-like reptiles, precursors to mammals; another group led to the evolution of the birds; and yet another developed into the dinosaurs, or "terrible lizards." Reptiles also invaded the sea and the air, but these kinds of reptiles eventually died out.

The Mesozoic is divided into three periods: the **Triassic,** the **Jurassic,** and the **Cretaceous.** These names refer to geological events or formations of the eras and do not chronicle the evolution of life during this time. During the Triassic Period, the earliest period of the Mesozoic, small dinosaurs and primitive mammals appeared. Using fossil evidence regarding changes in the jaws and teeth of these organisms, some scientists think that mammals evolved from reptiles of the Permian Period (Paleozoic Era) that were distinguished by "sails" on their backs. Scientists speculate that their upright fans of tissue collected the sun's heat when it was cold and radiated excess body heat when it was hot.

During the Jurassic Period, large dinosaurs dominated the Earth, and birds first appeared. The earliest known bird, *Protoavis* ("first bird") (Figure 26-11), was reported in 1986 by Sankar Chaterjee of Texas Tech University. The fossil bones of this organism were found in rocks dated to be 225 million years old, very early in the Mesozoic Era. Until this find, the oldest known fossil bird was *Archaeopteryx*, estimated to be about 150 million years old (Figure 26-12).

FIGURE 26-11 **Protoavis, a 225 million-year-old fossil bird, from west Texas.**
This remarkable find, which, if it proves to be a true ancestor of birds, will extend the history of birds back 75 million years, was announced by Sarkar Chaterjee of Texas Technological University in 1986.

FIGURE 26-12 **Protoavis.**
About the size of the crow, *Protoavis* lived in the forests of central Europe 150 million years ago. The teeth and long, jointed tail are features not found in any modern birds. Discovered in 1862, this bird was cited by Darwin in later editions of *On the Origin of Species* in support of his theory of evolution and was originally called *Archaeopteryx*.

Was *Protoavis* the first bird?

Those who study the history of life on Earth by examining fossils are paleontologists, and in June 1991 a paleontologist named Sankar Chaterjee published a detailed description of a fossil that has created a furor among other paleontologists. The fossil he described was 225 million years old, from the late Triassic Period, found by Chaterjee and colleagues 5 years earlier as a jumble of bones in an ancient west Texas mudbed (Figure 26-A). After 5 years of careful study, Chaterjee had assembled the bones into a nearly complete skeleton—and he claimed it was the skeleton of a bird! He gave it the name *Protoavis* ("first bird").

Why all the fuss? Chaterjee's claim is in effect a frontal assault on one of the most widely known animals in all paleontology—the fossil bird *Archaeopteryx* ("ancient wing"). *Archaeopteryx* dates back to the late Jurassic Period, about 150 million years ago, during the middle of the dinosaurs' reign on earth, whereas Chaterjee's fossil is as old as the oldest dinosaur, from the time when they first appeared on Earth. Most paleontologists believe that birds evolved from dinosaurs in the mid-Jurassic Period (in effect, that birds are dinosaurs with a feather coat) and that *Archaeopteryx* represents a transitional form, having the feathers and breastbone of a bird and the teeth and tail of a dinosaur.

Chaterjee bases his claim on the fact that the skeleton shows bird-like elements, such as a wishbone, a shoulder modified for flying, and a keeled sternum (which serves as an attachment point for flight muscles in modern birds). He claims the skull has no fewer then 23 features that are fundamentally birdlike. In particular, the creature's skull appears to lack holes that are present on the skulls of dinosaurs of the period; in *Protoavis,* these holes have merged with the eye sockets, making it similar to modern birds. The jaws of *Protoavis* are able to slide forward, and the upper jaw can elevate. Modern birds possess both these features, but ancient reptiles had neither.

The mud in which *Protoavis* fossilized was incapable of preserving any feathers, however, and indeed most of the *Protoavis* bones are poorly preserved. Other paleontologists are not convinced by Chaterjee's reconstruction, complaining that the poor state of the fossils makes it extremely difficult to identify many of the features important to Chaterjee's argument. Also, because the fossil was not articulated (that is, its bones were not attached to one another but were found lying in a jumble), they worry that Chaterjee's skeleton may in fact contain the bones of several different kinds of animals. If Chaterjee is right, birds are not the descendants of dinosaurs but their cousins. If a vote were taken today among paleontologists, his claim would probably lose. The world of paleontology awaits a better fossil.

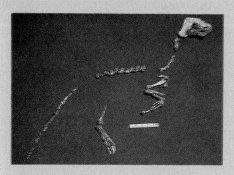

FIGURE 26-A The bones of *Protoavis.* Was *Protoavis* the first bird?

Whether *Archaeopteryx* arose from the same evolutionary line as *Protoavis* or independently from another dinosaur is not known. Some scientists question the validity of Chaterjee's "find" (see the boxed essay).

Archaeopteryx had feathers but apparently not enough of them to fly very effectively. Feathers, thought to have evolved from reptilian scales, probably helped "early" birds glide from tree to tree and provided them with insulation. Ultimately, feathers made possible the subsequent evolution of birds that could fly well. This ability allowed birds to inhabit unoccupied habitats and resulted in the evolution of a large and diverse class of organisms. Today, there are about 9000 different species of birds.

> **During the Mesozoic Era, the reptiles, which had evolved earlier from amphibians, became dominant and in turn gave rise to mammals (about 200 million years ago) and birds (at least 150 million years ago).**

During the Cretaceous Period, about 127 million years ago, flowering plants (angiosperms) began to appear, becoming the dominant form of plant life about 100 million years ago. Seed-bearing plants with fern-like leaves, similar to the living cycads, were abundant at that time. As flowering plants became more diverse, so did insects, which had feeding habits that were closely linked with the characteristics of flowering plants. Insects and flowering plants have **coevolved** with one another. Indeed, all groups of terrestrial organisms, including mammals, birds, and fungi, have evolved characteristics largely in relation to those of the flowering plants. Today, with about 240,000 species of flowering plants, the angiosperms still dominate the plant kingdom, greatly outnumbering all other kinds of plants.

Around 65 million years ago, as the Cretaceous Period (and the Mesozoic Era) ended, sudden shifts occurred in the kinds of marine organisms that existed. Some became extinct and others began to flourish. Scientists can infer these changes in the populations of marine organisms living at that time by studying marine fossils exposed in certain European geological strata. For example, many of the larger plankton (free-drifting microscopic protists) disappeared about 65 million years ago, and a fewer number of smaller ones took their place. The same rapid changes occurred in at least some nonplanktonic marine animal groups, such as the bivalve mollusks (clams and their relatives). The ammonites, a large and diverse group related to octopuses but with shells, abruptly disappeared. And, of course, a major extinction occurred on land: the dinosaurs disappeared, although at a pace much slower than that of the plankton.

Scientists have long discussed hypotheses regarding the extinction event of the late Cretaceous. Some propose that the climate cooled. Others have suggested that the sea level changed or that the vegetation cover of the earth changed. These reasons, however, would not account for the change in the plankton community. In 1980 a group of scientists headed by Luis W. Alvarez of the University of California, Berkeley, presented a dramatic hypothesis about the reasons for these changes. Alvarez and his associates observed that the usually rare element iridium was abundant in a thin layer in the geological strata that marked the end of the Cretaceous Period. Alvarez and his colleagues proposed that if a large, iridium-rich meteorite or asteroid had struck the surface of the earth, a dense particulate cloud would have been thrown up. This cloud would have darkened the earth for a time, greatly slowing or temporarily halting photosynthesis and driving many kinds of organisms (particularly photosynthetic plankton) to extinction. By disrupting and killing plant life, other organisms dependent on the plants would die out as well. Then, as its particles settled, the iridium in the cloud would have been incorporated in the layers of sedimentary rock that were being deposited at that time. Although the "meteorite hypothesis" is the most plausible one put forth to date, scientists are not certain that the impact happened, nor that it caused any of the extinctions of that time.

Changes in the Earth: The impact on evolution

During the Mesozoic Era, the continents did not exist as they do today. Instead, they were all joined in one giant continent scientists call **Pangea,** meaning "all Earth." Pangea began to break up into smaller pieces during the Mesozoic Era, with this movement continuing during the following era, the Cenozoic. These changes greatly affected evolution.

After the Earth coalesced approximately 4.5 billion years ago, its surface was hot and violent for about 600 to 700 million years. At first, these conditions were too inhospitable for living things or even for biochemical evolution to take place. Scientists hypothesize that geological activity eventually resulted in the formation of fragments of continents. Some geologists think that these land masses were formed as melted rock material rose to the surface of the Earth, pushed upward by other heavier, sinking, melted metals such as iron and nickel. Others hold that the highly active volcanoes prevalent at that time spewed so much lava into the seas that it eventually accumulated, forming land masses.

Approximately 200 million years ago, small land masses united to form the single, large "supercontinent" Pangea. Its northern half was called **Laurasia** and consisted of the present-day North America, Europe, and Asia. Its southern half was called **Gondwana** and was a combination of present-day South America, Africa, India, Antarctica and Australia. Pangea remained as a supercontinent for approximately 100 million years. However, forces were at work beneath it that eventually divided it—once again—into smaller land masses.

Scientists now have some insight into the movement of land masses over time, such as those that resulted in the formation of Pangea and its subsequent division. Today, the outer shell of the Earth is made up of six large "plates" and several smaller ones (Figure 26-13). According to the **plate tectonics theory,** these plates are rigid pieces of the Earth's crust, which are from 75 to 150 kilometers (approximately 50 to 100 miles) thick. The rocks beneath these plates are less rigid than the plates they support. The surface plates move, or drift, over this underlying semiliquid rock. Scientists think that this "liquid" rock flows slowly, rising as it heats, and sinking as it cools (Figure 26-14). At the mid-oceanic ridges—long, narrow mountain ranges under the sea—the plates move away from one another in a process called *seafloor spreading*. Molten rock wells up from deep within the Earth, pushing the plates apart and filling in the space between. At other places the plates move toward each other. As they meet, one plate sinks below the other, pushing up mountain ranges in the process. The Himalayas, for example, have been thrust up to the highest elevations on Earth as a result of the grinding, prolonged collision of the Indian subcontinent with Asia. Other types of violent geological activity are also associated with the movements of the crustal plates. The earthquakes that destroyed San Francisco in 1906 and the one that occurred in the same region in 1989 resulted from two plates sliding past each other.

Figure 26-15 summarizes the movements of the continents as they gradually moved apart from their positions as parts of Pangea 200 million years ago. The South Atlantic Ocean opened about 125 to 130 million years ago as Africa and South America moved apart from one another. South America then moved slowly toward North America and, approximately 3.1 to 3.6 million years ago, became connected to it by the Isthmus of Panama. Changes occurred in the positions of all the continents, which played a major role in the distribution of organisms that are seen today.

As land masses move away from one another, populations inhabiting those lands become separated from one another. Geographically isolated, populations of the same species often undergo speciation as differing pressures of natural selection act on them. Similarly, as land masses collide, species are thrust together and must compete with one another to survive. Shorelines disappear and with them go a variety of habitats and species. And as continents move, ocean currents change, causing climatic changes and yet new selection pressures.

The Cenozoic Era: The age of mammals

The mass extinctions of the Cretaceous marked the end of the Mesozoic Era and heralded in the Cenozoic Era about 65 million years ago. Extending to the present, the Cenozoic Era has two periods—the **Tertiary** and the **Quaternary**—

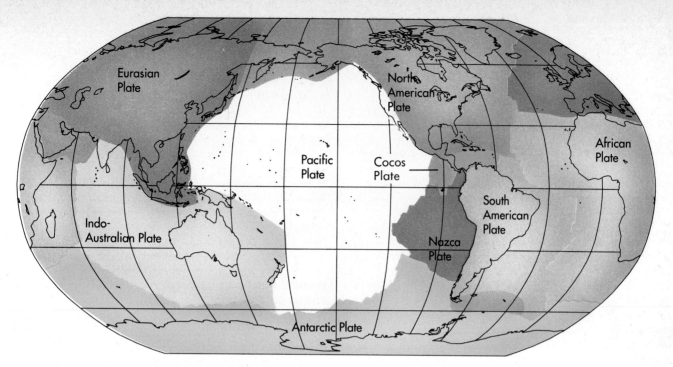

FIGURE 26-13 Plate tectonics theory.
Plate tectonics describes and explains the movement of the continents. According to this theory, Earth is made up of six large plates and several smaller plates. The rocks beneath these plates are semiliquid. The plates drift over this semiliquid rock, occasionally colliding. When this occurs, one plate is usually pushed below the other, a process called *subduction*. When plates move apart, the semiliquid rock wells up in the space created, a process called *seafloor spreading*.

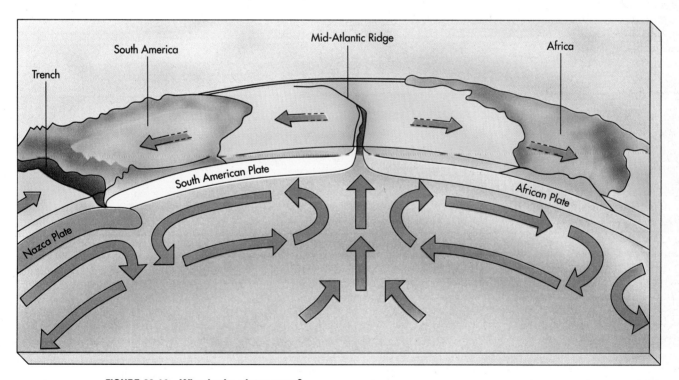

FIGURE 26-14 Why do the plates move?
The semiliquid rock on which the plates rest rises as it heats and sinks as it cools, causing movement.

THE EVOLUTION OF THE FIVE KINGDOMS OF LIFE

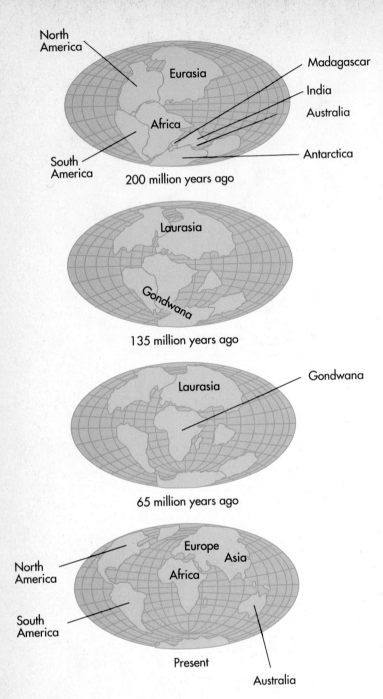

FIGURE 26-15 Movement of the continents.
The changes that occurred in the position of all the continents as a result of shifting plates have played a major role in the distribution of organisms seen today.

which are subdivided into many smaller time units called *epochs*. The epochs of the Tertiary Period are listed in Table 26-1. Notice also in Table 26-1 that most of the "year" has passed on the geologic calendar. Although 65 million years may seem like an incredibly long time, in terms of the history of life, it is incredibly short. During this very short time, life as it is known has evolved.

At the beginning of the Cretaceous during the **Paleocene Epoch,** the small mammals that survived the extinctions of the Mesozoic underwent adaptive radiation, quickly filling the habitats vacated by the dinosaurs. By the **Eocene Epoch,** many species of mammals were no longer small. Some of the mammals appearing at this time are organisms that were ancestors to present-day whales and bats. Eohippus, a small mammal (about the size of a dog) that gave rise to the horse, also appeared at that time. Eohippus evolved from an ancestor that gave rise to several groups of mammals, including tapirs and rhinoceroses, in addition to the horses. Monkey-like primates, the highest order of mammals, first appear as fossils about 36 million years ago in the early **Oligocene Epoch.** These **anthropoid primates**—monkeys, apes, and human beings and their direct ancestors—are the focus of Chapter 27, a continuation of the story of the evolution of life on Earth.

EVOLUTION: HOW LIVING THINGS CHANGE OVER TIME

Summary

1. Approximately 15 billion years ago, scientists theorize, all the material of the universe condensed into a single small space and then exploded in an event called the "big bang," forming an early generation of stars. Then, approximately 4.6 billion years ago, these stars exploded, forming the Earth and the other planets of the solar system.

2. When the Earth cooled, a hardened outer crust formed. This time is marked by the formation of the rocks, the oldest of which have been dated at about 3.8 billion years old. Fossil bacteria about 3.5 billion years old are the oldest direct evidence of life on Earth. These fossil bacteria, or cyanobacteria, are found in the massive limestone deposits of stromatolites.

3. Somehow, in the 300 million years between the the time when the Earth cooled and the first bacteria appeared, biochemical evolution became biological evolution. The long-held "primordial soup" theory states that life arose in the sea as elements and simple compounds of the primitive atmosphere reacted with one another to form simple organic molecules such as amino acids and sugars. Recently, scientists have begun to reexamine the premises of this theory and have proposed alternate hypotheses.

4. After the synthesis of complex organic molecules had taken place, self-replicating systems of RNA molecules may have begun the process of biological evolution. To be considered life, however, this "early" genetic material must have been organized within a cellular structure. Although scientists have shown that membrane-bounded droplets can form spontaneously in solutions, they recognize that such droplets are devoid of cellular components and doubt their role as the "first cells."

5. Scientists speculate that the first cells to evolve fed on organic materials in the environment. Because of limited organic material, the evolution of a pigment system that could capture the energy from sunlight and store it in chemical bonds most likely evolved relatively quickly. Then, oxygen-producing bacteria evolved, gradually oxygenating the atmosphere.

6. The first unicellular eukaryotes appeared about 1.4 billion years ago during the Proterozoic Era. Eukaryotic cells likely arose as "pre-eukaryotic" cells became hosts to endosymbiotic prokaryotes. Multicellular organisms first appeared about 630 million years ago; the earliest, soft-bodied forms are poorly represented in the fossil record.

7. Many of the phyla of organisms in existence today, except the chordates and the land plants, appear to have evolved during the Cambrian Period (590 to 500 million years ago) of the Paleozoic Era. These organisms evolved in the sea. Chordates and then vertebrates evolved later in the Paleozoic Era, with the first vertebrates appearing about 430 million years ago.

8. Terrestrial plants appeared about 410 million years ago and flourished during the Carboniferous Period (370 to 285 million years ago). Fishes became abundant and diverse during the Devonian Period (410 to 370 million years ago). The amphibians, organisms that spend part of their time on land and part in the water, evolved from the fishes, appearing about 360 million years ago. Then, about 300 million years ago, the amphibians gave rise to the first reptiles.

9. Insects evolved on land during the late Carboniferous Period. The abundant plant life of that time provided them with food and shelter. The fungi, a kingdom of organisms that live on dead and decaying matter, also evolved on land during the late Carboniferous Period.

10. A major extinction event occurred during the Permian Period, wiping out many of the species living at that time. The Mesozoic Era (250 to 65 million years ago) was a time of the adaptive radiation of the organisms surviving the Permian extinction, dominated by the evolution of the reptiles. The reptiles gave rise to the mammals (about 200 million years ago) and the birds (at least 150 million years ago).

11. The evolution of the five kingdoms of life has been influenced by movement of the land masses that has occurred over time.

REVIEW QUESTIONS

1. What is the "primordial soup" theory, and what does it attempt to explain? Describe another theory that explains the same event.
2. What is dehydration synthesis? What role might it have played in the evolution of life?
3. Explain the importance of a cell membrane to the evolution of early cells.
4. Create a table that lists the major eras and periods of geological time.
5. What are cyanobacteria, and when did they first appear? Explain their evolutionary significance.
6. Is the term *Proterozoic* accurate? Explain your answer.
7. When was the Cambrian Period, and why was it important?
8. Which period has been called *the age of fishes*, and why? What other evolutinary event occurred during this period?
9. Describe the major events of the Carboniferous Period. What were the living conditions on Earth during this time?
10. Summarize the major extinction events that have occurred on Earth. When did each happen?
11. Fill in the blanks: During the Mesozoic Era, one group of _____ evolved into precursors of _____, another led to the evolution of _____, and a third group developed into the large _____.
12. What might have caused the extinction event of the late Cretaceous Period? What organism(s) became extinct at that time?
13. What is the plate tectonics theory? Relate it to the evolution of living organisms.
14. Summarize the events of the Cenozoic Era. When did it begin?
15. Fill in the blanks: Monkeys, apes, and human beings are all examples of _____ _____, a particular order of _____.
16. Many movies have shown early human beings battling ferocious dinosaurs. Is this accurate? Explain.

KEY TERMS

adaptive radiation 455
anthropoid primate 460
archaebacteria 450
Archean Era 448
Cambrian Period 452
Carboniferous Period 454
Cenozoic Era 448
chordate 454
coevolution 457
Cretaceous Period 456
cyanobacteria 450
Devonian Period 454
endosymbiotic theory 452
Eocene Epoch 460
epoch 448
era 448
eubacteria 452
Gondwana 458
Jurassic Period 456
Laurasia 458
mass extinction 455
Mesozoic Era 448
notocord 454
Oligocene Epoch 460
Ordovician Period 454
ostracoderm 454
Paleocene Epoch 460
Paleozoic Era 448
Pangea 458
period 452
Permian Period 455
plate tectonics theory 458
Precambrian 453
proteinoids 447
Proterozoic Era 448
protocell 448
Quaternary Period 458
Silurian Period 454
stromatolite 450
template 447
Tertiary Period 458
Triassic Period 456

THOUGHT QUESTIONS

1. The Miller-Urey experiments suggest that the key building blocks of biological macromolecules could have been present in a "primordial soup," but the experiments do not suggest a way that the building blocks could have assembled themselves. The problem is that RNA monomers join in a dehydration synthesis to form RNA chains (that is, water is a *product* of the reversible reaction). So how could the first RNA molecules have formed in a watery primordial soup?
2. The most diverse time in the history of life was the pre-Cambrian Era, when creatures with many different body plans lived. Only a few of these basic body plans survived into the post-Cambrian Era, the other body plans were weeded out by evolution. Stephen Gould has speculated that evolution does not foster ever-greater complexity, more complex forms evolving from simpler primitive ones, but rather acts in the opposite fashion, tending to eliminate more and more options. Discuss his suggestion in light of what you have learned about evolution.
3. Archaebacteria have introns and in many other ways seem to be the direct ancestors of eukaryotes. Eubacteria, on the other hand, seem the direct ancestors of mitochondria. Which transition must have happened first?

FOR FURTHER READING

Alvarez, W., & Asaro, F. (1990, October). What caused the mass extinction?—an extra-terrestrial impact. *Scientific American*, pp. 78-84.
This article features the debate of the cause of the dinosaurs' extinction. The discoverers of iridium deposited at the time argue that a giant meteorite collided with the Earth.

Cortillot, V. (1990, October). What caused the mass extinction?—a volcanic eruption. *Scientific American*, pp. 85-92.
In a companion article, geologists argue that the iridium came not from a meteorite but from within the Earth, blown into the sky by enormous volcanoes produced when plumes in the Earth's mantle reached the surface 65 million years ago.

Silk, J. (1992, April). Cosmology back to the beginning. *Nature*, pp. 741-742.
Using data drom NASA's COBE satellite, scientists have detected temperature ripples in the cosmological microwave background, which they believe are remnants of the big bang that spawned the universe. This short article describes the findings in a simple, straightforward way.

Zimmer, C. (1992, May). Ruffled feathers. *Discover*, pp. 44-54.
Sakar Chaterjee, the Texas Tech University professor who claims he has discovered the fossil of the first bird, has been hit with charges from other scientists that his work is misleading and his conduct unprofessional. This article delineates the Protoavis controversy and provides a look into the sometimes contentious process of scientific discovery.

Human evolution

27

Footprints in the sand of a parent and child walking hand in hand on the wet edge of a beach might leave a permanent imprint of their presence. But the story behind these footprints is more complex than it might appear. Not surprisingly, these footprints were made near a lake. However, they are not embedded in sand but in volcanic ash that has preserved them for over 3½ million years. Although human in appearance, these footprints are not human. They were made by non-human ancestors, *Australopithecus afarensis,* and give scientists important clues to our heritage. One clue, for example, is the size of the footprints. Some scientists suspect that they were made by a male and a female rather than by a parent and child and therefore may reflect our ancestors' sexual dimorphism, or size difference with gender. Another clue to our heritage is that these organisms are walking erect on two legs instead of four, a characteristic called *bipedalism*. Bipedalism is only one of the many evolutionary changes that led to the appearance of modern humans.

HIGHLIGHTS

▼

The story of human evolution began over 500 million years ago with the appearance of the first chordates, organisms with a stiff internal rod that runs down the back beneath a central nerve.

▼

Vertebrates evolved from the chordates and include a class of organisms called *mammals*, animals that are warm blooded and have hair and whose females secrete milk from mammary glands to feed their young.

▼

The primates, the order of mammals to which humans belong, reflect an arboreal heritage having excellent depth perception and flexible, grasping hands; the earliest primates arose about 75 million years ago and lived on the ground before moving into the trees.

▼

The first human fossils date about 2 million years ago and are of the extinct species *Homo habilis*, a species that is considered human because they exhibited an intelligence far greater than their ancestors by making tools and clothing.

OUTLINE

Evolution of the vertebrates
 Evolution of the fishes
 Evolution of the amphibians, reptiles, and birds
 Evolution of the mammals
Evolution of the primates
 The prosimians
 The anthropoids
Evolution of the anthropoids
Evolution of the hominids
 The first hominids: *Australopithecus*
 The first humans: *Homo habilis*
 Human evolution continues: *Homo erectus*
 Modern humans: *Homo sapiens*

Evolution of the vertebrates

The beginning of the human evolution story can be traced all the way back to Cambrian times (see Chapter 26) and the appearance of the first chordates. This phylum of animals is named for the presence of a notochord: a stiff, internal rod that runs down the back beneath the central nerve cord. One of the earliest known chordate fossils is shown in Figure 27-1, A. This photo is of *Pikaia*, a small fishlike organism that resembles today's lancelets, shown in Figure 27-1, B.

Fossil evidence suggests that ancient flatworms are the ancestors to all the more complex animals, including the chordates (and *Pikaia*). As the flatworms exploited the variety of habitats in the sea, the processes of adaptive radiation resulted in the evolution of various phyla. Remarkably well-preserved fossils exist from this period of 100 million years ago, known as the Precambrian and early Cambrian Periods. The large numbers suggest an "explosion" of animal diversity during which time all the major phyla of animals originated with the exception of chordates. Included in this diversity are the roundworms, the arthropods, the mollusks, and the echinoderms (Figure 27-2). Chordate fossils became abundant in the Ordovician Period.

The approximately 42,500 species of chordates include fishes, amphibians, reptiles, birds, and mammals (and therefore humans). Chordates are distinguished by three principal features, two of which have already been mentioned: (1) a single hollow **nerve cord** located along the back; (2) a rod-shaped **notochord**, which forms between the nerve cord and the developing gut (stomach and intestines); and (3) **pharyngeal (gill) arches and slits,** which are located at the

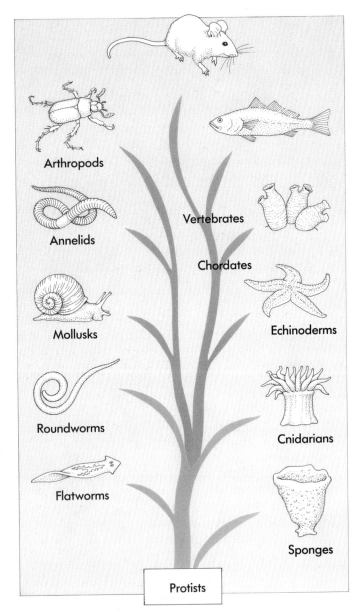

FIGURE 27-1 The earliest known chordate.
A *Pikaia gracilens* is a small fish-like animal, and it is one of the first organisms with a notochord. This particular fossil was found in the Burgess Shale.
B A lancelet. Note the resemblance to *Pikaia*.

FIGURE 27-2 Evolution of the major phyla.
During the Precambrian and Cambrian Periods, an explosion of animal diversity took place, resulting in the origination of all the major phyla with the exception of the chordates.

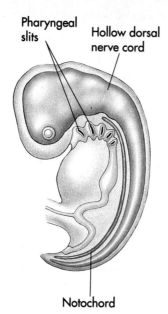

FIGURE 27-3 Chordate features. Embryos reveal the three principal features of the chordates: a nerve cord, a notochord, and pharyngeal (gill) arches and slits.

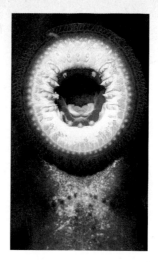

FIGURE 27-5 A lamprey, member of class Agnatha. Lampreys attach themselves to fish by means of their sucker-like mouths, rasp a hole in the body cavity, and suck out blood and other fluids from within. Lampreys do not have a distinct jaw.

throat (pharynx). (These three chordate features are shown in Figure 27-3 as they appear in the embryo.) All the features of chordates are evident in their embryos, even if they are not present in the adult form of the organism.

> Human evolution can be traced back to the appearance of the first chordates, which were small, fish-like organisms. Today, chordates include fishes, amphibians, reptiles, birds, and mammals.

Each of the three chordate characteristics played an important role in the evolution of the chordates and the vertebrates. The dorsal nerve cord increased the animals' responsiveness to the environment. In the more advanced vertebrates, it became differentiated into the brain and spinal cord. The notochord was the starting point for the development of an internal stabilizing framework, the backbone and skeleton, which provided support for locomotion. The bony structures that support the pharyngeal arches evolved into jaws with teeth, allowing these organisms to feed differently than their ancestors (Figure 27-4).

With the exception of two groups of relatively small marine animals, the tunicates and lancelets (see Figure 27-1, B), all chordates are vertebrates. Vertebrates (a subphylum of the chordates) differ from these two chordate groups in that the adult organisms have a **vertebral column,** or backbone, that develops around and replaces the embryological notochord. In addition, most vertebrates have a distinct head and a bony skeleton, although the living members of two classes of fishes, Agnatha (lampreys and hagfishes) (Figure 27-5) and Chondrichthyes (sharks and rays), have a cartilaginous skeleton.

> Vertebrates evolved from chordates. Vertebrates are characterized by a vertebral column surrounding a nerve cord located along the back. In addition, most vertebrates have a distinct head and a bony skeleton.

Evolution of the fishes

The first vertebrates to evolve were jawless fishes, members of the class Agnatha, about 500 million years ago (see Figures 27-4 and 27-5). Although traces of agnathan fossils are found in Cambrian strata, most date back to the Ordovician and Silurian Periods, about 400 to 500 million years ago.

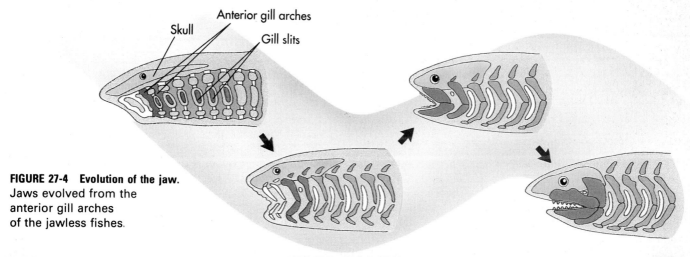

FIGURE 27-4 Evolution of the jaw. Jaws evolved from the anterior gill arches of the jawless fishes.

HUMAN EVOLUTION

FIGURE 27-6 Sharks and rays.
A Blue shark.
B Manta ray.
C Diamond sting ray.

Jaws first developed among vertebrates that lived about 410 million years ago, toward the end of the Silurian and beginning of the Devonian Periods. The first jawed fishes that evolved were the placoderms, ancestors to today's bony fishes and cartilaginous fishes—the sharks, skates, and rays (Figure 27-6). They were also one of the first groups of fishes to have paired fins. During the Devonian Period, fishes having a bony skeleton appeared: the "ray-finned" fishes and the "fleshy-finned" fishes.

The ray-finned fishes are the ancestors of the variety of bony fishes that exists today, such as the salmon, trout, cod, and tuna. (Others in this group are shown in Figure 27-7.) These fishes are the most successful class of vertebrates, accounting for about half of all living vertebrate species. One of the reasons for the success of the bony fishes is the development of the swim bladder, a gas-filled sac that allows these fishes to regulate their buoyant density; for this reason they can remain suspended at any depth in the water.

In the late Devonian Period, the "lungfishes" and the "lobe-finned" fishes appear in the fossil record. Fleshy-finned fishes are ancestors to both. Only a few species of lungfishes exist today (Figure 27-8). The lungs of these fishes are supplementary to their gills but allow them to breathe air when they are buried in the mud of dried-up lakes and streams.

Most of the Devonian Period lobe-finned fishes also had primitive lungs similar to the lungfishes. Interestingly, these fishes were thought to be extinct until the discovery of a living member of this group in 1938. Since then, other specimens of this same species have been found.

> The first vertebrates to evolve were the jawless fishes, about 500 million years ago. Jawed fishes evolved toward the end of the Silurian Period, about 410 million years ago.

Evolution of the amphibians, reptiles, and birds

One group of the lobe-finned fishes is believed to be the ancestor of the land-living **tetrapods,** or four-limbed vertebrates. The first land vertebrates were **amphibians,** animals able to live both on land and in the water. Amphibians are a class of vertebrates that lay their eggs in water or moist places and live in water during their early stages of development. Although a mature amphibian spends time on land, it must be able to return to the water frequently to keep its thin skin moist. The early amphibians had fish-like bodies,

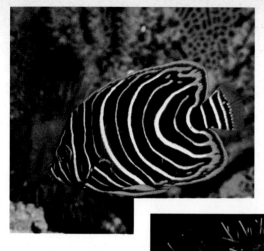

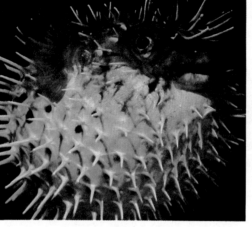
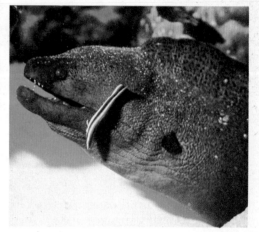

FIGURE 27-7 The bony fishes.
The bony fishes (class Osteichthyes) are extremely diverse.
A Koran angelfish. *Pomacanthus semicircularis,* in Fiji.
B Puffer. Puffers avoid being eaten by inflating themselves when attacked, thus projecting their spines outward.
C A sea horse. Sea horses move slowly and are difficult to see among marine algae and sea grasses.
D Moray eel, in Rangiroa, French Polynesia.

FIGURE 27-8 A lungfish.
Lungfishes have supplemental lungs that allow them to breathe air when they are buried in mud.

HUMAN EVOLUTION

FIGURE 27-9 Reconstruction of *Ichthyostega*, one of the early amphibians. *Ichthyostega* had efficient limbs for crawling on land, an improved olfactory sense associated with a lengthened snout, and a relatively advanced ear structure for picking up airborne sounds. Despite these features, *Ichthyostega*, which lived about 350 million years ago, was still quite fish-like in overall appearance.

short stubby legs, and lungs (Figure 27-9). Abundant about 300 million years ago during the Carboniferous Period, these amphibians probably gave rise to the other tetrapod groups.

The moist climate of the Carboniferous Period fostered the dominance of the amphibians during this time. However, following the Carboniferous Period, the climate changed dramatically as the Permian Period began. The amphibians, the dominant land vertebrates for about 100 million years, were gradually replaced by the reptiles as the dominant organisms. In contrast to amphibians, reptiles have water-resistant skins and more efficient lungs. These adaptations to dry climates gave reptiles a survival advantage over the amphibians. In addition, reptiles developed a reproductive advantage—the amniotic egg. An **amniotic egg** has a thick shell that encloses the developing embryo and a nutrient source within a watery sac (see Chapter 33). This type of egg protects the embryo from drying out, nourishes it, and enables it to develop away from water. Birds, too, have amniotic eggs.

> One group of jawed, bony fishes, the lobe-finned fishes, are thought to be the ancestors of the amphibians. Amphibians, animals that live on land but return to water to breed, arose over 300 million years ago and became dominant during the warm, moist climate of the Carboniferous Period. As the climate changed during the Permian Period, however, reptiles took over the land.

Birds (class Aves) evolved from reptiles during the Mesozoic Era, although the entire story of the evolution of birds is somewhat of a mystery due to gaps in the fossil record. Scientists have fossils of *Protoavis* and *Archaeopteryx*, two of the earliest known birds (see Chapter 26). No fossils have been found of birds that lived after *Archaeopteryx* until about 10 million years later in the Cretaceous Period. However, scientists think that all modern birds evolved from animals like *Archaeopteryx*.

Evolution of the mammals

The fossil record clearly shows that **mammals**, too, evolved from reptiles. Mammals are warm-blooded vertebrates that have hair and whose females secrete milk from mammary glands to feed their young. Organisms having *both* mammalian and reptilian characteristics, called **transitional forms**, first appear in the fossil record approximately 245 million years ago. Then, about 200 million years ago in the early Mesozoic Era, the first known mammals appear. These early mammals resemble today's shrews (Figure 27-10). They were small, fed on insects, and were probably **nocturnal**, or active at night. About 65 million years ago, following the extinction of the dinosaurs (see Chapter 26), mammals became abundant.

Various natural selection pressures resulted in the emergence of a number of significant anatomical and physiological changes in the mammals as they evolved from their reptilian ancestors. These changes resulted in a class of organisms that not only survived but flourished in a wide variety of habitats. Changes occurred in the reptilian arrangement of limbs as the mammals evolved, raising them high off the ground and allowing them to walk quickly and to run. A hinge developed between the lower jaw and the

FIGURE 27-10 A shrew.
Shrews are small nocturnal animals that feed on insects. They resemble the first known mammals that appeared about 200 million years ago.

All about Eve, or how scientists are attempting to construct the human family tree

Have you ever wanted to trace your "roots" and discover who your long-lost relatives were? Scientists, too, have this same curiosity but are much more ambitious. Ever since Darwin published his great works *On the Origin of Species* in 1860 and *The Descent of Man* in 1871, scientists have been searching for the "missing links" that link modern humans to their ancestors.

This search for missing links officially began when Dutch anatomist Eugene Dubois travelled to Java, a small island situated off the coast of Indonesia, and unearthed the first evidence of a hominid species called *Homo erectus* in 1877. Other fossil finds followed, and paleontologists, such as Richard and Mary Leakey, their son Louis Leakey, and Donald Johansen, built their careers out of the elusive hunt for hominid remains. Based on this fossil evidence, most scientists agree that *H. sapiens* evolved from *H. erectus* about 500,000 years ago. Fossil remains of *H. erectus* have been found throughout Europe, Asia, and Africa, leading many scientists to conclude that the races of *H. sapiens* evolved independently from separate isolated groups of *H. erectus* living in these different locations.

Recently, however, new molecular data has challenged this view, using DNA within mitochondria of living humans rather than fossils. Mitochondria are the cellular organelles that provide energy for the cell. Like tiny bacteria, mitochondria also contain their own DNA. Because mitochondrial DNA is passed from the mother only, looking at mitochondrial DNA made the genetic sleuthing much easier.

Scientists examined samples of blood from 189 people of different races and origins, including the !Kung people of Botswana, the Eastern pygmies of Zaire, and the Yorubans of Nigeria, as well as Europeans, Asians, and African-Americans. Feeding this information on DNA into a computer, they then constructed a family tree and reported that there was a single point of origin—in Africa (Figure 27-A)! The oldest fragment of mitochondrial DNA present in all humans today was first carried by an early human female in Africa, perhaps as recently as 200,000 years ago. This report argues strongly that our species has a single origin in Africa and that this African group of humans emigrated out of Africa and completely replaced the groups of *H. erectus* living in Europe, Asia, and other locations.

However, the methodology used to generate this now-famous family tree has been questioned, and it appears that many other family trees are as likely—the data do not allow a clear conclusion about human origins. Because African mitochondrial DNA is far more diverse, scientists are still betting on an African origin for *H. sapiens*, but the matter is far from proven.

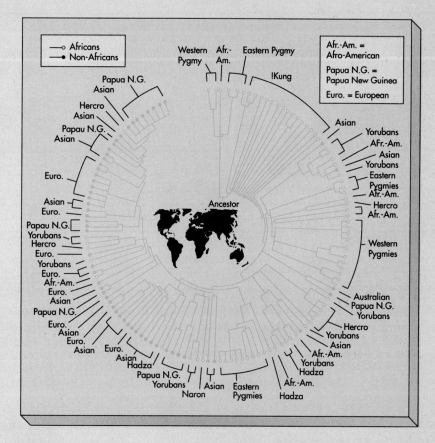

FIGURE 27-A This genetic tree is based on analysis of genetic sequences in 189 people of diverse races and shows that a common female ancestor existed in Africa.

FIGURE 27-11 Convergent evolution.
This illustration demonstrates the convergent evolution of marsupials in Australia and placental mammals in the rest of the world.

skull, and the teeth became differentiated, allowing mammals to eat a wide variety of food. The reptilian heart, having two ventricles with an incomplete separation, developed a complete wall in the mammals, preventing the mixing of oxygenated and deoxygenated blood. This change was significant in that mammals had also developed **warm bloodedness:** a constant internal body temperature. Warm-blooded animals are also called **endotherms,** meaning "within" (*endo-*) "temperature" (*-therm*). Their ancestors were **cold blooded,** having internal body temperatures that followed the temperature of their environments. Cold-blooded animals are also called **ectotherms,** meaning "outside" (*ecto-*) "temperature" (*-therm*). Endotherms have a higher rate of metabolism than do ectotherms and need more oxygen for the increase in the rate of cellular respiration.

Important changes also occurred in reproduction as the mammals evolved from the reptiles. The most "reptilian" type of reproduction is seen in the **monotremes,** one of the three subclasses of mammals living today. Monotremes lay eggs with leathery shells and incubate these eggs in a nest. The underdeveloped young that hatch from these eggs feed on their mother's milk until they mature. The ability of a mother to feed her young with milk she produces in mammary glands is another characteristic that has been advantageous to the adaptive radiation of the mammals. Present-day examples of monotremes are platypuses and spiny anteaters of Australia. These organisms are thought to have arisen on Gondwana, which is today's South America, Africa, and Australia (see Chapter 26), and were then isolated as the continents drifted apart.

The **marsupials,** a second subclass of mammals, do not lay eggs but give birth to immature young. These blind, embryonic-looking creatures crawl to the mother's pouch and nurse until they are mature enough to venture out on their own. Like the monotremes, marsupials were present in Gondwana; today most marsupials, such as kangaroos, wombats, and koalas, are found in Australia. Interestingly, the marsupials of Australia resemble the placental mammals that are present on the other continents. Figure 27-11 compares individual members of these two sets of mammals. The members of each pair have similar habitats and find their food in similar ways. These characteristics, along with their strikingly similar anatomy, suggest that the evolution of these two subclasses of mammals is a product of **convergent evolution:** the development of similar structures having similar functions in different species as the result of the same kinds of selection pressures.

> Mammals evolved from the reptiles approximately 200 million years ago. Changes in the structure and placement of their limbs, the structure of the heart, and reproductive strategies resulted in their ability to survive in a variety of climates and to eventually become abundant.

In **placental mammals,** the young develop to maturity within the mother. They are named for an organ formed during the course of their embryonic development, the placenta. The placenta is located within the walls of the uterus, or womb. Composed of both maternal and fetal tissues, the placenta is connected to the fetus by the umbilical cord. At the placenta, fetal wastes pass into the bloodstream of the mother, and oxygen and nutrients in the mother's bloodstream pass into the bloodstream of the fetus, a mechanism that allows the fetus to develop within the mother until it reaches a certain stage of maturity.

Evolution of the primates

There are 14 orders of placental mammals, which include many animals familiar to you, such as dogs, cats, horses, whales, squirrels, rabbits, bats, and a variety of others. The order of placental mammals that humans belong to is the **primates.** Primates are mammals that have characteristics reflecting an arboreal, or tree-dwelling, lifestyle. Among these characteristics are hands and feet that are able to grasp objects (such as tree branches), flexible limbs, and a flexible spine. Table 27-1 lists primate characteristics. (Figure 27-12 diagrammatically represents the taxonomic relationships of organisms living today.)

The earliest fossils that resemble any living primates belong to a group of organisms that look very much like today's insect-eating shrews (Figure 27-13). (Once classified as primates, today's tree shrews are now considered a separate order of placental mammals.) These primate ancestors are thought to have evolved in the late Mesozoic Era, about 75 million years ago, filling new habitats created by the appearance of the flowering plants, shrubs, trees, and the insects that fed on these plants. These early primates lived in the underbrush, hiding as they competed with the dinosaurs and other reptiles that were dominant at that time. As thousands of years passed, humans' shrew-like ancestors probably moved into the trees, eating insects that fed on the fruits and flowers growing on tree branches.

TABLE 27-1 Primate characteristics

- Ability to spread toes and fingers apart
- Opposable thumb (thumb can touch the tip of each finger)
- Nails instead of claws
- Omnivorous diet (teeth and digestive tract adapted to eating both plant and animal food)
- A semi-erect to an erect posture
- Binocular vision (overlap of visual fields of both eyes) resulting in depth perception
- Well-developed eye-hand coordination
- Bony sockets protecting eyes
- Flattened face without a snout
- A complex brain that is large in relation to body size

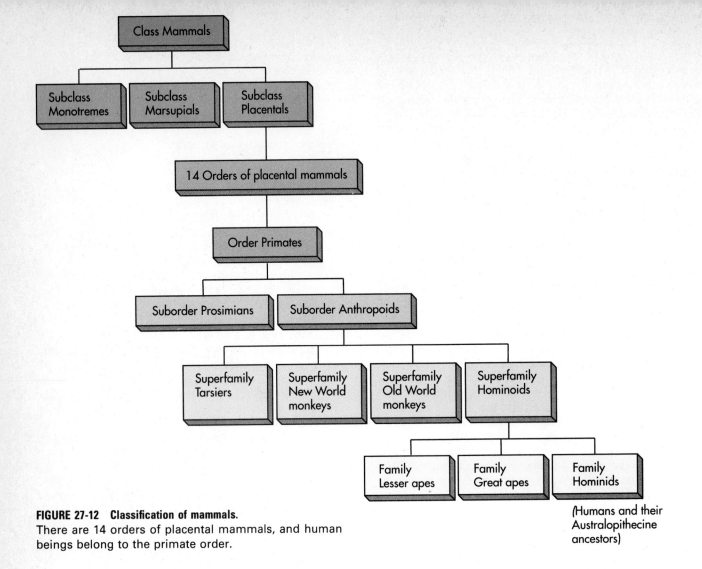

FIGURE 27-12 Classification of mammals.
There are 14 orders of placental mammals, and human beings belong to the primate order.

FIGURE 27-13 Where primates came from.
Reconstruction of an insect-eating mammal from the late Mesozoic Era, a mouse-sized animal that probably resembled the common ancestor of the primates and insectivores, two contemporary orders of mammals.

Selection pressures must have been great for these arboreal creatures. Those that survived in this habitat developed excellent **depth perception,** the ability to see in more than one plane as they moved from tree to tree. Depth perception is a function of **stereoscopic vision,** vision created by two eyes focusing on the same object. As primates evolved from their shrew-like ancestors, the eyes moved closer together from their placement on either side of the head (Figure 27-14). With this new placement, each eye can focus on the same object but from a slightly different angle. The brain puts both sets of messages together, resulting in the perception of three-dimensional form and shape. For example, eye placement that results in stereoscopic vision is apparent in predatory birds, such as owls, but is absent in nonpredatory birds, such as robins.

The primates also developed long limbs with flexible hands and feet adapted to grasping and swinging from branch to branch. Primates have two bones in the lower part of a limb that enable the wrists and ankles to rotate. In addition, the hands and feet of primates have digits that can be spread apart from one another, helping primates balance themselves when walking or running or helping

EVOLUTION: HOW LIVING THINGS CHANGE OVER TIME

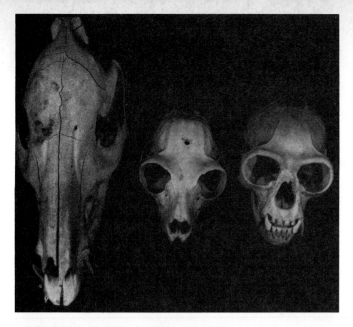

FIGURE 27-14 Evolution of depth perception.
As primates evolved, their eyes moved closer together from their placement on either side of the head. Because the eyes are closer together, each eye focuses on the same object but with a slightly different angle, resulting in greater depth perception.

them grasp objects. An **opposable thumb,** one that can touch the tip of each finger, helps in grasping. Most primates also developed flattened nails at the end of the digits, replacing the claws of their mammalian relatives.

> The primates, an order of mammals that humans belong to, reflect an arboreal heritage. The earliest primates arose about 75 million years ago and lived on the ground, feeding on plants and insects. Over time, the primates moved into the trees, developing excellent depth perception and flexible, grasping hands.

The prosimians

The primates are divided into two suborders: the **prosimians** and the **anthropoids** (the suborder that includes humans). The prosimians (meaning "before ape") are small animals, such as lemurs, indris, aye-ayes, and lorises; they range in size from less than a pound to approximately 14 pounds—about the size of a cat or a small dog. Most are nocturnal and have large ears and eyes to help them see and hear in the night. Prosimians have elongated snouts, reflecting their highly developed sense of smell. These characteristics are clearly seen in the two prosimians pictured in Figure 27-15, in addition to their elongated rear limbs, which help them leap from tree to tree in their tropical rain forest habitat. By the end of the Eocene Epoch (38 million years ago), prosimians were abundant in North America and Eurasia and were probably present in Africa also. However, their descendants now live only in the tropics of Asia, in tropical Africa, and on the island of Madagascar.

FIGURE 27-15 Prosimians.
A Ringtail lemur, *Lemur catta.* All living lemurs are restricted to the island of Madagascar, and all are in danger of extinction as the rain forests are destroyed.
B Tarsier, *Tarsius syrichta,* tropical Asia. Note the large eyes of the tarsier, an adaptation to nocturnal living.

FIGURE 27-16 Facial differences between an Old World monkey and a New World monkey.
A This macaque has a nose in which the nostrils are next to each other and point downward. Such noses are typical of Old World monkeys.
B This marmoset has a nose in which the nostrils face outward. This trait is characteristic of New World monkeys.

The anthropoids

The *anthropoids* (meaning "human-like") include the monkeys, apes, gorillas, chimpanzees and humans. These primates differ from the prosimians in the structure of the teeth, brain, skull, and limbs. The prosimians have pointed molars and horizontal lower front teeth. These horizontal teeth are used to comb the coat or to get at food. The anthropoids have more rounded molars and no horizontal lower front teeth. The brain of the prosimian is much smaller in relation to body size than the brain of an anthropoid. The face of an anthropoid is somewhat flat. In addition, the eyes of an anthropoid are closer together than those of a prosimian. Lastly, a prosimian's front limbs are short in relation to its long hind limbs, whereas both the front and hind limbs of an anthropoid are long. Compare these features in the prosimians (see Figure 27-15) and the anthropoids (see Figures 27-16 and 27-17).

In addition to these structural differences, prosimians and anthropoids also exhibit behavioral differences. Anthropoids are **diurnal,** that is, active during the day, whereas the prosimians are nocturnal. The anthropoids have evolved color vision, probably in relationship to their diurnal existence. Also, the anthropoids live in groups in which complex social interaction occurs. In addition, they tend to care for their young for prolonged periods.

The **hominoids** (also meaning "human-like") are one of four superfamilies of the anthropoid suborder. This superfamily includes apes and humans. The other three anthropoid superfamilies include (1) tarsiers (formerly classified with the prosimians, see Figure 27-15), (2) New World monkeys, and (3) Old World monkeys. In general, New World monkeys are found in the New World—South and Central America and they have flat noses whose nostrils face outward (Figure 27-16, *B*). Old World monkeys have noses similar to humans in which the nostrils are next to each other and point downward (Figure 27-16, *A*); these monkeys are found in the Old World—Africa, southern Asia, Japan, and Indonesia.

The hominoid superfamily is divided into three families: the lesser apes, the great apes, and the hominids, or humans. A variety of characteristics put the apes in a different superfamily from the monkeys: most apes are bigger than monkeys, have larger brains than monkeys, and lack tails.

The **lesser apes** include the gibbons, which are the smallest hominoids and closest in size to monkeys. They are about the size of a small dog, weighing 4 to 8 kilograms (9 to 18 pounds). As all monkeys and apes do, they live in tropical rain forests. Here, they leap and swing from tree to tree with their long arms.

The **great apes** include the orangutans, gorillas, and chimpanzees. These apes are much larger than the gibbons (Figure 27-17). The orangutans are about the size of humans, weighing between 50 and 100 kilograms (110 to 220 pounds). These apes exhibit sexual dimorphism; the females weigh about half that of the males. Like the gibbons, the orangutans have long arms, but they walk—they do not swing—from branch to branch of the rain forest trees. The

FIGURE 27-17 The apes.
A Gorilla, *Gorilla gorilla.*
B Mueller gibbon, *Hylobates muelleri.*
C Chimpanzee, *Pan troglodytes.*
D Orangutan, *Pongo pygmaeus.*

gorillas are the largest of the apes; they weigh about 160 kilograms, or 350 pounds. The females are smaller, ranging in weight from 165 to 240 pounds. The gorillas spend most of their time on the ground. When alarmed, male gorillas beat on their chests in a behavioral display. However, they are not the fierce animals humans think they are. In fact, gorillas are quite peaceable. The smallest of the great apes are the chimpanzees—humans' closest relatives. These animals weigh between 40 and 50 kilograms, or 90 to 110 pounds; females are slightly lighter. Like the other apes (except the gorillas), they spend their time in the trees.

The **hominids**, the family that includes humans of today, are the most intelligent of the hominoids. They are distinguished from the other families of hominoids in that they walk upright on two legs; they are said to be **bipedal**. In addition, hominids communicate by language and exhibit culture—a way of life that is passed on from one generation to another. The only living hominids are humans—*Homo sapiens*.

HUMAN EVOLUTION

Evolution of the anthropoids

Scientists have found jaw fragments that suggest that the ancestors of monkeys, apes, and humans began their evolution approximately 50 million years ago. The tarsioids may have begun their evolution even earlier—about 54 million years ago. Scientists do not know what adaptive pressures resulted in the appearance of these early anthropoid primates, nor do they know which prosimian was their ancestor.

Although the only fossil evidence of the emergence of the anthropoids consists of pieces of jaw, biochemical studies complement the knowledge gained from the fossil record. Together, these techniques tell scientists a great deal about the evolution of the anthropoids and the relationships among humans, apes, and monkeys. The basis for biochemical studies lies in the concept that as organisms change over time, their genetic material, or DNA, also changes. Each evolutionary change involves the formation of "new" genes from "old" by mutation; favorable new genes endure because of natural selection. Thus evolutionary change involves a progressive accumulation of genetic changes between organisms and their ancestors. Using biochemical techniques, scientists can determine the relatedness of species that have evolved from a common ancestor by comparing their DNA molecules. Scientists compare DNA indirectly by determining the similarity between comparable protein molecules or directly by measuring the degree of "mismatch" between two comparable stretches of DNA. Organisms that are more distantly related will have accumulated a greater number of genetic differences than organisms that more recently evolved from a common ancestor. In this way, scientists can estimate the time since two species diverged from a single ancestor, thereby determining their evolutionary relatedness. Relatedness can then be depicted in a **phylogenetic tree,** or family tree.

The phylogenetic tree of the anthropoids is shown in Figure 27-18. Fossils and biochemical studies show that the New World monkeys branched from the line leading to the Old World monkeys and the hominoids about 45 million years ago, in the mid-Eocene Epoch. In addition, scientists have discovered many fossils in North Africa that date from 25 to 30 million years ago, in the Oligocene Epoch. One of the earliest of these fossils, which scientists have named *Parapithecus,* probably led to the line of Old World monkeys,

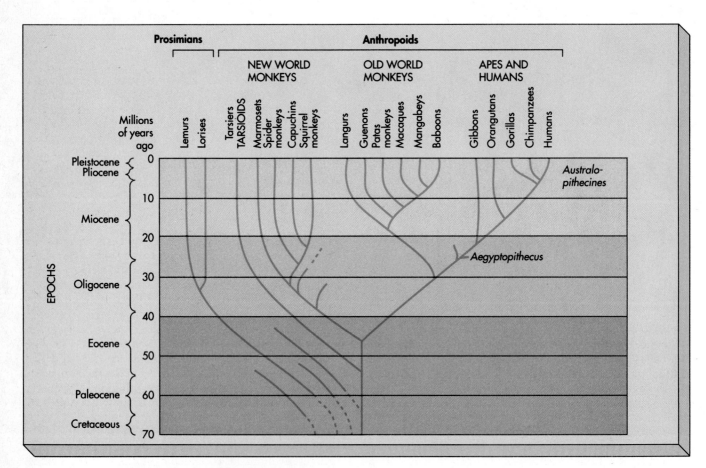

FIGURE 27-18 The primate family tree.
This tree is based on DNA comparisons between living species. The dating of the branches is based on the ages of fossils.

FIGURE 27-19 *Aegyptopithecus zeuxis.*
This primate fossil dates to the late Oligocene Era and is thought to be the ancestor of the hominoids that lived in Africa during the early Miocene Era.

splitting from the hominoids approximately 30 million years ago. The others, scientists think, are probably members of the evolutionary line that leads to the hominoids.

The most well-known of the Oligocene fossils found in North Africa is called *Aegyptopithecus*. This fossil is from the late Oligocene Epoch and is thought to be the ancestor to the early Miocene hominoids of Africa. Looking at the partially restored skull in Figure 27-19, you can see that *Aegyptopithecus* had some prosimian characteristics. It had a pronounced snout, leading scientists to believe that its sense of smell was still highly specialized, much like that of the prosimians. In addition, like the prosimians, *Aegyptopithecus* lived singly rather than in social groups.

> Fossil evidence and biochemical studies show that the anthropoids, a suborder of primates that includes humans, monkeys, and apes, first appeared about 50 million years ago. The New World monkeys branched from the evolutionary line leading to the Old World monkeys and the hominoids about 45 million years ago. The Old World monkeys split from the hominoids approximately 30 million years ago.

Fossil evidence is still being accumulated that will help tell the story of hominoid evolution. From mid-Miocene times on, the hominoid fossil record is quite extensive but consists primarily of skull and teeth fragments. Investigators have calculated that the evolutionary line leading to gibbons diverged from the line leading to the other apes about 18 to 22 million years ago, the line leading to orangutans split off roughly 13 to 16 million years ago, the line leading to gorillas diverged 8 to 10 million years ago, and the split between hominids and chimps occurred approximately 5 to 8 million years ago. This last statement suggests that chimpanzees and gorillas are humans' closest relatives, with a common relative alive 5 to 8 million years ago.

Evolution of the hominids

The first hominids: *Australopithecus*

The two critical steps in the evolution of humans were the evolution of bipedalism (walking on two feet) and the enlargement of the brain. For many years, scientists have hypothesized how and when bipedalism arose. The current theory is that bipedalism arose as a *preadaptation* because of the way humans' arboreal, ape-like ancestors moved through the trees. In other words, the skeletons and muscles of human ancestors were structured in a way that these hominoids were able to walk bipedally even though they lived in the trees. Interestingly, these structural adaptations developed as a part of their arboreal life and the types of locomotion they exhibited in the trees. Figure 27-20 shows the movements of gibbons and other related brachiators—those hominoids that move through trees by hanging from the branches with their arms, "walking" themselves along. As the weather in Africa cooled somewhat and became seasonal, patches of savannah grasslands began to invade former areas of tropical forest. The hominoids best able to walk efficiently on the ground survived, and bipedalism evolved.

Although hominids may have first appeared as long ago as 5 million years, the oldest undisputed evidence of the hominids is 3.6 to 3.8 million years old. This find was made relatively recently, in 1976, by Mary D. Leakey and an international team of scientists. The team discovered the fossil footprints shown on p. 463. Anthropologists think that the individuals who made the footprints are or are very closely related to human ancestors. Fossil bones discovered a few years earlier, at a site in Ethiopia about 2000 kilometers (1250 miles) north of the footprints, support this hypothesis. At 3.5 million years old, these hominid bones are somewhat younger than the footprints but are the oldest hominid bones ever found. These fossils were named *Australopithe-*

FIGURE 27-20 Brachiation and bipedalism.
Brachiators, such as gibbons and siamangs, locomote by hanging from branches with their arms and reaching from hold to hold. Brachiation has preadapted these animals to bipedal walking, which they do on broad branches (as the siamang in the illustration is doing) or on the ground.

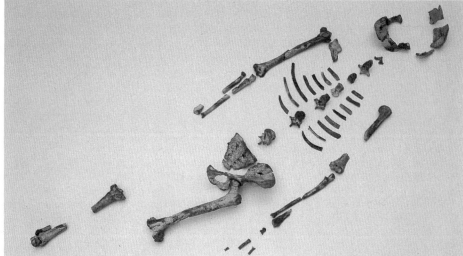

FIGURE 27-21 Lucy, from Ethiopia, is the most complete skeleton of *Australopithecus* discovered so far.
The reconstruction was made by a careful study of muscle attachments to the skull.

EVOLUTION: HOW LIVING THINGS CHANGE OVER TIME

cus afarensis, meaning "southern ape of Afar," by Donald C. Johanson and an international team of scientists who found them in the Afar region of Ethiopia. In 1974, this team also found and pieced together one of the most complete fossil skeletons of *A. afarensis*, which has since become famous (Figure 27-21). This "first" hominid was named Lucy after the Beatles song "Lucy in the Sky with Diamonds," which was popular at that time.

Although the australopithecines are considered to be the first hominids, they are not humans (members of the genus *Homo*). Their brains were still small in comparison to present-day human brains, and they had long, monkey-like arms. In addition, their faces were ape-like as shown in the photo of a reconstruction of Lucy. Lucy probably weighed about 25 kilograms (55 to 60 pounds) and was about 1 meter tall (slightly over 3 feet). Males were larger than females, however; they stood up to 1.2 meters tall (4 feet) and weighed as much as 45 kilograms (100 pounds).

A. afarensis evolved into two (and possibly more) lineages, including the species *A. africanus, A. robustus,* and *A. boisei*. These australopithecines lived on the ground in the open savannah of eastern and southern Africa. Their diets consisted primarily of plants. At night they probably slept in the few trees that existed in these grasslands, much like the savannah baboons do today to protect themselves from predators.

> The oldest evidence of the first appearance of the hominids is 3.6 to 3.8 million year old footprints. The oldest hominid fossil skeleton is 3.5 million years old and is not classified as human. This hominid is of the species *Australopithecus afarensis* and is nicknamed Lucy.

The phylogenetic relationships among these australopithecines is not clear; a family tree that is widely accepted at this time is shown in Figure 27-22. Notice from the diagram that no australopithecines are alive today; the last ones disappeared about 1 million years ago.

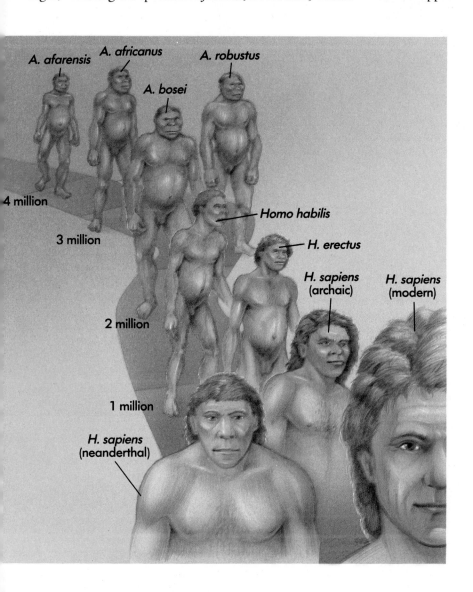

FIGURE 27-22 **The path of human evolution.** *Australopithecus robustus* and *A. boisei* seem to represent evolutionary dead ends, with no living descendants. Whether *A. africanus* was ancestral to *Homo habilis* or to *A. robustus* is currently in dispute, as is whether Neanderthals are the ancestors of modern humans or a parallel line that did not survive.

The first humans: *Homo habilis*

Climatic changes during the Pleistocene Epoch (2 million to 10,000 years ago) may have contributed to the disappearance of the australopithecines and the evolution of a new, more intelligent genus of hominids: the human (genus *Homo*). During the Pleistocene Epoch, the earth cooled, repeatedly undergoing **ice ages** during which vast regions of the earth (from the poles to a latitude of about 40 degrees) were covered by massive glaciers. For example, ice covered the land from the north pole southward over Canada and parts of the northern United States. Organisms such as the australopithecines had a low level of intelligence and no innate cold weather survival behaviors, and they died. Apparently, however, a species of hominid more intelligent than the australopithecines, was evolving from (most likely) *A. afarensis*. These hominids protected themselves from the cold by building shelters and wearing clothes. They are thought to have been the first humans and were given a name to emphasize their intelligence—*Homo habilis,* or "skillful human." Scientists have fossil bones and tools of *H. habilis* dating back about 2 million years. The first *H. habilis* fossils were discovered in 1964 by Louis Leakey, Philip Tobias, and John Napier in the Olduvai Gorge of eastern Africa.

Judging from the structure of the hands, *H. habilis* regularly climbed trees as did their australopithecine ancestors, although the *H. habilis* people spent much of their time on the ground and walked erect on two legs. Skeletons found in 1987 indicate that the *H. habilis* people were small in stature like the australopithecines, but fossil skulls and teeth reveal that the diet of *H. habilis* was more diverse than that of the australopithecines, including meat as well as plants. The tools found with *H. habilis* were made from stones fashioned into implements for chopping, cutting, and pounding food. The use of stone tools by these early humans marks the beginning of the **Stone Age,** a time which spans approximately 2 million to 35,000 years ago.

> The first human fossils date back about 2 million years and are of the extinct species *H. habilis,* meaning "skillful human." This species is considered human because they exhibited an intelligence far greater than their ancestors by making tools and clothing.

Human evolution continues: *Homo erectus*

All of the early evolution of the genus *Homo* seems to have taken place in Africa. There, fossils belonging to the second, also extinct species of *Homo*—*Homo erectus*—are widespread and abundant from 1.6 million to about 300,000 years ago. By 1 million years ago, however, *H. erectus* had migrated into Asia and Europe.

The *H. erectus* people were about the size of modern-day humans, were fully adapted to upright walking, and had brains that were roughly twice as large as those of their ancestors. However, they still retained prominent brow ridges, rounded jaws, and large teeth. The tools of *H. erectus* were much more sophisticated than those of *H. habilis* and were used for hunting, skinning, and butchering animals. These people were hunter gatherers, which means that they collected plants, small animals, and insects for food, while occasionally hunting large mammals. Researchers have found the first evidence of the use of fire by humans at 1.4 million-year-old campsites of *H. erectus* in the Rift Valley of Kenya, Africa. Fire is characteristically associated with populations of this species from that time onward. All of these activities: tool making, hunting, using fire, and building shelters are signs of **culture,** or a way of life that depends on intelligence and the ability to communicate knowledge of the culture to succeeding generations. Inherent in the concept of culture, then, is the concept of language. The ability to communicate enhances the ability of a species to survive, especially in harsh conditions, such as an ice age, by sharing survival tactics and warning each other of danger. Anthropologists think that the development of language was probably one of the most important factors in the appearance of *Homo sapiens*.

> Fossils of a second extinct species of human, *H. erectus,* date back to 1.6 million years ago. These humans made sophisticated tools, built shelters, used fire, and probably communicated with language.

Modern humans: *Homo sapiens*

The earliest fossils of *Homo sapiens* (meaning "wise humans") are dated to be about 200,000 years old and most likely evolved from the *H. erectus* species in Africa. The oldest *H. sapiens,* however, are not considered to be anatomically modern, that is, to have the same anatomical features of today's humans. This species is therefore referred to as an early or archaic form (see Figure 27-22). In general, these early *H. sapiens* had larger brains, flatter heads, more sloping foreheads, and more protruding brow ridges and faces than today's humans.

The fossil record shows gradual change of the species *H. sapiens,* with the early form evolving over a 75,000 year span to a subspecies of *H. sapiens* called **Neanderthal.** This subspecies was named after the Neander Valley in Germany where their fossils were first found. The Neanderthals lived from about 125,000 to 35,000 years ago in Europe and the Middle East.

Compared with modern humans, the Neanderthal people were powerfully built, short, and stocky. Their skulls were massive, with protruding faces, projecting noses, and rather heavy bony ridges over the brows. Their brains were even larger than those of modern humans, a fact that may have been related to their heavy, large bodies. The Neanderthals made diverse tools, including scrapers, borers, spearheads, and hand axes. Some of these tools were used for scraping hides, which they used for clothing. They lived in hutlike structures or in caves. Neanderthals took care of their injured and sick and commonly buried their dead, often placing food and weapons and perhaps even flowers with the bodies. Such attention to the dead strongly suggests that they believed in a life after death. For the first time, the

FIGURE 27-23 Cave painting.
Cave paintings, almost always showing animals and sometimes hunters, were made by Cro-Magnon people, our immediate ancestors. These paintings are found primarily in Europe and were made for about 20,000 years, until 8000 to 10,000 years ago.

kinds of thought processes characteristic of modern *H. sapiens*, including symbolic thought, are evident in these acts.

Approximately 10,000 years before the Neanderthal subspecies died out, modern *H. sapiens* made their appearance. This "modern" subspecies (our subspecies) is called **Homo sapiens sapiens.** The early members of this subspecies are the Cro-Magnons and are named after a cave in southwestern France where scientists found some of their fossils.

The **Cro-Magnons** had a stocky build, much like the Neanderthals, but their heads, brow ridges, teeth, jaws, and faces were much smaller than the Neanderthals and were more similar to today's humans. However, just as modern humans show variation among races, so, too, did the Cro-Magnons. The Cro-Magnons used sophisticated tools that were made not only from stone but from bone, ivory, and antler—materials that were not used by earlier peoples. Hunting was an important activity for the Cro-Magnons, evidenced by the abundance of animal bones found with human bones and elaborate cave paintings of animals and hunt scenes (Figure 27-23). The paintings appear to have been part of a ritual to ensure the success of the hunt.

The subspecies of *H. sapiens* that preceded us showed a gradual development of culture and society, which was the foundation for the development of "modern" culture and society. About 10,000 years ago, the last ice age came to a close, the global climate began to warm, and various groups of *H. sapiens sapiens* began to cultivate crops and breed animals for food. Archeologists have uncovered the remains of small, ancient cities, such as those of Jericho shown in Figure 27-24, which give evidence that by 9000 years ago humans had developed complex social structures. By 5000 years ago the first large cities and "great civilizations" appeared, such as those in Egypt (3100-1090 BC) and Mesopotamia (3100-1200 BC), and the final break occurred with the hunter-gatherer way of life.

The rest of the story of human development is one of history and is best left to the historians, anthropologists, and sociologists to tell. From a biologist's perspective, however, the spread of modern humans throughout the world has a dark side. As people inhabited new lands and increased their numbers, they destroyed many populations of animals and plants. Today, the growing human population is hastening global pollution, destruction, and devastation. Although humans have made incredible progress in the ability to survive, they still need to make progress in living in harmony with the rest of the living world and its resources.

FIGURE 27-24 The beginnings of civilization.
This photo shows the remains of a house in Jericho, which dates to about 7000 BC. The ruins of Jericho also contain the remnants of city walls and towers, demonstrating that by about 9000 years ago, many of our ancestors had moved away from the hunting-gathering life-style into an agricultural life-style.

Summary

1. The story of human evolution can be traced back to the appearance of the first chordates about 500 million years ago. Chordates are characterized by a single, dorsal, hollow nerve cord; a flexible rod, the notochord, which forms between the nerve cord and the developing gut in the early embryo; and pharyngeal (gill) arches located at the throat.

2. The vertebrates evolved from the chordates. Today, most chordates are vertebrates. Vertebrates differ from other chordates in that they usually possess a vertebral column, a distinct and well-differentiated head, and a bony skeleton.

3. The first vertebrates to evolve were the jawless fishes. Once abundant and diverse, they are represented among the living vertebrates only by the lampreys and the hagfishes. Jaws first developed among vertebrates that lived about 410 million years ago. Jawed fishes make up about half of the species of vertebrates and are dominant in fresh waters and saltwaters everywhere. The two classes of jawed fishes are the cartilaginous fishes, such as the sharks, rays, and skates, and the bony fishes, such as salmon, cod and tuna.

4. One ancient group of bony fishes, the lobe-finned fishes, is believed to be the ancestors of the first land vertebrates, the amphibians. Amphibians depend on water and lay their eggs in moist places. The amphibians arose over 300 million years ago. They gave rise to the reptiles.

5. The reptiles were the first vertebrates that were fully adapted to life on land. Amniotic eggs, which evolved in this group but are also characteristic of the birds and the few egg-laying mammals, represent a significant adaptation to the dry conditions that are widespread on land. The birds and mammals evolved from the reptiles.

6. Mammals, vertebrates that have hair and whose females secrete milk from mammary glands to feed their young, evolved from the reptiles approximately 200 million years ago. Mammals are warm blooded, or able to maintain a constant internal body temperature. Other than a few organisms such as the birds, all other living animals are cold blooded; their body temperatures vary with the temperature of the environment. Changes in the structure and placement of the mammalian limbs, the structure of the heart, and reproductive strategies resulted in mammals' ability to survive in a variety of climates and to eventually become abundant.

7. Primates, one of the 14 orders of mammals, first appeared 75 million years ago. Primates have large brains in proportion to their bodies, binocular vision, and five digits, including an opposable thumb; they exhibit complex social interactions.

8. The primates are divided into two suborders: the prosimians (small animals such as lemurs, indris, and aye-ayes) and the anthropoids (a group that includes monkeys, apes, and humans).

9. The hominoids (meaning "human-like") are one of four superfamilies of the anthropoid suborder. This superfamily includes apes and humans. The other three anthropoid superfamilies include tarsiers, New World monkeys, and Old World monkeys. The New World monkeys branched from the evolutionary line leading to the Old World monkeys and the hominoids about 45 million years ago. The Old World monkeys split from the hominoids approximately 30 million years ago. Ancestors to the apes gave rise to the gibbons, orangutans, chimpanzees, gorillas, and hominids.

10. The two critical steps in the evolution of humans were the evolution of bipedalism (walking on two feet) and the enlargement of the brain. The earliest hominids and the direct ancestors of humans belong to the genus *Australopithecus*. They appeared in Africa about 5 million years ago. They were small hominids, standing 3 to 4 feet tall and weighing from 50 to 100 pounds.

11. The genus *Australopithecus* gave rise to humans belonging to the genus *Homo*. The first species of this genus, *H. habilis*, appeared in Africa about 2 million years ago. Now extinct, the people of this species are considered human because they exhibited an intelligence far greater than their ancestors by making tools and clothing.

12. The second species of *Homo*, *H. erectus*, appeared in Africa approximately 1.6 million years ago. These people used fire, built shelters, fashioned sophisticated tools, and exhibited culture. *H. sapiens* probably evolved from *H. erectus* about 200,000 years ago.

REVIEW QUESTIONS

1. Describe the three features that distinguish chordates from other organisms.
2. Distinguish between chordates and vertebrates.
3. Fill in the blanks: The first land vertebrates were _____, organisms, which can live both on land and in water. They were gradually replaced as dominant organisms by the _____, animals with more efficient lungs and water-resistant skins.
4. Choose the correct answer: Birds evolved from: a) winged insects; b) flying mammals; c) reptiles; d) amphibians with webbed feet.
5. Fill in the blanks: Mammals are _____-blooded vertebrates with body _____. Female mammals secrete _____ from _____ _____ to feed their young.
6. What is warm bloodedness, and why is it important?
7. What do monotremes, placental mammals, and marsupials have in common? How do they differ?
8. What are primates? What two characteristics have helped them to be successful?
9. Distinguish between the prosimians and the anthropoids. Give an example of each.
10. Match each family with the appropriate animal(s):
    1. Hominids   a. Gibbon
    2. Hominoids   b. Chimpanzee
    3. Lesser apes   c. You, the
    4. Greater apes   chimp, and
                          the gibbon
                       d. You
11. Which statement is true? (Or are they both false)? Explain your answer:
    a. All hominids are hominoids.
    b. All hominoids are hominids.
12. Explain the significance of each of the following: *Australopithecus afarensis*, *Homo habilis*, and *Homo erectus*. Which is most like you?
13. "Neanderthal" is sometimes used as a derogatory term, meaning someone who is brutish and unsophisticated. Is this accurate? Explain your answer.
14. How did *Homo sapiens sapiens* differ from earlier *Homo* species?

KEY TERMS

amniotic egg   468
amphibian   466
anthropoid   473
Australopithecus afarensis   477
bipedal   475
convergent evolution   471
Cro-Magnon   481
culture   480
depth perception   472
diurnal   474
ectotherm   471
endotherm   471
great ape   474
hominid   475
hominoid   474
Homo erectus   480
Homo habilis   480
Homo sapiens   475
Homo sapiens sapiens   481
ice age   480
lesser ape   474
mammal   468
marsupial   471
monotreme   471
Neanderthal   480
nerve cord   464
nocturnal   468
notochord   464
opposable thumb   473
pharyngeal (gill) arches   464
phylogenetic tree   476
placental mammal   471
primate   471
prosimian   473
stereoscopic vision   472
Stone Age   480
tetrapod   466
vertebral column   465

THOUGHT QUESTIONS

1. If the oldest fossils of *Homo sapiens* are from Africa, then *Homo* must have migrated out of Africa twice, first as *Homo erectus* and later as *H. sapiens*. *H. erectus* survived in Africa until 500,000 years ago, in Europe until 400,000 years ago, and in Asia until about 250,000 years ago. Do you think *H. erectus* was driven to extinction by *H. sapiens*? Discuss.

FOR FURTHER READING

Thorne, A. G., & Wolpoff, M. (1992, April). The multiregional evolution of humans. *Scientific American*, pp. 66-83.

Wilson, A. C., & Cann, R. (1992, April). The recent African genesis of humans. *Scientific American*, pp. 66-83.

These two articles outline polar positions on the origins of humans: are humans descended from a single African ancestor, as postulated by researchers investigating DNA, or are modern humans related to a number of interconnected lineages scattered throughout the world, as suggested by fossil remains?

Walker, A., & Teaford, M. (1989, January). The hunt for proconsul. *Scientific American*, pp. 76-82.

This is an exciting account of the last common ancestor of great apes and human beings.

PART SEVEN

THE DIVERSITY AND UNITY OF LIVING THINGS

28 Viruses, bacteria, and genetic engineering

HIGHLIGHTS

▼
Viruses are noncellular infectious agents capable of causing certain diseases and cancers.

▼
Bacteria have a prokaryotic cell structure, occur in many habitats, play key ecological roles, are important in producing food, and can cause disease.

▼
Some bacteria can transfer genetic material, naturally recombining genes.

▼
Gene transfer is used to improve the characteristics of plants and animals and to produce medically important products. Scientists are beginning to treat genetic disorders in humans.

OUTLINE

Viruses
   The discovery of viruses
   The structure of viruses
   Viral replication
      The lytic cycle
      The lysogenic cycle
   Viruses and cancer
   The classification of viruses

Bacteria
   Bacterial reproduction
   Bacterial diversity and classification
      Archaebacteria
      Eubacteria
   Bacteria as disease producers

Genetic engineering
   Natural gene transfer among bacteria
   Human-engineered gene transfer among bacteria

Applications of recombinant DNA technology
   Medically important products
   Gene therapy
   Agriculture

The statistics are astounding! A recent study revealed that nearly one out of four college women tested in Seattle, Washington are harboring a sexually transmitted virus that is strongly associated with cancer of the cervix. Computer color-enhanced, these deadly virus particles (formally known as *human papillomavirus 16*) are shown in the photo, as the green spheres speckled with red. The human cell they are destroying is yellow.

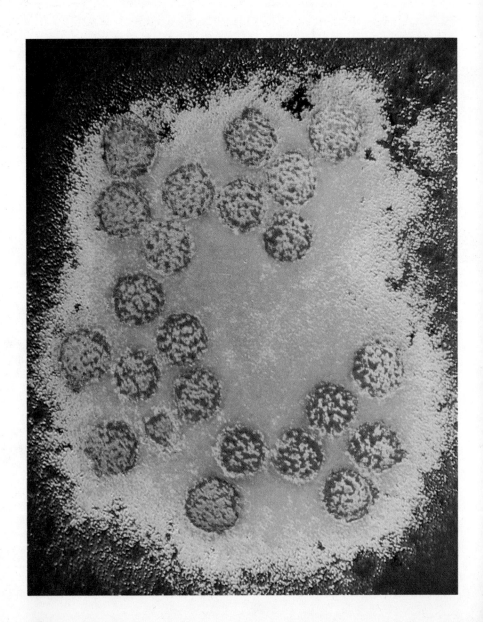

The human papillomavirus (HPV) causes a sexually transmitted disease commonly known as *genital warts*. HPV comes in 60 different varieties. Infection with any one of the 60 types can lead to an outbreak of warts, although some are more likely to cause warts than others. Ironically, the types that most often result in warts are the least dangerous. HPV types such as HPV 16 rarely cause warts on infection but are those most likely to be associated with cancer and the least likely to be detected.

Two other viral diseases transmitted by sexual contact are genital herpes and AIDS. Unlike sexually transmitted diseases (STDs) spread by bacteria, viral STDs cannot be cured. The drugs available to treat these diseases only help ease the symptoms but cannot destroy the viruses. Why are these diseases different in that regard? What makes a virus so hard to kill?

Viruses

Viruses are infectious agents. They enter living organisms and cause disease. But although viruses invade living things and cause cells to make more viruses, the viruses themselves are not living! They do not have a cellular structure, which is the basis of all life. They are nonliving **obligate parasites,** which means that viruses cannot reproduce outside of a living system. They must exist in association with and at the expense of other organisms. Unfortunately, that "other organism" may be you!

The discovery of viruses

At the end of the nineteenth century, several groups of European scientists working independently first realized that viruses existed. As they filtered fluids derived from plants with tobacco mosaic disease and cattle with hoof-and-mouth disease, the scientists discovered that the infectious agents passed right through the fine-pored filters they used, which were designed to hold back bacteria. They concluded that the infectious agents associated with these diseases were *not* bacteria—they were too small. As they studied the filtrate containing these mysterious agents, the scientists also discovered that the disease-causing agents could multiply only within living cells. These infection agents, they hypothesized, must lack some of the critical "machinery" cells use to reproduce.

For many years after their discovery, viruses were regarded as very primitive forms of life, perhaps the ancestors of bacteria. Today, scientists know that this view is incorrect—viruses are not living organisms. The true nature of viruses became evident in the 1930s after the ground-breaking work of an American scientist, Wendell Stanley. Stanley prepared an extract of tobacco mosaic virus (TMV), purified it, and studied its chemical composition. His conclusion: TMV was a protein—and he was partially right. Scientists later discovered that TMV also contains ribonucleic acid, or RNA. In the late 1930s, with the development of the electron microscope, scientists were able to see the virus that Stanley purified (Figure 28-1, *A*).

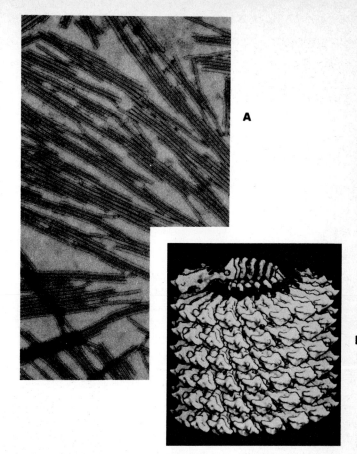

FIGURE 28-1 Tobacco mosaic virus.
A An electron micrograph of purified tobacco mosaic virus.
B Computer-generated model of a portion of tobacco mosaic virus. An entire virus consists of 2130 identical protein molecules—the yellow knobs—which form a cylindrical coat around a single strand of RNA (colored *red*).

The structure of viruses

Viruses primarily infect plants, animals, and bacteria. A specific virus can only infect a certain species of organisms. So you cannot be infected by a bacterial virus, nor can your dog catch your cold. Some viruses, however, can infect more than a single species of organism; for example, the human immunodeficiency virus (HIV virus that causes acquired immunodeficiency syndrome [AIDS]) is thought to have been introduced to humans from African monkeys.

Each virus has its own unique shape (Figure 28-2), but all contain the same basic parts: a nucleic acid **core** (either DNA or RNA) and a protein "overcoat" called a **capsid.** The structure of the TMV is shown in Figure 28-1, *B* and illustrates one way that a virus is put together. This virus is helical, with its single strand of RNA coiled like a spring, surrounded by a spiraling capsid of protein molecules. Many viruses have another chemical layer over the capsid called the **envelope,** which is rich in proteins, lipids, and carbohydrate molecules. Figure 28-3, *A* is an electron micrograph of a typical enveloped virus, the causative agent of herpes. Its structure is illustrated in Figure 28-3, *B*.

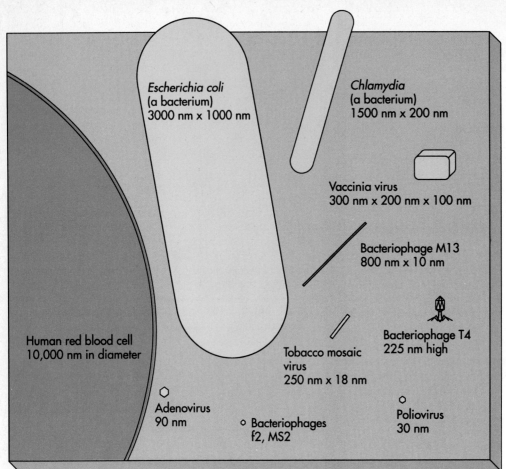

FIGURE 28-2 The shapes and sizes of viruses.
The size of various viruses are shown here in relation to a bacterium and a human red blood cell. Dimensions are given in nanometers.

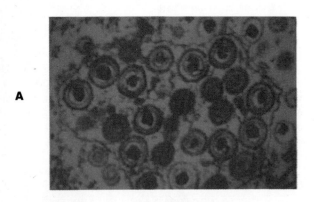

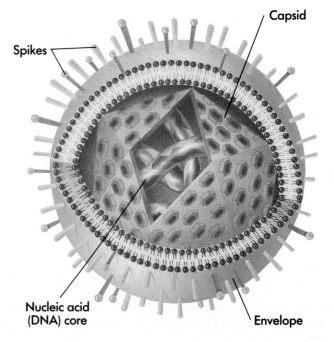

FIGURE 28-3 Structure of a typical virus.
A Electron micrograph of the herpesvirus.
B Structure of the herpesvirus.

It is hard to conceptualize how small viruses are, but Figure 28-2 helps by showing the size of a few viruses relative to the size of a bacterium and a human red blood cell. Some viruses, such as the poliovirus, are as small as the width of the plasma membrane on a human cell. The largest viruses are barely visible with a light microscope. Most viruses can be seen only by using an electron microscope.

> Viruses are made up of a nucleic acid core surrounded by a protein covering called a *capsid*. Some viruses also have an additional covering, or envelope.

Viral replication

Viruses cannot multiply on their own. They must enter a cell and use the cell's enzymes and ribosomes to make more viruses. This process of viral multiplication within cells is called **replication.** Various patterns of viral replication exist.

Some viruses enter a cell, replicate, and then cause the cell to burst, releasing new viruses. This pattern of viral replication is called the **lytic cycle.** Other types of viruses enter into a long-term relationship with the cells they infect, their nucleic acid replicating as the cells multiply. This pattern of viral replication is called the **lysogenic cycle.**

The lytic cycle

The process of viral replication has been studied most extensively in bacteria because bacteria are easier to grow in the laboratory and infect with viruses than plant or animal cells. Many bacterial viruses (usually called **bacteriophages** or simply **phages**) follow a lytic cycle pattern of replication. As shown in Figure 28-4, a bacteriophage first attaches to a receptor site on a bacterium. The virus then injects its nucleic acid into the host cell, while its protein capsid is left outside the cell. Next, the viral genes take over cellular processes and direct the bacterium to produce viral "parts" that will be used to assemble whole viruses. After their manufacture, these strands of nucleic acid and proteins are assembled into mature viruses that **lyse,** or break open, the host cell. Each bacterium releases many virus particles. Each "new" virus is capable of infecting another bacterial cell.

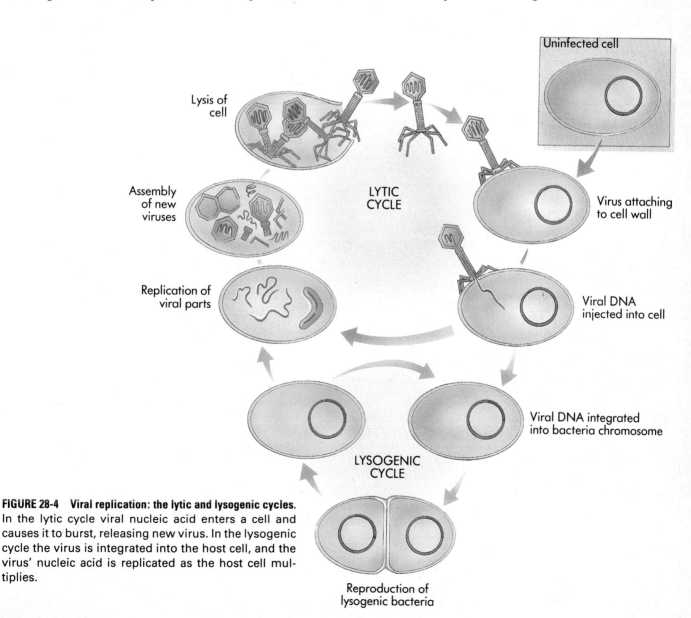

FIGURE 28-4 Viral replication: the lytic and lysogenic cycles. In the lytic cycle viral nucleic acid enters a cell and causes it to burst, releasing new virus. In the lysogenic cycle the virus is integrated into the host cell, and the virus' nucleic acid is replicated as the host cell multiplies.

Some animal viruses infect cells in a manner similar to bacterial virus infection, but they enter animal cells by endocytosis and must be uncoated before they can cause the cell to manufacture viruses. These cells may die as new virus particles are released in a lytic infection, or they may survive if virus particles are slowly budded from the cell by a process similar to exocytosis. This "slow budding" causes a persistent infection.

> During a lytic pattern of viral infection, a bacterial virus injects its nucleic acid into a host cell, whereas an animal virus enters by endocytosis. The viral nucleic acid directs the cell to produce "new" viral nucleic acid and protein coats. After assembly of these parts, the bacterial virus particles cause the cell to burst open, releasing them. Animal viruses often leave the cell by slow budding.

Plant cells are somewhat protected from viral infection by their rigid cell walls and protective outer, waxy cuticles. Viruses can only enter plants if they are damaged or if other organisms, such as sucking insects or fungi, assist them.

The lysogenic cycle

Some viruses, instead of killing host cells, integrate their genetic material with that of the host. Then, each time the host cell reproduces, the viral nucleic acid is replicated as if it were a part of the cell's genetic makeup (see Figure 28-4). In this way, the virus is passed on from cell to cell. Infection of this sort is called a **latent infection.** These integrated latent genes may not cause any change in the host for a long time. Then, triggered by an appropriate stimulus, the virus may enter a lytic cycle and produce symptoms.

The herpes simplex virus causes latent infections of the skin. Herpes nucleic acid remains in nerve tissue (sensory ganglia) without damaging the host until a cold, a fever, or other factor such as ultraviolet radiation from the sun acts as a trigger, and the cycle of cell damage begins. This "damage" is manifested as cold sores or fever blisters. The herpes zoster virus (the chickenpox virus) can also act in the same way. This virus may remain latent in the nerve tissue of a person having had chickenpox, only to be triggered at a later time to cause the painful nerve disorder shingles.

> During a lysogenic pattern of viral infection, a virus integrates its genetic material with that of a host and is replicated each time the host cell reproduces.

Viruses and cancer

Depending on which genes a virus carries, a virus can often seriously disrupt the normal functioning of the cells it infects. For thousands of years, diseases caused by viruses have been known and feared. Among the diseases caused by viruses are smallpox, chickenpox, measles, German measles (rubella), mumps, influenza, colds, infectious hepatitis, yellow fever, polio, rabies, and AIDS. Additionally, certain animal viruses can disrupt the normal functioning of cells so much that they change normal cells into cancer cells.

TABLE 28-1 Human cancers that may be caused by viruses

CANCER	VIRUS	FAMILY
Adult leukemia (cancer of the white blood cells)	Human T cell leukemia virus I and II	Retrovirus
Burkitt's syndrome (cancer of the white blood cells, often accompanied by a large facial tumor)	Epstein-Barr virus	Herpes
Cervical cancer	Herpes simplex II virus	Herpes
Liver cancer	Hepatitis B virus	Not classified
Nose and throat cancer	Epstein-Barr virus	Herpes
Skin and cervical cancer	Papillomavirus	Papova

Infection with certain viruses can lessen the growth requirements of the invaded cells, resulting in their growing faster than other cells—often uncontrollably. Such uncontrolled growth leads to the formation of large masses called **tumors.** If the body successfully walls off the tumor so that it is unable to spread, it is called a **benign** tumor and its cells are not considered cancerous. The human papillomavirus, for example, causes the growth of small benign tumors commonly known as warts.

Certain viruses can also transform normal cells into rapidly-growing but highly invasive cells. Along with viruses, other stimuli, such as chemicals called **carcinogens** and certain types of radiation such as ultraviolet rays or x-rays, can transform cells in this way. These highly invasive cells are cancer cells. Cancer cells grow out of control, forming tumors that invade and destroy body tissues. Such tumors are called **malignant** (cancerous) tumors. Human cancers that may be induced by viruses are listed in Table 28-1.

> Infection with certain viruses can transform normal cells into cancer cells.

The classification of viruses

Because viruses are not living things, they are not included in the five kingdoms of life. Scientists have, however, devised a classification scheme for viruses based on the host they infect. Viruses are first grouped according to whether they infect plants, animals, or bacteria. Further classification usually focuses on differences in morphology (shape and structure), type of nucleic acid, and manner of replication.

Bacteria

Bacteria and viruses both have "reputations" as being agents of disease. But the role of bacteria in the world of living things is much broader than that of a pathogen. Bacteria make life on Earth possible because they perform integral functions as decomposers of organic material and are natural recyclers of nitrogen and other inorganic compounds in ecosystems (see Chapter 38). They are also used to produce certain foods, such as yogurt, sauerkraut, dill pickles, and olives.

> Bacteria have a cell structure different from other organisms: the cytoplasm contains no internal compartments or organelles, the hereditary material is not enclosed by a membrane to form a nucleus, and the cell is bounded by a membrane encased within a cell wall. For these reasons, bacteria are classified as a separate kingdom of organisms, Monera.

Bacteria are the oldest, most abundant, and simplest organisms. Bacteria were abundant for well over 2 billion years before eukaryotes appeared in the world. They were largely responsible for creating the properties of the atmosphere and the soil during the long ages in which they were the only form of life on Earth (see Chapter 26). Bacteria are present on and in virtually everything you eat and you touch. Bacteria are also the only organisms with a prokaryotic cellular organization. (This type of cellular organization is described in Chapter 3.) It differs from eukaryotic cellular organization in primarily two ways: (1) the prokaryotic cell has no membrane-bounded nucleus, and (2) it contains no membrane-bounded organelles that compartmentalize the cell. This structural uniqueness places bacteria in a kingdom all their own: **Monera,** meaning "alone."

Bacterial reproduction

Reproduction among bacteria is asexual; one cell divides into two with no exchange of genetic material among cells. This process is called **binary fission.** Before fission, or division of the cell, the genetic material replicates and divides.

Bacterial DNA exists as a single, circular molecule that is attached at one point to the interior surface of the cell membrane. As eukaryotic cells do, bacteria make a copy of their genetic material before cell division. The bacterium also grows in size and manufactures sufficient ribosomes, membranes, and macromolecules for two cells before dividing. When the cell reaches an appropriate size and the synthesis of cellular components is complete, binary fission begins.

The first step of binary fission is the formation of a new cell membrane and cell wall between the attachment sites of the two DNA molecules (Figure 28-5). As the new membrane and wall are added, the cell is progressively constricted in two. Eventually the invaginating membrane and wall reaches all the way into the cell center, forming two cells from one.

Most bacteria reproduce every 1 to 3 hours. Some bacteria take a great deal longer. However, bacteria having conditions favoring their growth could produce a population of billions in little more than a day! This type of growth

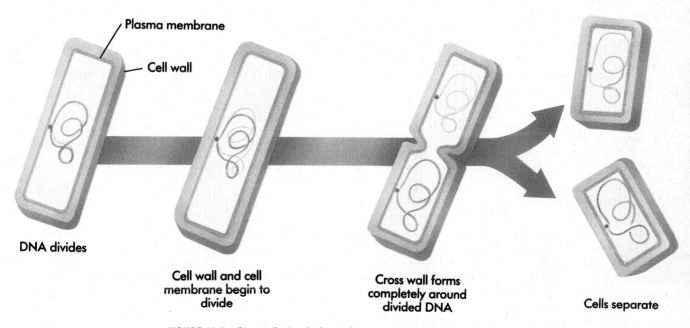

FIGURE 28-5 Binary fission in bacteria.
During binary fission, one bacterium divides into two bacteria.

is referred to as **exponential growth,** a period of rapid doubling of cells. However, the population cannot grow unchecked indefinitely. Many cells compete for food and other growth factors they need to survive. With resources limited, the entire population cannot be maintained, and cells begin to die. Wastes also accumulate, poisoning some of the cells. The growth of the population "levels out." Eventually, if the growth requirements of the bacterial population are no longer met, the population may begin to die as rapidly as it once grew.

> Bacteria reproduce asexually by binary fission. First they produce sufficient cell parts for two cells and replicate their single, circular molecule of DNA. Then the cell wall and membrane grow inward, between the attachment sites of the two DNA molecules, and literally split the cell in two.

Bacterial diversity and classification

Bacteria occur in a wide range of habitats and play key ecological roles in each of them. Some thrive in hot springs where the water temperature can be as high as 78° C (172° F). Others live more than a quarter of a mile beneath the surface of the ice in Antarctica. Still other bacteria, capable of dividing only under high pressures, exist around deep-sea vents formed by undersea volcanoes. Bacteria live practically everywhere, even in ground water where they were once thought to be absent. They are able to play many ecological roles because they make up a kingdom of organisms that is extremely diverse in its physiology.

This diverse kingdom of organisms is classified according to criteria listed in Table 28-2. They are not grouped into phyla but into four divisions that reflect differences in the chemistry of their cell walls. Each division is further split into sections rather than classes of organisms. Classification from this point on is similar to the classification of the eukaryotes in that further subdivisions use the titles of order, family, and genus. Within this classification scheme, most of the bacteria are considered **eubacteria,** or "true" bacteria.

TABLE 28-2 Criteria for classifying bacteria

- Shape of individual cells
- Arrangements of groups of cells
- Presence of flagella
- Staining characteristics
- Nutritional characteristics
- Temperature, pH, and oxygen requirements
- Biochemical nature of cellular components, such as RNA and ribosomes
- Genetic characteristics, such as percentages of DNA bases

But one section of organisms is different from all the rest. These are the **archaebacteria,** or "ancient bacteria." Many taxonomists consider archaebacteria to be a separate, sixth kingdom (see Chapter 26).

> The kingdom of bacteria, Monera, is divided into four divisions, which are further subdivided into sections, orders, families, genera, and species.

Archaebacteria

The cell walls and the cell membranes of the archaebacteria are chemically different from those of the eubacteria, as are certain of their physiological processes. These bacteria have such unusual chemical processes that they live in quite unusual places!

One group of archaebacteria produces methane (also known as marsh gas) from carbon dioxide and hydrogen. These bacteria can be found in places such as the gut of cattle and the depths of landfills. The methane they produce unfortunately adds to the blanket of greenhouse gases surrounding the Earth and therefore to the problem of global warming (see Chapter 40). Another group of archaebacteria live only in areas having high concentrations of salt, such as in salt marshes in the intertidal zone, where fresh water meets seawater in stagnant, concentrated salt pools. A third group of archaebacteria were already mentioned: those that live in the incredible heat and pressure of the deep-sea vents.

> The archaebacteria are a taxonomic section of bacteria that are chemically different in certain structures and metabolic processes from all other bacteria.

Eubacteria

The eubacteria make up all the rest of the bacteria—quite a diverse collection. However, these thousands of species can be placed into one of three groups according to their mode of nutrition.

One of these groups is the photoautotrophic (photosynthetic) bacteria. Like plants, photosynthetic bacteria use energy from the sun to produce "food" in the form of carbohydrates. They are therefore called **photoautotrophs,** which means *light* (photo) *self* (auto) *feeders* (trophs). Photosynthesis takes place in green bacteria, purple bacteria, and cyanobacteria (formerly known as blue-green algae). These bacteria contain chlorophyll (chemically different from the chlorophyll in plants), which is found within a system of membranes that ring the interior periphery of the cell. In the green and purple bacteria, photosynthesis does not take place exactly as it does in plants, and oxygen is not a by-product. In fact, oxygen tends to break down their chlorophyll, so these species usually live in polluted water that contains little oxygen. The cyanobacteria, however, release oxygen during photosynthesis as plants do and can be found living near the surface of lakes and ponds (Figure 28-6). The cyanobacteria were among the first cells to evolve and probably oxygenated Earth's primitive environment.

A second group of eubacteria are the chemoautotrophic bacteria. "Chemical self-feeders" or **chemoautotrophs,** derive the energy they need from inorganic molecules such as ammonia (NH_3), methane (CH_4), and hydrogen sulfide

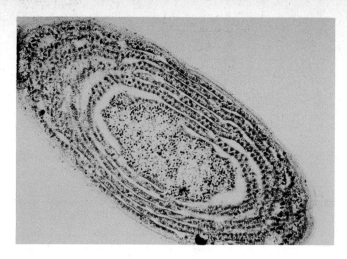

FIGURE 28-6 A cyanobacterium.
The outer regions of the cell are filled with photosynthetic membranes. The dark spots between the membranes are storage areas for the carbohydrates produced by photosynthesis.

FIGURE 28-7 Soybean root nodules.
The nodules growing on the roots of this soybean plant contain nitrogen-fixing bacteria. These bacteria convert atmospheric nitrogen (N_2) to ammonia (NH_3), which can then be used as a nitrogen source by the plants.

(H_2S) gases. With this energy and carbon dioxide (CO_2) as a carbon source, they can manufacture all their own carbohydrates, fats, proteins, nucleic acids, and other growth factors. The archaebacteria that live near deep-sea vents are chemoautotrophs, as are eubacteria called **nitrifying bacteria.** Nitrifying bacteria play an important role in the cycling of nitrogen between organisms and the environment. These bacteria live in nodules (spherical swellings) in the roots of legumes, such as beans, peas, and clover (Figure 28-7). They convert ammonia to nitrates, a form of nitrogen used by plants. Other species of chemoautotrophs play a key role in the sulfur cycle, using H_2S or elemental sulfur for energy and converting it in the process to sulfates, other plant nutrients.

The third group of eubacteria are heterotrophic bacteria. Most bacteria are "other feeders" or **heterotrophs,** obtaining their energy from organic material that enters these cells by diffusion and active transport. Humans are heterotrophs, too, eating plants and animals—organic material that once lived. Heterotrophic bacteria are considered **decomposers** and play a key role in the carbon cycle. They break down large organic compounds, such as proteins and carbohydrates, into small compounds, such as CO_2, which is released into the atmosphere to be recycled as it is "fixed" by plants during the process of photosynthesis.

Bacteria as disease producers

Most plant diseases are not caused by bacteria but by fungi (see Chapter 29). However, a few genera of bacteria do infect plants, primarily causing types of plant rot and wilt. Wilt occurs when bacteria block water from moving up the xylem vessels in the plant. In addition, some genera of bacteria cause tumorlike growths called *galls* in plants. Figure 28-8 is a photograph of a plant with crown gall disease caused by bacteria of the genus *Agrobacterium*. Interestingly, scientists have learned ways to use these disease-causing bacteria in productive ways in the genetic engineering of plants.

FIGURE 28-8 Crown gall disease on a tobacco plant.
This disease is caused by bacteria of the genus *Agrobacterium*.

VIRUSES, BACTERIA, AND GENETIC ENGINEERING

In contrast to plant diseases, many human diseases are caused by bacteria, including the diseases listed in Table 28-3 with which you may be familiar. The bacteria that cause disease are all heterotrophic. Most disease-producing bacteria use their hosts for food, but some poison their hosts. To cause disease, bacteria or the poisons they produce must first get into the body. Usually this happens if you eat contaminated food or drink contaminated water. You may inhale bacteria present in the air after an infected person coughs or sneezes, touch a contaminated object, or have sexual intercourse with an infected partner. Sometimes bacteria enter the body through broken skin as the result of an injury or injection with a contaminated needle.

After entering the body, bacteria then attach to body cells and cause various types of tissue damage. The chemicals bacteria produce digest the tissues so that the breakdown products can be taken into the bacteria and metabolized. In addition, certain bacteria, such as the bacteria that cause staphylococcal food poisoning, gas gangrene, tetanus, or cholera, produce **toxins,** or poisons. These toxins can cause powerful effects, such as high fevers, violent muscle spasms, vomiting, diarrhea, heart damage, and respiratory failure. All the while, your body combats these cells using its nonspecific and specific defense systems (see Chapter 13). The ability of your body to wage this war affects the degree and length of the illness and its eventual outcome.

TABLE 28-3 Some diseases caused by bacteria

RESPIRATORY DISEASES
Bacterial pneumonia
Bacterial meningitis
Strep throat
Tuberculosis
Legionnaire's disease

GASTROINTESTINAL DISEASES
Cholera
Dysentery
Typhoid fever

GASTROINTESTINAL POISONING
Staph food poisoning
Botulism

SEXUALLY TRANSMITTED DISEASES
Syphilis
Gonorrhea
Chancroid

SKIN DISEASES
Staph infections
Leprosy
Yaws

Genetic engineering

Producing disease is only one of the varied roles that bacteria play in the web-like interactions of living things. In recent decades, scientists have added a new role—bacteria are now being used as microscopic factories, manufacturing hormones and a variety of medically useful products and helping to produce crop plants and feed animals with desired characteristics. These new techniques of molecular biology that involve the manipulation of genes are called **genetic engineering** or **recombinant DNA technology.**

Natural gene transfer among bacteria

Scientists have long known that certain bacteria can transfer genetic material from one cell to another. Genes move from one prokaryote to another by three methods: **transformation, transduction,** and **conjugation** (Figure 28-9).

In transformation, free pieces of DNA move from a donor cell to a recipient cell. The pieces of DNA come from bacteria that have lysed and released their DNA. Certain cells, called **competent cells,** are able to "take up" this DNA and incorporate pieces into their genomes, thereby becoming transformed, or changed. Not all bacteria are able to be transformed; the ability to take up DNA fragments is an inherited characteristic.

During transduction, DNA from a donor cell is transferred to a recipient cell by a virus. For this transfer to occur, the virus must first merge its genetic material with that of a bacterium it has infected. Only certain lysogenic viruses or damaged viruses are capable of incorporating bacterial genes with their genome. Also, not all bacteria are capable of being infected by a virus carrying bacterial genes. But this process does occur in certain viral-bacterial interactions and results in genes from one bacterium being transferred to another bacterium.

During conjugation, a donor and a recipient cell make contact, and the DNA from the donor is injected into the recipient cell. This transfer takes place in bacteria having extrachromosomal pieces of DNA called **plasmids** that have genes for chromosome transfer. Plasmids are fragments of DNA separate from the main circular molecule that occurs in bacteria. They replicate independently of the main chromosome and make up about 5% of the DNA of many bacteria. (Figure 28-10 shows these small circles of DNA.) Certain plasmids are called **F plasmids** because they have a fertility factor—several special genes that promote the transfer of the plasmid to other cells. These genes code for proteins that form a tube called a **pilus** on the surface of the cell. When the pilus of one cell makes contact with the surface of another cell that lacks pili (and therefore does not contain the F plasmid), the replication and then transfer of the plasmid occurs (Figure 28-11). If the F plasmid has been incorporated into the larger bacterial chromosome, as sometimes happens, the cell will copy and transfer the entire bacterial chromosome to the recipient cell.

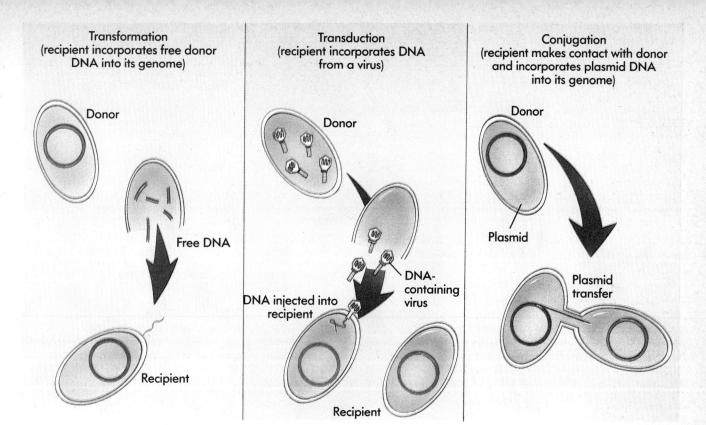

FIGURE 28-9 Natural gene transfer among bacteria.
Transformation, transduction, and conjugation.

FIGURE 28-10 A ruptured bacterial cell *(Escherichia coli)* has released its chromosomes and plasmids.
Plasmids are the small circles of "extra" DNA that replicate independently of the main chromosome.

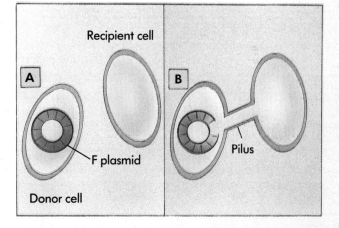

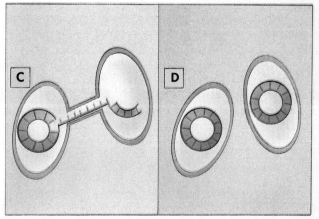

FIGURE 28-11 Conjugation.
A-C During conjugation, a pilus is formed between two bacterial cells across which the F plasmid is transferred. **D** If the F plasmid becomes incorporated into the larger bacterial chromosome, the cell will copy and transfer the entire bacterial chromosome to the recipient cell.

BIOLOGY TECHNOLOGY & society

The new agriculture

Genetic engineering is revolutionizing how farmers grow their crops. In the past, much of a farmer's efforts were devoted to cultivating soil to eliminate weeds and spraying crops with pesticides to rid them of crop-destroying insects. Over 40% of the chemical insecticides used in the world today, for example, are sprayed on cotton plants to kill plant-eating insects.

Now weeds can be eliminated from fields of many commercially important broadleaf plants in a new way. These plants have been genetically engineered to be resistant to the herbicide glyphosate, a powerful biodegradable chemical that kills most growing plants. The herbicide blocks an enzyme that plants need to make aromatic amino acids. (Humans do not make them but get them from their diet, so the herbicide is harmless to humans.) Genetic engineers have added a gene encoding a glyphosate-resistant enzyme to the crops. Now when fields are treated with glyphosate, all the plants in the field die—except the genetically engineered crop. Instead of cultivating and applying many different kinds of herbicides (most herbicides kill only a few kinds of weeds), only a single herbicide is required, one that readily breaks down in the environment.

Genetic engineering is also making many crops immune to attack by insects. In one approach, tomatoes have been made resistant to the caterpillar-like tomato hornworm. Scientists inserted a gene isolated from soil bacteria (a kind called *Bacillus thuringiensis*) into the tomato's chromosomes. This gene encodes a protein that converts enzymes in the caterpillar's stomach into a toxin causing paralysis and death in insects. Because the necessary enzyme is not found in mammals, the protein is harmless to humans. The genetically engineered tomato plants are resistant to the hornworm, without the application of any pesticide (Figure 28-A).

Many plant pests attack roots, and to counter this threat biologists have introduced the gene encoding the *B. thuringiensis* protein into root-colonizing bacteria. Insects that eat the roots convert the protein into toxin and die.

A major effort is underway to transfer into crop plants the genes that enable soybeans and other legume plants to "fix" atmospheric nitrogen, which all other plants must obtain from the soil. Worldwide, farmers applied over 60 million tons of nitrogen fertilizer in 1990, an expensive undertaking. The genes needed to obtain the necessary nitrogen directly from the air are present only in certain symbiotic bacteria living in the legume plant's roots. These genes have been isolated and introduced into crop plants, but they do not seem to function properly in their new eukaryotic environment. Experiments are being pursued actively because perfecting the ability to grow crops without nitrogen fertilizers would greatly expand the world's ability to feed a hungry population.

FIGURE 28-A A successful experiment in plant resistance. The two tomato plants were exposed to destructive caterpillars under laboratory conditions. The nonengineered plant on the left has been completely eaten, but the engineered plant shows no signs of damage.

Human-engineered gene transfer among bacteria

Within the past couple of decades, scientists have learned how to manipulate these natural mechanisms of gene transfer to produce bacteria with desired genomes. One use of this technology has been to insert *human genes* into bacteria that instruct the bacteria to produce human proteins. For example, the genes that code for the production of human insulin have been inserted into bacteria, causing the bacteria to produce this hormone. Bacteria grown in large numbers, such as shown in Figure 28-12, produce large quantities of this hormone, which can then be purified and used in the treatment of diabetes.

How do scientists put genes into bacteria? Continuing with the example of the production of human insulin, scientists first cut the pieces of DNA they need—in this case the gene that codes for insulin—from human DNA. To do this job, researchers extract human DNA from cells and treat it with "chemical scissors" called **restriction enzymes.** Restriction enzymes work by "recognizing" certain base sequences in DNA molecules and breaking the bonds between the bases at that point. Such a sequence is called a **recog-**

FIGURE 28-12 A pharmaceutical engineer checks on the purification process for Humulin, a human insulin created from recombinant DNA technology.
Humulin is used to manage diabetes in humans.

nition sequence. In nature, restriction enzymes are produced by bacteria to protect themselves from bacteriophage invasion. They protect bacteria by "cutting" viral DNA into pieces before it can take over the cell. And now these naturally occurring bacterial enzymes have become one of the many tools of the genetic engineer.

A myriad of restriction enzymes exist. Different restriction enzymes recognize different recognition sequences, so scientists can pick the enzyme that will cut DNA in a spot they choose. The DNA sequences that restriction enzymes recognize are typically four to six nucleotides long. Because the nucleotides on double-stranded DNA are complementary to one another, the restriction enzyme recognizes and cuts both strands. However, the bases of the recognition sequence, being complementary to one another, run in opposite directions on the two strands. Therefore the restriction enzyme cuts both strands but cuts one at each end of the recognition segment as shown in Figure 28-13. The result is that the cut piece of DNA has ends in which one strand is longer than the other. These ends are "sticky ends" because they can bond with another piece of DNA cut with the same restriction enzyme. Therefore fragments of human DNA and bacterial DNA cleaved by the same restriction enzyme have the same complementary nucleotide sequences at their ends and can be joined to one another. A "sealing" enzyme called a **ligase** helps re-form the bonds. In this example, insulin genes and bacterial plasmids that have been cut with the same restriction enzyme can be joined.

The next step is to insert "hybrid" plasmids into bacteria. If these plasmids are mixed under the proper conditions with competent bacteria, the bacteria will take up the plasmids from the mixture. These bacteria now have the ability to produce human insulin. (This procedure of genetically engineering bacteria is summarized in Figure 28-14.)

Inserting fragments of foreign DNA into bacterial cells using plasmids is a technique that scientists use frequently. Scientists also use certain viruses to insert DNA fragments into bacteria. First, they insert the DNA into the genome of a virus instead of into a plasmid. The virus then carries the DNA fragment into a cell it is capable of infecting. Such viruses are called **vectors.** Animal viruses can also be used as vectors. In fact, scientists have used animal viruses to carry bacterial genes into monkey cells and animal genes into plant cells.

There has been considerable discussion about the potential danger of inadvertently creating an undesirable life form in the course of a recombinant DNA experiment. What if someone fragmented the DNA of a cancer cell and then incorporated the fragments at random into viruses that are propagated within bacterial cells? Might there not be a danger that one of the resulting bacteria could be capable of infecting humans and causing an infective form of cancer?

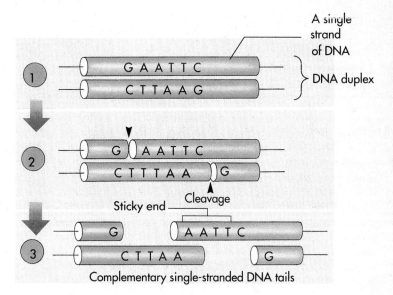

FIGURE 28-13 The method in which restriction enzymes produce DNA fragments with "sticky ends."
The restriction enzyme EcoRI always cleaves sequence GAATTC at the same spot, after the first G. Because the same sequence occurs on both strands, both are cut. But the position of the G is not the same on the two strands; the sequence runs the opposite way on the other strand. As a result, single-stranded tails are produced. The single-stranded tails are complementary to each other, or sticky.

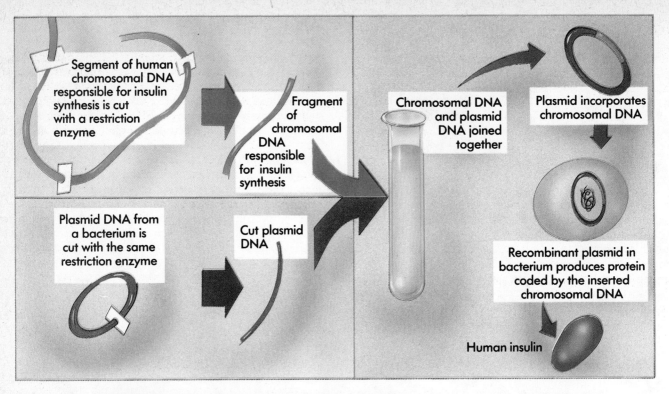

FIGURE 28-14 How insulin is made through genetic engineering.
DNA is inserted into plasmids, and the plasmids are mixed with bacteria. The bacteria will take up the plasmids through transformation and will produce the proteins directed by the inserted DNA.

Even though most recombinant DNA experiments are not dangerous, such concerns should be taken seriously. Both scientists and governments monitor experiments in genetic engineering to detect and forestall any such hazard. In addition, researchers have established appropriate experimental safeguards. For example, the bacteria used in many recombinant DNA experiments are unable to live outside of laboratory conditions; many of them can live only in an atmosphere that is free of oxygen and therefore cannot survive in the Earth's atmosphere. Experiments that are clearly dangerous are prohibited.

Applications of recombinant DNA technology

The newly found ability to isolate individual genes and transfer them from one kind of organism to another has revolutionized scientist's ability to improve the characteristics of useful plants and animals and to avoid disease. These techniques offer enormous potential for the agriculture and medicine of the future and have already produced important applications. Genetic engineering is also a useful tool that helps scientists learn about gene structure, function, and regulation. So these powerful techniques are used not only to produce new products but also to generate new knowledge.

Medically important products

Using gene technology to produce proteins useful in treating human disorders or diseases involves the process described earlier: introducing human genes that code for clinically important proteins into bacteria. Along with producing human insulin, this technique has been used to produce more than 150 medically important products. Some of them are human growth hormone, used to treat growth disorders in children; interferon, used to fight cancer, viral diseases, and rheumatoid arthritis; and interleukin-2, used to treat certain deficiencies of the human immune system. Previously, these medically important products were unavailable or were obtained from animals at great expense. Humans often had allergic reactions to proteins obtained from other animals, but the genetically engineered products do not produce side effects because they have the human protein structure.

Genetic engineering techniques have also been used to produce vaccines, such as hepatitis B vaccine that helps protect against the leading cause of liver cancer. In one approach, scientists use gene-splicing techniques to produce a noninfective virus whose protein coat will stimulate the body's immune system to produce antibodies against the "real" virus. Scientists use the DNA from a harmless virus and add to it the genes that code for the protein coat of the infective virus. When injected into humans, these viruses do not cause disease but stimulate the immune system to produce antibodies specific for the infective form of the virus.

Gene therapy

Recombinant DNA technology entered a new phase of application on September 14, 1990, when a 4-year-old girl became the first person to undergo **gene therapy.** Gene therapy is the treatment of a genetic disorder by the insertion of "normal" genes into the cells of a patient. The young girl that was treated suffered from the rare genetic disorder severe combined immune deficiency. Key immune system cells called *T cells* (see Chapter 13) were not working because they lacked the enzyme ADA (adenosine deaminase). Without her T cells, this girl had no defense against infection.

To treat her, doctors removed some of her blood and cultured her defective T cells. They then added viruses that contained working copies of the ADA gene to the cell culture. When the viruses infected the T cells, they inserted these ADA genes into the cells. Doctors grew these altered T cells in the laboratory until they numbered in the billions and then injected them into the child's blood. The genetically repaired cells multiplied, populating the girl's immune system with gene-corrected T cells that produce the necessary enzyme.

Agriculture

Another important application of genetic engineering is the manipulation of the genes of crop plants. In broadleaf plants, such as tomatoes, tobacco, and soybeans, a plasmid of the bacterium *Agrobacterium tumefaciens*, which causes crown gall in plants, is the main vehicle that has been used to introduce foreign genes. A part of this plasmid integrates into the plant DNA, carrying its spliced genes. (Figure 28-15 diagrams the steps of this process.) Through the use of this technique, scientists are working to improve crops and forest trees by making them more resistant to disease, frost, and herbicides (chemicals that kill weeds).

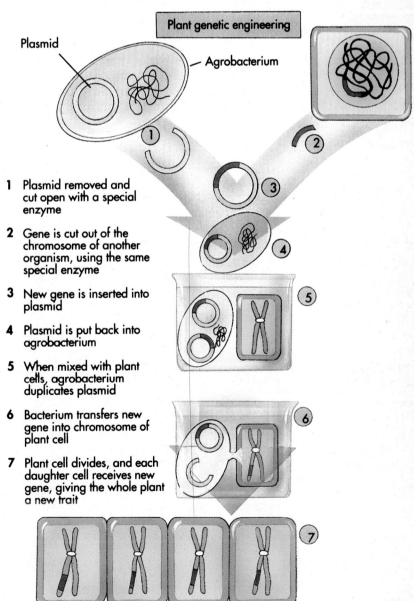

FIGURE 28-15 Use of a plasmid of *Agrobacterium tumefaciens* in genetic engineering. These genetic engineering procedures induce resistance to viruses, herbicides, and frost in broadleaf plants.

1. Plasmid removed and cut open with a special enzyme
2. Gene is cut out of the chromosome of another organism, using the same special enzyme
3. New gene is inserted into plasmid
4. Plasmid is put back into agrobacterium
5. When mixed with plant cells, agrobacterium duplicates plasmid
6. Bacterium transfers new gene into chromosome of plant cell
7. Plant cell divides, and each daughter cell receives new gene, giving the whole plant a new trait

Summary

1. Viruses were discovered at the end of the nineteenth century. For many years after their discovery, viruses were regarded as primitive forms of life. Today, scientists know that viruses are not living organisms because they are not cells. Viruses are protein-coated nucleic acids that replicate (multiply) within living cells.

2. Viruses primarily infect plants, animals, and bacteria, replicating within their cells. Various patterns of viral replication exist. In the lytic cycle a virus enters a cell and causes it to produce viral nucleic acid and protein coats. After these viral parts are assembled, the new virus particles may burst from the host cell or may leave the host cell by budding. In the lysogenic cycle, viruses enter into a long-term relationship with the cells they infect, their nucleic acid replicating as the cells multiply.

3. Some viruses can seriously disrupt the normal functioning of the cells they infect, transforming them into rapidly growing, invasive cells. These cells grow out of control, forming cancerous tumors that destroy body tissues.

4. Bacteria (kingdom Monera) have a prokaryotic cell structure. They differ from eukaryotic cells in many ways but primarily in that they have no membrane-bounded nucleus or membrane-bounded cellular organelles.

5. Bacteria reproduce asexually by binary fission, a splitting in two, after replication of the gentic material takes place. The numbers of bacteria within a population increase rapidly when growth conditions are favorable. These growth conditions vary among bacteria, since they occur in a wide range of habitats.

6. Most bacteria are considered eubacteria, meaning "true bacteria." These thousands of genera of bacteria can be placed into one of three groups according to their mode of nutrition: photoautotrophs, which make their own food by photosynthesis using the energy of the sun; chemoautotrophs, which make their own food by deriving energy from inorganic molecules; and heterotrophs, which obtain energy from organic material.

7. The archaebacteria differ from the eubacteria in their structural and physiological chemistry. Archaebacteria inhabit harsh environments, such as land fills, salt marshes, and deep-sea vents.

8. Bacteria play diverse ecological roles and are extremely important as decomposers. Bacteria are also important in the manufacture of certain foods. A few genera of bacteria cause diseases in plants, and many genera cause diseases in animals, including humans.

9. Scientists have learned to use the techniques of natural gene transfer in bacteria to direct gene transfer in ways that are useful to humans. Bacteria are now being used to manufacture hormones and a variety of medically useful products and to help produce crop plants and animals with desired characteristics. Scientists are also beginning to learn to use these techniques to treat genetic disorders.

REVIEW QUESTIONS

1. What are viruses? What do scientists mean when they say that viruses are not alive?
2. Diagram the structure of a generalized virus. Label the core, capsid, and envelope.
3. Distinguish between the lytic cycle and the lysogenic cycle.
4. One of your friends usually develops a cold sore on his mouth when he gets a cold. Why? What pattern of viral replication is involved?
5. What is the difference between a benign tumor and a malignant tumor? Which is more dangerous?
6. Why are bacteria classified into a kingdom that is unique from all other organisms?
7. Although many bacteria can cause dangerous diseases, in general, bacteria make life on earth possible. Why?
8. Fill in the blanks: Bacteria reproduce by a process called _____ _____. This can lead to a period of rapid doubling of cells called _____ _____.
9. What group(s) of bacteria would you expect to find living in each of the following locations: a) in nodules in the roots of bean plants; b) on a fallen log on the forest floor; c) in a shallow freshwater pond; d) in a highly concentrated salt pool?
10. What is meant by the terms *genetic engineering* and *recombinant DNA technology*?
11. What do transformation, transduction, and conjugation have in common? How are they different?
12. Summarize how researchers insert human genes into bacteria. Why is this technique useful?
13. What is gene therapy?
14. Summarize some of the potential risks and benefits of genetic engineering experiments.

KEY TERMS

archaebacteria 492
bacteriophage 489
benign 490
binary fission 491
capsid 487
carcinogen 490
chemoautotroph 492
competent cells 494
conjugation 494
core 487
decomposer 493
envelope 487
eubacteria 492
exponential growth 492
F plasmid 494
gene therapy 499
genetic engineering 494
heterotroph 493
infectious agent 487
latent infection 490
ligase 497
lyse 489
lysogenic cycle 489
lytic cycle 489
malignant 490
Monera 491
nitrifying bacteria 493
obligate parasite 487
photoautotroph 492
pilus 494
plasmids 494
recognition sequence 496
recombinant DNA technology 494
replication 489
restriction enzyme 496
toxin 494
transduction 494
transformation 494
tumor 490
vector 497
virus 487

THOUGHT QUESTIONS

1. Most plant diseases are caused by fungi and few by bacteria; most human diseases are caused by bacteria and few by fungi. Why the difference?
2. Why would you use genetic engineering to make a crop resistant to a herbicide?

FOR FURTHER READING

Mee, C.L., Jr. (1990, February). How a mysterious disease laid low Europe's masses. *Smithsonian*, pp 66-79.
This is the story of the bubonic plague, the Black Death, which changed the character of Europe in the fourteenth century.

Pennisi, E. (1992, February 22). Making sense of the disorder inside viruses. *Science News*, 141(8):116-117.
Scientists peer into the interior of tumor viruses using a new, unusual technique.

Radetsky, P. (1989, April). Taming the wily rhinovirus. *Discover*, pp. 38-43.
This article examines contemporary approaches to preventing colds, which affect about 1 in 20 people at any given time.

Resenberg, Z., & Fauci, A. (1990, February 10). Inside the AIDS virus. *New Scientist*, pp. 51-54.
Understanding the mechanisms by which the AIDS virus destroys the immune system is of central importance in developing a cure.

29 > Protists and fungi

OUTLINE

Protists
    Animal-like protists: Protozoans
        Amebas
        Flagellates
        Ciliates
        Sporozoans
    Plant-like protists: Algae
        Dinoflagellates
        Golden algae
        Brown algae
        Green algae
        Red algae
    Fungus-like protists
        Cellular slime molds
        Plasmodial (acellular) slime molds
        Water molds
Fungi
    Zygote-forming fungi
    Sac fungi
    Club fungi
    Imperfect fungi
    Lichens

"Please don't drink the water!" may seem like an unnecessary warning if you were camping along this pristine-looking mountain stream. But after an invigorating hiking trip, you could return home with nausea, cramps, bloating, and diarrhea—all symptoms of giardiasis, or "hiker's diarrhea." The culprit of this uncomfortable ailment is a single-celled organism no larger than a red blood cell: *Giardia lamblia*.

HIGHLIGHTS

▼

Protists make up an extremely diverse kingdom of eukaryotic organisms that are primarily single-celled but that include certain phyla containing multicellular forms.

▼

Protists can be grouped according to their mode of nutrition into those that are animal-like heterotrophs, plant-like photosynthetic autotrophs, and fungus-like saprophytes, although taxonomists differ in the classification of many forms.

▼

Fungi make up a kingdom of multicellular eukaryotic organisms that are primarily saprophytic; that is, they feed on dead or decaying organic material.

▼

Lichens are symbiotic associations between fungi and algae that allow both to survive in extremely harsh environments.

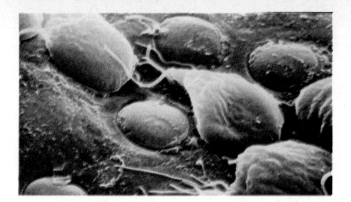

FIGURE 29-1 *Giardia lamblia.*
These single-celled organisms live in water. When humans drink water infected with *Giardia* organisms, they can experience nausea, cramps, bloating, vomiting, and diarrhea. It is important not to drink *any* untreated water, even though it may look safe.

The organism *G. lamblia* is found throughout the world, including all parts of the United States and Canada. It occurs in water, including the clear water of mountain streams and the water supplies of some cities. In addition to humans, it infects at least 40 species of wild and domesticated animals. These animals can transmit *G. lamblia* to humans by contaminating water with their feces.

Flagella protude from one end of *G. lamblia*, allowing it to move along the intestinal wall of its host. When feeding, it attaches to the wall, sucking blood for nutrition and leaving the suction-cup–shaped marks shown in Figure 29-1. This motile form of the protist exists only while inside the body of its victim. Dormant, football-shaped cysts are expelled in the feces, which can survive for long periods of time outside of their hosts—especially in the cool water of mountain streams. When ingested by other hosts, the cysts develop into their motile, feeding form.

What should you do to prevent infection by *G. lamblia* when hiking or camping? First, *never* drink untreated water, no matter how clean it looks. Because *G. lamblia* is resistant to usual water treatment agents, such as chlorine and iodine, you should boil the water you drink for at least a minute. Better still, *do not* drink the water—bring your own bottled water to be safe.

Protists

The **protists** (kingdom Protista) are a varied group of eukaryotic organisms. Many are single celled, although some phyla of protists include multicellular or colonial forms (single cells that live together as a unit). Within this kingdom are animal-like, plant-like, and fungus-like organisms (Figure 29-2)—quite a diverse array! The animal-like protists are called **protozoans** and are considered animal-like because they are heterotrophs: they take in and use organic matter for energy. *G. lamblia*, for example, is a protozoan. The plant-like protists are the **algae** (including **diatoms**).

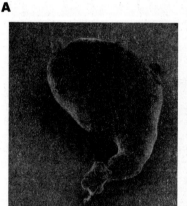
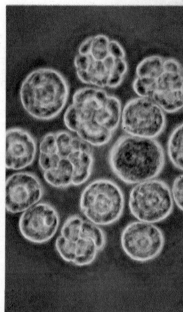

FIGURE 29-2 Single-celled protists.
A Animal-like protist *Pelomyxa palustris,* a unique, single-celled organism that lacks mitochondria and does not divide mitotically.
B Plant-like protist *Eudorina elegans,* a type of green alga.
C Fungus-like protist, the plasmodial slime mold. The individual cells cannot be distinguished in such a plasmodium.

These organisms are plant-like because they are eukaryotic photosynthetic autotrophs: they manufacture their own food using energy from sunlight. The fungus-like protists consist of two phyla of **slime molds** and one of **water molds**. They secrete enzymes onto food sources to predigest the food before they absorb it as the fungi do.

Although many protists are single celled, these organisms are incredibly different from prokaryotic single-celled organisms, the bacteria. The single-celled protists are much larger than bacteria, having approximately 1000 times the volume, and they contain typical eukaryotic cellular organelles. This compartmentalization of eukaryotic cells by membrane-bounded organelles increases their organization to a level much more complex than that of bacteria and allows single-celled eukaryotic organisms to carry out cell functions that support their larger cell volume.

> Most of the protist phyla are unicellular eukaryotes, although some phyla contain multicellular forms. Protists can be grouped as animal-like (protozoans), plant-like (algae), or fungus-like according to their mode of nutrition.

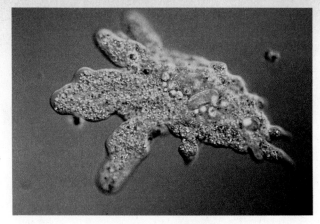

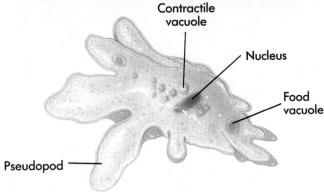

FIGURE 29-3 Structure of an ameba.
The pseudopods, literally "false feet," allow the ameba to move and also trap food particles.

Animal-like protists: Protozoans

The animal-like protists, or protozoans, obtain their food in diverse ways. These heterotrophic characteristics are linked to the ways that these organisms move and provide a means by which they can be grouped. There are four protozoan phyla: the amebas, flagellates, ciliates, and sporozoans.

Amebas

Amebas appear as soft, shapeless masses of cytoplasm. Within the cytoplasm lies a nucleus and other typical eukaryotic organelles. The cytoplasm streams, pushing out certain parts of the cell while retracting others. These cytoplasmic extensions are called **pseudopods** (from the Greek meaning "false feet") and are a means of both locomotion and food procurement. In fact, these cell extensions give the amebas their phylum name, *Rhizopoda*, which means "root-like feet." Shown in Figure 29-3, the pseudopods simply stream around the ameba's prey, engulfing it within a vacuole. Enzymes digest the contents of the vacuole, which are then absorbed into the cytoplasm to be further broken down for energy. Amebas reproduce by binary fission: the nucleus reproduces by mitosis and then the cell splits in two.

Amebas are abundant throughout the world in fresh water and saltwater, as well as in the soil. Many species are parasites of animals, including humans, and can cause diseases such as amebic dysentery, an infection of the digestive system that produces a diarrhea containing blood and mucus. Although amebic dysentery is a disease associated with poor sanitation and is found primarily in the tropics, medical researchers estimate that about 2 million Americans are infected with the causative agent, *Entamoeba histolytica*.

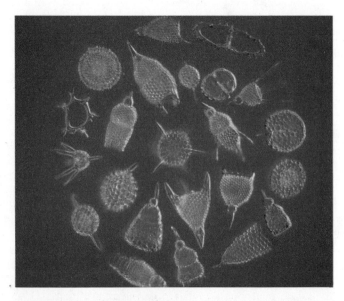

FIGURE 29-4 Radiolarian shells.
The shells of radiolarians resemble delicate glass sculptures.

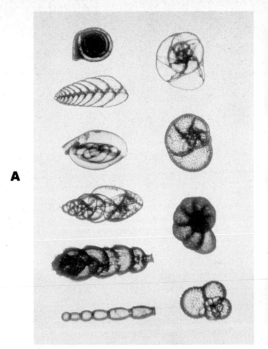

FIGURE 29-5   Foraminiferan shells.
A These beautiful foraminiferan shells are made out of limestone.
B The shells of foraminifera are punctured with tiny holes through which the foraminiferan pseudopods poke.

Certain groups of ameba secrete shells that cover and protect their cells. The **radiolarians** secrete shells made of silica that are glass-like and delicate (Figure 29-4). The **foraminifera** (or forams) secrete beautifully sculpted shells made out of calcium carbonate ($CaCO_3$), or limestone (Figure 29-5, A). The name *foraminifera* means "hole bearers" and refers to the microscopic holes in their shells through which their pseudopods poke (Figure 29-5, B). Food particles stick to these cellular extensions and are then absorbed into the cell.

Forams are abundant in the sea—so abundant, in fact, that their shells litter the sea floor. When studying geological strata, scientists often use the forams as indicators of geological age by noting the types of forams present in ancient rock. Interestingly, the white cliffs of Dover (Figure 29-6) are actually masses of foram shells, uplifted millions of years ago with the sea floor in an ancient geological event.

> Amebas are protozoans that have changing shapes brought about by cytoplasmic streaming, which forms cell extensions called *pseudopodia*.

FIGURE 29-6   The white cliffs of Dover.
The picturesque white cliffs are actually composed of masses of foram shells, which were uplifted from the sea floor millions of years ago.

PROTISTS AND FUNGI

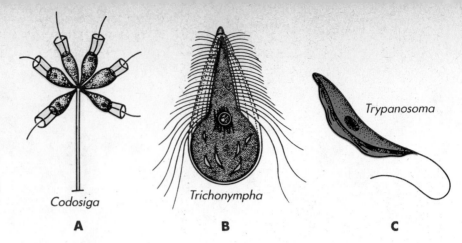

FIGURE 29-7 The flagellates.
A *Codosiga*, a colonial flagellate that remains attached to its substrate.
B *Trichonympha*, one of the flagellates that inhabit the gut of termites and wood-feeding cockroaches. *Trichonympha* ingests wood cellulose, which it finds in abundance in the digestive tracts of its hosts.
C *Trypanosoma* causes sleeping sickness in humans. It has a single flagellum attached to its head.

Flagellates

Flagellates are an interesting group because they are so diverse; a few representative genera of flagellates are shown in Figure 29-7. Although they all have at least one **flagellum** (a long, whip-like organelle of motility), some members of this phylum have many flagella, and some have thousands! Their phylum name is *Zoomastigina*, which means "animal whip."

All flagellates have a relatively simple cell structure. They do not have cell walls or protective outer shells as some of the amebas or ciliates do. They also have no complex internal digestive system of organelles as the ciliates do. A flagellate simply absorbs food through its cell membrane, sometimes using its flagella to ensnare food particles.

The flagellates are generally found in lakes, ponds, or moist soil where they can absorb nutrients from their surroundings. The colonial flagellate *Codosiga* shown in Figure 29-7, *A*, for example, consists of groups of cells that are often found anchored to the bottom of a lake or pond by a cellular stalk. The flagella create currents in the water that draw food toward the cells. The flagellate *Trichonympha*, shown in Figure 29-7, *B*, lives a protected life in the gut of termites, digesting wood particles the termite eats. Many flagellates are found living within other organisms; some of these relationships are not harmful to the hosts but other relationships are. Figure 29-7, *C* shows the flagellate *Trypanosoma* that lives in a parasitic relationship with certain mammals, including humans, causing the disease sleeping sickness.

▶ Flagellates are protozoans characterized by fine, long hair-like cellular extensions called *flagella*. ◀

One class of flagellates, the **euglenoids**, generally have chloroplasts and make their own food by photosynthesis. Each euglenoid has two flagella, a short one and a long one that are located on the anterior end of the cell. They move by whipping the long flagellum (Figure 29-8). The euglenoids reproduce asexually by transverse fission, a process in which the parent cell divides across its short axis (see Figure 29-11, *A*). No sexual reproduction is known among this group, which is named after its most well-known member: *Euglena*.

Euglena lives in ponds and lakes and can withstand stagnant water. It has a hard yet flexible covering beneath its plasma membrane called a *pellicle*, with ridges spiraling around its body. These ridges can be seen clearly in Figure 29-8. Two organelles, an **eyespot** and a **photoreceptor**, help *Euglena* stay near the light. (The name *Euglena*, in fact, means "true eye.") The photoreceptor, located near the base of its longer flagellum, is shaded by the nearby eye spot. As light filters through the pigment of the eyespot, the receptor senses the direction and intensity of the light source. Information from the receptor assists the movement of *Euglena* toward the light, a behavior known as **positive phototaxis**.

Interestingly, *Euglena* can survive without light—the chloroplasts simply become small and nonfunctional, and the organism begins to absorb food like a heterotroph. *Euglena* is a good example of an organism that is difficult to classify because of its ability to change its mode of nutrition. In fact, some euglenoids never have chloroplasts and live a totally heterotrophic existence.

▶ Euglenoids are flagellates, most of which have chloroplasts. ◀

FIGURE 29-8 *Euglena*, a euglenoid.
Euglenoids make their own food by photosynthesis and swim by means of flagella.

THE DIVERSITY AND UNITY OF LIVING THINGS

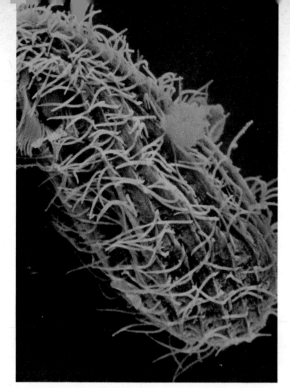

FIGURE 29-9 The ciliates.
Stentor, a funnel-shaped ciliate, showing spirally arranged cilia.

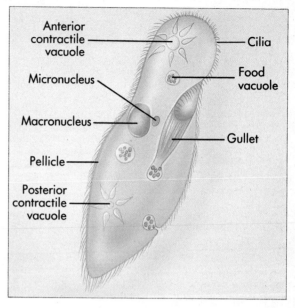

FIGURE 29-10 *Paramecium*.
This diagram shows the location of the contractile vacuoles, the pellicle, micronucleus and macronucleus, cilia, and food vacuole.

Ciliates

Ciliates (phylum Ciliophora) get their name from a Latin word meaning "eyelash"—a name that is descriptive of the fact that all or parts of these cells are covered with hair-like extensions called *cilia* (Figure 29-9). These cilia beat in unison, moving the cell about or creating currents that move food particles toward the gullet of the cell.

Ciliates possess a wide array of cellular organelles that perform functions similar to the organs of multicellular organisms. An example of this interesting cellular organization is shown in the diagram of the paramecium in Figure 29-10. The paramecium is a ciliate that is classically used as one example of this group. The beating cilia of the paramecium sweep food into its gullet. From the gullet, food passes into a vacuole where enzymes and hydrochloric acid aid in digestion. After absorption of the digested material is complete, the vacuole empties its waste contents into the **anal pore,** located in a special region of the pellicle. The waste then leaves the cell by a process similar to exocytosis.

Paramecia reproduce asexually by **transverse fission** (Figure 29-11, *A*.) In addition, paramecia undergo a type of sexual reproduction shown in Figure 29-11, *B* called **conjugation.** However, conjugation is not really a reproductive process. Instead, two cells exchange genetic material. So although conjugation does not produce offspring cells, it does promote genetic variability among cells that normally produce clones of identical cells when they reproduce. Genetic variability enhances the ability of the population to survive. Some algae, fungi, and bacteria also exchange genetic material in similar processes, which are also termed conjugation.

Most ciliates live in fresh water or salt water and do not infect other organisms. However, one species called *Balantidium coli* inhabits the intestinal tracts of pigs and rats.

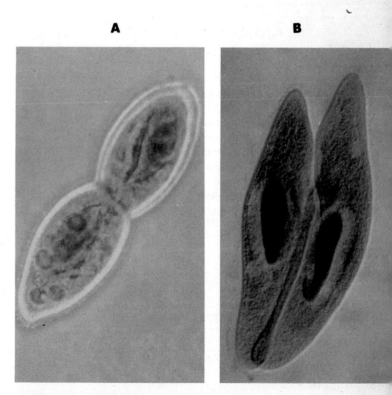

FIGURE 29-11 Reproduction in *Paramecium*.
A Transverse fission. When a mature *Paramecium* divides, two complete individuals result.
B Conjugation. In this process, two *Paramecia* exchange genetic material to promote genetic variability.

PROTISTS AND FUNGI

BIOLOGY & you

Malaria today

Malaria, caused by the sporozoan *Plasmodium,* is one of the most serious diseases in the world. As many as 250 million people are infected with it at any one time, and approximately 1 million of them, mostly children, die of malaria each year.

Malaria occurs primarily in the tropics, where the female *Anopheles* mosquito, the carrier of the disease, is prevalent. When an infected mosquito bites a person, *Plasmodium* sporozoans are injected into the bloodstream. As Figure 29-A shows, these sporozoans are carried to the liver where they undergo several cycles of asexual reproduction. In a week or two, these cells emerge from the liver, invade red blood cells, and undergo additional stages of asexual division and development. After these stages of development, the sporozoans are called *merozoites*. These cycles of development cause cycles of symptoms. The host often feels well during development of the parasite within the blood cells. Then the infected red blood cells burst, releasing the merozoites, which enter other red blood cells. This event produces chills and fever, accompanied by nausea, vomiting, and severe headache. These symptoms cycle every few days as cells are infected and then burst. During their development with the blood cells, some of the merozoites develop into gametes (sex cells). If the host is again bitten by an *Anopheles* mosquito, these gametes are taken up by the mosquito, and the sexual stages of reproduction take place.

Scientists are currently focusing on developing a vaccine against malaria and are starting to produce promising results. Each of the three different stages of the life cycle of *Plasmodium* sporozoans (the asexual stages in the liver, the stages in the red blood cells, and the sexual stages) produces different antigens and is sensitive to different antibodies. Because the numbers of sporozoans injected by the *Anopheles* mosquito are so high (about a thousand are injected!) and because they can travel to the liver within a few minutes and increase eightfold within 24 hours, a vaccine against these sporozoans is not likely to be effective. Scientists are therefore searching for a compound vaccine that will be effective against the sporozoans, merozoites, and gametes. Human trials are underway and promise to eventually control the disease, despite the great complexity of the *Plasmodium* sporozoan life cycle.

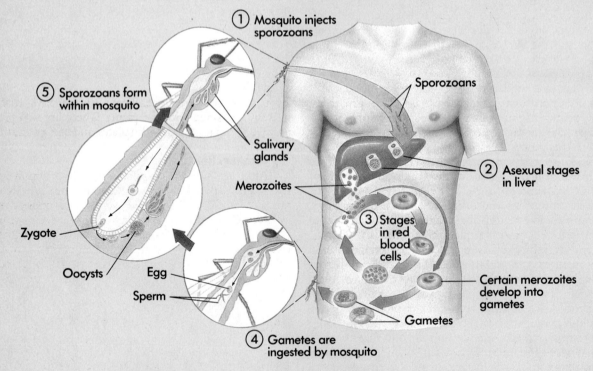

FIGURE 29-A The life cycle of *Plasmodium,* a parasite that causes malaria.

People who come in contact with this protozoan on farms or in slaughterhouses can become infected. The ciliates embed themselves into the lining of the intestines, producing sores and causing dysentery, similar to amebic dysentery. Occasionally, epidemics of balantidiasis occur in areas having poor sanitation.

> Ciliates are protozoans characterized by fine, short, hair-like cellular extensions called *cilia*.

Sporozoans

All **sporozoans** (phylum Sporozoa) are nonmotile, spore-forming parasites of vertebrates, including humans. Sporozoans have complex life cycles that involve both asexual and sexual phases. Their spores are small, infective bodies that are transmitted from host to host by various species of insects.

The sporozoan used classically to represent this phylum of protists is *Plasmodium,* the causative agent of malaria. Approximately 1 million people die from this disease each year, so it is considered one of the most serious diseases in the world. The boxed essay describes this disease and the life cycle of *Plasmodium* in more detail.

> Sporozoans are nonmotile protozoans that are parasites of vertebrates, including humans. They undergo complex life cycles in which they are passed from host to host by various species of insects.

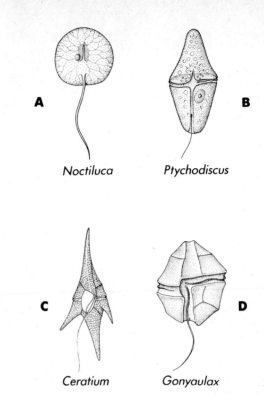

FIGURE 29-12 The dinoflagellates.
A *Noctiluca* lacks the heavy cellulose armor characteristic of most dinoflagellates.
B *Ptychodiscus.*
C *Ceratium.*
D *Gonyaulax.*

Plant-like protists: Algae

Algae are eukaryotic organisms that contain chlorophyll and carry out photosynthesis, so they can be thought of as plant like. However, algae lack true roots, stems, leaves, and vascular tissue (an internal water-carrying system). Algae are widely distributed in the oceans and lakes of the world, floating on or near the surface of the water no lower than the sun's rays can reach.

Some systems of classification place the unicellular algae in the protist kingdom and the multicellular algae in the plant kingdom. However, current thinking favors placing all the algae with the protists, broadening the scope of this kingdom. Formerly, the protist kingdom included only the protozoans and the unicellular algae. It now includes multicellular algae and the fungus-like protists, each of which lacks some of the characteristics of either plants or fungi. These changes reflect the ongoing debate among scientists regarding the evolutionary relationships among organisms and how organisms should be classified to reflect these relationships. Classifying all algae as protists suggests closer evolutionary relationships among the various phyla of algae and other protists than between the algae and the plants. The green algae, thought to be the ancestors to the land plants, are sometimes still placed in the plant kingdom.

There are three phyla of multicellular algae: the **brown algae,** the **green algae** (which includes many unicellular forms), and the **red algae.** These are the phyla that were formerly classified with the plants. There are two phyla of unicellular algae: the dinoflagellates and the **golden algae.**

Dinoflagellates

Although the **dinoflagellates** (phylum Dinoflagellata) have flagella, they look nothing like the flagellates or euglenoids. Many dinoflagellates have outer coverings of stiff cellulose plates, which give them very unusual appearances (Figure 29-12). Their flagella beat in two grooves, one encircling the cell like a belt and the other perpendicular to it. As they beat, the encircling flagellum causes the dinoflagellate to spin like a top; the perpendicular flagellum causes movement in a particular direction. This spinning is a characteristic for which this phylum was named. The word *dinoflagellate* comes from Latin words meaning "whirling swimmer."

Most dinoflagellates live in the sea and carry on photosynthesis. Their photosynthetic pigments are usually golden brown, but some are green, blue, or red. The red dinoflagellates are also called *fire algae.* In coastal areas, these organisms often experience population explosions or "blooms," causing the water to take on a reddish hue referred to as a *red tide.* Red tides destroy other living things because many species of dinoflagellates produce powerful toxins.

PROTISTS AND FUNGI

FIGURE 29-13 Red tide.
Red tides are caused by a population explosion of certain dinoflagellates. The release of toxins from these dinoflagellates can poison marine life.

These poisons kill fishes, birds, and marine mammals (Figure 29-13). In addition, shellfish strain these dinoflagellates from the water and store them in their bodies. Although the shellfish are not harmed, they are poisonous to humans and other animals that eat them.

Dinoflagellates reproduce primarily by longitudinal cell division, but sexual reproduction has also been shown in more than 10 genera of dinoflagellates.

> Dinoflagellates, a type of flagellated unicellular red algae, are characterized by stiff outer coverings. Their flagella beat in two grooves, one encircling the cell like a belt and the other perpendicular to it.

Golden algae

The three types of organisms found in phylum Chrysophyta are the **yellow-green algae,** the **golden-brown algae,** and the diatoms. These organisms are named after the gold-green color of the photosynthetic pigments in their chloroplasts. (The prefix *chryso-* means "color of gold.") An unusual characteristic of these organisms is that they store food as oil. Some forms smell "fishy," giving unpleasant odors to the fresh water lakes they inhabit.

The most well-known members of this group, abundant in both the ocean and in fresh water, are the diatoms—microscopic aquatic pillboxes. Diatoms are made up of top and bottom shells that fit together snugly (Figure 29-14, *A*). These organisms reproduce asexually by separating their top from their bottom, each half then regenerating another top or bottom shell within itself. The shells are composed of silica, so like the radiolarians, the diatoms look somewhat glass like. However, the shells of the diatoms are so characteristically striking and intricate (Figure 29-14, *B*) that it would be hard to confuse them with any other group of protists. Interestingly, the shells of fossil diatoms often form very thick deposits on the sea floor, which are sometimes mined commercially. The resulting "diatomaceous earth" is used in water filters, as an abrasive, and to add the sparkling quality to products, such as the paint used on roads and frosted fingernail polish.

> The golden algae have gold-green photosynthetic pigments and store food as oil. They are often represented by diatoms, organisms that look like microscopic pillboxes.

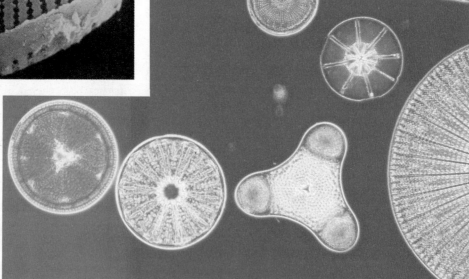

FIGURE 29-14 Diatoms.
A Diatoms are composed of a top and bottom shell that fit together.
B Several different types of diatoms.

Brown algae

The brown algae (phylum Phaeophyta) are the dominant algae of the rocky, northern shores of the world. The types of brown algae that grow attached to rocks at the shoreline are known as *rockweed* (Figure 29-15, *A*). Their puffy air bladders keep the plant afloat during high tide. One type of rockweed is also called *Sargasso weed* and gave the Sargasso Sea its name. The Sargasso Sea is an area of ocean in the mid-Atlantic, east of Bermuda, with unusual water and current patterns that cause it to be quite calm. This calmness allows floating species of the Sargasso weed to proliferate and dominate the area.

The large brown alga with enormous leaf-like structures is kelp (Figure 29-15, *B*). These algae are an important source of food for fish and invertebrates, as well as some marine mammals and birds that live among these seaweeds. Some genera of the kelps are among the longest organisms in the world (rivaling the height of the giant sequoia trees), reaching lengths of up to 100 meters (328 feet)! These algae usually have a structure in which a root-like portion, descriptively termed a **holdfast,** anchors the seaweed to the ocean floor or to rocks. A stem-like **stipe** carries the leaf-like **blades** of the seaweed, which float on or in the water, capturing the sun's rays (Figure 29-15, *C*).

Most brown algae have life cycles that parallel the generalized life cycle of plants (see Chapter 30 and Figure 30-1). The other multicellular algae (brown algae and red algae) also have plant-like life cycles. However, the life cycles of the red algae are quite complex.

> Brown algae are large, multicellular algae found predominantly on northern, rocky shores. Some grow to enormous sizes.

FIGURE 29-15 The brown algae.
A Rockweed. The air bladders attached to the algae help keep the plant afloat.
B Kelp. Kelp have large blades (leaf-like structures) and are an important food source for marine fishes, birds, mammals, and invertebrates.
C A kelp forest. The blades of kelp capture sunlight, which they use for photosynthesis. Kelp forests can be quite large.

PROTISTS AND FUNGI

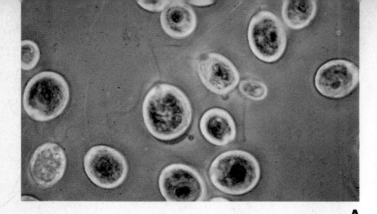

Green algae

The green algae, or Chlorophyta (literally, "green plants") are an extremely varied phylum of protists. In fact, more than 7000 species exist! Of these species, most are aquatic (as are other algae), but some are **semiterrestrial**. Semiterrestrial algae live in moist places on land, such as on tree trunks, on snow, or in the soil. These algae are primarily unicellular microscopic forms, but some are multicellular. Green algae show many similarities to land plants: they store food as starch, they have cell walls composed of cellulose, they have a similar chloroplast structure, and their chloroplasts contain chlorophylls *a* and *b*. For these reasons, scientists think the green algae were ancestors to the plant kingdom.

Well known among the unicellular green algae is the genus *Chlamydomonas* (Figure 29-16, *A*). Individuals are microscopic, green, and rounded and have two flagella at their anterior ends. They are aquatic and move rapidly in the water as a result of the beating of their flagella in opposite directions. They have eyespots that help direct the algae to the sunlight. The life cycle of *Chlamydomonas* is very simple. This haploid organism reproduces asexually by cell division and sexually by the functioning of some of its cells as gametes. After gametes fuse, the diploid zygote divides by meiosis, restoring the haploid state.

Some genera of green algae live together in groups. *Volvox* is one of the most familiar of these colonial green algae. Each colony is a hollow sphere made up of a single layer of individual, biflagellated cells (Figure 29-16, *B*). The flagella of all of the cells beat in such a way as to rotate the colony in a clockwise direction as it moves through the water. In some species of *Volvox* there is a division of labor among the different types of cells, making them truly multicellular organisms.

The multicellular forms of green algae grow in either fresh water or saltwater, such as the sea lettuce *Ulva* shown in Figure 29-16, *C*. Sea lettuce is extremely plentiful in the ocean and is often found clinging to rocks or pilings. Interestingly, this alga consists of sheets of tissue only two cells thick. Another familiar multicellular green alga is the filamentous *Spirogyra*. This alga is interesting because it has spiral chloroplasts (Figure 29-16, *D*).

> Green algae include both unicellular and multicellular forms. Most forms are aquatic, as are other algae, but some species live in moist places on land.

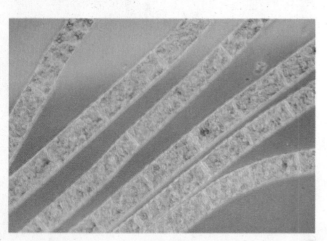

FIGURE 29-16 The green algae.
A *Chlamydomonas* is a unicellular green alga that has two flagella and an eyespot sensitive to sunlight.
B *Volvox* is a colonial green alga. Each colony is made up of individual cells.
C *Ulva,* or sea lettuce.
D *Spirogyra,* a filamentous green alga.

Red algae

Almost all red algae, or Rhodophyta ("red plants"), are multicellular, and the majority of their species are marine. Their color comes from the types and amount of photosynthetic pigments present in their chloroplasts. Many species have a predominance of red pigments in addition to chlorophyll and so are red (as their name suggests). Some red algae have a predominance of other pigments so they look green, purple, or greenish black (Figure 29-17). The red pigment, however, is especially efficient in absorbing the green, violet, and blue light that penetrates into the deepest water. (It *reflects* red light.) This enables some of the red algae to grow at greater depths than other algae and inhabit areas in which most algae cannot exist.

The red algae produce substances that make them interesting both ecologically and economically. The coralline algae, for example, deposit calcium carbonate (limestone) in their cell walls. Along with coral animals, these red algae play a major role in the formation of coral reefs. Also, all red algae have glue-like substances in their cell walls: agar and carrageenan. Agar is used to make gelatin capsules, the material for making dental impressions and a base for cosmetics. It is also the main ingredient in the laboratory media on which bacteria, fungi, and other organisms are often grown. Carrageenan is used mainly as a stabilizer and thickener in dairy products, such as creamed soups, ice cream, puddings, and whipped cream, and as a stabilizer in paints and cosmetics. Some of the red algae are used as food in certain parts of the world, such as in Japan.

> The red algae play a major role in the formation of coral reefs and produce glue-like substances that make them commercially useful.

FIGURE 29-17 The red algae.
Red algae do not always appear red. Some algae show different colors depending on their daily exposure to light.
A *Ahnfeltia plicata* growing on rocks. This type of alga is edible and is considered a delicacy in many Asian cultures.
B *Bossiella* is a coralline red alga, which means that it secretes the hard substance calcium carbonate.
C *Ahnfeltia concinna,* a common red alga.

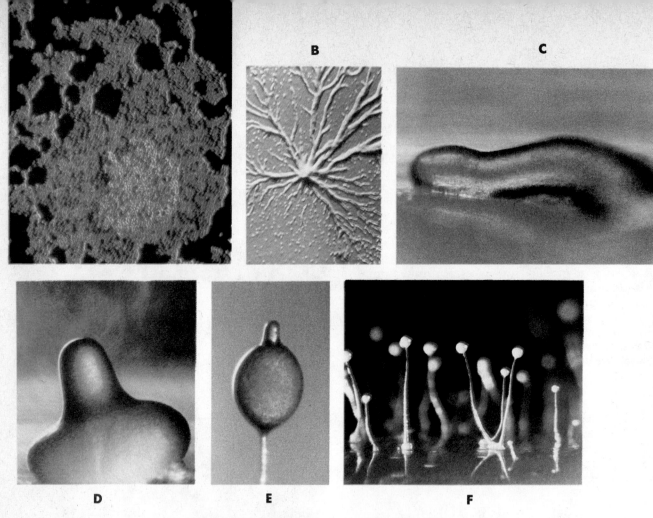

FIGURE 29-18 Development in a cellular slime mold.
A The ameba stage, in which the mold looks and behaves like an ameba.
B The amebas aggregate and move toward a fixed center.
C They form a slug that migrates towards light.
D The slug stops moving and begins to form into a fruiting body.
E The head of a fruiting body.
F Many fruiting bodies together.

Fungus-like protists

The slime molds make up two unique and interesting phyla of protists. They are called molds because they give rise to mold-like (fungus-like) spore-bearing stalks during one stage of their life (see p. 518). However, slime molds look very much like amebas at other times in their life cycles, forming visible "slimy" masses. In fact, scientists now think slime molds may be most closely related to the amebas.

Slime molds are heterotrophs and feed on bacteria, which they find in damp places rich in nutrients, such as rotting vegetation (especially rotting logs), damp soil, moist animal feces, and water. The slime molds are subdivided into two classes: the **cellular slime molds** and the **plasmodial (acellular) slime molds**.

> Slime molds are organisms that are fungus-like in one phase of their life cycles and ameba-like in another phase of their life cycles.

Cellular slime molds

Most of their lives, the cellular slime molds look and behave like amebas, moving along and capturing bacteria by means of pseudopods (Figure 29-18, *A*). At a certain phase of their life cycle, which is often triggered by a lack of food, the individual organisms move toward one another (Figure 29-18, *B*) and form a large, moving mass called a *slug* (Figure 29-18, *C*). (Their phylum name, Acrasiomycota, comes from the name of the chemical attractant, acrasin, the cellular form produces, which causes the cells to aggregate.) The slug eventually transforms into a fruiting body (Figure 29-18, *D, E, F*). The tips of the fruiting bodies contain dormant, cyst-like forms of the ameba-like cells and are called *spores*. Some of these spores fuse and undergo a type of sexual reproduction before being released; others do not. Each spore becomes a new "ameba" if it falls onto a suitably moist habitat. The amebas begin to feed and continue the life cycle.

FIGURE 29-19 The plasmodial slime mold.
Plasmodial slime molds move about as a plasmodium, a multinucleated mass of cytoplasm.

Plasmodial (acellular) slime molds

The plasmodial slime molds also have an ameboid stage to their life cycle, but these "amebas" are quite unusual. These bizarre organisms stream along as a **plasmodium**—a nonwalled, multinucleate mass of cytoplasm—which resembles a moving mass of slime (Figure 29-19). Their phylum name, Myxomycota, literally means "mucous mold." The plasmodia engulf and digest bacteria, yeasts, and other small particles of organic matter as they move along. At this stage of its life cycle, a plasmodium may reproduce asexually; the nuclei undergo mitosis simultaneously, and the entire mass grows larger.

When food or moisture is in short supply, the plasmodium moves to a new area and forms spores. These spores are held in spore cases, which have a characteristic look for each genera of acellular slime mold. Figure 29-20 shows three different types of spore cases. The spores are resistant to unfavorable environmental influences and may last for years if dry. Meiosis occurs in the spores. When conditions are favorable, the spore cases open and release spores that germinate into flagellated, haploid cells called *swarm cells*. The swarm cells can divide, producing more swarm cells, or can act as gametes. Gametes can fuse and form a new plasmodium by repeated mitotic divisions.

Water molds

Taxonomists disagree as to whether the water molds should be considered protists or fungi. A primary issue is that water molds have flagellated spores, which are not characteristic of fungi. In this textbook, these organisms are classified with the fungus-like protists.

If you have an aquarium and have seen white fuzz on any of your fish, you have been introduced to the water molds. They live not only in fresh water and saltwater and on aquatic animals but also in moist soil and on plants. Some of the plant diseases caused by this group are late blight and downy mildew. Although this group gets the name *water molds* because many species thrive in moisture, they are sometimes called *egg fungi* because of the large egg cells present during their sexual reproduction.

> Water molds thrive in moist places and aquatic environments, parasitizing plants and animals. During sexual reproduction, they produce large egg cells.

FIGURE 29-20 Spore cases of three types of plasmodial slime molds.
A *Arcyria.*
B *Fuligo.*
C *Tubifera.*

PROTISTS AND FUNGI

Fungi

Fungi are a separate kingdom of mostly multicellular eukaryotic organisms that are **saprophytic;** that is, they feed on dead or decaying organic material. To do this, fungi secrete enzymes onto a food source to break it down and then absorb the breakdown products. Some fungi are **parasites** and feed off living organisms in the same way (as happens in ringworm and athlete's foot). Most fungi are multicellular. Unlike plants, fungi have no chloroplasts and do not produce their own food by photosynthesis. They are composed of slender filaments that may form cottony masses or that may be packed together to form complex structures, such as mushrooms.

The slender filaments of fungi are barely visible to the naked eye. Termed **hyphae** (singular, hypha), they may be divided into cells by cross walls called **septa** (sing., septum) or may have no septa at all. The septa rarely form a complete barrier, however, so cytoplasm streams freely throughout either type of hyphae. Because of this streaming, proteins made throughout the hyphae may be carried to their actively growing tips. As a result, the growth of fungal hyphae may be very rapid when food and water are abundant and the temperature is optimum.

A mass of hyphae is called a **mycelium** (plural, mycelia). Part of the mycelial mass grows above the food source (substrate) and bears reproductive structures. The rest grows into the substrate (Figure 29-21). The part of the mycelium embedded in the food source secretes enzymes that digest the food. For this reason, many fungi are harmful because their mycelia grow into and decay, rot, and spoil foods (and sometimes other organic substances, such as leathers). In addition, some fungi cause serious diseases of plants and animals, including humans.

In their roles as decomposers, fungi may seem troublesome to humans, but they are essential to the cycling of materials in ecosystems (see Chapter 38). In addition, many antibiotics (drugs that act against bacteria) are produced by fungi. Yeasts, which are single-celled fungi, are also useful in the production of foods such as bread, beer, wine, cheese, soy sauce, and yogurt. These foods all depend on the biochemical activities of yeasts, in which they produce certain acids, alcohols, and gases as by-products of their metabolism of sugar. Yeasts can also cause infections in humans, such as vaginal and oral yeast infections.

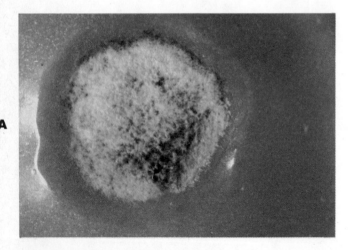

FIGURE 29-21 The structure of mold.
A A cottony mold growing on a tomato.
B This "falsely color" micrograph shows the fungus *Erisyphepisi,* which causes powdery mildew on grasses, magnified 45 times. The thread-like mycelium *(red)* covers the surface of the grass leaf *(greenish yellow)*. The mycelium is made up of original hyphae.

Most fungi reproduce both asexually and sexually. (Figure 29-22 shows a generalized life cycle for many fungi.) During sexual reproduction, hyphae of two genetically different mating types (called + and - strains) come together, and their haploid nuclei fuse and produce a diploid zygote. This zygote germinates into diploid hyphae, which bear **spores.** Spores are reproductive bodies formed by cell division (mitosis or meiosis) in the parent organism. These spores are formed by meiosis in a diploid parent and by mitosis in a haploid parent; the spores are always haploid. They germinate into haploid hyphae after being released and finding appropriate growth conditions. The cycle comes full circle when their hyphae fuse with those of a genetically different mating type, forming zygotes. Some fungi may reproduce asexually by budding, by growing new hyphae from fragments of parent hyphae, or by producing spores by mitosis.

Fungi are classified into divisions rather than into phyla as plants are. Three divisions of fungi include those organisms that have a distinct sexual phase of reproduction: the **zygote-forming fungi,** the **sac fungi,** and the **club fungi.** A fourth group is a "catch-all" division in which organisms are placed because they appear to have no sexual stage of reproduction. This division is termed the **imperfect fungi.**

Fungi are eukaryotic organisms that feed on dead or decaying organic material or parasitize living organisms. Although some fungi cause disease in plants and animals, some are important in the production of food, and most are ecologically important decomposers.

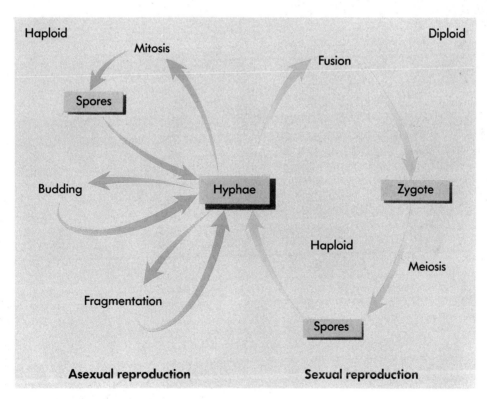

FIGURE 29-22 A generalized life cycle for fungi.
Fungi alternate between sexual and asexual reproductive stages.

PROTISTS AND FUNGI

Zygote-forming fungi

You may have seen black mold growing on bread or other food (Figure 29-23, *A*), but you may not have seen this mold under a microscope (Figure 29-23, *B*). Members of this group are found on decaying food and other organic material and are characterized by their formation of sexual spores called **zygospores**. For this reason, this group of fungi is called the *zygote-forming fungi* (division Zygomycota). The life cycle of these fungi (Figure 29-23, *C*) parallels the generalized life cycle shown in Figure 29-22 quite closely.

> Zygote-forming fungi, such as black bread mold, live on decaying organic material and are characterized by their formation of sexual spores called *zygospores*.

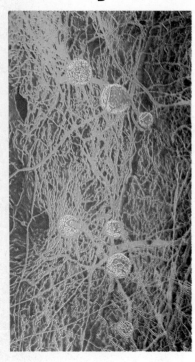

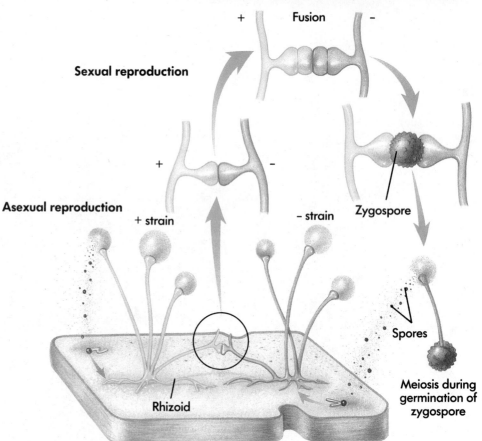

FIGURE 29-23 Black bread mold.
A Black bread mold growing on food.
B A microscopic view of black bread mold. This scanning electron micrograph has been color-enhanced to bring out contrasts. The hyphae (the *green strands*) and the sporangia *(yellow balls)* can be clearly seen.
C The life cycle of black bread mold.

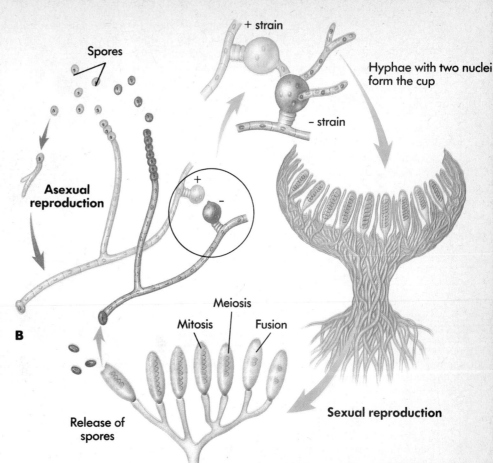

FIGURE 29-24 Sac fungi (division Ascomycota).
A A cup fungus, *Cookeina tricholoma,* from the rain forest of Costa Rica.
B The life cycle of a cup fungus.

Sac fungi

The sac fungi are the largest division of fungi, having at least 30,000 named species. They live in a wide variety of places, such as in the soil, in saltwater and fresh water, on dead plants and animals, and on animal feces. They are called sac fungi because their sexual spores, which occur in groups of eight, are enclosed in sac-like structures called **asci** (sing., ascus). Their life cycles are similar to the club fungi.

Among the sac fungi (division Ascomycota) are such familiar and economically important fungi as yeasts, cup fungi (Figure 29-24), and truffles. Unfortunately, this class of fungi also includes many of the most serious plant pathogens, including the causative agent of Dutch elm disease, which has killed millions of elms in North America and Europe (Figure 29-25). The spores of the mold causing this disease are carried from tree to tree by a bark beetle.

FIGURE 29-25 Dutch elm disease.
Millions of elms in North America and Europe have been killed by Dutch elm disease, which is caused by an ascomycete, *Ceratocystis ulmi.* The spread of Dutch elm disease can be controlled by a combination of sanitation (removing dead and dying trees promptly), killing beetles in traps baited with chemicals, and treating the trees in advance with fungicides and bacteria that inhibit the growth of the fungus.

Although single cells, the yeasts are also sac fungi and are one of the most economically important of the class because of their use in the production of various foods. Most of the reproduction of the yeasts is asexual and takes place by binary fission or by budding (the formation of a smaller cell from a larger one). Sometimes, however, whole yeast cells may fuse, forming sacs with zygotes. These cells divide by meiosis then mitosis, forming eight spores within each sac. When the spores are released, each functions as a new cell.

> The sac fungi live in both aquatic and terrestrial environments and are characterized by sexual spores borne in sac-like structures. Familiar examples of this group are cup fungi and yeasts.

Club fungi

This division of fungi includes the mushrooms, toadstools, puffballs, jelly fungi, and shelf fungi (Figure 29-26). Some species of club fungi are commonly cultivated; for example, the button mushroom—commonly served in salad bars—is grown in more than 70 countries, producing a crop with a value of over 15 billion dollars. Other kinds of club fungi are represented by the rusts and smuts, which are devastating plant pathogens. Wheat rust, for example, causes enormous economic losses to wheat wherever it is grown.

FIGURE 29-26   Club fungi (division Basidiomycota).
A An earth star, *Geastrum saccatus*. Earth stars are a kind of puffball. Basidia form within puffballs, which may have spores. They release hundreds of thousands—or even millions—of basidiospores when mature.
B A shelf fungus, *Trametes versicolor*, growing on a tree trunk.
C A jelly fungus growing in the Amazonian rain forest of Brazil. Although many familiar basidiomycetes form their basidia within mushroom-like basiocarps, many do not.

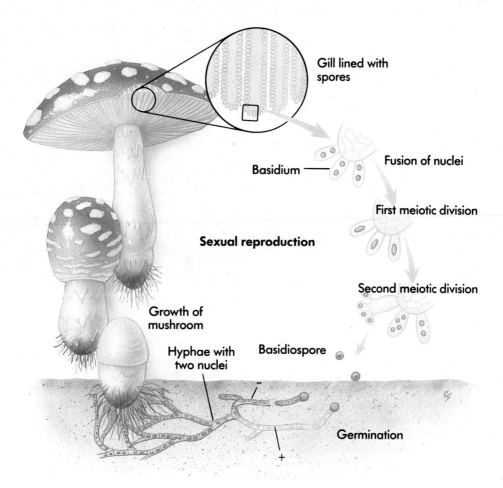

FIGURE 29-27 Life cycle of a common club fungus—a mushroom.
This class of fungi rarely undergoes an asexual stage of reproduction.

The club fungi are so-named because they have club-shaped structures from which unenclosed spores are produced. These structures are called **basidia** (sing., basidium) and give the division name of Basidiomycota. In mushrooms, basidia line the gills found under the cap (Figure 29-27). In rusts and smuts, basidia arise from hyphae at the surface of the plant.

As shown in Figure 29-27, a mushroom is formed when pairs of hyphae (often of two different mating strains) fuse and intermingle their cytoplasm and haploid nuclei. The fused hyphae develop into the mushroom, including the gills on the underside of its cap. The basidia develop as single cells on the free edges of the gills. Within the cells that develop into basidia, the nuclei present from the fusion of hyphae now fuse to form zygotes. Each zygote undergoes meiosis, producing four haploid spores in each basidia. The zygotes within some basidia undergo mitosis, producing basidia having eight haploid spores. Turgor pressure (water build-up within the basidia) bursts the basidia and hurls the spores from mushroom—at an average rate of 40 million per hour! These spores germinate to produce hyphae, which fuse to begin a sexual cycle of reproduction once again. Asexual reproduction is rare in this class of fungi.

> The club fungi, named because they have club-shaped structures from which unenclosed spores are produced, include the mushrooms, puffballs, and shelf fungi.

PROTISTS AND FUNGI

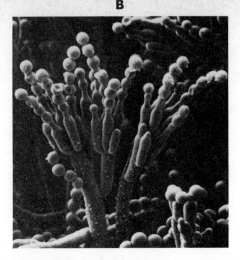
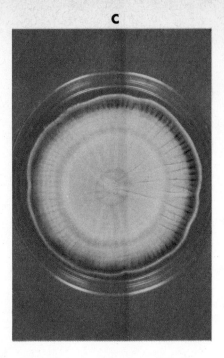

FIGURE 29-28 Imperfect fungi.
A *Aspergillis.*
B *Penicillium* growing on an orange.
C *Penicillium* growing in a petri dish.

Imperfect fungi

This division of fungi includes members whose hyphal structure and asexual reproduction suggest that they are zygote-forming fungi, sac fungi, or club fungi. However, the fungi in this group have lost the ability to reproduce sexually, or their sexual stages of reproduction have not been observed. Therefore these genera cannot be placed in their appropriate division because placement is based on features related to sexual reproduction. They are therefore referred to as *imperfect*—the Fungi Imperfecti. Although these fungi do not reproduce sexually, hyphae of different mating types often fuse, providing some genetic recombination. This characteristic is important to the survival of each species, especially in the plant pathogens because mutation and recombination produce new strains that can still infect plants bred for resistance to these fungal diseases.

The imperfect fungi reproduce asexually by spores. An enormous range of diversity occurs in the structure of the spore cases found at the tips of their hyphae. These variations represent adaptations of the fungi to dispersal of their spores. For example, spores distributed by insects usually have sticky, slimy spore cases with odors attractive to insects. Those dispersed by the wind have dry spore cases.

Among the economically important genera of imperfect fungi are *Penicillium* and *Aspergillus* (Figure 29-28). Some species of *Penicillium* are sources of the well-known antibiotic penicillin, and other species of the genus give the characteristic flavors and aromas to cheeses such as Roquefort and Camembert. Species of *Aspergillus* are used for fermenting soy sauce and soy paste, processes in which certain bacteria and yeasts also play important roles.

Most of the fungi that cause diseases in humans are members of the imperfect fungi. Some of these fungi cause infections of the skin, such as athlete's foot (Figure 29-29) and other forms of ringworm. The term *ringworm* refers to any fungal infection of the skin, including the scalp and nails. There is also a small group of fungi that cause diseases of various organs, including the brain. Some of these diseases are spread by birds, bats, or contaminated soil. Such diseases are usually contracted by breathing air heavily contaminated with spores of the causative agent. Other fungal diseases occur only in persons weakened by other diseases, making them more suseptible to infection by pathogenic fungi.

> The imperfect fungi have no sexual stage of reproduction and reproduce asexually by spores. Some members of this division are sources of antibiotics or are important in food production, but many cause disease in humans.

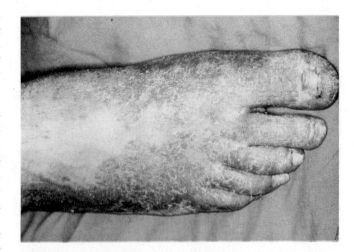

FIGURE 29-29 Athlete's foot.
This itchy, painful condition is caused by a member of the imperfect fungi.

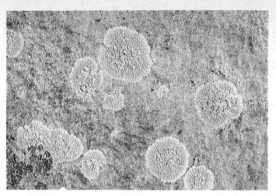

Lichens

Lichens (Figure 29-30) are associations between fungi and photosynthetic partners and are classified with the fungi. They provide an example of mutualism, a living arrangement in which both partners benefit. Most of the visible body of a lichen consists of fungus, but cyanobacteria or green algae (or sometimes both) live within the fungal tissues (Figure 29-31). The photosynthetic organism provides nutrients for itself and the fungus. Specialized fungal hyphae penetrate or envelop the photosynthetic cells and transfer water and minerals to them, although this part of the mutualistic relationship has been questioned recently. Researchers now suspect that the fungus acts more like a parasite than a mutualistic partner.

Lichens are able to invade the harshest of habitats at the tops of mountains, in the farthest northern and southern latitudes, and on dry, bare rock faces in the desert. In such harsh, exposed areas, lichens are often the first colonists, breaking down the rocks and setting the stage for the growth of other plants. These amazing algae-fungus partnerships are able to dry or freeze and then recover quickly and resume their normal metabolic activities. The growth of lichens may be extremely slow in harsh environments—so slow, in fact, that many small lichens appear to be thousands of years old and are among the oldest living things on Earth.

FIGURE 29-30 Three types of lichens.
A Crustose (encrusting) lichens growing on a rock in California.
B A fruticose (shrubby) lichen, *Cladina evansti,* growing on the ground in Florida. Fruticose lichens predominate in deserts because they are more efficient in capturing water from moist air than either of the other two types.
C A foliose (leafy) lichen, *Parmotrema gardneri,* growing on the bark of a tree in the mountain forest in Panama.

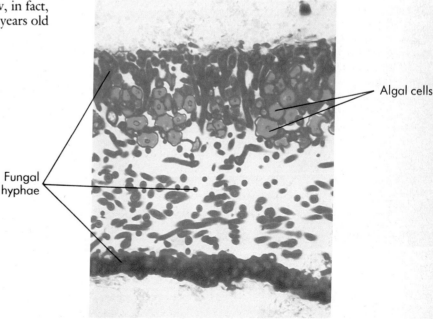

FIGURE 29-31 Section of a lichen.
The fungal hyphae are more densely packed into a protective layer on the top and bottom layers of the lichen. The cells (shown as blue) near the surface of the lichen are a green alga. This alga supplies carbohydrates to the fungus.

Summary

1. The protists (kingdom Protista) are a varied group of eukaryotic organisms. Many are single celled, although some phyla of protists include multicellular or colonial forms. Within this kingdom of organisms are animal-like cells, which take in organic matter for nutrition; plant-like cells, which manufacture their own food; and fungus-like cells, which live on dead or decaying organic matter.

2. Animal-like protists, the heterotrophs, are usually referred to as protozoans. The protozoans are classified according to the means by which they move and feed. The amebas move and eat by means of cell extensions, or pseudopods. The flagellates move by means of long whip-like cellular extensions, their flagella. The ciliates move and sweep food toward themselves by means of short hair-like cellular processes, their cilia. And the sporozoans are nonmotile parasites of vertebrate animals, carried from one host to another by insects.

3. The algae contain chlorophyll, carry out photosynthesis, but lack true roots, stems, leaves, and vascular tissue. There are five phyla of plant-like protists: the dinoflagellates, the golden algae, the brown algae, the green algae, and the red algae. The dinoflagellates and the golden algae are unicellular organisms, whereas brown algae, green algae, and red algae are multicellular. However, the green algae has certain unicellular species.

4. The slime molds have unique life cycles that have fungus-like spore-bearing stages and "slimy" ameba-like stages. The water molds grow in fresh water and saltwater, on aquatic animals, in moist soil, and on plants. These protists produce large egg cells during sexual reproduction.

5. Fungi are multicellular, eukaryotic organisms, most of which feed on dead or decaying organic material. They are important decomposers that are essential to the cycling of materials in ecosystems, are important in the production of certain foods, and also cause certain diseases. Most fungi are composed of slender microscopic filaments, or hyphae, that form cottony masses or compact plant-like structures.

6. Three divisions of fungi have sexual stages of reproduction. These are the zygote-forming fungi (such as black bread mold), the sac fungi (such as cup fungi and yeasts), and the club fungi (such as mushrooms, puffballs, and shelf fungi).

7. The imperfect fungi are a fourth division of fungi. Members of this division have no known sexual stages of reproduction. The imperfect fungi include organisms that produce antibiotics and are used in the production of certain foods but that also cause certain diseases in humans.

8. Lichens are associations of fungi with green algae and/or cyanobacteria. The photosynthetic organism provides nutrients for itself and the fungus. Researchers now question whether the fungus helps the photosynthetic organism; it has been thought that fungal hyphae penetrated or enveloped the photosynthetic cells and transferred water and minerals to them. Lichens are able to exist in harsh environments and live for long periods of time. Some lichens may be among the oldest organisms on Earth.

KEY TERMS

algae 503
ameba 504
anal pore 507
asci 519
basidium (pl., basidia) 521
blade 511
brown algae 509
cellular slime mold 514
ciliate 507
club fungi 517
conjugation 507
diatom 503
dinoflagellate 509
euglenoid 506
eyespot 506
flagellum 506
foraminiferan (pl., foraminifera) 505
golden algae 509
green algae 509
holdfast 511
hyphae 516
imperfect fungi 517
mycelium (pl., mycelia) 516
plasmodial (acellular) slime mold 514
plasmodium 515
positive phototaxis 506
protist 503
protozoan 503
pseudopod 504
radiolarian 505
red algae 509
sac fungi 517
saprophytic 516
semiterrestrial 512
septum (pl., septa) 516
slime mold 504
spore 517
sporozoan 509
stipe 511
swarm cell 525
transverse fission 507
water mold 504
zygospore 518
zygote-forming fungi 517

REVIEW QUESTIONS

1. Explain how the following terms are related: protists, protozoa, algae, diatoms, slime molds.
2. Describe an ameba. How does it move and take in food?
3. What do flagellates and ciliates have in common? How do they differ?
4. What are sporozoans? Briefly summarize a sporozoan's life cycle.
5. What do euglenoids, dinoflagellates, and golden algae have in common? Describe each one.
6. Identify the phylum or phyla to which the following belong: yellow-green algae, diatoms, golden-brown algae.
7. Fill in the blanks: The _____ _____ are protists that are _____-like in one phase of their life cycles, and _____-like in another phase of their life cycles.
8. Fill in the blanks: A fungus that feeds on a fallen tree in a forest is _____. A fungus that invades the skin between your toes, creating the uncomfortable condition known as athlete's foot, is a(n) _____.
9. Fungi are both helpful and harmful to human beings. Give some examples of each.
10. Match each term with the most appropriate statement:
    _____ Club fungi
    _____ Zygote-forming fungi
    _____ Sac fungi
    _____ Water molds
    _____ Imperfect fungi
    a. Sexual spores borne in sac-like structures
    b. Have no sexual stage of reproduction
    c. Sexual spores are produced by basidia
    d. Sexual spores are called zygospores
    e. During sexual reproduction, they produce large egg cells
11. What are lichens, and where are they found?
12. Fill in the blanks: Amebas that secrete delicate shells made of silica are called _____; amebas that secrete shells made of calcium carbonate or limestone are called _____.

THOUGHT QUESTIONS

1. Some single-celled protists like *Volvox* form spherical colonies of hundreds of cells. Within the colony, some of the cells become specialized for sexual reproduction. Why is such a colony not considered to be a multicellular organism?
2. If the protist *Euglema* can act "plant-like" by photosynthesizing in the presence of light, and "animal-like" by injesting its food in the absence of light, in which group does it belong? What does this say about the usefulness of this approach to classifying protists?
3. Many fresh water lakes, when polluted by high-phosphate detergents, become overgrown with mats of photosynthetic green algae. If the pollution continues, all other life in the lake soon dies. Why? Why can you not avoid this lethal result by simply poisoning the algae?

FOR FURTHER READING

Barron, G. (1992, March). Jekyll-Hyde mushrooms, *Natural History*, pp. 47-52.
Many fungi belie their benign appearance and actively trap and consume a variety of organisms.

Cochran, M.F. (1990, February). Back from the brink: Chestnuts. *National Geographic*, pp. 128-140.
Nearly destroyed by the chestnut blight, the American chestnut is being rescued by dedicated scientists and volunteers.

McKnight, K.H., & McKnight, V. (1987). *A field guide to mushrooms of North America.* Peterson Field Guide Series. New York: Houghton Mifflin.
This is an excellent guide that includes most of the common and edible species of North America.

Saffo, M.B. (1987). New light on seaweeds. *BioScience*, 37, 654-664.
Seaweeds occur at different depths in the ocean, their photosynthetic pigments being efficient in harvesting energy for the particular light waves that reach that depth.

30 Plants
Reproductive patterns and diversity

Although this does not look like a family photo of two generations of individuals—it is! The leaf-like greenery are plants of one generation and the stalks are of another generation. But interestingly, both are the same plant—the hairy-cap moss *Polytrichum*.

The green plants are the gametophyte generation. Gametophytes do as their name suggests: they produce gametes, or sex cells. These male and female gametes fuse during fertilization to form zygotes.

HIGHLIGHTS

▼

Plants are multicellular, eukaryotic organisms that typically live on land and make their own food by the process of photosynthesis.

▼

Plants, unlike animals, have two different multicellular phases that occur alternately within their sexual life cycles.

▼

The differences between the life cycles of the lower, nonvascular plants and the higher, vascular plants reflects an adaptation of the vascular plants for life on land.

▼

Plants are also able to reproduce asexually by vegetative propagation, a process in which a new plant develops from a portion of a parent plant.

OUTLINE

Characteristics of plants

The general pattern of reproduction in plants

Patterns of reproduction in nonvascular plants

   The bryophytes

      Mosses

      Liverworts

      Hornworts

Patterns of reproduction in vascular plants

   Seedless vascular plants

   Vascular plants with naked seeds

   Vascular plants with protected seeds

   Mechanisms of pollination

   Fruits and their significance in sexual reproduction

   Seed formation and germination

   Types of vegetative propagation in plants

Regulating plant growth: Plant hormones

The generation of plants that develops from the zygotes, the stalk-like plants called *sporophytes*, looks very different from its gametophyte parents. The sporophyte generation is named because it produces spores. And the gametophyte generation is in turn produced from these spores, resulting in a cycling of two very different generations of the same plant.

Characteristics of plants

Because plants have the ability to reproduce sexually, a characteristic they have in common with most living things, other characteristics differentiate them from other organisms and place them in their own kingdom. Plants are multicellular, eukaryotic, photosynthetic autotrophs. That is, they produce their own food (which most plants store as starch) by using energy from the sun and carbon dioxide from the atmosphere (a process described in Chapter 8). However, all organisms fitting this description are *not* plants. The multicellular algae, at one time classified as plants and now classified by most systematists (taxonomists) as protists, also fit this description. Today, plants are often defined as multicellular, photosynthetic eukaryotes that *live on land*, although some botanists disagree with this definition. The few species of plants that live in the water have returned to the water after evolving on land.

Another characteristic that differentiates the multicellular algae from the plants is their type of chlorophyll, the pigment that absorbs energy from the sun during the process of photosynthesis. Plants have the pigments chlorophyll *a* and chlorophyll *b*. The red and brown algae do not have chlorophyll *b* and so are thought not to have evolved from the same ancestors as the plants. The green algae do have chlorophyll *a* and *b* but, because they have many single-celled forms, are also classified as protists. The green algae are thought by most scientists to be the evolutionary predecessors of the land plants.

All plants can be placed into one of two groups: the nonvascular plants and the vascular plants. Most plants (over 80% of all living plant species) are **vascular plants,** those having specialized tissues to transport fluids. There are three major groups of vascular plants: seedless vascular plants (such as ferns), vascular plants with naked seeds (such as pine trees), and vascular plants with protected seeds (such as flowering plants). Plants lacking these specialized transport tissues are called **nonvascular plants.** There is one major group of nonvascular plants: the bryophytes (such as "true" mosses).

Traditionally, plants are taxonomically separated into **divisions** rather than into phyla, but these categories are basically equivalent to one another. In this classification scheme, the plant kingdom has ten divisions. Table 30-1 lists these divisions, including examples. It shows how the divisions fit into the groupings of plants previously described.

TABLE 30-1 The ten divisions of plants

GROUP	DIVISION	EXAMPLES
Nonvascular plants	Bryophyta	"True" mosses, liverworts, hornworts
Seedless vascular plants	Psilophyta Lycophyta Sphenophyta Pterophyta	Whisk ferns Club mosses Horsetails Ferns
Vascular plants with naked seeds	Coniferophyta Cycadophyta Ginkgophyta Gnetophyta	Conifers Cycads Ginkgos Gnetae
Vascular plants with protected seeds	Anthophyta Class Monocotyledons Class Dicotyledons	Flowering plants Grasses, irises Flowering trees, shrubs, roses

The general pattern of reproduction in plants

It is necessary to take a broader look at the patterns of reproduction of all living things to investigate patterns of reproduction in specific divisions of plants. The series of events that take place from one stage during the life span of an organism, through a reproductive phase, and until a stage similar to the original is reached in the next generation is called a **life cycle.** For example, a bacterium's life cycle can be described as beginning when it (along with another daughter cell) splits from a mother cell during binary fission. This life cycle is complete when the bacterium itself splits into two new cells. This is an example of an asexual life cycle.

Animals, plants, some protists, and some fungi have sexual life cycles, which are characterized by the alternation of meiosis and fertilization. For example, your life cycle could be described as all of the events of growth and development that began at fertilization and continued after your birth through the time of your sexual maturation. This is the time you began producing sex cells by the process of meiosis. The cycle would begin again, or have come "full circle," when you have a child—the next generation.

PLANTS: REPRODUCTIVE PATTERNS AND DIVERSITY

Figure 30-1, *A* diagrams this type of sexual life cycle, which is common to most animals. Notice in the diagram that during the animal life cycle, the multicellular organism has a double set of hereditary material; it is diploid. (One set was contributed by a father and one set was contributed by a mother.) The gametes, however, are haploid single cells.

> Animal life cycles are characterized by a multicellular diploid phase and a unicellular haploid phase. The multicellular diploid organisms produce haploid gametes by meiosis. These gametes fuse during fertilization, forming the first cell of a new diploid multicellular organism.

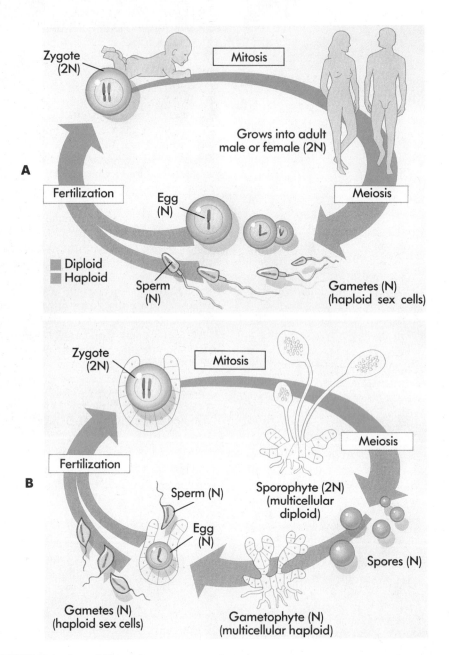

FIGURE 30-1 Sexual life cycles.
A The animal sexual life cycle. The organism that results from the union of haploid gametes is diploid.
B A generalized plant life cycle showing the alternation of generations. The sporophyte generation (diploid) alternates with the gametophyte generation (haploid). Haploid spores are produced by the sporophyte generation by meiosis. When dispersed, these haploid spores grow into gametophytes.

In some protists and many fungi, the haploid phase of the life cycle is multicellular. Gametes are produced from these organisms by mitosis rather than by meiosis. Only the zygotes are diploid (see Chapter 29).

Plants and some species of algae have life cycles that differ from the two sexual life cycles already described. They have both a multicellular haploid phase and a multicellular diploid phase. Because both phases of the life cycle are multicellular, this type of life cycle is called **alternation of generations** (Figure 30-1, *B*). The alternating generations of plants are called the **sporophyte** (spore-plant) generation and the **gametophyte** (gamete-plant) generation. As previously mentioned, gametophytes form gametes, or sex cells. Because they are **haploid** individuals, having *half* the usual number of chromosomes for that species of organism, they form gametes (which are haploid) by the process of mitotic cell division. These gametes fuse during fertilization, a process that depends on the presence of water in which sperm swim to the egg. The plants produced from fertilized eggs are **diploid** plants, having a full complement of genetic material. These diploid plants are the sporophytes. Because sporophytes are diploid, they use the process of meiosis to produce haploid spores. When dispersed, spores grow into gametophyte plants.

> **Plant life cycles are marked by an alternation of generations of diploid sporophytes with haploid gametophytes. As a result of meiosis, sporophytes produce spores, which grow into gametophytes. Gametophytes produce gametes as a result of mitosis. These gametes fuse during fertilization and grow into sporophytes.**

Looking at general characteristics of the bryophytes and the vascular plants and their patterns of reproduction, you will notice that each group has its own variation of a life cycle. However, the gametophyte (haploid) generation often dominates the life cycles of the nonvascular plants, whereas the sporophyte (diploid) generation dominates the life cycles of the vascular plants. This difference reflects an adaptation of the vascular plants for life on land as they evolved from earlier, nonvascular forms. These adaptations included the development of spores with protective walls able to tolerate dry conditions, efficient water and food-conducting systems, and gametophyte generations that became protected by and nutritionally dependent on the sporophyte generation.

Plants produce gametes and spores in specialized structures. In the bryophytes and several divisions of vascular plants, eggs are formed in structures called **archegonia** (singular, **archegonium**). Sperm are produced in structures called **antheridia** (sing., **antheridium**). In the more specialized vascular plants, except in a few species of seed-bearing plants, these specialized gamete-producing structures have been reduced in size during the course of evolution as the sporophyte generation began to dominate the life cycle. Eggs and sperm differentiate from a small number of haploid cells within the sporophyte. These haploid cells are actually the gametophyte generation.

FIGURE 30-2 Bryophytes.
A A liverwort, *Marchantia*. The sporophytes are carried within the tissues of the umbrella-shaped structures that arise from the surface of the flat, green, creeping gametophyte.
B *Antheros*, a hornwort. This photo shows the long, slender sporophyte with the gametophyte below.

Patterns of reproduction in nonvascular plants

The bryophytes

Bryophytes are a single division of small, low-growing plants that are commonly found in moist places. The three classes of bryophytes are mosses (such as the hairy-cap moss shown on p. 526), liverworts (Figure 30-2, *A*), and hornworts (Figure 30-2, *B*).

FIGURE 30-3 Peat mosses, *Sphagnum*.
A A peat bog being drained by the construction of a ditch. In many parts of the world, accumulations of peat, which may be many meters thick, are used as fuel.
B The glistening black, round objects are the spore cases, which contain spores. The spore cases of *Sphagnum* have a lid that blows off explosively, releasing the spores.

Mosses

The largest class of bryophytes and probably the one most familiar to you is the **mosses**. The gametophytes of most mosses have small, simple leaf-like structures often arranged in a spiral around stem-like structures. These bryophyte structures are different from the stems and leaves of vascular plants; see Chapter 31. Many small, carpet-like plants are mistakenly called mosses. For example, *Spanish moss* is actually a flowering plant, a relative of the pineapple. Even eliminating these impostors, however, there are still some 10,000 species of true mosses, and they are found almost everywhere on Earth.

One kind of moss that is most important economically is *Sphagnum*. (Figure 30-3). This moss grows in boggy places (low-lying, wet, spongy ground), forming dense and deep masses that are often dried and sold as peat moss. Peat moss is used in gardening as a mulch, layered around trees or plants to protect the roots from temperature fluctuations, retain moisture, control weeds, and enrich the soil. It functions well as a mulch and a soil additive because its tissues have special water storage cells. These cells allow the peat to absorb and retain up to 90% of its dry weight in water, whether or not the moss is alive.

Figure 30-4 shows the life cycle of the hairy-cap moss. Flask-shaped archegonia are found among the top leaves of the female gametophytes. Each archegonium produces one egg. Antheridia are found in a similar place in the male gametophytes. Each antheridium produces many sperm. The flagellated sperm swim through drops of water from rain, dew, or other sources into the neck of the archegonium and then to the egg. After fertilization takes place, the zygote develops into a young sporophyte within the archegonium. It then grows out of the archegonium and differentiates into a slender stalk. This stalk is initially green, but its chlorophyll disintegrates as it matures, leaving the stalk yellow or brown. At this stage the sporophyte stalk derives its nourishment from the gametophyte. Each sporophyte stalk bears a spore capsule near its tip. Haploid spores are produced by meiosis within this capsule. When the top of the spore capsule pops off, the spores are freed. Under the proper conditions, these spores germinate into thread-like filaments; the characteristic leafy gametophytes arise from buds that form on these filaments.

Liverworts

Liverworts were given their name in medieval times when people believed that plants resembling particular body parts were good for treating disease of those organs. Some liverworts are shaped like a liver and were thought to be useful in treating liver ailments. The ending *-wort* simply means "herb."

A well-known example of a liverwort is *Marchantia*, shown in Figure 30-2, *A*. It has green, leafy gametophytes that grow close to the ground. Antheridia and archegonia develop within the umbrella-like portion of the stalks that grow up from the leaf-like gametophyte. The sporophytes develop encased within these tissues; spores are freed from these structures.

Hornworts

The **hornwort** is named because it has elongated sporangia that protrude like horns from the surface of the creeping gametophytes (Figure 30-2, *B*). Its life cycle parallels quite closely those of the mosses and liverworts.

> The life cycles of all three classes of bryophytes—the mosses, liverworts, and hornworts—are somewhat uniform, all having two distinct phases. The sporophyte generation grows out of or is embedded in the tissues of the gametophyte generation and depends on the gametophyte generation for nutrition. Although the gametophyte plant is the most familiar, neither the sporophyte nor gametophyte generation tends to dominate the life cycle.

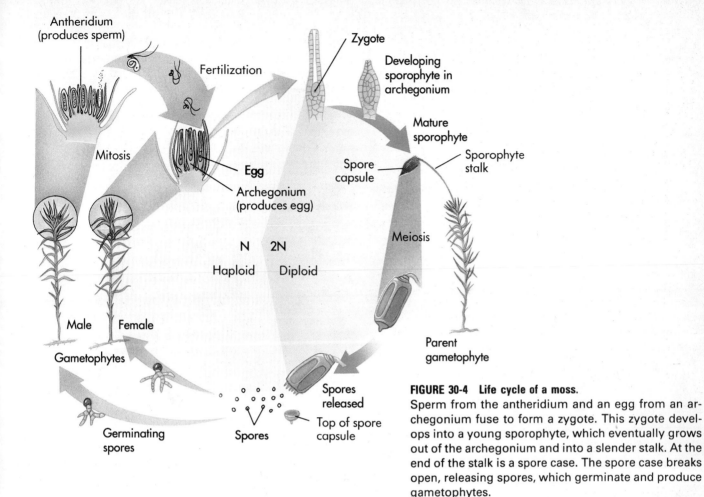

FIGURE 30-4 Life cycle of a moss.
Sperm from the antheridium and an egg from an archegonium fuse to form a zygote. This zygote develops into a young sporophyte, which eventually grows out of the archegonium and into a slender stalk. At the end of the stalk is a spore case. The spore case breaks open, releasing spores, which germinate and produce gametophytes.

Patterns of reproduction in vascular plants

Seedless vascular plants

As listed in Table 30-1, the members of four divisions of vascular plants do not form seeds. **Seeds** are structures from which new sporophyte plants grow; they protect the embryonic plant from drying out or being eaten when it is at its most vulnerable stage. Seeds also contain stored food for the new plant. The seedless plants overcome these problems in interesting ways. Figure 30-5 diagrams the life cycle of a fern—a familiar member of the **seedless vascular plants.** Its life cycle is representative of this group.

When a fern plant is mature, it produces spores by meiosis. The spore cases of ferns look like dots on the underside of the fern leaf, or **frond.** Because it produces spores, the fern is the sporophyte generation—the dominating form in the life cycle of the seedless vascular plants. After its spores are dispersed, those that settle in a moist environment will germinate into haploid plants that look very *unlike* ferns.

These plants are small, ground-hugging, heart-shaped gametophytes. Their antheridia and archegonia are protected somewhat by being located on the underside of the plant. Sperm, when released, swim through moisture collected on the underside of the leaf to the archegonia. Each fertilized egg (zygote) starts to grow within the protection of the archegonium. After this initial protected phase of growth, the fern sporophyte is able to grow on its own and becomes much larger than the gametophyte.

> The life cycles of seedless vascular plants are similar to one another. The sporophyte (diploid) generation is dominant and lives separately from the gametophyte. The gametophytes produce motile sperm that need water to swim to the eggs. The fertilized eggs produce young sporophytes that grow protected within gametophyte tissues but eventually become free living.

PLANTS: REPRODUCTIVE PATTERNS AND DIVERSITY

Many of the seedless vascular plants that lived about 300 million years ago were converted long ago to a fuel used today—coal. Club mosses that grew on trees, horsetails, ferns, and tree ferns made up great swamp forests during the Carboniferous Period (see Chapter 26). Areas of New York State, Pennsylvania, and West Virginia, for example, were lying near the equator at that time (see Chapter 26). Dead plants did not completely decay in the stagnant, swampy waters, and they accumulated. These swamps were later covered by ocean waters. Marine sediments piled on top of the plant remains. Pressure and heat acted on the layers of dead plant material beneath the ocean floor and converted the remains to coal. When you burn fossil fuels such as coal, you are burning a resource that was formed under special conditions that have not been repeated in the last 300 million years. Coal is therefore called a *nonrenewable resource*. Burning coal adds carbon dioxide to the atmosphere that had been locked in this fossil fuel for millions of years. This gas adds to the amount of carbon dioxide in the atmosphere and contributes to the problems of global warming and acid rain (see Chapter 40).

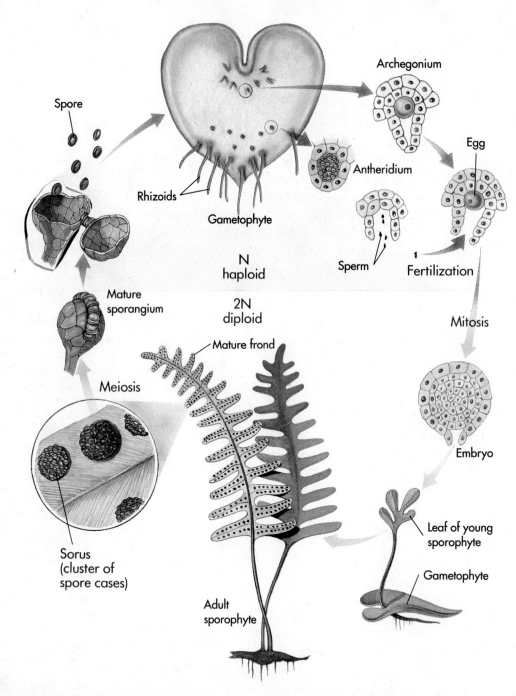

FIGURE 30-5 Life cycle of a fern.
The dark dots that you have seen on the underside of a fern frond are actually clusters of spore cases. The spore case breaks open, releasing spores, which develop into a heart-shaped gametophyte. The archegonia and antheridia produce eggs and sperm, which fuse to form a zygote. The zygote develops into a sporophyte, which becomes much larger than the gametophyte.

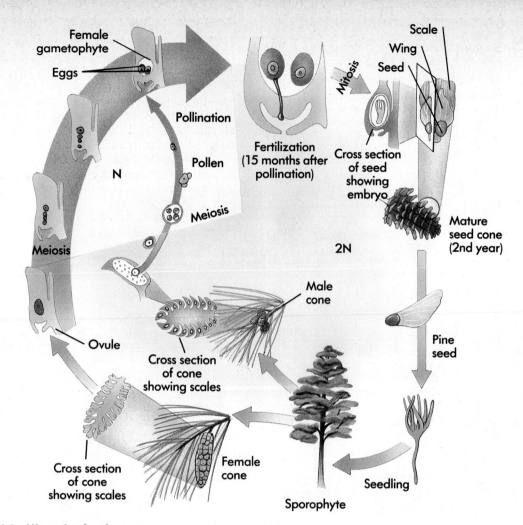

FIGURE 30-6 Life cycle of a pine.
Start at the sporophyte, which is a full-grown pine tree. The gametophyte generation of a pine is so small that it is not visible to the naked eye! The male gametophyte is the pollen. The female gametophyte (archegonium) is multicellular and contains two eggs, of which only one will develop into an embryo. The female gametophyte is depicted as the white area around the eggs and is multicellular tissue.

Vascular plants with naked seeds

Four divisions of vascular plants fall in the category of vascular plants with naked seeds: the conifers, cycads, ginkgo, and gnetophytes. This group is called the **gymnosperms**, a name derived from Greek words meaning "naked seed." The term *naked* refers to the seeds of gymnosperms, which are not completely enclosed by the tissues of the parent at the time it is pollinated.

The most familiar division of gymnosperm are the conifers, or cone-bearing trees. The conifers include pines, spruces, firs, redwoods, and cedars. Figure 30-6 diagrams the life cycle of a pine. The pine tree is the sporophyte (diploid) generation. Interestingly, the gametophyte generation is so reduced that it is not visible to the naked eye and is comparatively short lived.

Pines and most other conifers bear both male cones and female cones on the same tree. You can tell them apart because the female cones are the larger of the two. These cones produce spores that undergo meiosis, producing the male and female gametophytes. Male gametophytes are pollen grains, each consisting of four cells. The male gametophytes produce sperm, which remain located in the pollen grains. The multicellular female gametophytes each produce two to three eggs. The eggs develop within protective structures called **ovules** within the scales of the female pine cones. In the spring, the male cones release their pollen, which is blown about by the wind. (Those allergic to pine pollen know this fact quite well!) As some of this pollen passes by female cones, it gets trapped there by a sticky fluid produced by the now open female cones. As this fluid evaporates, the pollen is drawn further into the cone. When the pollen comes into contact with the outer portion of the ovule, it germinates and forms a pollen tube that slowly makes its way into the ovule—to the egg. After 15 months the tube reaches its destination and discharges its sperm. Fertilization takes place, producing a zygote. The development of the zygote into an embryo takes place within the ovule, which matures into a seed. Eventually the seed falls from the cone and germinates, and the embryo resumes growing and becomes a new pine tree.

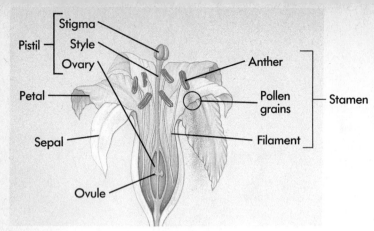

FIGURE 30-7 An angiosperm flower.
The main parts of a flower, which contain both male and female structures.

Vascular plants with protected seeds

Flowers are the organs of sexual reproduction in the vascular plants with protected seeds—the **angiosperms.** These plants bear seeds in a **fruit.** The flower is structured to promote sexual reproduction—the union of gametes that ultimately develops into an embryo within the seed.

Developmentally, flower parts are actually modified leaves. The outermost whorl, or ring, of modified leaves are the **sepals** (Figure 30-7). Sepals enclose and protect the growing flower bud. In some flowers such as roses, the sepals remain small, green, and somewhat leaflike. In other flowers, such as tulips, the sepals become colored and look like the **petals,** the next whorl of flower parts.

Flower petals are frequently prominent and colorful and attract pollinating animals, especially insects. Although the sepals and petals are the dominant outward features of flowers, they are not the organs of sexual reproduction. Many flowers either do not have sepals or petals or have inconspicuous sepals or petals—particularly flowers that are pollinated by the wind. The sex organs of the flower are the innermost modified leaves, the male **stamens** and, at the center of the flower, the female **pistil.**

Each stamen consists of an **anther,** a compartmentalized structure where haploid **pollen grains** are produced. Each pollen grain is a male gametophyte, enclosed within a protective outer covering. This gametophyte will produce sperm by mitosis, which will be enclosed within the pollen grain. A long, thin filament bears and supports the anther, exposing the pollen to wind or pollinating animals.

The pistil consists of three parts. At its tip is a sticky surface called the **stigma** to which pollen grains can adhere. At the base of the pistil is the **ovary,** a chamber that completely encloses and protects the ovules. Within each ovule, a single mother cell develops and then divides meiotically, producing four cells. One of these cells develops into a female gametophyte. When mature, the female gametophyte is called an *embryo sac*. Within this sac are typically eight cells, one of which is the egg. The ovary will become a seed when its eggs are fertilized by the male gamete. The **style** is a narrow stalk arising from the top of the ovary that bears the stigma. The style may be either long or short to facilitate the best exposure of the stigma to the method of pollination that characterizes the plant (Figure 30-8). Pollination in the flowering plants refers to the transfer of pollen from the anther to the stigma.

Figure 30-9 diagrams the life cycle of an angiosperm. After a pollen grain lands on the stigma, it produces a long tube that grows from the pollen grain down the **style** and penetrates the ovary, entering an ovule. One of the two haploid nuclei (sperm) in the pollen tube fertilizes the haploid egg nucleus in the ovule, producing a diploid zygote.

FIGURE 30-8 Different modes of pollination.
A Pollination by a bumblebee. As this bumblebee, *Bombus,* collects nectar from the flame azalea, the stigma contacts its back and picks up any pollen that the bee might have acquired there during a visit to a previous flower.
B Pollination by a butterfly. This copper butterfly, *Lycaena gorgon,* is probing a flower with its proboscis, a coiled, tongue-like organ, to extract its nectar.
C Pollination by a hummingbird. Long-tailed hermit hummingbird extracting nectar from the flowers of *Heliconia imbricata* in the forests of Costa Rica. Notice the pollen grains on the bird's beak.

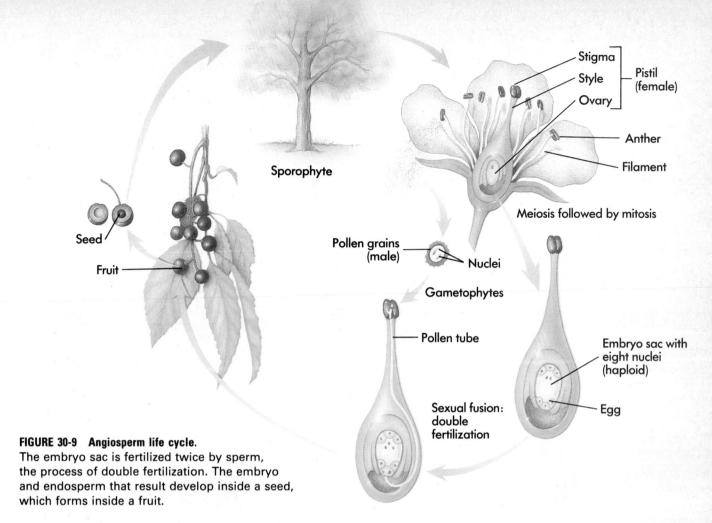

FIGURE 30-9 Angiosperm life cycle.
The embryo sac is fertilized twice by sperm, the process of double fertilization. The embryo and endosperm that result develop inside a seed, which forms inside a fruit.

The zygote will become a new plant embryo. The other sperm nucleus fuses with two other nuclei of the embryo sac, producing a *triploid* endosperm nucleus. The endosperm nucleus develops into tissue that will feed the embryo as it grows into a plant. The ovule becomes the seed within which the embryo develops, and the ovary ripens into a fruit. The fruit is a food that attracts animals, which play a role in seed dispersal.

> Sexual reproduction in gymnosperms and angiosperms produces seeds by the union of male gametes released from pollen grains and female gametes within ovules. Germinating pollen tubes convey sperm to eggs without a watery environment. The seeds, formed within the ovary after the union of sperm and eggs, protect and nourish the young sporophytes as they begin to develop into new plants.

Mechanisms of pollination

Pollination takes place in many ways, such as by insects, animals, and the wind. Insects and other animals often visit the flowers of angiosperms for a liquid called **nectar**. Nectar is a food rich in sugars, amino acids, and other substances.

In certain angiosperms and in most gymnosperms, pollen is blown about by the wind and reaches the stigmas passively. However, wind does not carry pollen far or precisely compared with insects or other animals. Therefore plants pollinated by the wind usually grow close together. These plants typically produce large quantities of pollen.

In some angiosperms, the pollen does not reach other individuals at all. Instead, it is shed directly onto the stigma of the same flower, sometimes in bud. This process is termed **self-pollination**.

The relationship between animals known as **pollinators** and flowering plants has been one of the most important features in the evolution of both groups. For plants to be effectively pollinated by animals, a particular insect or other animal must visit many plants of the same species. Flowers have evolved various colors and forms that attract certain pollinators, thereby promoting effective pollination. Yellow flowers, for example, are particularly attractive to bees, whereas red flowers attract birds but not insects. Insects in turn have evolved a number of special traits that enable them to obtain food efficiently from the flowers of the plants they visit.

> Plants can be pollinated by animals, insects, or the wind. The evolution of both flowering plants and their animal pollinators is closely linked.

PLANTS: REPRODUCTIVE PATTERNS AND DIVERSITY

Fruits and their significance in sexual reproduction

Parallel to the evolution of the angiosperms' flowers and nearly as spectacular has been the evolution of their fruits. Fruits have evolved a diverse array of shapes, textures, and tastes and exhibit many differing modes of dispersal.

Fruits that have fleshy coverings—often black, bright blue, or red—are normally dispersed by birds and other vertebrates. Just as red flowers attract birds, the red fruits signal an abundant food supply. By feeding on these fruits, birds and other animals carry seeds from place to place and thus transfer the plants from one suitable habitat to another (Figure 30-10, *B*). Other fruits, such as snakeroot, beggar ticks, and burdock (Figure 30-10, *A*), have evolved hooked spines and are often spread from place to place because they stick to the fur of mammals or the clothes of humans. Others, like the coconut and those that occur on or near beaches, are regularly spread by water (Figure 30-10, *C*). Some fruits have wings and are blown about by the wind. The dandelion is a familiar example (Figure 30-10, *D*).

> Seeds are dispersed by sticking to or being eaten by animals, by being blown by the wind, or by floating to new environments across bodies of water.

FIGURE 30-10 How seeds are dispersed.
A Animal dispersing seeds. This English setter is covered with hound's tongue weeds seeds that have stuck to his coat.
B Bird dispersing seeds. This cedar waxwing feeds berries to the waiting young in the nest. Birds that eat seeds digest them rapidly so that much of the seed is left intact. What is excreted from the bird can grow into a mature plant.
C The seeds of a coconut, *Cocos nucifera,* sprouting on a sandy beach. One of the most useful plants for humans in the tropics, coconuts have become established even on the most distant islands by drifting in the waves.
D The seeds of a dandelion, *Pyropappus caroliniana,* are dispersed by the wind. The "parachutes" disperse the fruits of dandelions widely in the wind, much to the gardener's despair.

Finding new food plants

Just three species of plants—rice, wheat, and corn—supply half of all human energy requirements, and only about 150 kinds of plants are used extensively. Among the 250,000 known species of plants, there may be tens of thousands of additional kinds of plants that could be used for human consumption if their properties were known and they were cultivated. Finding new food plants in the tropics is especially crucial, since the population in these areas is expected to climb steadily. Making these tropical countries less dependent on imported food is a goal that researchers are striving to fulfill. Unfortunately, efforts to find new food plants in the tropics are hampered by deforestation. Cutting tropical rain forests to make way for more traditional food crops will greatly reduce the opportunity to uncover new food plants that may reside in the rain forest.

One of the most fertile areas for biological exploration is the study of plant use by people who live in direct contact with natural plant communities. These people know a great deal about the plants with which they come into contact, and the screening that they and their ancestors have performed over many generations can, if the results are understood and cataloged, help scientists recognize new plants. For example, the herbal curer in the jungles of southern Surinam has a profound knowledge of uses of plants that grow in his region, but this knowledge is being lost as civilization and modern medicine move into the area.

Some recent triumphs in the search for new, useful plants are the following:

Jojoba (*Simmondsia chinensis*) (Figure 30-A, *1*) is native to the deserts of northwestern Mexico and the southwestern United States. It has been an important crop in several parts of the world since the late 1970s because the liquid waxes in its seeds have characteristics similar to sperm whale oil, which is important for certain kinds of fine lubrication. Jojoba has the added advantage of being easy to grow: it can be cultivated in lands too dry for most other crops to grow. Industries once dependent on sperm whale oil for lubrication have found the jojoba oils cheaper to use. This new availability of jojoba oils has made the hunting of sperm whales uneconomical, helping to save them from extinction.

Grain amaranths (*Amaranthus* species) were important grain crops of the Latin American highlands in the days of the Incas and Aztecs, but are little-used now (Figure 30-A, *2*). Their use was suppressed because they played a role in pagan ceremonies of which the Spanish conquerors disapproved. The grain amaranths, fast-growing plants that produce abundant grain, are rich in lysine, an amino acid rare in most plant proteins but essential in animal nutrition. These ancient crop plants are currently being investigated for widespread development.

Winged bean (*Psophocarpus tetragonolobus*) is a tropical vine that produces highly nutritious seeds, pods, and leaves. Its tubers are eaten like potatoes, and its seeds produce large quantities of an edible oil much like safflower oil. First cultivated locally in New Guinea and Southeast Asia, the winged bean has spread since the 1970s throughout the tropics, where it holds great promise as a source of food. (Figure 30-A, *3*).

Perhaps the strangest of all the new food crops is not a plant at all, but a bacterium. *Spirulina* is a photosynthesizing bacterium that is used as a traditional food source in Africa, Mexico, and other regions. It is currently being studied as a possible food source for other countries. *Spirulina* has a higher protein content than soybeans and is easy to grow—the bacterium thrives in very alkaline water and is cultivated easily in large ponds. However, people must overcome certain psychological barriers to eating this kind of food if the efforts to transplant *Spirulina* out of its traditional cultures are to be successful.

FIGURE 30-A
1 Jojoba.
2 Grain amaranth.
3 Winged bean.

Seed formation and germination

The long chain of events between fertilization and maturity is **development**. During development, cells become progressively more specialized, or differentiated. Development in seed plants results first in the production of an embryo, which remains dormant within a seed until the seed germinates, or sprouts.

Seed **germination** depends on a variety of environmental factors, the most important of which is water. However, the availability of oxygen (for aerobic respiration in the germinating seed), a suitable temperature, and sometimes the presence of light are also necessary. The first step in germination of a seed occurs when it imbibes, or takes up, water. Once this has taken place, metabolism within the embryo resumes.

Germination and early seedling growth require the mobilization of food storage reserves within the seed. A major portion of almost every seed consists of food reserves. Angiosperms fall into two groups regarding the placement of stored food in their seeds: the **monocots** and the **dicots**. In dicots, most of the stored food is in the **cotyledons**, or seed leaves. In monocots, most of the food is stored in extraembryonic tissue called **endosperm** (Figure 30-11). (Although dicots store some food in endosperm, the amount of endosperm varies among species of plants.) Its single cotyledon absorbs food from the endosperm and shuttles it to the embryo. Monocots and dicots also differ from one another in a number of features. Monocots usually have parallel veins (fluid-carrying tissues) in their leaves, and their flower parts are often in threes. Among the monocots are the lilies, grasses, cattails, orchids, and irises (Figure 30-12, *A* to *C*). Dicots usually have netlike veins in their leaves and their flower parts are in fours or fives. The dicots include the great majority of familiar plants: almost all kinds of flowering trees and shrubs and most garden plants, such as snapdragons, chrysanthemums, roses, and sunflowers (Figure 30-12, *D* to *F*).

Usually the first portion of the embryo to emerge from the germinating seed is the young root, or **radicle,** which anchors the seed and absorbs water and minerals from the soil. Then, the shoot of the young seedling elongates and emerges from the ground.

As the shoot emerges from the soil, the first true leaves are protected by a straight sheath or a curved stem. Multiplication of the cells in the tips of the stem and roots, along with their elongation and differentiation, initiates and continues the growth of the young seedling.

> The embryo of a plant remains dormant within the seed until the seed germinates. Germination begins when the seed takes up water and begins to sprout.

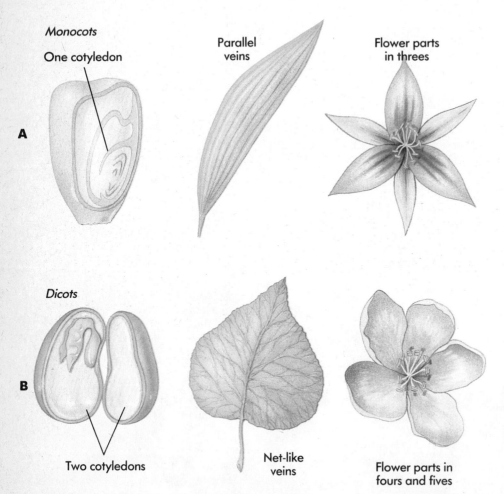

FIGURE 30-11 Monocots and dicots.
A In monocots, most of the food for the embryo is stored in endosperm. Other characteristics of monocots are the parallel veins in their leaves and the occurrence of their flower parts in threes.
B Dicots store their food in cotyledons, or seed leaves. They have net-like veins in their leaves, and their flower parts occur in fours or fives.

FIGURE 30-12 Examples of monocots and dicots.
Monocots include the cattail (**A**), the crested dwarf iris (**B**), and the pink lady's slipper (**C**). All of these plants show the parallel leaf veination typical of monocots. Dicots include the rose (**D**), the sea grape (**E**), and the bloodroot (**F**). These examples show the net-like veination typical of dicots.

FIGURE 30-13 Clones of aspen.
The contrast between the golden quaking aspens *(Populus tremuloides)* and the dark green Engelmann spruces *(Picea engelmannii)* evident in this autumn scene near Durango, Colorado, makes it possible to see their clones, large colonies produced by single individuals spreading underground and sending up new shoots periodically.

Types of vegetative propagation in plants

Vegetative propagation is an asexual reproductive process in which a new plant develops from a portion of a parent plant. Some plants, such as irises and grasses, produce new plants along underground stems called **rhizomes.** Other plants, such as strawberries, have horizontal stems that grow above the ground called **runners** or **stolons.** These plants produce new roots and shoots at nodes (places where one or more leaves are attached) along these stems. New plants can also arise vegetatively from specialized underground storage stems called **tubers.** A white potato, for example, is a tuber and can grow a new plant from each of its eyes.

In some plants, new shoots can arise from roots and grow up through the surface of the soil. For example, a group of aspen trees often consists of a single individual that has given rise to a colony of genetically identical trees by producing new shoots from its horizontal roots (Figure 30-13).

In a few species of plants, even the leaves are reproductive. The plant in Figure 30-14 is commonly called the *maternity plant* because small plants arise in the notches along the margins of the leaf. When mature, they drop to the soil and take root. Gardeners commonly propagate African violets from leaf cuttings and many other plants from stem cuttings.

A major breakthrough in the asexual propagation of plants has been the development of **cell culture techniques** (Figure 30-15). Using these techniques, scientists are able to remove individual cells from a parent plant and grow these cells into new individuals. In this way, botanists are able to produce virtually unlimited numbers of genetically identical offspring. Cell culture techniques have been particularly useful in propagating plants that are slow to multiply on their own, such as coconut palms and redwoods, and in cultivating varieties of individual plants with special characteristics, such as large flowers. Award-winning varieties of orchids, for example, are often produced in this way. Cell culture is also used for the commercial production of tremendous numbers of plants in a short period of time—such as the chrysanthemums you may buy at the grocery store. The timber industry uses cell culture techniques in developing rapidly growing conifers such as the Douglas fir.

The use of cell culture techniques in agriculture will revolutionize the way in which farmers produce some crops. In the future, farmers will be able to grow crop plants that have been mass-produced by cell culture and are genetically superior in some way, such as disease-resistance, high yield per plant, or desirable taste. In addition, the techniques of genetic engineering allow scientists to insert genes that express favorable characteristics into the chromosomes of culture cells—another potential revolution in worldwide agriculture.

> Many plants can reproduce vegetatively by growing new plants from their roots, stem, or leaves. Scientists, using cell culture techniques, can also propagate many plants from a single cell of the parent. Use of these techniques along with the genetic engineering of plants will revolutionize the way that farmers produce some crops.

FIGURE 30-14 *Kalanchoe daigremontiana.*
The small plants growing in the notches along the leaf margins will drop to the soil and take root. For this reason, *Kalanchoe daigremontiana* is called the "maternity plant."

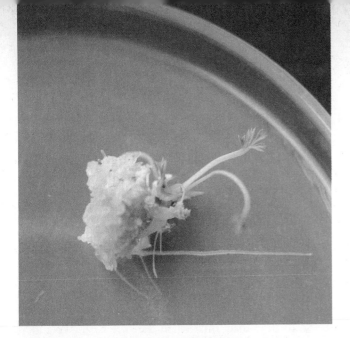

FIGURE 30-15 A plant growing in a cell culture.
The small shoots growing in the petri dish are genetic clones of carrots. Researchers use clones and tissue cultures to try to find hardier and more insect-resistant varieties of plants.

Regulating plant growth: Plant hormones

As a plant grows, it is influenced by environmental factors such as the amount of water and light it receives. However, a plant's growth, differentiation, maturation, flowering, and many other activities are also regulated by chemicals called **hormones.** Hormones are chemical substances produced in small, often minute, quantities in one part of an organism and then transported to another part of the organism where they bring about physiological responses.

There are at least five major kinds of hormones in plants. **Auxins, gibberellins,** and **cytokinins** promote and regulate growth. Cytokinins stimulate cell division, and auxins and gibberellins promote growth through cell elongation (Figure 30-16). The differences in concentration of auxin from one side of the stem or root to another, for example, controls the bending of plants toward or away from light (phototropism) (Figure 30-17) and toward or away from gravity (gravitropism). In contrast, **abscisic acid** is a growth inhibitor that induces and maintains dormancy or otherwise opposes the three growth-promoting hormones. **Ethylene** is released by plants as a gas and effects the ripening of fruit and leaf drop in nearby plants.

Other chemicals, such as plant pigments called **phytochromes,** also affect a plant's response to its environment. Phytochromes change form in response to day length and thereby stimulate or inhibit flowering. The study of plant hormones and other regulatory chemicals, especially how they produce their effects, is an active and important field of research today.

FIGURE 30-16 The effect of the plant hormone gibberellin.
This cabbage plant has been treated with gibberellin, which promotes growth through cell elongation.

FIGURE 30-17 Phototropism.
Plants grow toward the light because auxin, a plant hormone that controls plant growth, is produced in higher levels in darkness. Thus the side of the plant away from the light will grow faster, causing the plant to bend toward the light. The stems of this blood sorrel are oriented toward the window, which lets in plenty of sunlight.

Summary

1. Plants are multicellular, eukaryotic, photosynthetic autotrophs that live on land: the bryophytes (nonvascular plants) and the vascular plants. Current classification schemes usually classify the multicellular algae, once classified as plants, with the protists.

2. Animals, plants, some protists, and some fungi have sexual life cycles that are characterized by the alternation of the processes of meiosis and fertilization. Plants and some species of algae have life cycles that have both a multicellular haploid phase and a multicellular diploid phase. Because both phases of the life cycle are multicellular, this type of life cycle is called *alternation of generations*.

3. In plant life cycles, multicellular haploid plants produce gametes by mitosis and are gametophytes. These gametes fuse during fertilization and grow into the multicellular, diploid spore-producing plants, or sporophytes. These individuals form spores by meiosis, which grow into gametophytes.

4. In general, the gametophyte generation often dominates the life cycles of the nonvascular plants, whereas the sporophyte generation dominates the life cycles of vascular plants. This difference reflects an adaptation of the vascular plants for life on land.

5. The bryophytes include three divisions of nonvascular plants: the mosses, liverworts, and hornworts. The life cycles of most have two distinct phases, with neither the sporophyte nor the gametophyte dominating the life cycle, although the gametophyte plants are the ones usually more familiar. The sporophyte plants live on and derive nutrients from the gametophyte plants.

6. The sporophyte phase of the life cycle is the dominant phase of the life cycle of vascular plants. The seedless vascular plants produce motile sperm that swim through moisture on the gametophyte to fertilize the egg. The sporophytes that develop are protected initially by the gametophyte tissues. The gametophytes in seed-bearing plants consist of only a few cells. These cells produce eggs and sperm. After fertilization the embryonic sporophyte is protected and nourished within a seed.

7. The organs of sexual reproduction in the flowering plants, or angiosperms, are in the flower. Pollen grains are male gamete–producing cells. The female gamete–producing cells are in the ovules in the ovary.

8. The flowers of angiosperms make possible the precise transfer of pollen from the anther to the stigma, an event that precedes fertilization and formation of a seed. Pollen is transferred by insects, particularly bees; birds and other animals; and the wind.

9. The seeds of angiosperms remain within the ovary, which develops into a fruit. Many fruits are fleshy and often sweet; animals that consume fruit may carry the seeds for long distances before excreting them as waste. The seeds can then germinate, or sprout, in the new location.

10. Seeds germinate only when they receive water and appropriate environmental cues. In the germination of seeds, mobilization of the food reserves stored in the cotyledons and in the endosperm is critical.

11. The change of a zygote into a mature individual, initiated immediately after fertilization, is development. Development is a process of progressive specialization that results in differentiation, the production of highly individual tissues and structures.

12. Plants can also reproduce asexually by vegetative propagation; a new plant grows from a portion of another plant.

KEY TERMS

abscisic acid 541
alternation of generations 529
angiosperm 534
anther 534
antheridium (pl., antheridia) 529
archegonium (pl., archegonia) 529
auxin 541
cell culture techniques 540
cotyledon 538
cytokinin 541
dicot 538
divisions 527
endosperm 538
ethylene 541
frond 531
fruit 534
gametophyte 529
germination 538
gibberellin 541
gymnosperm 533
hornwort 530
liverwort 530
monocot 538
moss 530
nectar 535
nonvascular plant 527
ovary 534
ovule 533
petal 534
phytochrome 541
pistil 534
pollen grain 534
pollinator 535
radicle 538
rhizome 540
runner/stolon 540
seed 531
seedless vascular plant 531
self-pollination 535
sepal 534
sporophyte 529
stamen 534
stigma 534
style 534
tuber 540
vascular plant 527
vegetative propagation 540

REVIEW QUESTIONS

1. Fill in the blanks: _____ plants contain specialized tissues within them to transport fluids. _____ plants lack these specialized tissues.
2. Fill in the blanks: _____ produce spores, which grow into _____. These _____ produce gametes, which fuse during fertilization and grow into _____.
3. Name the three classes of bryophytes and summarize their life cycles.
4. What are seeds? Explain their function(s).
5. Summarize the life cycle of seedless vascular plants. Give an example of this type of plant.
6. Match each of the following with the most appropriate term:
   Gymnosperms
   Angiosperms
   Bryophytes
   Ferns
   a. Seedless vascular plants
   b. Nonvascular plants
   c. Vascular plants with naked seeds
   d. Vascular plants with protected seeds
7. Draw a generalized diagram of a flower. Label the following: sepals, petals, stamens, pistil, anther, pollen grains, stigma, ovary, ovules, and style.
8. Summarize the sexual reproduction of angiosperms and gymnosperms.
9. The evolution of flowering plants and their animal pollinators is closely linked. Explain why.
10. Summarize the significance of fruits in sexual reproduction.
11. Place the following events in the correct sequence:
    a. The radicle emerges from the seed.
    b. Fertilization occurs.
    c. The shoot emerges from the ground.
    d. An embryo develops.
    e. The seed germinates.
12. What do monocots and dicots have in common? How do they differ?
13. What is vegetative propagation? Give an example.
14. What are hormones? Summarize the function(s) of the five major kinds of plant hormones mentioned.

THOUGHT QUESTIONS

1. The incredible success of flowering plants argues that increasing the efficiency of sexual reproduction has been of great evolutionary advantage to the angiosperms. Why then are so many angiosperms self-pollinators?
2. You probably know what a dandelion looks like, a delicate white sphere at the tip of a slender stalk—blowing on dandelion flowers is something almost all children do. Dandelions are obligatorily asexual, which is to say they never practice sexual reproduction. So why do they have flowers?
3. Why do you think angiosperm seed tissue (endosperm) has evolved to be triploid when there is no triploid tissue in the seeds of gymnosperms?

FOR FURTHER READING

Batra, S.W.T. (1984, February). Solitary bees. *Scientific American*, pp. 120-127.
This is an excellent article on these diverse and fascinating pollinators, some of which are of great commercial importance.

Cook, R.E. (1983). Clonal plant populations. *American Scientist, 71*, 244-253.
This is an excellent discussion on the role of asexual reproduction among plants in natural populations.

Schaller, G. B., Jinchu, H., Wenshi, P., & Jing, Z. (1985). *The giant pandas of Wolong*. Chicago: University of Chicago Press.
A fascinating account of the mutual adaptations of pandas and bamboo, this book offers the first glimpse of the life of pandas in their remote mountain home in western China.

31 Plants
Patterns of structure and function

HIGHLIGHTS

Vascular plants are organized along a vertical axis with underground roots and aboveground shoots made up of stems and leaves.

Vascular plants have specialized tissues that make up the structure of roots, stems, and leaves, enabling them to absorb nutrients, transport fluids, store food, carry on photosynthesis, and produce new cells for plant growth.

Water and dissolved minerals move upward from root to stem and leaves by transpirational pull; water and dissolved sugar are transported throughout the plant by mass flow.

The nonvascular plants have various levels of organization within their groups; all have a much simpler organization than the vascular plants.

OUTLINE

The organization of vascular plants

Tissues of vascular plants
  Fluid movement: Vascular tissue
  Food storage: Ground tissue
  Protection: Dermal tissue
  Growth: Meristematic tissue

Organs of vascular plants
  Roots
  Shoots
    Stems
    Leaves

Movement of water and dissolved substances in vascular plants

The organization of nonvascular plants

Although the photo looks like the discarded shell of a huge turtle, it is actually a plant. In fact, one nickname for this plant is the "turtleback." The turtleback's "shell" is really a modified stem that stores water for the plant. This helps the turtleback survive the dry season in its tropical environment. During the rainy season, a system of green branches grows out its top, manufacturing food and bearing leaves, flowers, and fruits. Through the years these branches die, covering the turtleback with an intertwining mass.

Turtlebacks commonly grow in Africa. The native people, however, call this plant *Hottentot's bread* because its fleshy storage organ can be eaten and has a texture similar to a giant turnip. This organ commonly grows a meter (3 feet) in diameter. Some grow as tall as 2⅓ meters (approximately 7 feet) and weigh up to 318 kilograms (700 pounds)—rather large for the produce section of the grocery store!

The organization of vascular plants

For any plant to grow as tall as the turtleback, it must have a means to transport substances to and from its various parts. Most plants (over 80% of all living species) have a system of specialized tissues that performs this function. Plants with these tissues are called **vascular plants** (see Chapter 30 for discussion of their reproductive cycles). Despite their reproductive diversity, however, vascular plants all have the same basic architecture.

A vascular plant is organized along a vertical axis (Figure 31-1). The part below the ground is called the **root.** The part above the ground is called the **shoot.** The root penetrates the soil and absorbs water and various ions crucial for plant nutrition. It also anchors the plant. The shoot consists of a **stem** and **leaves.** The stem serves as a framework for the positioning of the leaves, where most photosynthesis takes place. The arrangement, size, and other characteristics of the leaves are critically important in the plant's production of food. Flowers and ultimately fruits and seeds are also formed on the shoot.

> Vascular plants are those having a system of tissues that transports water and nutrients. Vascular plants have the same basic architecture: they are made up of underground roots and aboveground shoots. The shoot consists of a stem and leaves.

Tissues of vascular plants

The organs of a vascular plant—the leaves, roots, and stem—are made up of different mixtures of tissues, just as your legs are composed of different tissues such as bone and muscle. A tissue is a group of cells that works together to carry out a specialized function. Vascular plants have three

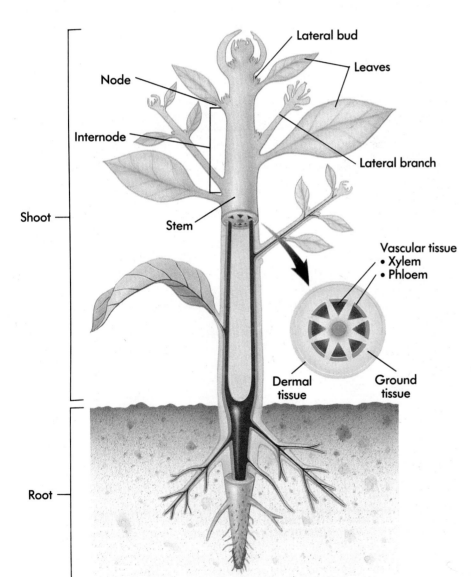

FIGURE 31-1 Structure of a typical plant. The root anchors the plant while the shoot supports the leaves, structures in which photosynthesis takes place. The vascular tissue is the circulatory system of plants and transports water and nutrients.

PLANTS: PATTERNS OF STRUCTURE AND FUNCTION

types of differentiated tissues, groups of cells having specific structures to perform specific functions: **vascular tissue, ground tissue,** and **dermal tissue** (see Figure 31-1).

The word *vascular* comes from a Latin word meaning "vessel." Thus vascular tissue forms the "circulatory system" of a plant, conducting water and dissolved minerals up the plant and the products of photosynthesis throughout the plant. Ground tissue stores the carbohydrates the plant produces. It is the tissue in which the vascular tissue is embedded and forms the substance of the plant. Dermal tissue covers the plant, protecting it—like skin—from water loss and injury to its internal structures. Plants also contain **meristematic tissue,** or growth tissue, an undifferentiated tissue in which cell division occurs. This tissue, often just called *meristem,* is considered undifferentiated because the cells it produces will eventually become one of the other three cell types.

> The three types of differentiated tissues in vascular plants are (1) dermal tissue, which protects the plant; (2) vascular tissue, which conducts water, minerals, carbohydrates, and other substances throughout the plant; and (3) ground tissue, which stores food the plant manufactures. In addition, an undifferentiated type of tissue called *meristem* produces new plant cells during growth.

Fluid movement: Vascular tissue

Vascular tissue is of two principal types: **xylem,** which conducts water and dissolved minerals, and **phloem,** which conducts carbohydrates the plant uses as food along with other needed substances. Each type of vascular tissue is made up of different, specialized conducting cells.

Figure 31-2, *A* and 31-2, *B* show two different types of xylem cells. Both types of cells conduct water and dissolved minerals only after they die and lose their cytoplasm, becoming hollow and thick walled. Stacked end to end, they form pipelines that extend throughout the plant. The cells pictured in Figure 31-2, *A,* called *tracheids,* show the pores these two cell types have along their length, like soaking hoses used in a garden. These pores allow some water to move laterally to cells surrounding the xylem pipelines. Figure 31-2, *C* shows a sieve tube member and companion cell.

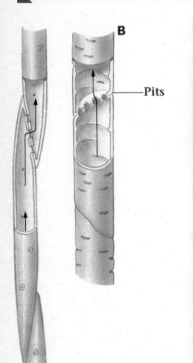
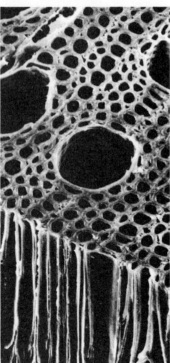
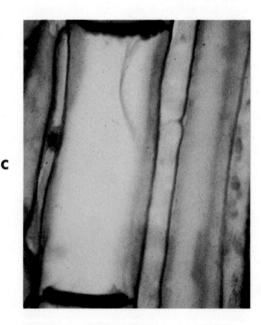
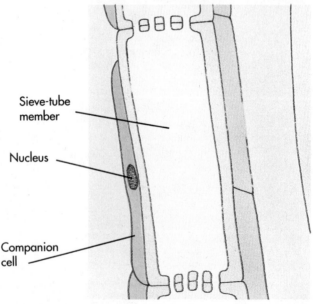

FIGURE 31-2 Vascular tissues of a plant.
A Tracheid. In tracheids, the water passes from cell to cell by means of pits.
B Vessel elements. In vessel elements, water passes from cell to cell by means of pits. Both tracheids and vessel elements make up xylem.
C Seive tube. Individual cells, or seive tube members, are stacked end to end to form a tube. Seive tube members are a part of the phloem and transport nutrients.

The conducting cells (sieve tube members) are alive and contain cytoplasm but are not typical cells. They lack nuclei, for example. The ends of these elongated cells have pits that allow the easy passage of sugar-filled water from where it is produced—in the leaves, for example—to where it is used rapidly—at the reproductive structures, for example. Companion cells, which contain all of the organelles commonly found in plant cells (including nuclei), secrete substances into and remove substances from the sieve tube members. These substances include sugars produced during photosynthesis.

Food storage: Ground tissue

Ground tissue contains **parenchyma cells,** which function in photosynthesis and storage. These thin-walled cells contain large vacuoles and may also be packed with chloroplasts. They are the most common of all plant cell types and form masses in leaves, stems, and roots. Fleshy storage roots, such as carrots or sweet potatoes, contain a predominance of storage parenchyma. The flesh of most fruits is also made up of these cells. **Sclerenchyma cells,** also found in the ground tissue, are hollow cells with strong walls. These cells help support and strengthen the ground tissue. The gritty texture of a pear is due to schlerenchyma cells that are dispersed among the softer parenchyma cells. Both types are shown in Figure 31-3.

Protection: Dermal tissue

Dermal tissue covers the outside of a plant with the exception of woody shrubs and trees, which have protective bark in its place. Bark is made up of other tissue types (see p. 552). **Epidermal cells** are the most abundant type of cell found in the dermal tissue. These cells are often covered with a thick, waxy layer called a **cuticle** that protects the plant and provides an effective barrier against water loss (Figure 31-4, *A*). Other types of cells found in the dermis

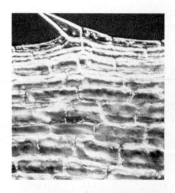

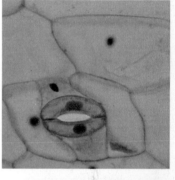

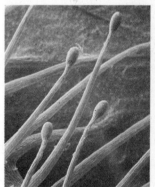

FIGURE 31-4 Dermal tissue.
A Epidermal cells. These cells are covered with a waxy substance and thus protect the plant from water loss.
B Guard cells. These cells regulate the movement of water vapor and gases into and out of the plant.
C Trichomes. Trichomes perform various functions: they increase surface area, defend the plants against insects and other predators, and reflect the sunlight away from the plant, reducing water loss.

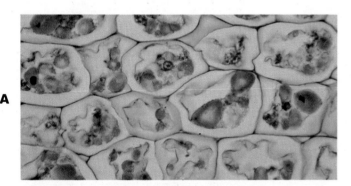
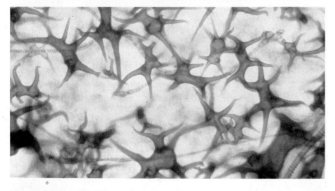

FIGURE 31-3 Ground tissue.
A Parenchyma cells. Parenchyma cells store starch and contain chloroplasts.
B Sclerenchyma cells. These cells are hollow, with strong walls. Sclerenchyma cells help support the plant.

are guard cells, which surround openings in the leaves through which gases and water vapor enter and leave (Figure 31-4, *B*), and trichomes (Figure 31-4, *C*), which are outgrowths of the epidermis (much like the hairs on your body) that have various functions. For example, on "air plants," trichomes help provide a large surface area for the absorption of water and minerals. On leaves or fruits, fuzz-like trichomes reflect sunlight, which helps control water loss. In some desert plants, white multicellular trichomes reflect enough sunlight to reduce temperatures in internal tissues. Some trichomes defend a plant against insects or larger animals with their sharpness or by secreting chemicals.

Growth: Meristematic tissue

Plants contain meristems, areas of undifferentiated cells that are centers of plant growth. Every time one of these cells divides, one of the two resulting cells remains in the meristem. In this way, meristem cells remain "forever young," capable of repeated cell division. The other cell goes on to differentiate into one of the three kinds of plant tissue, ultimately becoming part of the plant body.

Plants can grow only in relationship to where their meristematic tissue is located. Located at the tips of the roots and the tips of the shoots, this tissue is called **apical meristem** (*apex* meaning "tip"). Apical meristem allows plants to grow taller and their roots to grow deeper into the ground. This type of plant growth, which occurs mainly at the tips of the roots and shoots, is called **primary growth**.

Some plants not only grow taller but grow thicker as well. This type of growth is called **secondary growth** and occurs in all woody trees and shrubs such as pines, oaks, and rhododendrons. These plants have a cylinder of meristematic tissue along the length of their stems and branches, which is called **lateral meristem.** Herbaceous (nonwoody) plants, such as tulips, have only primary growth. However, their stems do grow somewhat thicker as they develop. One reason for this growth is that plant development involves both cell division *and* cell growth; as cells enlarge, plant parts enlarge. Additionally, some cell division occurs in the cells differentiating from meristematic tissue.

Organs of vascular plants

Roots, stems, and leaves are the organs of vascular plants. Together, the stems and leaves make up the shoot. The roots and shoots of all vascular plants share the same basic architecture, but there are differences. This chapter discusses the structure of the plants that dominate the plant world today: the flowering plants, or **angiosperms.** The principal differences in the structures of the roots, stems, and leaves in the seedless vascular plants and those with naked seeds lie primarily in the relative distribution of the vascular and ground tissue systems.

Roots

The function of a root system is to anchor a plant in the ground and to absorb water and minerals from the soil.

TABLE 31-1 Inorganic nutrients important to plants

NUTRIENT	RELATIVE ABUNDANCE IN PLANT TISSUE (ppm)
MACRONUTRIENTS (nutrients required in greater concentration)	
Hydrogen	60,000,000
Carbon	35,000,000
Oxygen	30,000,000
Nitrogen	1,000,000
Potassium	250,000
Calcium	125,000
Magnesium	80,000
Phosphorus	60,000
Sulfur	30,000
MICRONUTRIENTS (nutrients required in minute quantities)	
Chlorine	3000
Iron	2000
Boron	2000
Manganese	1000
Zinc	300
Copper	100
Molybdenum	1

ppm, Parts per million. Parts per million equals units of an element by weight per million units of oven-dried plant material.

(The minerals [essential inorganic nutrients] that plants use are listed in Table 31-1.) During the process of photosynthesis, plants use the water they absorb along with carbon dioxide they capture from the air to produce carbohydrates (see Chapter 8). But primarily, water absorbed at the roots replaces the water released by the plant into the air in a process called *transpiration* (see p. 555). Leaves are the principal organs of transpiration. But why do plants need to absorb minerals? The answer is simple. Although plants manufacture carbohydrates during photosynthesis, these sugars are not the only substances that plants need to live. Plants need proteins, fats, and vitamins. These substances are formed from the carbohydrates plants manufacture and from the minerals that plants take in from the soil and concentrate. Many plants therefore are an important source of minerals in the human diet. Broccoli and cabbage, for example, are excellent sources of calcium. Bananas provide you with potassium.

> **Roots, the part of a plant usually found below ground, absorb water and minerals as well as anchor the plant.**

Structurally, the roots of dicots (plants having netlike veins in their leaves, see Chapter 30) have a central column of xylem with radiating arms. Between these arms are strands of phloem (Figure 31-5). Ringing this column of vascular

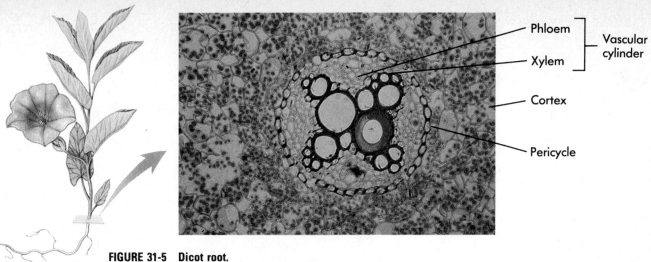

FIGURE 31-5 Dicot root.
This typical dicot root from a buttercup has a central column of xylem with radiating arms.

tissue (often called the *vascular cylinder*) and forming its outer boundary is another cylinder of cells called the **pericycle.** This tissue is made up of parenchyma cells able to undergo cell division to produce branch roots (roots that arise from other older roots). Surrounding the pericycle is a mass of parenchyma called the **cortex.** These cells store food for the growth and metabolism of the root cells. The innermost layer of the cortex is called the **endodermis,** which consists of specialized cells that regulate the flow of water between the vascular tissues and the outer portion of the root. The outer layer of the root is the **epidermis,** which absorbs water and minerals. These cells have extensions called root hairs that provide the epidermal cells with a larger surface area over which absorption can take place—like the function of microvilli in the intestinal walls of animals.

Monocot roots (plants having parallel veins in their leaves, see Chapter 30) are similar to dicot roots with one important exception: monocot roots often have centrally located parenchyma (storage) tissue called **pith.** The xylem and phloem are arranged in rings around the pith (Figure 31-6).

> Most dicot roots have a central column of xylem with radiating arms and strands of phloem between these arms. Surrounding this vascular tissue is a layer of cells called the *pericycle* that is capable of cell division. Parenchyma (storage) cells of the cortex surround the pericycle. The entire root is covered with a protective epidermis. Monocot roots are structured similarly but often have an additional storage tissue called *pith,* which is located in the center of the root.

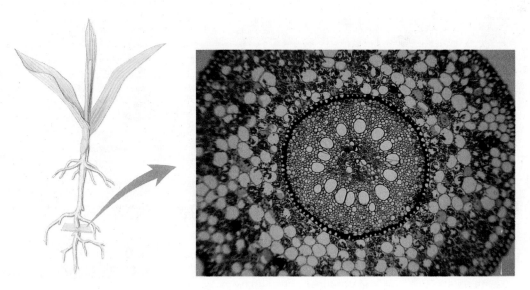

FIGURE 31-6 Monocot root.
The important difference between a dicot and monocot root is that the monocot root has a centrally located storage tissue, or pith. The xylem and phloem are arranged in rings around the pith.

PLANTS: PATTERNS OF STRUCTURE AND FUNCTION

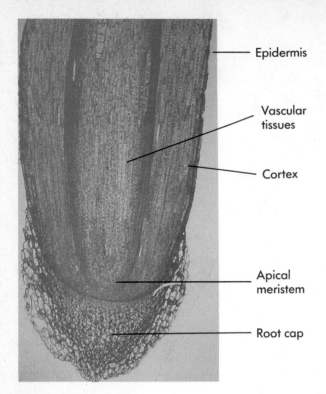

FIGURE 31-7 A root tip.
Section of a root tip in corn, *Zea mays*, showing the differentiation of epidermis, cortex, and column of vascular tissues.

The end of a root is tipped with apical meristem. These growth cells divide and produce cells inwardly, back toward the body of the plant, and outwardly. Outward cell division results in the formation of a thimble-like mass of relatively unorganized cells, the **root cap,** which covers and protects the root's apical meristem as it grows through the soil. Just behind its tip, the root cells elongate. Velvety root hairs, the tiny projections from the epidermis, form above this area of elongation (Figure 31-7). Here, too, the cells differentiate to produce specialized cell types.

If you have ever pulled a plant up by its roots, you will have noticed that the roots of one type of plant may look different from those of another type of plant. These differences in appearance are linked to differences in function. Many dicots, like dandelions, for example, have a single root called a **taproot,** which is sometimes the only major underground structure (Figure 31-8, *A*). Taproots grow deep into the soil, firmly anchoring the plant. Some taproots, like carrots and radishes, are fleshy because they are modified for food storage. The plant draws on these food reserves when it flowers or produces fruit, thus taproot crops are harvested before that time.

FIGURE 31-8 Types of roots.
A Taproots in a dandelion, *Taraxacum officinale*. Even a small section of these taproots can regenerate a new plant, which is one reason why dandelions are so difficult to eliminate from lawns and gardens.
B Prop roots in corn, *Zea mays,* are adventitious—they arise from stem tissue and take over the function of the main root.
C Fibrous roots in grass. These roots have no central, predominant root. Fibrous roots are especially efficient in achoring the plant and absorbing nutrients.

The taproot that develops in monocots, on the other hand, often dies during the early growth of the plant and new roots develop from the lower part of the stem. These roots are called **adventitious roots,** roots that develop from an aboveground structure. Often, adventitious roots help anchor a plant, such as "prop" roots in corn (Figure 31-8, *B*). Certain dicots, such as ivy plants, also develop adventitious roots. The adventitious roots of ivy plants help them cling to walls.

Have you ever pulled up a clump of grass and looked at its roots? Grass has **fibrous roots,** a type of root system that has no predominant root (Figure 31-8, *C*). Most monocots have fibrous roots. They work well to anchor the plant in the ground and absorb nutrients efficiently because of their large surface area. They also help prevent soil erosion by holding soil particles together.

Shoots

Plant shoots grow aboveground and are made up of stems and leaves. As with roots, the growing end of a shoot is tipped with apical meristematic tissue. Young leaves cluster around the apical meristem, unfolding and growing as the stem itself elongates (see Figure 31-1).

Leaves form on the stem at locations called **nodes.** The portions of the stem between the nodes are called the **internodes**. As the leaves grow, tiny undeveloped side shoots called **lateral buds** develop at the angles between the leaves and the stem. These buds, which contain their own embryonic leaves, may elongate and form lateral branches given the proper environmental conditions. However, most of these branches remain small and dormant.

Stems

One purpose of stems is to support the parts of plants that carry out photosynthesis. This process takes place primarily in the leaves, which are arranged on the stem so that light will fall on them. In addition, stems conduct water and minerals from the roots to all plant parts and bring the products of photosynthesis to where they are needed or stored (Figure 31-9).

The stem "transportation system" is made up of strands of xylem and phloem tissue that are positioned next to each other, forming cylinders of tissue called **vascular bundles** (often referred to as the *veins* of leaves). The xylem tissue

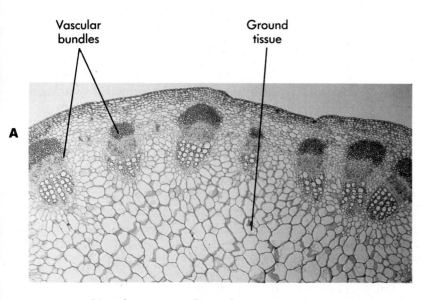

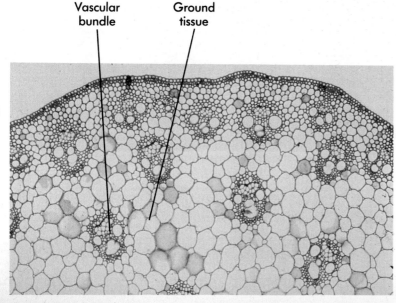

FIGURE 31-9 Stems.
A A dicot stem from the common sunflower, in which the vascular bundles are arranged around the outside of the stem.
B A monocot stem from corn, showing the scattered vascular bundles characteristic of monocots.

551

characteristically forms the part of each bundle closer to the interior of the stem. The phloem lies closer to the epidermis. In herbaceous dicots (those with soft stems rather than woody, tree-like stems), the vascular bundles are arranged in a ring near the periphery of the stem (Figure 31-9, *A*). In monocots, the vascular bundles are scattered throughout the stem (Figure 31-9, *B*).

Because of the arrangement of their vascular bundles, dicot stems have a mass of ground tissue in the center of the stem called the pith and a ring of ground tissue between the epidermis and the vascular bundles called cortex. Monocot stems also have ground tissue, but because it surrounds scattered vascular bundles, ground tissue does not form areas of pith or cortex. The epidermis of both monocot and herbaceous dicot stems is covered with a protective, waxy coating called the *cuticle*.

> Stems support the photosynthetic structures—the leaves—and transport water and dissolved substances throughout plants. Herbaceous dicot stems are characterized by an inner cylinder of ground tissue called *pith* surrounded by a ring of vascular bundles. Encircling the ring of vascular bundles is additional ground tissue called *cortex*. Monocot stems are characterized by scattered vascular bundles embedded in ground tissue.

Dicots and gymnosperms (such as pine trees) with woody stems (or trunks), such as flowering trees, have lateral meristems called **cambia**. One type of cambium in woody stems is called **vascular cambium**. As the name suggests, this growth tissue lies between the vascular tissue—the xylem and phloem—connecting the bundles to form a ring (Figure 31-10). As the cambial cells divide during secondary growth, one of the resulting daughter cells remains as a cambial cell, and the other differentiates into either a xylem or a phloem cell. This new xylem and phloem is called *secondary xylem* and *secondary phloem* (31-11).

The wood of trees is actually accumulated secondary xylem. The wood of dicot trees (such as cherry, hickory, oak, and walnut) is commonly referred to as *hardwood*, whereas the wood of conifers (such as fir, cedar, pine, and spruce) is called *softwood*. However, these names are not accurate descriptions of each group. Each contains trees having woods of varying hardness. The hardness of wood relates to its density, which depends on its proportion of wall substance to the space bounded by the cell wall. The denser a wood, the more wall substance it has in relation to the space it bounds and the stronger it is. When used for building, denser woods are harder to nail and machine, but they generally shrink and swell less than less dense, softer woods. Denser woods are also better fuel woods.

When growth conditions are favorable, as in the spring and early summer in most temperate regions, the cambium divides most actively, producing large, relatively thin-walled cells. During the rest of the year, the cambium divides more slowly, producing small, thick-walled cells. This pattern of growth results in the formation of rings in the wood. These rings are called **annual rings** and can be used to calculate the age of a tree (Figure 31-12).

Other lateral meristem tissue called **cork cambium** lies just under the epidermis. The outermost cells of this growth tissue produce densely packed cork cells. Cork cells have thick cell walls that contain fatty substances, making these cells waterproof and resistant to decay. When mature, the cork cells lose their cytoplasm and become hardened in a manner similar to the epidermal cells on your skin. The innermost cells of the cork cambium produce a dense layer of parenchyma cells. The cork, the cork cambium, and the parenchyma cells make up the outer protective covering of the plant called the **bark**.

> Woody stems have lateral meristem tissue that provides for secondary growth, which increases the diameter of the stem. Other lateral meristem tissue produces a dense layer of cells called *bark* that protects the plant.

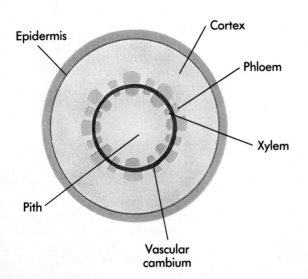

FIGURE 31-10 The vascular cambium in a stem.
The vascular cambium is the growth tissue in woody stems and lies between the xylem *(green)* and the phloem *(blue)*.

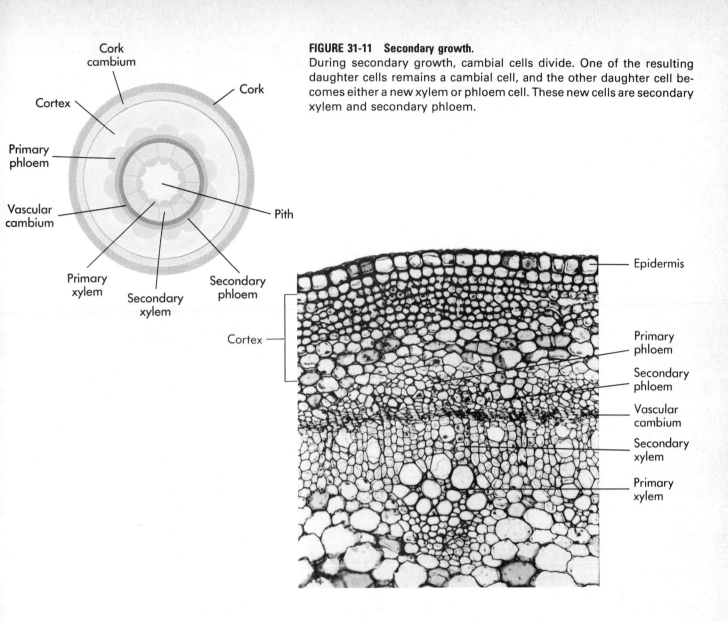

FIGURE 31-11 **Secondary growth.**
During secondary growth, cambial cells divide. One of the resulting daughter cells remains a cambial cell, and the other daughter cell becomes either a new xylem or phloem cell. These new cells are secondary xylem and secondary phloem.

FIGURE 31-12 **The structure of a tree trunk.**
The yearly growth of the vascular cambium produces growth rings.

PLANTS: PATTERNS OF STRUCTURE AND FUNCTION

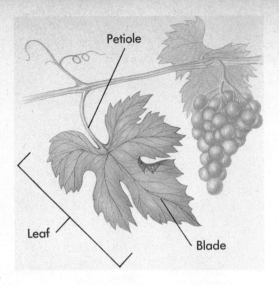

FIGURE 31-13 Structure of a leaf.
A leaf consists of a petiole and blade.

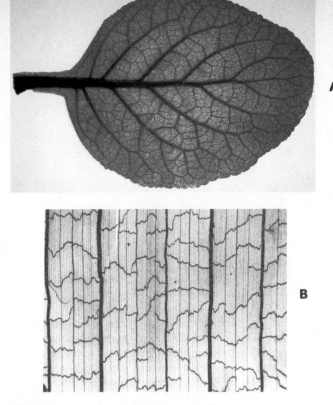

FIGURE 31-14 Dicot versus monocot leaves.
The leaves of dicots, such as this African violet relative (A), have net veination; those of monocots, like this palm (B), have parallel veination.

Leaves

Leaves, outgrowths of the shoot apex, are the light-capturing photosynthetic organs of most plants. Some exceptions exist; a major exception is found in most cacti, in which the stems are green and have largely taken over the function of photosynthesis for the plants.

Most leaves have a flattened portion, the **blade,** and a slender stalk, the **petiole** (Figure 31-13). Veins, consisting of both xylem and phloem, run through the leaves. In monocots, veins are usually parallel, and in dicots, they are usually net-like (Figure 31-14). Many conifers (vascular plants with naked seeds) have needle-like leaves suited for growth under dry and cold conditions. These modified leaves have thick, waxy coverings beneath, which are compactly arranged, thick-walled cells.

Microscopically, a cross section of a typical leaf looks somewhat like a sandwich: parenchyma cells in the middle, bounded by epidermis. The vascular bundles, or veins, run through the parenchyma. The leaf parenchyma is appropriately called the **mesophyll,** or "middle-leaf" (Figure 31-15).

The mesophyll of most dicot leaves is divided into two layers: the **palisade layer** and the **spongy layer.** The palisade layer lies beneath the upper epidermis of the leaf and consists of one or more layers of closely packed, column-like cells. These cells and the cells of the spongy layer contain chloroplasts, the organelles in which photosynthesis takes place. The spongy layer lies beneath the palisade layer and is made up of irregularly shaped cells. These cells are not closely packed like those in the palisade layer but instead have many air spaces between them. These spaces are connected, directly or indirectly, with openings to the outside called **stomata** (singular, **stoma**). Each stoma is bracketed by two **guard cells** that regulate its opening and closing (Figure 31-16).

The stomata open and close because of changes in the water pressure of their guard cells. As Figure 31-16 shows, guard cells are thicker on the side next to the stomatal opening and thinner on their other sides and ends. When the guard cells are plump and swollen with water, the cells bow. This change in the shape of the guard cells opens the stomata. When photosynthesis is taking place, water enters the guard cells, opening the stomata because the guard cells actively transport potassium ions to their interior. The water follows the potassium ions by osmosis because of the osmotic gradient the ions create. The oxygen produced by photosynthesis diffuses into the atmosphere through the stomatal openings, while the carbon dioxide needed for pho-

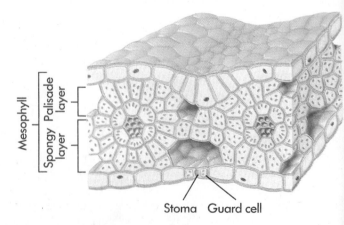

FIGURE 31-15 Internal structure of a leaf.
This diagram shows the internal structure of a typical leaf.

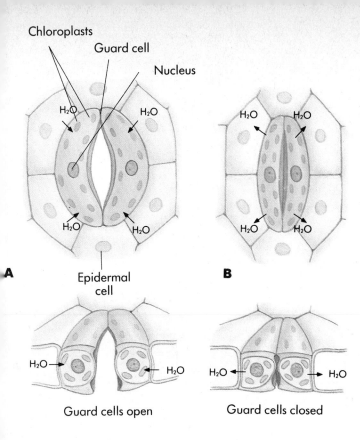
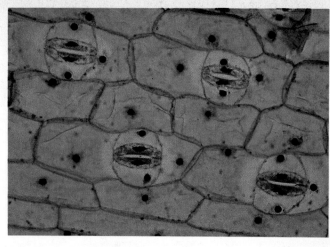

FIGURE 31-16 Guard cells.
A When solute pressure is high within the guard cells, the two guard cells bow outward, opening the stomata.
B When solute pressure is low, the guard cells become limp and close the stomata.

tosynthesis to take place diffuses in. In addition, water leaves the leaf through these openings in the form of water vapor, a process called **transpiration**.

> Leaves, the light-capturing organs of most vascular plants, are made up of parenchyma cells bounded by epidermis. The parenchyma cells contain chloroplasts, the organelles of photosynthesis.

Movement of water and dissolved substances in vascular plants

Did you ever wonder how trees manage to move water to their uppermost leaves? Osmosis alone cannot account for this amazing feat. In fact, many forces are at work. To understand these forces and their interactions, try a little experiment. Fill a glass jar with water—a jar that apple or cranberry juice comes in, for example. Now fill a dishpan about three-fourths full with water. Put a piece of cardboard over the mouth of the jar, quickly turn it upside down as you place the mouth of the jar under water, and then remove the cardboard. What happens? Unless some water escaped from the jar before the mouth was submerged, the water stayed in the jar well above the level of the water in the dishpan. But why? The answer to this question is one of the reasons plants can move water up their vascular system from the roots to the leaves. Interestingly, the answer is air pressure.

The force of gravity pulls on the air that is over the water in the dishpan creating a force called *air pressure*. The air pushing down on the water in the dishpan causes water to be pushed up into the jar and helps it stay there. At the same time, however, gravity pulls down the water within the jar. The interaction of these two forces (the push up and the pull down) are both dependent on gravity and determine the level of water in the jar. At sea level, air pressure overcomes the pull of gravity on a column of water in a microscopically thin tube (such as xylem in plants), pushing it to a height of about 10.4 meters (about 31 feet or 3½ stories)! Two other forces, adhesion and cohesion, help prevent the collapse of this column. The term **adhesion** means that water molecules adhere, or stick to, the walls of the very narrow xylem of plants. **Cohesion** means that water molecules tend to stick to one another. (See Chapter 2 for a more complete discussion of the properties of water molecules.) These two forces link water molecules together and to the sides of the xylem with weak chemical bonds called *hydrogen bonds*.

Where does the water come from that is in the xylem of plants? Most of the water absorbed by a plant comes in through its root hairs. Minerals and other nutrients also pass into the cells of the root hairs by means of special cellular "pumps." In this way, root cells maintain a higher concentration of dissolved minerals than the concentration of minerals in the water of the soil. Therefore water tends to steadily move into the root hair cells from an area of higher concentration (in the soil) to an area of lower concentration (in the root hairs) by osmosis,

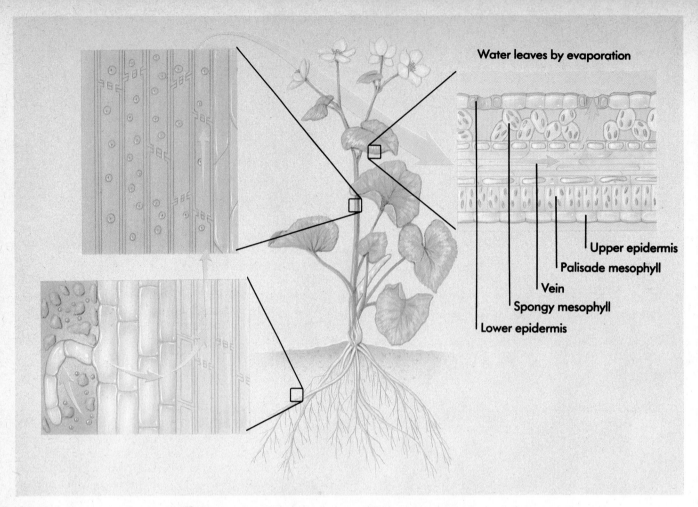

FIGURE 31-17 Water movement through a plant (transpiration).
Water is pulled up through the plant by a combination of forces.

developing a force called **root pressure.** Once inside the roots, the water and dissolved minerals pass inward to the conducting elements of the xylem.

In addition to air pressure, adhesion, cohesion, and root pressure, the water in the xylem is pulled upward by forces associated with transpiration. Transpiration is a process by which water vapor passes out of a plant through the stomatal openings in its leaves. The water moves into the leaves after travelling up the plant through the xylem. When it reaches the end of the xylem, it diffuses into the photosynthetic mesophyll cells. Many air spaces surround the mesophyll cells, and water molecules easily pass through the membranes of the cells and into the spaces. Notice the close placement of the xylem, the mesophyll cells, and the air spaces in the cross-sectional view of the leaf in Figure 31-15. From the air spaces, the water evaporates through open stomata. The evaporation of water from the surfaces of the leaves creates a concentration gradient of water molecules, which causes the water molecules in areas of higher concentration (the xylem and mesophyll cells) to move to areas of lower concentration (the air spaces), creating a tension, or pull, on the column of water. Figure 31-17 shows how the forces of transpiration, adhesion, cohesion, and root pressure work together to move water and minerals from the roots to the leaves of a plant.

> Water rises in a vascular plant beyond the point at which it would be supported by air pressure because evaporation from its leaves produces a force that pulls up on the entire water column all the way down to the roots. The forces of cohesion and adhesion work to maintain an unbroken column of water.

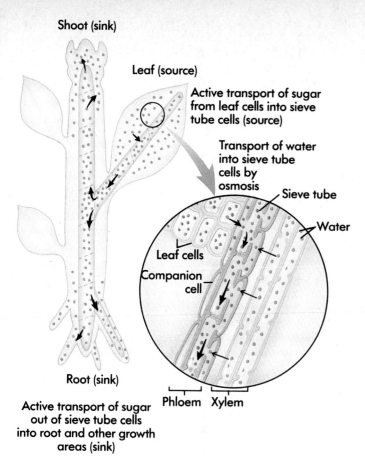

FIGURE 31-18 Mass flow.
Sucrose from the source is transported into sieve tube members by companion cells, and water follows by osmosis. At a sink—a place where sucrose is needed—sucrose is actively transported from the phloem, and water again follows by osmosis. The high water pressure in the phloem near the source and the low pressure near the sink cause this flow of sucrose, which can be very rapid.

Fluid is also transported in plants by the phloem. As mentioned previously, these tissues transport the products of photosynthesis—sugars—dissolved in water. This sugary solution is commonly called *sap*. Have you ever seen the collection of sap from maple trees? If so, you realize that plants can rapidly transport large volumes of fluid very quickly. Phloem sap contains 10% to 25% sucrose in addition to minerals, amino acids, and plant hormones and may travel as fast as one meter per hour!

The forces of diffusion and osmosis alone cannot account for this rapid movement of phloem sap. Instead, a pressure-flow or **mass flow** system performs this function. The mass flow system is shown in Figure 31-18 and works in the following way: Sucrose is produced at a **source**, such as a photosynthesizing leaf, and is actively transported into sieve tube members by companion cells. As the concentration of sucrose increases in the phloem, water follows by osmosis. In the roots below or at some other **sink** where sucrose is needed, companion cells actively transport sucrose out of the phloem. Water again follows by osmosis. The high hydrostatic (water) pressure in the phloem near the source and low pressure near the sink causes the rapid flow of the sap. After the sap reaches its destination in the plant, the water can be recycled by moving back to the source through the xylem.

The organization of nonvascular plants

The nonvascular plants, the bryophytes, are made up of three groups of plants: the liverworts, the hornworts, and the mosses (see Chapter 30). These plants are not organized the same as the vascular plants. Only certain genera have specialized vascular-like tissue. None have true roots, stems, or leaves. Bryophytes range from those forms that look like filamentous algae to those that look somewhat like certain vascular plants. But as a consequence of having less sophisticated transport systems than the vascular plants do, the bryophytes do not grow very tall. Some look as though they are creeping over their substrate, or food source.

Some bryophytes have distinct stems and leaf-like structures, such as most mosses (see Chapter 30), whereas other bryophytes do not. Bryophyte stems and leaf-like structures look outwardly different from the stems and leaves of vascular plants and are anatomically similar but simpler. These organs have an outer layer of epidermis made up of protective cells and growth cells. The cortex is made of parenchyma cells, much like those found in vascular plants. In addition, these stems and leaf-like structures have a central area of water-conducting tissue.

Bryophytes have no roots, but some have slender, usually colorless projections called **rhizoids** that anchor these simple plants to their substrate. Unlike roots, however, rhizoids consist of only a few cells and do not play a major role in the absorption of water or minerals. These substances often enter a bryophyte directly through its stems or leaves.

In general, the bryophytes have no specialized vascular tissues. The sporophytes of many moss species and the gametophytes of some moss species do have a central strand of somewhat specialized water-conducting tissue in their stems, and food-conducting tissue has been identified in a few genera. Even when such tissues are present, however, their structures are much less complex than those found in the vascular plants.

> The bryophytes—liverworts, hornworts, and mosses—are primarily low-growing plants. Some have stems and leaf-like structures, but these structures are anatomically much simpler than those of vascular plants. Although most bryophytes have no specialized vascular tissues, a few species have somewhat specialized water-conducting tissues. Food-conducting tissues are rare.

Summary

1. The body of a vascular plant has two parts, a root and a shoot. These parts are made up of three principal tissues: vascular tissue, ground tissue, and dermal tissue. Vascular tissue conducts water and dissolved minerals up the plant, ground tissue stores the carbohydrates the plant produces and forms the substance of the plant, and dermal tissue covers the plant, protecting it. In addition, plants contain meristematic tissue, an undifferentiated tissue in which cell division occurs.

2. Vascular tissue is of two types: xylem and phloem. The xylem conducts water and dissolved minerals from the roots through the stem and to the leaves. Phloem conducts water and dissolved sugars (sap).

3. Plants grow in length by cell elongation and apical meristems, zones of active cell division at the ends of the roots and the shoots. This type of growth is primary growth. Secondary growth in both stems and roots takes place in woody trees and shrubs by means of lateral meristems along the length of their stems and branches.

4. Roots, stems, and leaves are the organs of vascular plants. The root system anchors a plant in the ground and absorbs water and minerals from the soil. Stems support the leaves—the primary organs of photosynthesis. In addition, stems conduct water and minerals from the roots to all plant parts and bring the products of photosynthesis to where they are needed or stored.

5. Together, the xylem and phloem are called *vascular bundles*. They form different, distinct organizational patterns in the roots, stems, and leaves of monocots (plants with parallel vascular bundles, or veins, in their leaves) and the dicots (plants with a web-like pattern of veins in their leaves).

6. Dicots with woody stems have lateral meristem tissue, or cambium. The vascular cambium lies between the xylem and phloem; its dividing cells form xylem toward the interior (secondary xylem) and phloem (secondary phloem) toward the exterior. As a result, the diameter of a plant increases.

7. Wood is accumulated secondary xylem; it often displays rings because it exhibits different rates of growth during different seasons.

8. Leaves, the photosynthetic organs of most vascular plants, are made up of specialized ground tissue cells, or parenchyma, bounded by epidermis. Vascular bundles run through the parenchyma, and its cells contain chloroplasts, the organelles in which photosynthesis takes place. Openings in the epidermis—stomata—allow the carbon dioxide needed for photosynthesis to enter and allow the oxygen produced by photosynthesis to escape. Water vapor also evaporates from the plant through the stomata.

9. Water flows through plants in a continuous column, driven mainly by the evaporation of water vapor from the stomata and osmotic pressure at the roots. The force of air pressure helps maintain the height of the column of water in the xylem. In addition, the cohesion of water molecules and their adhesion to the walls of the narrow xylem through which they pass are important factors in maintaining the flow of water to the tops of plants.

10. Sucrose moves from where it is produced in a plant to where it is used by the process of mass flow. First, it is actively transported into phloem cells where it is produced and is actively transported out of the phloem cells where it is used. These active transport processes produce a sugar gradient and a water pressure gradient, which cause their movement.

11. The bryophytes are not organized the same as the vascular plants. Only large, multicellular forms of these plants have tissue differentiation and specialized transport cells. However, the organization and transport systems of these genera are much less complex than that of the vascular plants.

KEY TERMS

adhesion 555
adventitious root 551
annual ring 552
apical meristem 548
bark 552
blade 554
cambium 552
cohesion 555
cork cambium 552
cortex 549
cuticle 547
dermal tissue 546
endodermis 549
epidermal cell 547
epidermis 549
fibrous root 551
ground tissue 546
guard cell 554
internode 551
lateral bud 551
lateral meristem 548
leaf 545
mass flow 557
meristematic tissue 546
mesophyll 554
node 551
palisade layer 554
parenchyma cell 547
pericycle 549
petiole 554
phloem 546
pith 549
primary growth 548
rhizoid 557
root 545
root cap 550
root pressure 556
sclerenchyma cell 547
secondary growth 548
shoot 545
sink 557
source 557
spongy layer 554
stem 545
stomata (sing., stoma) 554
taproot 550
transpiration 555
vascular bundle 551
vascular cambium 552
vascular plant 545
vascular tissue 546
xylem 546

REVIEW QUESTIONS

1. What are vascular plants?
2. Fill in the blanks: In a vascular plant, the part below ground is called the _____. The part aboveground, called the _____, consists of _____ (structures in which most photosynthesis takes place) and a(n) _____ (which serves as a framework).
3. Match each type of tissue to its function:
   1. Meristem
   2. Dermal tissue
   3. Ground tissue
   4. Vascular tissue
   a. Stores food manufactured by the plant
   b. Protects the plant
   c. Produces new plant cells during growth
   d. Conducts water, minerals, carbohydrates, and other substances throughout the plant
4. What do xylem and phloem have in common? How do they differ?
5. Identify and give a function for each of the following: a) parenchyma cells; b) epidermal cells; c) stomata; d) root cap.
6. Distinguish between primary and secondary growth.
7. Parents are always telling their children to eat lots of vegetables. Using your knowledge of plant physiology, give one benefit of eating vegetables.
8. Draw two diagrams, one showing the root of a "typical" monocot, the other the root of a "typical" dicot. Label the pericycle, xylem, cortex, endodermis, epidermis, and pith.
9. What type of root system would you expect to find in a dandelion, an ivy plant, and a clump of grass? What are the advantages of each type of root?
10. How can annual rings help you estimate the age of a tree? What type of tissue is involved? In which type of growth does this result?
11. Describe the process by which water rises in a vascular plant. What forces are involved?
12. Diagram the movement of phloem sap in a mass flow system. Label the source and sink, and show the direction of flow.
13. In general, how does the appearance of nonvascular plants differ from that of vascular plants? Why?

THOUGHT QUESTIONS

1. In tropical climates, many tall plants are CAM plants—they shut their stomates during the hot days and open them at night (see the boxed essay in Chapter 8). If their stomates are closed during the day so that transpiration cannot suck water up the xylem, why doesn't the water within the plant fall down the stem?
2. The roots of many plants have permanent symbiotic associations with the fungi mycorrhizae. Among them are the most ancient fossil plants known. What might be the advantage of this association to the plant?

FOR FURTHER READING

Sandved, K. B., & Prance, G. T. (1984). *Leaves*. New York: Crown Publishers, Inc.
This book provides a beautiful introduction to the diversity of leaves.

Swain, R. B. (1988, November). Notes from the radical underground. *Discover*, pp. 16-18.
Cooperation and competition between tree roots—this article offers a new view of the relationships between plants.

Woodward, L. (1989, February 18). Plants, water and climate. Part I. *New Scientist*, (supplement), pp. 1-4.
This is an excellent account of the way water moves through plants.

32 Invertebrate animals
Patterns of structure, function, and reproduction

HIGHLIGHTS

▼

Animals are a diverse group of multicellular, heterotrophic, eukaryotic organisms that most likely evolved from protist ancestors.

▼

Other than the sponges, all animals exhibit either radial symmetry or bilateral symmetry.

▼

All bilaterally symmetrical organisms (except roundworms and flatworms) have coeloms, which are lined, fluid-filled enclosures that contain body organs suspended by thin sheets of connective tissue.

▼

The echinoderms have embryological developmental patterns suggesting that they are more closely related to the vertebrates than to the other invertebrates.

OUTLINE

Characteristics of animals
　Patterns in symmetry
　Patterns in body cavity structure
　Patterns in embryological development

Invertebrates: Diversity in symmetry and coelom
　Asymmetry: Sponges
　Radial symmetry: Hydra and jellyfish
　Bilateral symmetry: Variation on a theme in many phyla
　Flatworms
　Roundworms
　Clams, snails, and octopuses
　Earthworms and leeches
　Lobsters, insects, and spiders
　Sea urchins and starfish

Coral reefs are one of the most beautiful, exotic, and colorful living landscapes. They are only partially made up of corals, which are animals that grow on rocks in shallow, tropical waters (from 30 degrees north to 30 degrees south of the equator). In fact, the organisms that make up a coral reef are quite diverse. Many organisms such as red algae help build and strengthen coral reefs. Even the corals themselves show incredible diversity.

The term *coral* is a general term that refers to a variety of invertebrate animals (those without a backbone) of the phylum Cnidaria. Some corals are hard corals—those that have a hard, outer shell. Others are soft corals—those that lay down fragments of shell-like material within their tissues. The tissues of corals contain pigments that produce a wide range of hues, as do the microscopic algae that live within the coral animals. Corals therefore exist in many colors: brown, blue, green, yellow, pink, purple, red, and even black. Later in this chapter you will learn about the patterns of structure and function that corals and other invertebrate animals have in common despite their incredible diversity. But first you will look at the "bigger picture" of the animal kingdom to discover its unifying themes.

Characteristics of animals

How are animals different from the other kingdoms of living things? Like the plants, fungi, and protists, animals are eukaryotic organisms, having a cellular structure different from the prokaryotic structure of bacteria (see Chapter 28). Also like the plants, most fungi, and some protists, animals are multicellular. No single-celled animals exist. Only their gametes are single-celled, but these cells are not independently living organisms. As soon as fertilization takes place, the development of a new multicellular individual begins. (See Figure 30-1, which diagrams the sexual life cycle of animals.)

Animals are heterotrophs, unable to make their own food. Therefore animals must eat plants, other organisms, or organic matter for food. Some simple animals, such as the sponges, take organic matter directly into their cells. Most animals digest food within a body cavity. The resulting molecules are then taken into the body cells to be broken down further by the chemical reactions of cellular respiration (see Chapter 7). The end product of cellular respiration is energy, which is used to drive the activities of life, including growth, maintenance, reproduction, and response to the external environment. As part of this response, most animals are capable of movement to capture food or to protect themselves from injury.

The cells of all animals (except the sponges) are organized into tissues, which are groups of cells combined into structural and functional units (see Chapter 9). In most animals, the tissues are organized into organs, complex structures made up of two or more kinds of tissues. Organs that work together to perform a function are organ systems.

Animals are extraordinarily diverse in their forms and how they function. This diverse group is often divided into two subgroups: the invertebrates and the vertebrates. The invertebrates are animals without a backbone, a series of bones that surrounds and protects a dorsal (back) nerve cord. In some invertebrates, the dorsal nerve, as well as the backbone, does not exist. Examples of invertebrates are spiders, sponges, jellyfish, snails, and worms. The vertebrates (a group that includes you) have both a backbone and a dorsal nerve cord. Examples of vertebrates are bears, fish, dogs, cats, birds, and frogs. Interestingly, although the vertebrates are usually larger and more commonly known, the invertebrates make up over 95% of all animal species.

Most zoologists think that the animals arose from protist ancestors (see Chapters 26 and 27). (The evolutionary relationships among the animals and their ancestors are shown in Figure 32-1.) Each animal on the lower branches of the tree represents a present-day phylum in the animal kingdom. The animals on the upper, right branches—the mammals, birds, reptiles, amphibians, and fishes—are all classes of animals in the phylum Chordata, subphylum Vertebrata. The chordate phylum includes three subphyla: the lancelets, the tunicates, and the vertebrates. Lancelets and tunicates (sea squirts) are both groups of marine animals that have a cartilaginous notochord instead of a bony vertebral column. In addition, the lancelets and tunicates lack a brain, an organ common to all vertebrates. The animals on the upper, left branches—the crustaceans, insects (with the centipedes and millipedes), and spiders—are the three subphyla of the phylum Arthropoda. Arthropods are animals that have a hard outer shell, jointed legs, and a segmented body.

> Animals are a diverse group of eukaryotic, multicellular, heterotrophic organisms. This diverse group is often divided into two subgroups: the invertebrates and the vertebrates. The invertebrates are animals without backbones; the vertebrates are animals with a backbone that surrounds and protects a dorsal nerve cord. Most zoologists think that the animals evolved from protist ancestors.

Figure 32-1 shows that the sponges, cnidarians (jellyfish, corals), and flatworms are the present-day phyla of animals most closely related to the evolutionary ancestors of the animals. Likewise, animals higher on the tree represent present-day phyla, subphyla, or classes of organisms that are less closely related to the ancestors of the animals. Each branch of this phylogenetic tree represents an evolutionary pathway that diverged from an ancestral pathway.

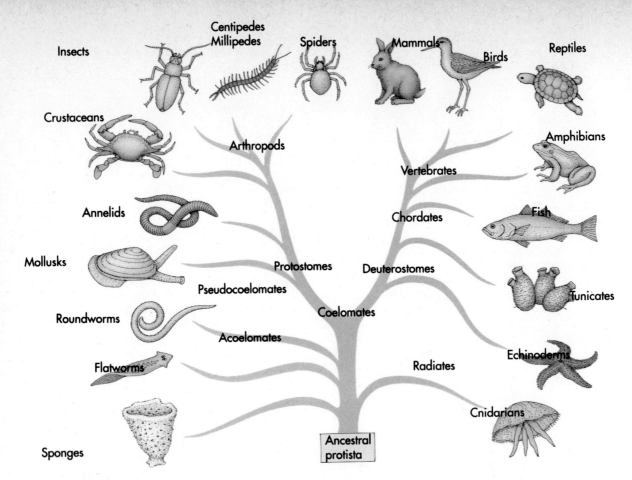

FIGURE 32-1 The animal ancestral tree.
All animals on the lower branches of the tree are phyla. The animals on the upper right branches are classes of phylum Chordata, subphylum Vertebrata. The animals on the upper left branches of the tree are the subphyla of phylum Arthropoda.

Patterns in symmetry

The word *symmetry* refers to the distribution of the parts of an object or living thing. All animals (except the sponges) exhibit either **radial symmetry** or **bilateral symmetry.** Sponges are asymmetrical, or without symmetry.

Radial symmetry means that body parts emerge, or radiate, from a central point, much like spokes on a wheel (Figure 32-2). As noted in Figure 32-1, the cnidarians and their relatives are radially symmetrical. These organisms have a top and a bottom, but they have neither a dorsal (back) side nor a ventral (belly) side (see Figure 32-6).

Animals exhibiting radial symmetry are aquatic organisms. Many are sessile; that is, they are anchored in one place. Their radially symmetrical body allows them to interact with the watery environment in all directions. Other organisms such as starfish (phylum Echinodermata) are also radially symmetrical but are not grouped with the cnidarians because their embryonic development and internal anatomy suggest that they are related to organisms having bilateral symmetry. In addition, their larvae are bilaterally symmetrical.

Bilateral symmetry means that the right side of an object or an organism is a mirror image of the left side. Animals with bilateral symmetry have a dorsal and a ventral side and a cephalic (head) end and a caudal (tail) end. All phyla of animals other than the sponges, jellyfish, and echinoderms exhibit bilateral symmetry.

Patterns in body cavity structure

A **coelom,** or body cavity, is a fluid-filled enclosure lined with connective tissue within most bilaterally symmetrical organisms and the echinoderms. Organs are located within this body cavity. For example, you have an abdominal cavity lined with a thin, nearly transparent sheet of connective tissue (peritoneum). Suspended by thin sheets of connective tissue arising from the peritoneum, the stomach and intestines hang in this coelom. All bilaterally symmetrical organisms have coeloms, except roundworms and flatworms.

Organisms having a coelom are called **coelomates.** Most of the organisms in Figure 32-1 are coelomates. Notice that the roundworms are labeled **pseudocoelomates.** These organisms (along with tiny aquatic organisms called *rotifers*) have a false body cavity. Although a fluid-filled cavity houses their organs, it is not lined nor are the organs suspended by thin sheets of connective tissue. This difference arises during the embryological development of the pseudocoelomates.

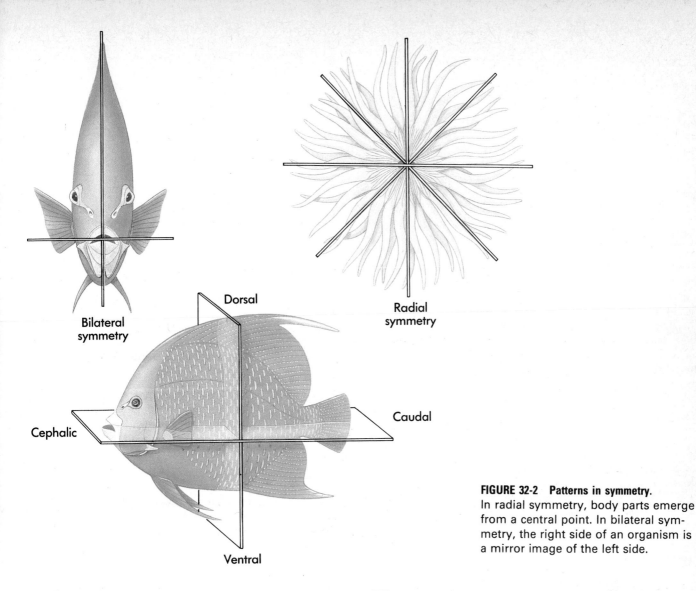

FIGURE 32-2 Patterns in symmetry. In radial symmetry, body parts emerge from a central point. In bilateral symmetry, the right side of an organism is a mirror image of the left side.

The **acoelomates,** the flatworms, have no body cavity. Embedded within the other tissues of the body, their organs are not protected from the movements and compressions of other body tissues.

The three patterns of body cavity structure are shown in Figure 32-3.

Coelomates have lined, fluid-filled body cavities in which organs are suspended by thin sheets of connective tissue. Pseudocoelomates have body cavities but they are not lined and do not have organs suspended in them. Acoelomates have no body cavities.

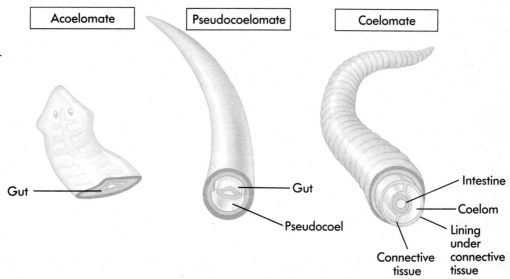

FIGURE 32-3 Three plans for construction of animal bodies. Animals can be coelomate (having a coelom), pseudocoelomate (having a false coelom), and acoelomate (having no coelom).

Patterns in embryological development

Two main branches arise from the trunk of the animal family tree in Figure 32-1. These two branches of coelomate animals represent two distinct evolutionary lines. One includes the mollusks (clams), annelids (segmented worms), and arthropods (lobsters, insects, and spiders). These organisms are called *protostomes*, meaning "first" (*proto*) "mouth" (*stome*). This strange name refers to events in the embryological development of these organisms.

During early embryological development (see Chapter 21), all organisms consist of a solid ball of cells usually called a *morula*. As development proceeds, the cells secrete fluid that fills the interior of the ball and pushes the cells to the edges of the sphere. This stage is called the *blastula*. This fluid-filled ball of cells then forms an indentation, assuming the shape of a blown-up balloon with a fist pushing in one side. This stage of development, the *gastrula*, gives rise to a two-layered embryo and begins the formation of the gut. In the **protostomes**, this first indentation becomes the mouth of the organism. In another group of organisms called **deuterostomes**, which includes the echinoderms (starfish) and chordates, the first indentation becomes the anus and a second one becomes the mouth. The term *deuterostome* means "second mouth." Figure 32-4 shows these differences.

The protostomes and the deuterostomes also show different cleavage patterns during embryological development. During the morula stage of development, the cells of the protostomes divide in a way that forms a spiral pattern. The cells of the deuterostomes cleave in a radial pattern. These patterns of embryological development (along with other developmental similarities) suggest close evolutionary relationships among the organisms within each group. In addition, because the protostome developmental pattern occurs in all acoelomates, scientists think that it was the pattern of development in the ancestors to the animals.

> The protostomes (mollusks, annelids, and arthropods) differ from the deuterostomes (echinoderms and chordates) with regard to certain embryological events. These differences suggest close evolutionary relationships among the organisms within each group.

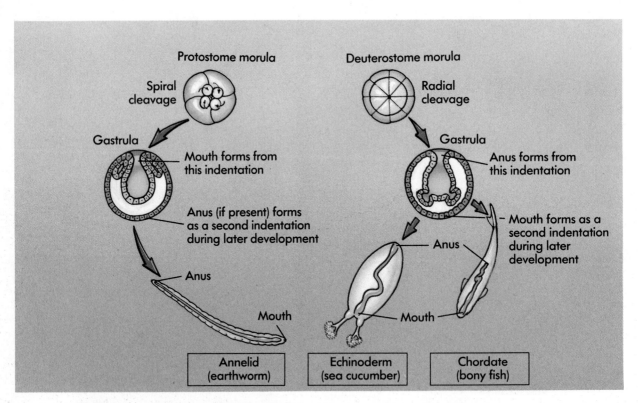

FIGURE 32-4 Protostomes and deuterostomes.
The differences between protostomes and deuterostomes arise during embryonic development. The mouth of the protostome develops from the first indentation of the gastrula. The mouth of a deuterostome develops from a second indentation. The deuterostome anus develops from the first indentation.

Invertebrates: Diversity in symmetry and coelom

Asymmetry: Sponges

Sponges (phylum Porifera) are aquatic organisms; most species live in the ocean rather than in fresh water. These sessile creatures are considered simpler than other animals in their organization because they have no tissues, no organs, and no coelom. The asymmetrical bodies of sponges consist of little more than masses of cells embedded in a gelatinous material, or matrix (Figure 32-5, *A*).

The body of a sponge is shaped like a sac or vase (Figure 32-5, *B, C*). The body wall is covered on the outside by a layer of flattened cells called the **epithelial wall.** Lining the inside cavity of the sponge are specialized, flagellated cells called **collar cells.** The matrix makes up the substance of the sponge, sandwiched between the outer, epithelial layer and the inner layer of collar cells. Within the matrix are ameboid-type cells, needle-like crystals of calcium carbonate or silica, and tough protein fibers. Pores, channel-like openings that span the matrix, are dispersed throughout the sponge. These pores are integral to the movement of water, dissolved substances, and particulate matter to the interior of the sponge. They give the sponges their phylum name *Porifera*, which means "pore bearers."

As they beat, the flagella of the collar cells create a current of water that flows from the outside of the sponge, through pores in the matrix, to the internal cavity of the

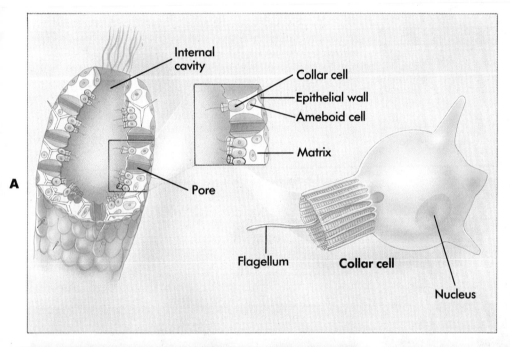

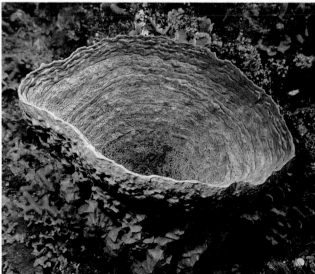

FIGURE 32-5 The sponge (phylum Porifera).
A Diagram of a sponge, with detail of a collar cell.
B A barrel sponge.
C Red boring sponge.

sponge, and then out again through the large opening at the top of the sponge. The circulation of water in this way brings the nutrients in the water to the collar cells.

Sponges frequently reproduce asexually by fragmentation; groups of cells become separated from the body of the sponge and develop into new individuals. In addition, sponges may develop branches that grow over the rocks on the sea floor, much like a plant develops underground runners. Colonies of sponges grow along these branches.

Sponges reproduce sexually and asexually. Most species of sponges produce both female sex cells (eggs) and male sex cells (sperm), which arise from cells in the matrix. Both types of sex cells are produced within the same organism. Such individuals are called **hermaphrodites,** after the Greek male god Hermes and the female goddess Aphrodite. The sperm are released into the cavity of the sponge and are carried out of the sponge with water currents and into neighboring sponges through their pores. Fertilization occurs in the gelatinous matrix where the eggs are held. There, the fertilized eggs develop into flagellated, free-swimming larvae that are released into the sponge's cavity. After the larvae leave the interior of the sponge, they settle on rocks and develop into adults.

> Sponges (phylum Porifera) are aquatic, asymmetrical, acoelomate organisms shaped somewhat like a vase. They are considered the simplest of animals because they have no tissues or organs.

Radial symmetry: Hydra and jellyfish

Animals other than the sponges have a definite shape and symmetry. Only two phyla are classified as radially symmetrical: the Cnidaria, which includes jellyfish, hydra, sea anemones, and corals (Figure 32-6), and Ctenophora, a minor phylum that includes the comb jellies (Figure 32-7).

FIGURE 32-6
Representatives of three classes of cnidarians (phylum Cnidaria).
A Hydra.
B Jellyfish.
C Yellow cup coral.

FIGURE 32-7 A comb jelly.
Note the comb-like plates and two tentacles.

There are two basic body plans exhibited by the cnidarians: **polyps** and **medusae** (Figures 32-6 and 32-8). Polyps are aquatic, cylindrical animals with a mouth at one end that is ringed with tentacles. (The name *polyp* actually means "many feet.") Polyps such as the sea anemones and corals live attached to rocks. Like the corals, many polyps build up a hard outer shell, an internal skeleton, or both. Some polyps are free floating, such as the fresh water *Hydra*. In contrast, most medusae are free floating and are often umbrella shaped. Commonly known as jellyfish, medusae have a thick, gelatinous interior. The mouth of a medusa is usually located on the underside of its umbrella shape, with its tentacles hanging down around the umbrella's edge.

Structurally, epithelial tissue covers the outside of cnidarians, and an inner tissue layer, the gastrodermis, lines the gut cavity (see Figure 32-8). The mesoglea (literally, "middle glue") lies between. This layer is quite thick within medusae and gives them their jelly-like appearance. A network of nerve cells extends through cnidarians, but they have no brain-like controlling center.

Some cnidarians such as the hydra, sea anemones, and corals occur only as polyps. Simple polyps such as hydra usually reproduce asexually by budding (see Figure 5-6). However, they do have some of their tissue organized into primitive ovaries that produce eggs and testes producing sperm, as do the more complex polyps such as sea anemones and corals. (Some species of these organisms *never* reproduce by budding.) Some hydra, like sponges, are hermaphrodites. Others exist as males and females. Eggs remain attached to the hydra but exposed to the water. Sperm are discharged from the testes and swim to the egg. After fertilization, developing hydras grow while attached to the parent.

Some cnidarians exist only as medusae. Medusae reproduce sexually. Ovaries hang from the underside of female

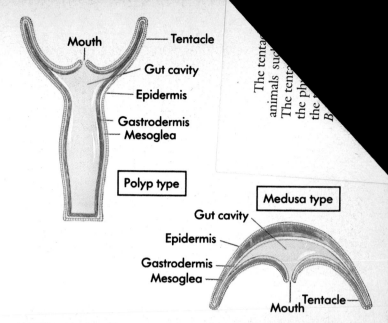

FIGURE 32-8 Body forms of cnidarians: the medusa and the polyp. These two phases alternate in the life cycles of many cnidarians, but a number—including the corals and sea anemones—exist only as polyps.

medusae, and testes hang from the males. Eggs and sperm are shed into the water, where fertilization takes place. The fertilized egg develops into a larva that never settles down to become a sessile polyp but develops directly into a medusa. However, most species have medusae that have a life cycle in which the larvae develop into polyps. These polyps develop into medusae. This type of alternating life cycle is shown in Figure 32-9.

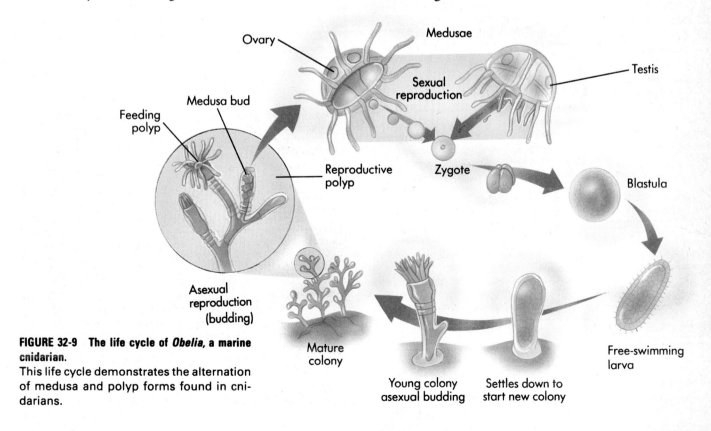

FIGURE 32-9 The life cycle of *Obelia*, a marine cnidarian.
This life cycle demonstrates the alternation of medusa and polyp forms found in cnidarians.

INVERTEBRATE ANIMALS: PATTERNS OF STRUCTURE, FUNCTION, AND REPRODUCTION

cles of a cnidarian help it capture prey—other as small fishes, shrimp, and aquatic worms. cles bear stinging cells called *cnidocytes*, which give ylum its name. You can see these cells as tiny dots in tentacles of the yellow cup coral shown in Figure 32-6. If you have ever been stung by a jellyfish, you know how powerful its sting can be. These stinging cells work much like harpoons. Powered by water pressure, the stinging cells are jettisoned out of the animal, spearing and immobilizing the prey. The tentacles then draw the prey back to the mouth.

> Cnidarians have two layers of tissues and a nerve net that coordinates cell activities. They exist either as polyps (corals and sea anemones), which are cylindrically shaped animals that anchor to rocks, or as medusae (jellyfish), which are free-floating, umbrella-shaped animals. In some cnidarians, these two forms alternate during the life cycle of the organism.

Bilateral symmetry: Variation on a theme in many phyla

Flatworms

The remaining phyla of animals are all bilaterally symmetrical or (in the case of the echinoderms) are thought to have evolved from bilaterally symmetrical forms. In addition, they all develop embryologically from three layers of tissue: an inner layer or **endoderm,** an outer layer or **ectoderm,** and a middle layer or **mesoderm.** Although cnidarians have a middle layer between their two tissue layers, it consists of a jelly-like material with only widely dispersed cells.

The bilaterally symmetrical animals with the simplest body plan are the flatworms (phylum Platyhelminthes). This phylum name comes from Greek words meaning "flat" (*platys*) "worm" (*helminthos*) and describes their flattened ribbon or leaf-like shapes. Although simple in structure, the flatworms have organs and some organ systems, but because they have no coelom, the organs are embedded within the body tissues (see Figure 32-3). Flatworms are also the simplest animals to have a distinct head, a characteristic common to many of the bilaterally symmetrical animals.

There are three classes of flatworms: the turbellarians, the flukes, and the tapeworms (Figure 32-10). The turbellarians are free living and found in fresh or salt water or in damp soil. The flukes and tapeworms are parasites and live on or in other animals, deriving nutrition from their hosts.

Free-living flatworms move from place to place, feeding on a variety of small animals and bits of organic debris. They move by means of ciliated epithelial cells that are concentrated on their ventral surfaces. In fact, the name *turbellarian* comes from a Latin word meaning "to bustle or stir" and refers to the water turbulence created by their movement. Sensory pits or tentacles along the sides of their heads detect food, chemicals, and movements of the fluid in which they are moving. They also have eyespots on their heads, which contain light-sensitive cells that enable the worms to distinguish light from dark. These organs are part of the flatworm

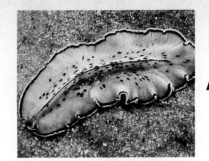

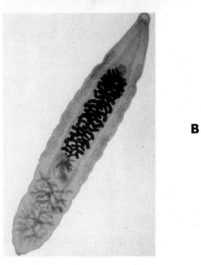

FIGURE 32-10 Flatworms (phylum Platyhelminthes).
A A marine, free-living turbellarian.
B The human liver fluke, *Clonorchis sinensis*.

nervous system and connect to a ladder-like paired nerve cord that extends down the length of the animal. Tiny swellings at the cephalic, or head end, of the organism are considered a primitive brain.

The flatworm has a digestive system consisting of a digestive sac, or gut, open only at one end (Figure 32-11). Muscular contractions in the upper end of the gut of flatworms cause a strong sucking force by which the flatworms ingest their food and tear it into small bits. The cells making up the gut wall engulf these particles; most digestion takes place within these cells. Wastes from within the cells diffuse into the digestive tract and are expelled through the mouth. In addition, the flatworm excretes excess water and some wastes by means of a primitive excretory system: a network of fine tubules that runs along the length of the worm. Specialized bulb-like cells, or flame cells, are located along these tubules (see Figure 32-11). As cilia within them beat (looking like a flickering flame), they move the water and wastes into the tubules and out excretory pores.

Reproduction in flatworms is much more complicated than in sponges or cnidarians. Although most flatworms are hermaphroditic, a characteristic exhibited by the sponges and cnidarians, the organs of reproduction are better developed. When flatworms mate, each partner deposits sperm in a copulatory sac of the other. The sperm travel along special tubes to reach the egg. In free-living flatworms, the fertilized eggs are laid in cocoons and hatch into miniature adults. In some parasitic flatworms, there is a complex

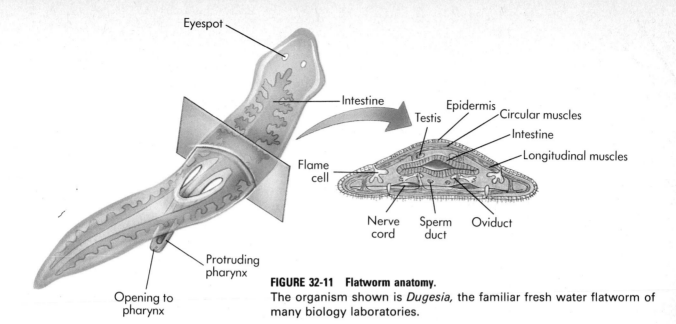

FIGURE 32-11 Flatworm anatomy.
The organism shown is *Dugesia,* the familiar fresh water flatworm of many biology laboratories.

succession of distinct larval forms. Flatworms are also capable of asexual reproduction. In some genera, when a single individual is divided into two or more parts, each part can regenerate an entire new flatworm.

The flukes and the tapeworms are two classes of flatworms that live within the bodies of other animals. The adult form of both classes parasitize humans. The life cycles of these organisms are discussed in the boxed essay.

> **The acoelomates, typified by the flatworms, are the most primitive bilaterally symmetrical animals. Although simple in structure, the flatworms have organs and some organ systems, including a primitive brain. Because flatworms do not have a coelom, their organs are embedded within the body tissues.**

Roundworms

Seven phyla have a pseudocoelomate body plan (see Figure 32-3)—a body cavity with no lining. Only one of the seven, the roundworms, includes a large number of species.

Roundworms are classified in the phylum Nematoda, a word that comes from a Greek word meaning "thread." These cylindrical worms are diverse in size, but some of them are so small and slender that they look like fine threads. In fact, they can be so microscopically small that a spadeful of fertile soil may contain millions of these worms (Figure 32-12). Some species are abundant in fresh or salt water.

Many members of this phylum are parasites of vertebrates. About 50 species of roundworms parasitize human beings, causing problems such as blockage of the lymphatic vessels (Figure 32-13) or intestines and infections of the

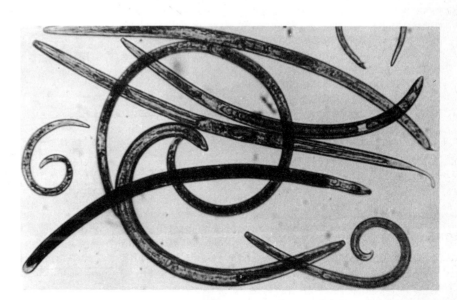

FIGURE 32-12 Living dirt.
One square meter of ordinary lawn soil teems with 2 to 4 million roundworms. Although most are similar in form, they range in length from about 0.2 millimeters to about 6 millimeters.

INVERTEBRATE ANIMALS: PATTERNS OF STRUCTURE, FUNCTION, AND REPRODUCTION

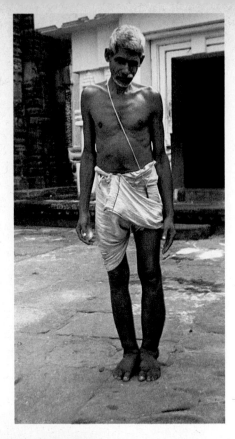

FIGURE 32-13 Elephantiasis.
This condition is caused by roundworms that live in the lymphatic passages and block the flow of lymph. As a result, fluids cannot drain, and swelling occurs.

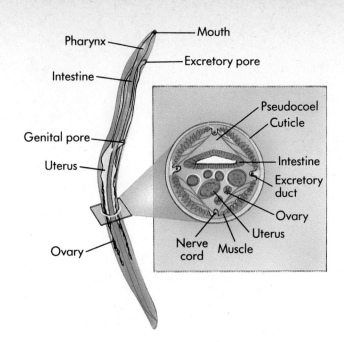

FIGURE 32-14 Anatomy of *Ascaris*, a parasitic roundworm of humans.
Note the pseudocoel.

muscles or lungs. Roundworms also parasitize invertebrates and plants. For this reason, some nematodes are being investigated as agents of biological control of insects and other agricultural pests.

Roundworms are covered by a flexible, tough, transparent multilayered tissue (cuticle), which is shed as they grow. A layer of muscle lies beneath this epidermal layer and extends lengthwise (Figure 32-14). These longitudinal muscles pull against both the cuticle and the firm, fluid-filled pseudocoelom, similar to how your muscles pull against your bones. All this effort gets them nowhere in clear water, but they can move in muddy water or soil, which provides surfaces against which the worm's body pushes. The movements do, however, push on the fluid-filled pseudocoelom and aid in the distribution of food and oxygen throughout the worm.

The roundworm digestive system has two openings—a mouth and anus. Most roundworms have raised, hair-like sensory organs near their mouths. The mouth itself often has piercing organs, or stylets. Food passes through the mouth as a result of the sucking action of a muscular pharynx. After passing through these organs, food continues through the digestive tract where it is broken down and then absorbed. The roundworms that parasitize animals take in digested food of the host; the cells lining the digestive system simply absorb these nutrients.

The roundworms also contain primitive excretory and nervous systems. The nervous system consists of a ring of tissue surrounding the pharynx and a solid dorsal and a ventral nerve cord (unlike the hollow nerve cord of a chordate [see p. 581]). The excretory system consists of two lateral canals that unite near the anterior end to form a single tube ending in an excretory pore.

> **The pseudocoelomates are typified by the roundworms. Muscles attached to their thick, outer cuticle push against the fluid-filled pseudocoelom, resulting in a whip-like movement. Many of the roundworms are parasites of invertebrates, vertebrates (including humans), and plants.**

Clams, snails, and octopuses

The rest of the invertebrate animals have a "true" coelom. The next three phyla of coelomates: the mollusks (clams, snails, and octopuses), the annelids (earthworms and leeches), and the arthropods (lobsters, insects, and spiders) are all protostomes. The acoelomate animals already discussed are not protostomes but exhibit similar developmental patterns to the protostomes. These similarities suggest an evolutionary closeness among these groups.

The mollusks are a large phylum of invertebrate animals having a muscular foot and a soft body covered by a mantle and are usually covered with a hard shell. The phylum name *Mollusca* comes from a Latin word meaning "soft bodied." The shelled mollusks include the snails, clams, scallops, and oysters. Unshelled mollusks are represented by the octopuses, squids, and slugs.

Mollusks are widespread and often abundant in marine and fresh water environments, and some, such as certain snails and the slugs, live on land. They range in size from

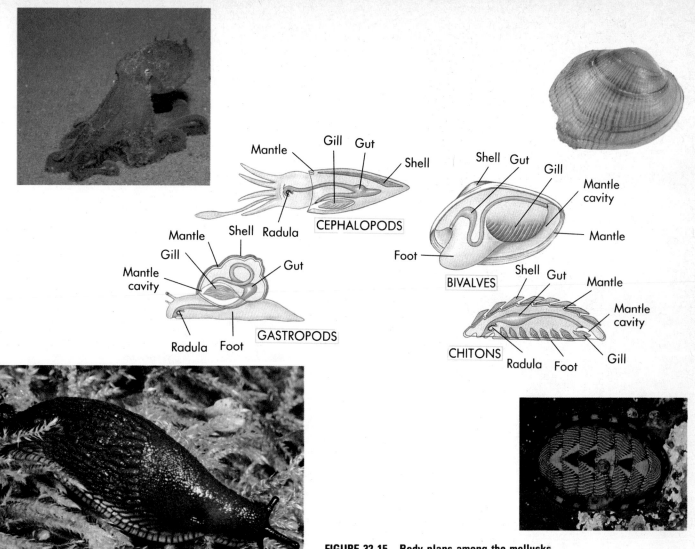

FIGURE 32-15 Body plans among the mollusks.
The name of each group describes its prominent features. The photos of each group are: slug, *Arion ater* (gastropod); a mussel of a nearly extinct species from the Wabash and upper Maumee river in Indiana (bivalve); octopus (cephalopod); and a chiton, *Tonicella lineata*.

being near microscopic to having the huge proportions of giant squid, which may be as long as 21 meters (almost 70 feet) and weigh up to 250 kilograms (550 pounds).

Mollusks exhibit four body plans: gastropod, cephalopod, bivalve, and chiton (Figure 32-15). The name of each group describes its prominent features. Although each group (with the exception of the bivalves) has a head end, the cephalopods (literally, "head-foot") have the most well-differentiated head and the most well-developed nervous system. The bivalves ("two-shelled") gastropods have the least well-developed nervous sytem of these groups. All four groups have a **visceral mass,** or group of organs, consisting of the digestive, excretory, and reproductive organs. The visceral mass is covered with a soft epithelium called the **mantle,** which arises from the dorsal body wall and encloses a cavity between itself and the visceral mass. This cavity is *not* the coelom; the coelom surrounds the heart only.

The mollusk's gills, the organs of respiration, lie within the mantle cavity. **Gills** are a system of filamentous projections of the mantle tissue that are rich in blood vessels. These projections greatly increase the surface area available for gas exchange. In land-dwelling mollusks, a network of blood vessels within the mantle cavity serves as a primitive lung.

Mollusks exhibit both open circulatory systems (in clams, for example) and closed circulatory systems (in squid, for example). Both types of circulatory system have a heart to pump blood. In a **closed circulatory system,** blood is enclosed within vessels as it travels throughout the body of the organism. In an **open circulatory system,** blood flows in vessels leading to and from the heart but through irregular channels called **blood sinuses** in many parts of the body.

In most coelomate animals (most of which have a closed circulatory system), the blood vessels are intimately associ-

BIOLOGY & you

Parasitic flatworms

Parasitic flatworms are an unusual and frightening group. Although tiny, their effects on humans and animals on which humans depend can be devastating.

Most parasitic flatworms have two hosts. Eggs or larvae grow and develop within an animal called the *secondary host*. Once the appropriate stage of development is reached (different for each species of flatworm), the parasites then migrate into a *primary host*, where they complete their maturation into adults and begin to reproduce. Humans most often serve as primary hosts to parasitic flatworms, and animals such as snails, fish, cattle, and pigs serve as secondary hosts. It is easy to see how humans can become infected with parasitic worms—the secondary hosts are sources of food.

Parasitic flatworms are of two types: the flukes and the tapeworms. Several species of the parasitic fluke *Schistosoma* are found in Africa, South America, the West Indies, the Middle East, and the Far East. Infection with *Schistosoma* organism ranks as one of the major infectious diseases of the world. The life cycle of *Schistosoma* flatworms begins when the eggs are shed into the water in the feces or urine of an infected individual. The eggs hatch in the water. The larvae must come into contact with a specific genus of snail that serves as their secondary host within a few hours, or they die. Each species of *Schistosoma* parasitizes a specific genus of snail. Once within the snail, the *Schistosoma* larvae develop into a form that is equipped to swim. They then migrate out of the snail and into the water. When they come into contact with humans, they penetrate the skin and burrow until they find a blood vessel. The young schistosomes migrate first to the liver, where they complete their development into adults, before migrating to various other parts of the body. The females produce prodigious amounts of eggs, and these eggs are shed in the feces and urine of the infected individual, thus completing the life cycle.

Other parasitic flukes include the liver fluke (*Clonorchis*), which uses the snail *Parafossarulus* and fishes of the family Cyprinidae as secondary hosts. Humans become infected when they eat raw or undercooked fish. The lung fluke uses fresh water crabs as secondary hosts and migrates to the lungs of its primary host, the human, after being ingested. Infection with the lung fluke causes severe respiratory problems.

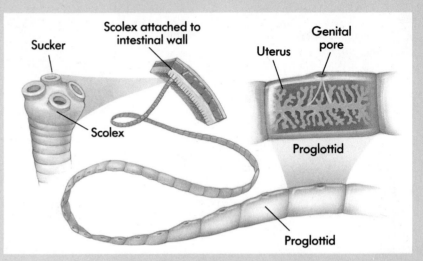

FIGURE 32-A Anatomy of the beef tapeworm.

Tapeworms are a second type of parasitic flatworm. The body of a tapeworm is long and flat (Figure 32-A). The head region is equipped with a scolex, an organ that contains suckers or hooks with which the tapeworm attaches itself to the intestinal wall of its primary host. The rest of the body is made up of repeating sections called *proglottids*. Each proglottid contains a full set of reproductive organs. Each proglottid, then, is capable of producing hundreds of offspring. No wonder tapeworms have been referred to as "reproduction factories!"

The most common tapeworm that affects humans is the beef tapeworm (*Taeniarhynchus saginatus*). The life cycle of the beef tapeworm begins when a few proglottids are broken off in the feces of an infected animal. The proglottids produce eggs as they lie in the vegetation or soil. The eggs are extremely hardy—they have been known to live up to 5 months as they lie on grass. When the eggs are ingested by grazing cattle, they hatch within the intestine, and the larvae burrow through the intestinal wall until they reach muscle. Here, the larvae encyst—they become encapsulated and quiescent. When a human eats raw or undercooked beef, the cyst dissolves, and the tapeworm attaches to the intestine with its scolex. New proglottids develop, and the tapeworm matures in 2 to 3 weeks. After maturation, the tapeworm produces numerous offspring. Proglottids are broken off in the feces of the affected individual, and the life cycle begins again.

Pork tapeworm (*Taenia solium*) has a life cycle similar to that of the beef tapeworm, except that humans can also serve as the secondary host in which the tapeworm encysts. Common sites for encystment are the brain and the eye, resulting in blindness or severe neurological problems.

While the thought of sharing your body with such creatures is quite unpleasant, the use of efficient sanitation procedures has greatly reduced the rate of parasitic worm infection in the United States. A further precaution against these parasites is to avoid eating raw or undercooked meat or fish—remember that cattle, fish, and other animals serve as these parasites' secondary hosts, and high cooking temperatures will kill the parasites.

ated with the excretory organs, making the direct exchange of materials between these two systems possible. Mollusks were one of the earliest evolutionary lines to develop an efficient excretory system. Wastes are removed from the mollusk by tubular structures called **nephridia.** In mollusks with open circulatory systems, wastes move from the coelom into the nephridia and are discharged into the mantle cavity. From there they are expelled by the continuous pumping of the gills. In animals with closed circulatory systems, such as annelids, some mollusks, and the vertebrates, the coiled tubule of a nephridium is surrounded by a network of capillaries. Wastes move from the circulatory system to the nephridium for removal from the body. All coelomates (except for arthropods and chordates) have basically similar excretory systems.

> Mollusks, widespread in marine and fresh water environments, exhibit four body plans as represented by snails, clams, squid, and chitons. These animals use gills for respiration in water; terrestrial species have adaptations for breathing on land. They exhibit both open and closed circulatory systems and well-developed excretory systems.

Earthworms and leeches

The other two major phyla of protostomes, the annelids and arthropods, have segmented bodies, whereas the mollusks do not. Segmentation underlies the organization of all of the more complex phyla of animals: the annelids (earthworms and leeches), arthropods (lobsters, insects, and spiders), echinoderms (sea urchins and starfish), and chordates (tunicates, fishes, amphibians, reptiles, birds, and mammals). In some adult arthropods the segments are fused; their segmentation is only apparent in their embryological development. So, too, embryological development reveals segmentation in the chordates. Segmentation in this phylum is exhibited only by the vertebrates in the repeating units of their backbones. Because segmentation in animals is different among phyla, scientists think it arose independently in more than one line of evolution.

The annelids (phylum Annelida) are worms characterized by a soft, elongated body composed of a series of ring-like segments. In fact, the word *annelid* means "tiny rings." Annelids are abundant in the soil and in both marine and fresh water environments throughout the world. Internally, their segments are divided from one another by partitions called **septa.** Digestive, excretory, and neural structures are repeated in each segment.

There are three classes of annelids: marine worms, fresh water and terrestrial worms, and leeches (Figure 32-16). Exhibiting unusual forms and sometimes iridescent colors, the marine worms live in burrows, under rocks, inside shells, and in tubes of hardened mucus they manufacture. Leeches occur mostly in fresh water, although a few are marine and some tropical leeches occur in terrestrial habitats. Most leeches are predators or scavengers, and some suck blood from mammals, including humans. The best-known leech is the medicinal leech, which was used for centuries to remove

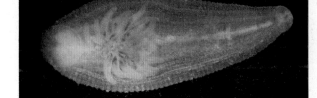

FIGURE 32-16 Representative annelids.
A Shiny bristleworm.
B Earthworms mating.
C A fresh water leech, with young leeches visible in the brood pouch.

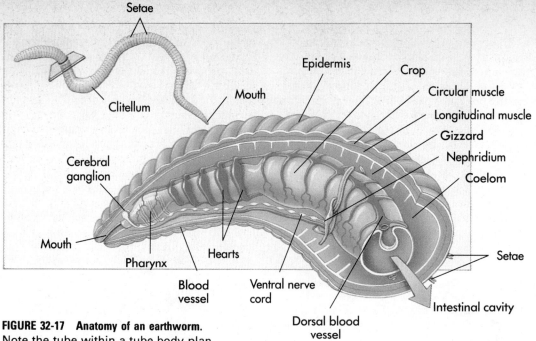

FIGURE 32-17 Anatomy of an earthworm.
Note the tube within a tube body plan.

what was thought to be excess blood responsible for certain illnesses. The medicinal leech is now used as a source of anticoagulant, which is used in research focusing on blood clotting. This animal is also used by some physicians to remove excess blood after surgery and to help restore circulation to severed body parts (such as fingers) after reattachment.

The earthworm exhibits the generalized body plan of this phylum: a tube within a tube (Figure 32-17). The digestive tract, a straight tube running from mouth to anus, is suspended within the coelom. An earthworm sucks in organic material by contracting its strong pharynx. It grinds this material in its muscular gizzard, aided by the presence of soil particles it takes in with its food.

The anterior segments of an earthworm contain a well-developed **cerebral ganglion,** or brain, and a few muscular blood vessels that act like hearts, pumping the blood through the closed circulatory system. Sensory organs are also concentrated near the anterior end of the worm. Some of these organs are sensitive to light, and elaborate eyes with lenses and retinas have evolved in certain members of the phylum. Separate nerve centers, or ganglia, are located in each segment and are connected by nerve cords. Each segment also contains both circular and longitudinal muscles, which annelids use to crawl, burrow, and swim. **Setae,** or bristles, help anchor the worms during locomotion or when they are in their burrows.

> **The annelids are characterized by serial segmentation and a tube within a tube body plan. The body is composed of numerous similar segments, each with its own circulatory, excretory, muscular, and neural structures.**

Reproduction differs among the annelid classes. In the marine worms, the sexes are usually separate, and fertilization is often external, occurring in the water and away from both parents. The earthworms and leeches, on the other hand, are hermaphroditic. When they mate, their anterior ends point in opposite directions, and their ventral surfaces touch. The **clitellum,** a thickened band on an earthworm's body, secretes a mucus that holds the worms together as they exchange sperm. Ultimately, the worms release the fertilized eggs into cocoons also formed by mucous secretions of the clitellum.

Lobsters, insects, and spiders

Lobsters (crustaceans), insects, and spiders (arachnids) are representatives of the diverse phylum Arthropoda (Figure 32-18). The name *arthropod* comes from the two Greek words "arthros" (jointed) and "podes" (feet) and describes the characteristic jointed appendages of all arthropods. The nature of the appendages differs greatly in different subgroups; appendages may take the form of antennae, mouthparts of various kinds, or legs.

The arthropods have a rigid external skeleton, or **exoskeleton,** which varies greatly in toughness and thickness among arthropods (Figure 32-19). The exoskeleton provides places for muscle attachment, protects the animal from predators and injury, and most importantly, protects arthropods from water loss. As an individual outgrows its exoskeleton, that exoskeleton splits open and is shed. A new soft exoskeleton lies underneath, which subsequently hardens. The animal then grows into its new "shell."

All arthropods can be placed into one of two groups: those with jaws and those without jaws. The crustaceans, insects, centipedes, millipedes, and a few other small groups of arthropods have jaws, or **mandibles.** These jaws are formed by the modification of one of the pairs of anterior appendages (but *not* the first pair). The appendages nearest the anterior end are sensory antennae (Figure 32-20, *A*). The remaining arthropods, which include the spiders, mites, scorpions, and a few other groups, lack mandibles. Their mouthparts usually take the form of pincers or fangs, which evolved from the appendages nearest the anterior end of the animal (Figure 32-20, *B*).

FIGURE 32-18 Representatives of phylum Arthropoda.
A A freshwater crayfish, *Procambarus*.
B Copulating grasshoppers.
C The arrowhead spider.

FIGURE 32-19 Exoskeletons.
Some arthropods have a tough exoskeleton, like this South American scarab beetle (**A**); others have a fragile exoskeleton, like the green darner dragonfly (**B**).

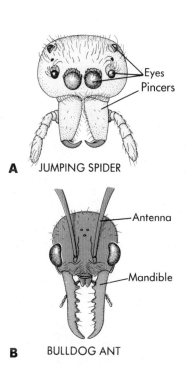

FIGURE 32-20 Arthropod mouthparts.
A The bulldog ant does have a jaw.
B The jumping spider lacks a jaw but does have pincers with which it catches prey.

INVERTEBRATE ANIMALS: PATTERNS OF STRUCTURE, FUNCTION, AND REPRODUCTION

FIGURE 32-21 The compound eye.
This compound eye is found on a robberfly.

An important structure of many arthropods such as bees, flies, moths, and grasshoppers is the **compound eye** (Figure 32-21). Compound eyes are composed of many independent visual units, each containing a lens. **Simple eyes,** composed of a single visual unit having one lens, are found in many arthropods with compound eyes and function in distinguishing light and darkness. In some flying insects, such as locusts and dragonflies, simple eyes function as horizon detectors and help stabilize the insects during flight.

In the course of arthropod evolution, the coelom has become greatly reduced, consisting only of the cavities that house the reproductive organs and some glands. Figure 32-22 illustrates the major structural features of a grasshopper as a representative of the arthropods. Like the annelids, the arthropods have a tubular gut that extends from the mouth to the anus. The circulatory system of arthropods is open; their blood flows through cavities between the organs. One longitudinal dorsal vessel functions as a heart, helping move the blood along.

Most aquatic arthropods breathe by means of gills. Their feathery-looking structure provides a large surface area over which gas exchange takes place between the surrounding water and the animal's blood. The respiratory systems of terrestrial arthropods generally have internal surfaces over which gas exchange takes place. The respiratory systems of insects, for example, consist of small, branched air ducts called **tracheae.** These tracheae, which ultimately branch into very small **tracheoles,** are a series of tubes that transmits oxygen throughout the body. The tracheoles are in direct contact with the individual cells, and oxygen diffuses from them to other cells directly across the cell membranes. Air passes into the trachea by way of specialized openings called **spiracles,** which can be closed and opened by valves in most insects.

Although there are various kinds of excretory systems in different groups of arthropods, a unique excretory system evolved in terrestrial arthropods in relation to their open circulatory system. The principal structural element is the **malpighian tubules,** which are slender projections from the digestive tract. Fluid is passed through the walls of the malpighian tubules to and from the blood in which the tubules are bathed. The nitrogenous wastes in it are precipitated and then emptied into the hindgut (the posterior part of the digestive tract) and eliminated. Most of the water and salts in the fluid is reabsorbed by the hindgut, thus conserving water.

▶ **Arthropods are a diverse phylum of organisms having jointed appendages and rigid exoskeletons.** ◀

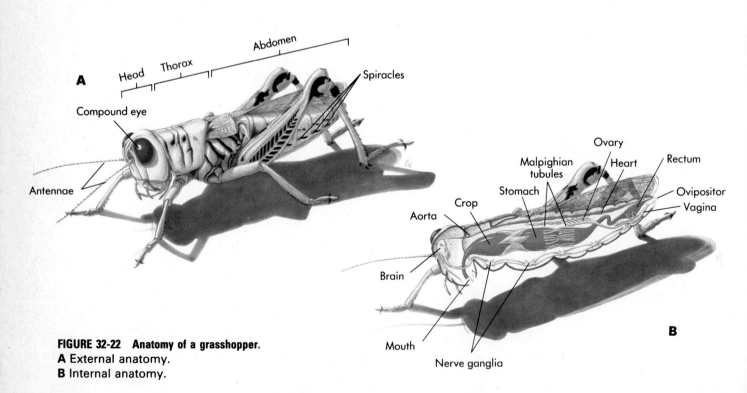

FIGURE 32-22 Anatomy of a grasshopper.
A External anatomy.
B Internal anatomy.

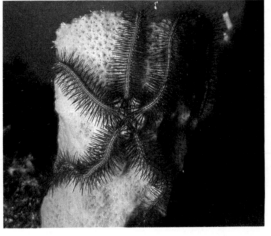
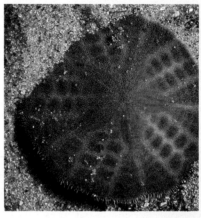

FIGURE 32-23
Representatives of the echinoderms.
A Sea star.
B Sea cucumber.
C Sea lilies.
D Brittle star.
E Sand dollar.

Sea urchins and starfish

Sea urchins and starfish are representatives of the phylum Echinodermata. The term *echinoderm* means "spine skin," an appropriate name for many members of this phylum (Figure 32-23). The five living classes of echinoderms are shown: the sea lilies, sea stars (starfish), brittle stars, sea urchins and sand dollars, and sea cucumbers. These marine animals live on the sea floor, with the exception of a few swimming sea cucumbers.

The echinoderms are different from the other invertebrates in that they are deuterostomes. The other invertebrates are protostomes. The embryological differences that distance them evolutionarily from the other invertebrates connect them with the chordates. It is thought that the echinoderms and the chordates, along with two smaller phyla not mentioned here, evolved from a common ancestor.

Echinoderms are bilaterally symmetrical as larvae but radially symmetrical as adults. They are evolutionary related to and grouped with the bilaterally symmetrical animals, however. Adult echinoderms have a five-part body plan corresponding to the arms of a sea star or the design on the "shell" of a sand dollar. Five **radial canals,** the positions of which are determined early in the development of the embryo, extend into each of the five parts of the body (Figure 32-24). This water vascular system is used for locomotion and is unique to echinoderms.

FIGURE 32-24 Anatomy of an echinoderm.
Notice the water vascular system of a sea star *(blue).*

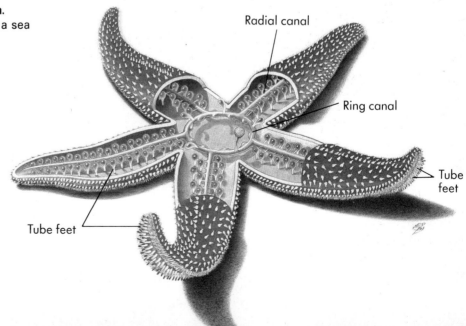

As adults, these animals have no head or brain. Their nervous systems consist of central **nerve rings** from which branches arise. The animals are capable of complex response patterns, but there is no centralization of function.

There is no well-organized circulatory system in echinoderms. Food from the digestive tract is distributed to all the cells of the body in the fluid that lies within the coelom. In many echinoderms, respiration takes place by means of skin gills, which are small, finger-like projections that occur near the spines. Waste removal also takes place through these skin gills. The digestive system is simple but usually complete, consisting of a mouth, gut, and anus.

Although the echinoderms are well-known for regenerating lost parts, most reproduction is sexual and external. The sexes in most echinoderms are separate, although usually little external difference exists between them. The fertilized eggs of echinoderms usually develop into free-swimming bilaterally symmetrical larvae that look quite different from the adults. They eventually change through a series of stages, or metamorphose, to adult forms.

> The echinoderms, or spiny-skinned animals, are marine organisms represented by sea urchins and starfish. They are closely related to the chordates.

Summary

1. Animals are a diverse group of multicellular, eukaryotic, heterotrophic organisms. The cells of all animals (with the exception of sponges) are organized into tissues, and in most animals the tissues are organized into organs. Most zoologists think that the animals arose from protist ancestors.

2. Other than the sponges, all animals exhibit either radial symmetry (the jellyfish and starfish) or bilateral symmetry (all other animals). Radial symmetry means that body parts emerge from a central point. Bilateral symmetry means that the right side of an organism is a mirror image of the left side.

3. A body cavity, or coelom, is a lined, fluid-filled enclosure, which contains body organs that hang suspended by thin sheets of connective tissue. All bilaterally symmetrical organisms and the echinoderms (starfish) have coeloms except roundworms and flatworms. Roundworms have pseudocoeloms, or false body cavities, which are not lined; roundworms' organs are not suspended by thin sheets of connective tissue. Flatworms have no coeloms.

4. Two main branches of coelomate animals, representing two distinct evolutionary lines, are the protostomes and the deuterostomes. The protostomes (mollusks, annelids, and arthropods) differ with regard to certain embryological events from the deuterostomes (echinoderms and chordates).

5. Sponges (phylum Porifera) are acoelomate, asymmetrical, aquatic animals. They are considered simple animals because they have no tissues or organs.

6. The hydra and jellyfish (phylum Cnidaria) are acoelomate, radially symmetrical, aquatic animals. They have two layers of tissues and exist either as cylindrically shaped polyps (corals and sea anemones) or medusae (jellyfish). In some cnidarians, these two forms alternate during the life cycle of the organism.

7. All animals other than the sponges and the cnidarians are bilaterally symmetrical or, in the case of the echinoderms, evolved from bilaterally symmetrical ancestors. In addition, they all develop embryologically from three tissue layers: endoderm, ecotoderm, and mesoderm.

8. The flatworms (phylum Platyhelminthes) are ribbon-like worms that live either in the soil or water or within other organisms. Having no coelom, they are considered to have the simplest body plan of the bilaterally symmetrical animals. However, the flatworms have organs and some organ systems, including a primitive brain.

9. The roundworms (phylum Nematoda) are pseudocoelomate, cylindrical worms that live in the soil, in water, or within other organisms. They have simple digestive, excretory, and nervous systems.

10. The mollusks (phylum Mollusca) are a large phylum of coelomates that exhibit four body plans. All four groups have a visceral mass, or group of organs, consisting of the digestive, excretory, and reproductive organs. They exhibit both closed and open circulatory systems.

11. The annelids (phylum Annelida) are worms characterized by a soft, elongated body composed of a series of ring-like segments, each with its own circulatory, excretory, muscular, and neural structures. They have a tube within a tube body plan.

12. The arthropods (phylum Arthropoda) are an extremely diverse phylum that includes organisms such as lobsters, insects, and spiders. Arthropods are characterized by a rigid external skeleton and jointed appendages.

13. The phylum Echinodermata, represented by sea urchins and starfish, are spiny-skinned marine animals that live on the sea floor. They are the only deuterostome invertebrates, a characteristic that shows relatedness with the chordates. Although adult echinoderms are radially symmetrical, their larvae are bilaterally symmetrical.

REVIEW QUESTIONS

1. What do you and jellyfish have in common? What is an important taxonomic difference between the two of you?
2. "Animals are a diverse group of eukaryotic, multicellular, heterotrophic organisms." Explain this statement.
3. Fill in the blanks: If an organism's body parts all emerge from a central point, it exhibits _____. If an organism has two sides that are a mirror image of one another, it shows _____ _____.
4. Distinguish among coelomates, pseudocoelomates, and acoelomates. Give an example of each.
5. Which organisms are considered the "simplest animals"? Why?
6. What two basic body plans are shown by cnidarians? Summarize the differences between these two plans.
7. Which organisms are the most primitive bilaterally symmetrical animals? Briefly describe these animals' structure.
8. People can become very ill if infested by roundworms. Describe these organisms and why they can be a health risk to humans.
9. Distinguish between open and closed circulatory systems. Which do you have?
10. Summarize the characteristics of mollusks.
11. If you've ever tried to swat a fly, you know that flies are quick to respond to even the slightest movements. What feature of a fly's body allows it to perceive motion so rapidly?
12. Briefly summarize the characteristics of arthropods.
13. In casual conversation, you refer to a spider on the wall as an insect. A friend who's studied biology informs you that spiders are not insects. Explain what she means.
14. How do echinoderms differ from all other invertebrates? What is the significance of this fact?

KEY TERMS

acoelomate 563
bilateral symmetry 562
blood sinus 571
cerebral ganglion 574
clitellum 574
closed circulatory system 571
coelom 562
coelomate 562
collar cell 565
compound eye 576
deuterostome 564
ectoderm 568
endoderm 568
epithelial wall 565
exoskeleton 574
gills 571
hermaphrodite 566
malpighian tubule 576
mandible 574
mantle 571
medusa 567
mesoderm 568
nephridium (pl., nephridia) 573
nerve ring 578
open circulatory system 571
polyp 567
protostome 564
pseudocoelomate 562
radial canal 577
radial symmetry 562
septa 573
setae 574
simple eye 576
spiracle 576
tracheae 576
tracheole 576
visceral mass 571

THOUGHT QUESTIONS

1. Coral reefs are among the most diverse communities of organisms on Earth. Why do you think this is so? Why aren't free-living marine communities as diverse?
2. Pseudocoelomates evolved before coelomates. Why do you think the coelomates have been so much more successful?

FOR FURTHER READING

Bavendam, F. (1989, August). Even for ethereal phantasms, it's a dog-eat-dog world. *Smithsonian*, pp. 94-101.
Aspects of the lives of some of the incredibly beautiful sea slugs, or nudibranchs, are explained.

Wilson, E.O. (1990, March). Empire of the ants. *Discover*, pp. 45-50.
This is an outstanding essay on the fascinating ways of ants.

33 Vertebrate animals
Patterns of structure, function, and reproduction

Although bats often conjure up images of blood-sucking vampires, these bats would be totally disinterested in your neck. Nicknamed "little red flying-foxes," they find the eucalyptus blossoms of their Australian homeland a delicacy. When this favorite food is in short supply, they often raid orchards. Although these bats can become pests, you can leave the vampire-fighting garlic and pointed stakes at home if you venture on a trip "Down Under." Little red flying foxes only bother the fruit-growers.

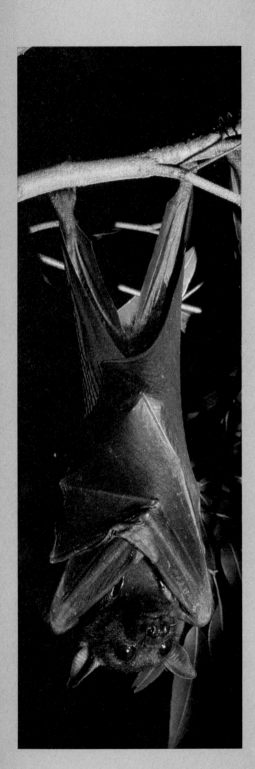

HIGHLIGHTS

▼

Chordates are organisms that, at some time during their development, have a single, hollow nerve cord along the back, a rod-shaped notochord between the nerve cord and the gut, and gill arches located at the throat.

▼

There are three subphyla of chordates: the tunicates, sac-like sessile marine organisms; the lancelets, tiny scaleless fish-like organisms, and the vertebrates.

▼

A vertebrate has a vertebral column surrounding the dorsal nerve cord, a distinct head with a skull that encases the brain, a closed circulatory system with a heart to pump the blood, and many other complex organs and organ systems.

▼

There are seven classes of living vertebrates: three classes are fishes and four classes are land-dwelling, four-footed animals.

OUTLINE

Chordates and vertebrates: Unity in symmetry and coelom
  Tunicates
  Lancelets
  Vertebrates
    Jawless fishes
    Cartilaginous fishes
    Bony fishes
    Amphibians
    Reptiles
    Birds
    Mammals

Interestingly, bats are mammals—the same class of animals to which humans belong. Bats are more like you than they are like birds! As this chapter progresses, you will come to understand those similarities as you explore the patterns of structure, function, and reproduction of the vertebrates.

Chordates and vertebrates: Unity in symmetry and coelom

The chordates and vertebrates are characterized by three principal features (see Chapter 27): (1) a single, hollow **nerve cord** located along the back; (2) a rod-shaped **notochord,** which forms between the nerve cord and the gut (stomach and intestines) during development; and (3) **pharyngeal (gill) arches,** which are located at the throat (pharynx). (These three chordate features are shown in Figure 27-3 as they appear in the embryo because they are present in the embryos of all chordates.) In addition, lancelets exhibit these characteristics as adults, and tunicates exhibit them as larvae. In vertebrates, the nerve cord differentiates into a brain and spinal cord. The notochord develops into a vertebral column that encases the nerve cord, protecting it. The pharyngeal arches develop into the gill structures of the fishes and into ear, jaw, and throat structures of the terrestrial vertebrates. The presence of the gill arches in all vertebrate embryos provides a clue to the aquatic ancestry of the subphylum Vertebrata.

Along with having these three traits, chordates have many other characteristics in common. All have a true coelom and bilateral symmetry. The embryos of chordates exhibit segmentation. Figure 33-1 shows segments of tissue called *somites* in the human embryo, which develop into the skeletal muscles. Most chordates have an internal skeleton to which their muscles are attached and work against, providing movement. (Larval tunicates and adult lancelets do not have internal skeletons; their muscles are attached to their notochords.) Finally, chordates have tails that extend beyond the anus, at least during embryonic development. Nearly all other animals have a terminal anus.

> Chordates are characterized by a single, hollow nerve cord located along the back, a rod-shaped notochord, and gill arches located at the throat. The vertebrates are an important subphylum of the chordates.

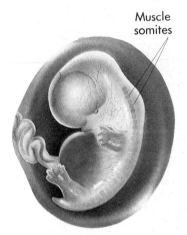

FIGURE 33-1   A chordate embryo.
All chordate embryos have segments of tissue called *somites.* The somites eventually develop into the skeletal muscles.

Tunicates

The **tunicates** are a group of about 2500 species of marine animals, most of which look like living sacs attached to the floor of the ocean (Figure 33-2, *A, B*). As shown in Figure 33-2, *C,* the tunicates are not much more than a large pharynx covered with a protective **tunic.** The tunic is a tough outer "skin" composed mainly of cellulose, a substance found in the cell walls of plants and algae but rarely found in animals. Some colonial tunicates live in masses on the ocean floor and have a common sac and a common opening to the outside (Figure 33-3). Colonial tunicates reproduce asexually by budding. Individual tunicates are hermaphodites, with each organism having both male and female sex organs.

The pharynx of a tunicate is lined with numerous cilia. As these cilia beat, they draw a stream of water through the incurrent siphon into the pharynx, which is lined with a sticky mucus. Food particles are trapped within the pharynx, and the filtered water flows out of the animal through the excurrent siphon. Because 90% of all species of tunicate have this structure and forcefully squirt water out their excurrent siphons when disturbed, they are also called *sea squirts.*

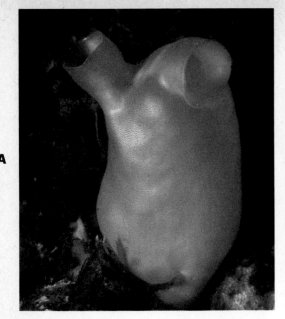
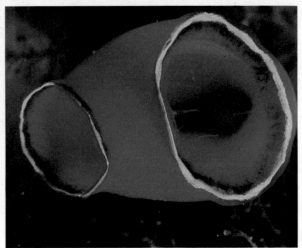
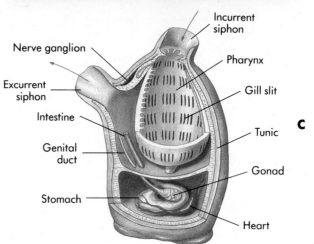

FIGURE 33-2 Tunicates (subphylum Urochordata).
A The sea perch *Halocynthia auranthium.*
B A beautiful blue and gold tunicate.
C The structure of a tunicate.

FIGURE 33-3 Colonial tunicates.
This colony was found at a depth of about 2 meters. The colony on the right is in the process of dividing.

As adults, tunicates lack a notochord and a nerve cord. The gill slits (which develop from the gill arches) are the only clue that adult tunicates are chordates. Only the larvae, which look like tadpoles (Figure 33-4), have notochords and nerve cords. The notochords are in their tails, and their nerve cords run dorsal to their notochords almost the entire length of their bodies. The subphylum name *Urochordate* comes from the placement of the larval notochord and literally means "tail chordate." The larvae remain free swimming for no more than a few days. Then they settle to the bottom and become sessile, attaching themselves to a suitable substrate by means of a sucker. As they mature, they adjust to a filter-feeding existence. Some tunicates that live in the world's warmer regions retain the ability to swim and never develop into a sessile form.

> Adult tunicates are sac-like, sessile, marine filter-feeders. Their only chordate characteristic is their gill slits. Their larvae, however, have notochords and nerve cords in the tails of their tadpole-like bodies.

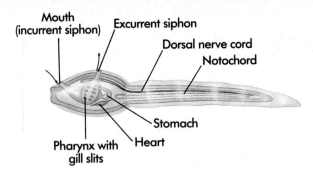

FIGURE 33-4 The structure of a larval tunicate.
The larvae of tunicates resemble tadpoles and contain notochords and nerve cords.

Lancelets

The **lancelets** are tiny, scaleless, fish-like marine chordates that are just a few centimeters long and pointed at both ends (Figure 33-5). They look very much like tiny surgical blades called *lancets,* from which they get their name. You may have had a few drops of blood taken in the doctor's office from a "fingerstick" done with a lancet.

Lancelets have a segmented appearance because of blocks of muscle tissue that are easily seen through their thin, unpigmented skin. Although they have pigmented light receptors, lancelets have no real head, eyes, nose, or ears. Unlike the the tunicates, the lancelet's notochord runs the entire length of its dorsal nerve cord. For this reason, lancelets are called *cephalochodates,* or *head chordates*. The lancelet retains its notochord throughout its lifespan.

The 23 species of lancelets live in the shallow waters of oceans all over the world. They spend most of their time partly buried in the sandy or muddy bottom with only their anterior ends protruding, feeding on plankton, floating microscopic plants and animals. In a manner similar to the tunicates, lancelets filter plankton from the water. Cilia line the anterior ends of their alimentary canals, and these beating cilia create an incoming current of water. The filtered water exits at an excurrent siphon. An oral hood projects beyond the mouth, or incurrent siphon, and bears sensory tentacles. The males and females are separate, but no obvious external differences exist between them.

> The lancelets are scaleless, fish-like marine chordates. The adult forms exhibit chordate characteristics.

Vertebrates

The **vertebrates** differ from the other chordates in that most have vertebral columns in place of notochords. A vertebral column, or backbone, is a stack of bones, each with a hole in its center, that forms a cylinder surrounding and protecting the dorsal nerve cord. (Each bone in the column is a vertebra.) One class of vertebrates that does not have a vertebral column is the jawless fishes. Present-day jawless

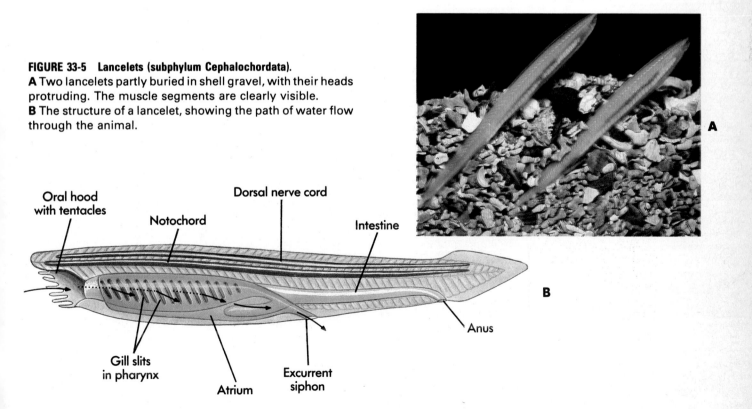

FIGURE 33-5 Lancelets (subphylum Cephalochordata).
A Two lancelets partly buried in shell gravel, with their heads protruding. The muscle segments are clearly visible.
B The structure of a lancelet, showing the path of water flow through the animal.

fishes have notochords, but their ancestors had bony skeletons and vertebral columns. Another class, the cartilaginous fishes, have skeletons and vertebral columns composed of cartilage (a tough yet elastic type of connective tissue) rather than bone.

In addition to having vertebral columns, vertebrates have distinct heads with skulls that encase their brains. They have closed circulatory systems (their blood flows within vessels) and a heart to pump the blood. Most vertebrates also have livers, kidneys, and endocrine glands. (Endocrine glands are ductless glands that secrete hormones, which play a critical role in controlling the functions of the vertebrate body.)

The human body plan is respresentative of the vertebrate body plan. Even though vertebrates have many similarities, they are a diverse group, consisting of animals adapted to life in the sea, on land, and in the air. There are seven classes of living vertebrates: three classes are fishes, and four classes are land-dwelling **tetrapods**, or four-footed animals.

> Vertebrates are a subphylum of chordates characterized by a vertebral column surrounding a dorsal nerve cord.

Jawless fishes

Other than the lamprey eels and hagfishes, the major groups of the **jawless fishes,** or agnatha (*a* meaning "not," *gnatha* meaning "jaws"), have been extinct for hundreds of millions of years. Only about 20 to 30 species of each of these two groups are alive today. Both groups are long, tube-like aquatic animals that usually live in the sea or in brackish (somewhat salty) water where the fresh water of a river meets the ocean. In addition to their lack of jaws, they have no paired fins to help them swim, and they have no scales. They do have notochords, however, and portions of cartilaginous skeletons that are remnants from their extinct ancestors.

Lampreys parasitize other fish. In fact, they are the only parasitic vertebrates. They have round mouths that function like suction cups (see Figure 27-5), attaching them to their prey—the bony fishes. When a lamprey attaches to a fish, it uses its spine-covered tongue like a grater, rasping a hole through the skin of the fish and then sucking out its body fluids. Sometimes lampreys can be so abundant that they are a serious threat to commercial fisheries, preying on salmon, trout, and other commercially valuable fishes. Entering the Great Lakes from the sea, they have become important pests there; millions of dollars are spent annually on their control. The hagfishes, although similar in size and shape to lampreys, are not parasitic. They are scavengers, often feeding on the insides of dead or dying fishes or large invertebrates.

> Lampreys and hagfish are tubular, scaleless, finless, jawless organisms that live in the sea or in brackish water. Lampreys parasitize bony fish and pose a serious threat to some commercial fisheries. Hagfish are scavengers.

To reproduce, jawless fishes **spawn** as do most fishes, amphibians, and shellfish. During spawning, the males and females deposit eggs and sperm directly into the water. Fertilization that takes place outside the body of the female is called **external fertilization.** Lampreys swim upstream to spawn as salmon do and create nests in which they deposit eggs and sperm. The fertilized eggs develop into larvae that feed on plankton. Over a period of years, the larvae mature and metamorphose (change) into parasitic adults. In contrast, the eggs of hagfishes do not develop into larvae. Completely formed hagfish hatch directly from fertilized eggs.

> Most fishes, amphibians, and shellfish have an external type of fertilization. Eggs are fertilized by sperm outside the body of the organism.

Cartilaginous fishes

The chondrichthyes (*chondri* meaning "cartilage," *ichthyes* meaning "fishes") include the sharks, skates, and rays (Figure 33-6). Hundreds of extinct species of **cartilaginous fishes** are known from the fossil record, but less than 800 species exist today.

FIGURE 33-6 Cartilaginous fishes (class Chondrichthyes).
Members of this class spend most of their time in graceful motion.
A Blue shark.
B Diamond sting ray.
C Manta ray.

Sharks, skates, and rays are all marine fishes. The skin of sharks (as well as the skates and rays) is covered with small, pointed, tooth-like scales called **denticles,** which give the skin a sandpaper texture. They have streamlined bodies and two pairs of fins: pectoral fins just behind the gills and pelvic fins just in front of the anal region. The dorsal (back) fins provide stability while motions of the other fins (including the asymmetrical tail fin) and sinuous motions of the whole body give the shark lift and propel it through the water.

Many sharks are predators and eat large fishes and marine mammals. Their sharp, triangular teeth saw and rip off pieces of flesh as the shark thrashes its head from side to side. Some sharks feed on plankton rather than prey on other animals. These sharks swim with their mouths open, and the plankton is strained from the water by specialized denticles on the inner surfaces of the gill arches.

Skates and rays are generally smaller than sharks and have flattened bodies with enlarged pectoral fins (see Figure 33-7, *B, C*) that undulate when these fishes move. Their tails, which are not the principal means of locomotion that they are in skarks, are thin and whip-like. They are sometimes armed with poisonous spines that are used as defense mechanisms rather than for predation. These animals have a mouth on their underside and feed mainly on invertebrates on or near the ocean floor.

> The cartilaginous fishes are all marine fishes and include the sharks, skates, and rays. The skin is covered with small, pointed, tooth-like scales that result in a sandpaper texture. Many sharks are predators, but skates and rays feed on invertebrates on the ocean floor.

The sensory systems of sharks, skates, and rays are quite sophisticated and diverse. Sharks have a lateral line system as all fishes and amphibians do. A **lateral line system** is a complex system of *mechanoreceptors* that lies in a single row along the sides of the body and in patterns on the head. These receptors can detect mechanical stimuli such as sound, pressure, and movement. Sharks often detect prey via their lateral line systems, but they can also detect electrical fields emitted by other fishes using their electroreceptors. Sharks have these receptors on their heads, and rays have them on their pectoral fins. Scientists think these fishes may use their electroreceptors for navigation as well. Chemoreception is another important sense in the chondricthyes. In fact, sharks have been described as "swimming noses" because of their acute sense of smell. Vision is also important to the feeding behavior of sharks; they have well-developed mechanisms for vision at low-light intensities.

> All fishes and amphibians have a complex system of mechanoreceptors called a *lateral line system*. These receptors, which lie in a row on the sides of the body and on the head, detect mechanical stimuli such as sound, pressure, and movement.

Another important characteristic of fishes is their ability to regulate their buoyancy, or ability to float at various depths in the water. Bony fishes have gas-filled *swim bladders* that regulate their buoyancy. Cartilaginous fishes do not have swim bladders but can adjust the size and oil content of their livers. Because oil is less dense than water and the liver has a high oil content, adjusting the oil content and size of the liver can regulate buoyancy to a certain degree. Most species of sharks, however, swim continually to keep from sinking and to keep water flowing over their gills.

Aquatic organisms show various adaptations regarding **osmoregulation,** the control of water movement into and out of their bodies. Marine organisms tend to lose water to their surroundings because their body fluids usually have a solute concentration (concentration of dissolved substances) lower than that of seawater. The cartilaginous fishes maintain solute concentrations close to that of seawater, however, because they change potentially toxic nitrogen-containing wastes (such as ammonia) into a less toxic compound called *urea* and retain it in their bodies rather than excreting it (Figure 33-7). (In fact, shark meat must be soaked in fresh water before it is eaten to remove most of the urea.) With similar solute concentrations both inside and outside the fish, water movement remains relatively equal in both directions.

Although most aquatic animals reproduce using external fertilization, the cartilaginous fishes have developed a method of internal fertilization. These fishes have *pelvic claspers,* which are rod-like projections between the pelvic fins of

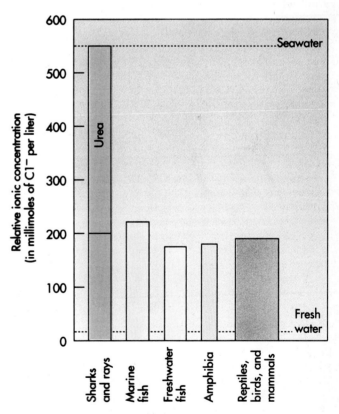

FIGURE 33-7 Ion concentrations for different classes of vertebrates. Ion concentrations in the bodies of different classes of vertebrates are roughly similar, with the exceptions of sharks and rays. Sharks and rays keep their ion concentrations close to that of seawater by adding urea to the bloodstream.

FIGURE 33-8 A viviparous fish giving birth.
Viviparous fishes such as this lemon shark carry live young within their bodies. The young complete their development inside their mother's body and are then released as small but competent adults.

the male fish. During *copulation,* or coupling of the male and female animals, the male inserts his clasper into the female's **cloaca.** The cloaca is the terminal part of the gut into which ducts from the kidney and reproductive systems open. The fish either swim side-by-side, or the male wraps around the female or holds on to her with his jaws. He then ejaculates into a groove in the clasper, which directs the sperm into the female's cloaca. The sperm then swim up the female's reproductive tract.

In addition to having a mechanism of internal fertilization, the cartilaginous fishes are ovoviviparous. **Ovoviviparous** organisms retain fertilized eggs within their oviducts until the young hatch (*ovum* meaning "egg," *vivus* meaning "alive," *pario* meaning "to bring forth"). However, although the young are born alive, they do not receive nutrition from the mother but from the yolk of the egg. The mother acts like an internal "nest" for the eggs until the young hatch. Certain fishes, some reptiles, and many insects are ovoviviparous. In contrast, **oviparous** organisms lay their eggs, and the young hatch from the eggs outside of the mother. Many skates and rays, for example, release eggs within protective egg cases. Birds and most reptiles, along with some of the cartilaginous fishes, are oviparous. **Viviparous** organisms such as mammals bear young alive as do ovoviviparous organisms, but the developing embryos primarily derive nourishment from the mother and not the egg. A few sharks, such as hammerhead sharks, blue sharks, and lemon sharks, are viviparous fishes (Figure 33-8).

> Most cartilaginous fishes fertilize their eggs internally and are ovoviviparous, retaining fertilized eggs within their oviducts until the young hatch.

Bony fishes

The vast majority of known species of fishes belong to the class Osteichthyes (*oste* meaning "bone"). **Bony fishes** live in both salt and fresh water, with many species spending a portion of their lives in fresh water and another portion in the sea.

The osteichthyes get their name from their bony internal skeletons. However, the skin of osteichthyes is also covered with thin, overlapping bony scales. Sometimes these scales have spiny edges (see Figure 27-7, *A, B*), which provide some protection for the animal. Bony fishes also have a protective flap that extends posteriorly from the head and protects the gills. This flap is called the **operculum.** Along with protecting the gills, the movement of the operculum enhances the flow of water over the gills, bringing more oxygen in contact with the gas-exchanging surfaces and allowing a fish to breathe while stationary. Unlike most of the cartilaginous fishes, bony fishes can remain motionless at various depths because of their ability to more finely regulate their buoyancy with their swim bladders.

The swim bladders of most bony fishes, the *ray-finned fishes* (the largest subclass of bony fishes), evolved from lunglike sacs of their ancestors. These sacs aided in respiration. In the ray-finned fishes, gas exchange between the swim bladder and the blood regulates its density and allows it to remain suspended in water without sinking to the bottom. The ray-finned fishes are the most familiar fish and include perch, cod, trout, tuna, herring, and salmon. Another subclass of bony fishes continues to use lungs to aid the gills in breathing: the *lobe-finned fishes*. Only four genera of lobe-finned fishes exist today; all are found in the southern hemisphere. Ancestors of this subclass of fishes gave rise to land-dwelling tetrapods.

> Bony fishes live in both salt and fresh water and get their name from their bony internal skeletons. By means of gas exchange between their swim bladders and their blood, bony fishes can regulate their buoyancy better than the cartilaginous fishes do. Cartilaginous fishes regulate buoyancy by controlling the amount of oil in their livers.

All chordates have closed circulatory systems with a system of blood vessels and a pump to push the blood through these vessels. The cartilaginous fishes and the bony fishes have tube-like hearts with four chambers, one right after the other (see Figure 33-12, *A*), but these four chambers really function as a two-chambered heart. The first two chambers (**sinus venosus** and **atrium**) collect blood from the organs. The second two (**ventricle** and **conus arteriosus**) pump blood to the gills. From the gills the blood moves to the rest of the body. Because of the great resistance in the narrow passageways of the capillaries at the gills, the movement of blood to the rest of the body is sluggish.

Although the cartilaginous fishes maintain a solute concentration of their body fluids near that of seawater, bony fishes do not. Only some bony fishes live in the sea; many groups live in fresh water. Marine and fresh water bony fishes have opposite situations regarding osmoregulation. The body fluids of marine fishes are *hypoosmotic* with respect to seawater. That is, their body fluids contain a lower concentration of dissolved substances than seawater does (see Figure 33-7). Consequently, water tends to leave these fishes (at the gill epithelium) by osmosis. To regulate water balance, marine fishes drink seawater and then excrete the salt by means of active transport at the gill epithelium. The fish kidney is not able to get rid of excess salt in the urine because it has no loop of Henle (see Chapter 14). This part of the kidney nephron works to concentrate urine and enable an organism to produce urine with a high solute concentration. Only birds and mammals have loops of Henle.

In contrast to marine fishes, the body fluids of fresh water fishes are *hyperosmotic* with respect to their fresh water environment. That is, their body fluids contain a higher concentration of dissolved substances than fresh water does. Consequently, water tends to enter fresh water fishes by osmosis. To regulate water balance, fresh water fishes do not drink water and excrete large amounts of very dilute urine. They reclaim some of the ions they lose in their urine by the uptake of sodium and chlorine ions by the gills and in the food they eat.

Because they tend to lose water by osmosis, marine fishes drink seawater and excrete salt at the gill epithelium. Because they tend to take on water, fresh water fishes do not drink water and excrete large amounts of very dilute urine.

Most osteichtheyes are oviparous, fertilizing their eggs externally. These eggs are often food for other marine organisms, however, and must survive other risks such as drying out. Most fishes (as well as other organisms that externally fertilize their eggs) lay large numbers of eggs, with some surviving these dangers. Many species of fishes build nests for their eggs and watch over them, which also enhances the chances of survival.

Amphibians

The word *amphibian* means "two lives" (*amphi* meaning "both," *bios* meaning "life") and refers to both the aquatic and terrestrial existence of this class of animals (Figure 33-

FIGURE 33-9 Amphibians (class Amphibia).
Examples of amphibians that have tails are the spotted salamander **(A)** and the Tennessee cave salamander **(B)**. Examples of amphibians without tails, the frogs and toads, are the pickeral frog, shown here in tadpole form **(C)**; the giant toad **(D)**; and the red-eyed tree frog **(E)**.

9). **Amphibians** (unlike reptiles, birds, and mammals) depend on water during their early stages of development. Many amphibians live in moist places like swamps and in tropical areas even when they are mature, which lessens the constant loss of water through their thin skins. The two most familiar orders of amphibians are those that have tails—the salamanders, mudpuppies, and newts—and those that do not have tails—the frogs and toads.

Most frogs and toads fertilize their eggs externally. The male grasps the female and sheds sperm over the eggs as they are expelled from the female. Most salamanders use internal fertilization but are still oviparous (lay their eggs). Because amphibian eggs have no shells or membranes to keep them from drying out, amphibians lay their eggs directly in water or in moist places. Figure 33-10, A shows the "nests" of foam that tropical tree frogs create to incubate their eggs. When the tadpoles develop, they drop from the tree branches into the water (Figure 33-10, B). Some amphibians protect their eggs by incubating them in their mouths, on their backs, or even in their stomachs! A few amphibian species are ovoviviparous and incubate the eggs within their reproductive organs until they hatch, and a few are viviparous.

The young of frogs and toads undergo *metamorphosis*, or change, during development from a larval to an adult form. The larvae are immature forms that do not look like the adult. The larvae of frogs and toads are tadpoles, which usually live in the water and have internal gills and a lateral line system like that of fishes. They feed on minute algae. These fish-like forms develop into carnivorous adults having legs and lungs; their gills and lateral line system disappear. The lungs of the adults are inefficient, however. Much of the gas exchange takes place across the skin and on the surfaces of the mouth. The skin of amphibians must therefore remain damp to allow gases to diffuse in and out.

The adults of certain salamanders (such as mudpuppies) live permanently in the water and retain gills and other larval features as adults. Other salamanders are terrestrial but return to water to breed. They, like frogs and toads, usually live in moist places such as under stones or logs or among the leaves of certain tropical plants.

> **Amphibians live both aquatic and terrestrial existences. Because they fertilize their eggs externally and their eggs have no shells or membranes, amphibians lay their eggs in water or in moist places. The young of frogs and toads change from a larval form to an adult form during their development.**

Along with having lungs rather than gills (in most cases), amphibians have a pattern of blood circulation different from that of the fishes. After the blood is pumped through the fine network of capillaries in the amphibian lungs, it does not flow directly to the body as it does in fishes. Instead, it returns to the heart. It is then pumped out to the body at a much higher pressure than if it were not returned to the heart. However, the oxygenated blood that returns to the heart from the lungs mixes with the deoxygenated blood returning to the heart from the rest of the body. Consequently, the heart pumps out a mixture of oxygenated and deoxygenated blood rather than fully oxygenated blood. Figure 33-12, B shows this pathway of blood flow. The blood from the lungs and the blood from the body enter the right and left atria of the heart, respectively. The blood in both these chambers flows into the single ventricle of the three-chambered amphibian heart and is pumped through two large vessels to both the lungs and the body.

> **Amphibians have a three-chambered heart rather than the two-chambered heart of fishes. However, oxygenated and deoxygenated blood mix in the heart.**

Reptiles

The three major orders of **reptiles** are the crocodiles and alligators, the turtles and tortoises, and the lizards and snakes (Figure 33-11). Reptiles have dry skins covered with scales that help retard water loss. As a result, reptiles can live in a wider variety of environments on land than amphibians can, but the crocodiles, alligators, and turtles are aquatic organisms.

The hearts of reptiles differ from amphibian hearts in that a partition called a *septum* subdivides the ventricle, the pumping chamber of the heart. The septum reduces the mixing of oxygenated and deoxygenated blood in the heart.

FIGURE 33-10 An amphibian egg-protection strategy. To keep their eggs from drying out, tropical tree frogs lay their eggs in nests of foam **(A)**. When the tadpoles hatch, they drop from their foam nest into a pond of water below **(B)**.

FIGURE 33-11 Reptiles (class Reptilia).
A The river crocodile. Crocodiles, like birds, are related to the dinosaurs.
B Red-bellied turtles. This attractive turtle is found frequently in the northeastern United States.
C An Australian skink. Although this species of burrowing lizard has legs, some species lack legs. Snakes evolved from one line of legless lizard.
D Smooth green snake.

In most crocodiles, the separation is complete. Figure 33-12, *C* shows that the reptilian heart closely resembles the four-chambered heart of birds and mammals shown in Figure 33-12, *D*. It is still considered to be a three-chambered heart, however.

One of the most critical adaptations of reptiles to life on land is the evolution of the shelled **amniotic egg.** Amniotic eggs are also characteristic of birds and egg-laying mammals (monotremes). The amniotic egg protects the embryo from drying out, nourishes it, and enables it to develop outside of water (Figure 33-13). This type of egg contains a yolk and albumin (egg white). The yolk is the primary food supply for the embryo, and the albumin provides additional nutrients and water. The embryo's nitrogenous wastes are excreted into the allantois, a sac that grows out of the embryonic gut. Blood vessels grow out of the embryo through the sac surrounding the yolk and through the allantois to the egg's surface, where gas exchange takes place. The amnion surrounds the developing embryo, enclosing a liquid-filled space within which the embryo develops and is protected. Lying just within the shell, a membrane called the *chorion* surrounds the embryo, amnion, yolk sac, and allantois and, along with the shell, controls the movement of gases into and out of the egg. In most reptiles, the eggshell is leathery, unlike the hard shell of bird eggs. Because of this difference, reptile eggs are somewhat permeable to water, whereas bird eggs are not.

Reptiles, like amphibians and fishes (but unlike birds and mammals), are **ectothermic** (*ectos* meaning "outside," *thermos* meaning "heat"). Ectothermic animals regulate body temperature by taking in heat from the environment. Even though ecotothermic animals are often called "cold blooded" (a misleading term), they often maintain body temperatures much warmer than their surroundings. Did you ever wonder why fish do not freeze in water near the freezing temperature? Interestingly, fishes (as well as certain lizards, invertebrates, and plants) produce their own internal antifreeze—chemical compounds that lower the freezing temperature of the body fluids of an organism. Along with such physiological adaptations, ectothermic animals protect themselves against the cold in behavioral ways. For example, frogs help protect themselves against freezing by spending the winter buried in the soil or in the mud at the bottom of ponds. Ectothermic animals also protect themselves from high heat by burrowing under rocks or remaining in shady, somewhat cooler areas. Desert tortoises, for example, construct shallow burrows to stay in during the summer and deeper burrows for hibernation in winter. Reptiles often

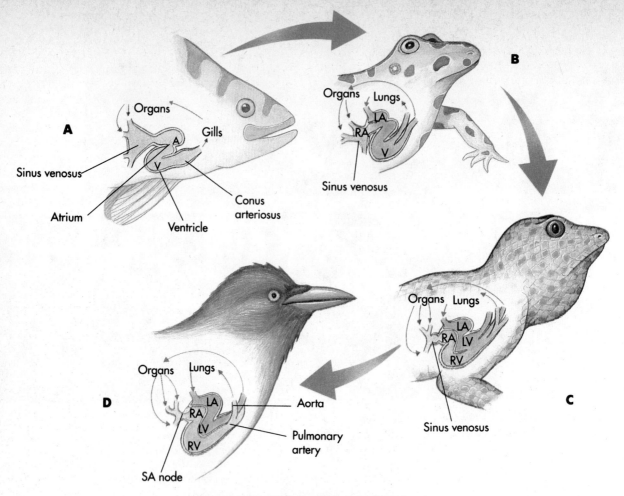

FIGURE 33-12 Evolution of the vertebrate heart.
A Fishes (two chambers).
B Amphibians (two chambers).
C Reptiles (three chambers).
D Birds and mammals (four chambers).

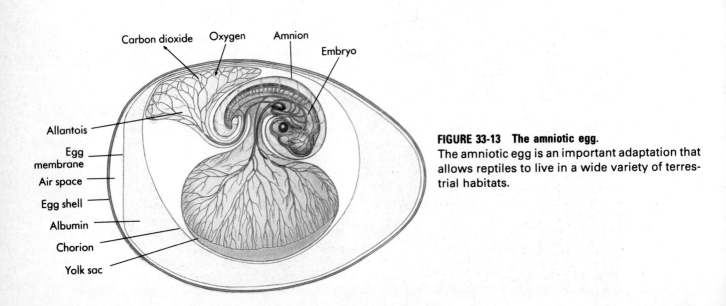

FIGURE 33-13 The amniotic egg.
The amniotic egg is an important adaptation that allows reptiles to live in a wide variety of terrestrial habitats.

THE DIVERSITY AND UNITY OF LIVING THINGS

bask in the sun, which raises their body temperature and their metabolic rate. When cold-blooded animals are cold, the metabolic rate slows down and they are unable to hunt for food or move about very quickly.

> **Reptiles have dry skins covered with scales that help retard water loss. Another critical adaptation to their life on land (although some reptiles are aquatic) is the development of the amniotic egg, which protects the embryo from drying out, nourishes it, and enables it to develop outside of water. Amniotic eggs are also characteristic of birds and egg-laying mammals.**

Birds

There are approximately 9000 species of birds living today (Figure 33-14). In birds, the wings are homologous to the forearms of other vertebrates. That is, they are derived from the same evolutionary origin but have been modified in the course of evolution. Birds have reptilian-like scales on their legs and lay amniotic eggs as reptiles do. They have hard, horny extensions of the mouth called **beaks** that tear, chisel, or crush their food. They also have digestive organs called **gizzards,** often filled with grit, that grind food. Beaks are not limited to birds. Many reptiles have beaks, including turtles, and so do some fishes, including the parrot fish, which uses its beak to rip away fragments of coral reef. Birds, however, are the only animals that have feathers.

FIGURE 33-14 Birds (class Aves).
The birds are a large and successful group of about 9000 species, more than any other class of vertebrates except the bony fishes.
A Ostrich. The group of flightless birds includes the ostrich (Africa), rhea (South America), emu (Australia), and kiwi (New Zealand). Flightless birds seem to represent an evolutionary line distinct from all other birds.
B A bee-eater in Tanzania. Bee-eaters nest communally, with many adult individuals participating in the care of the young.
C Northern saw-whet owl. Owls, like hawks, hunt other animals (prey) for food.
D A pair of wood ducks, showing the marked difference between males and females.
E Tufted puffin, one of many unique groups of sea birds.

VERTEBRATE ANIMALS: PATTERNS OF STRUCTURE, FUNCTION, AND REPRODUCTION

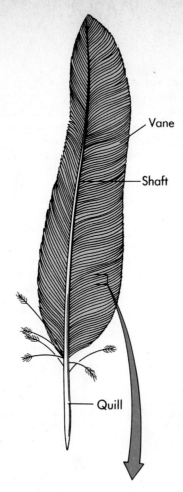

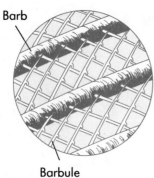

FIGURE 33-15 A feather.
The enlarged view shows the way in which the vanes are linked together by the microscopic barbules.

Feathers are flexible, light, waterproof epidermal structures. Several types of feathers form the body covering of birds, including contour feathers and down feathers. Contour feathers are flat (except for a fluffy, downy portion at the base) and are held together by tiny barbules (Figure 33-15). These feathers provide a streamlined surface for flight, and some are modified to reduce drag on the wings or act like individual propeller blades. In addition, birds can alter the area and shape of their wings by altering the positions of their feathers. Feathers also provide birds with waterproof coats and play an important role in insulating birds against temperature changes.

Along with feathers, birds have light, hollow bones adapted to flight. Birds also have highly efficient lungs that supply the large amounts of oxygen necessary to sustain muscle contraction during prolonged flight. Unlike fishes, amphibians, and reptiles (except for the crocodiles), birds and mammals have hearts that act as a double pump. The sides of the heart are completely separated with a septum (see Figure 33-12). The right side of the heart pumps blood to the lungs. The left side of the heart pumps blood to the body. Oxygenated blood and deoxygenated blood do not mix. As you compare the pathway of blood in the hearts of fishes, amphibians, reptiles, and birds in Figure 33-12, note that the four-chambered heart of birds and mammals evolved from only two of the chambers of the fish heart.

Birds and mammals are **endothermic** (*endo* meaning "within"); they regulate body temperature internally. The evolution of the four-chambered heart with separate pathways to the lungs and the body is thought to have been important in the evolution of endothermy in the birds and mammals. More efficient circulation is necessary to support the great increase in metabolic rate that is required to generate body heat internally. In addition, blood is the carrier of heat in the body, and an efficient circulatory system is required to distribute heat evenly throughout the body. Although endothermic animals are sometimes called "warm blooded," body temperature may often be cooler than that of the surroundings. Because endotherms maintain high internal body temperatures (37° C [98.6° F] in humans, for example), their internal temperatures are usually higher than that of the environment.

Endotherms maintain a constant high body temperature by adjusting heat production to equal heat loss from their bodies under various environmental conditions. The high metabolic rate of endotherms and the energy released during these chemical reactions produces much of this body heat. Increasing the action of skeletal muscles increases the metabolic rate and the amount of heat produced. Shivering in the cold, for example, is an action that produces body heat. But because endotherms usually live in environments cooler than body temperature, restricting heat loss is usually the concern. Animals that live in extremely cold temperatures, such as arctic birds and polar bears, are well-insulated with either feathers or hair that traps air. Raising or lowering the feathers or hair adjusts the insulating capacity. Getting "goosebumps" when you are cold is your body's reaction to raise your hairs and increase your insulation. Although humans no longer have substantial body hair, this mechanism is important in reducing heat loss in other endothermic vertebrates.

> Birds are winged vertebrates that are covered with feathers and have scales on their legs. They have beaks, which are horny extensions of the mouth. Their light, hollow bones are an adaptation to flight. They have a four-chambered heart as mammals do. Unlike the reptiles, amphibians, and fishes, birds can regulate body temperature internally.

Mammals

There are about 4500 species of living **mammals**, including human beings. Mammals are endothermic vertebrates that have hair and whose females secrete milk from mammary glands to feed their young. Mammals, like birds and most crocodiles, have a four-chambered heart with circulation to the lungs and separate circulation to the body. The locomotion of mammals is advanced over that of the reptiles, which in turn is advanced over that of the amphibians; the legs of mammals are positioned much farther under the body than those of reptiles and are suspended from limb girdles, which permit greater leg mobility.

The evolution of specialized teeth in mammals represents a major evolutionary advance. In fishes, amphibians, and reptiles, all teeth are essentially the same size and shape; in mammals, evolutionary specialization has resulted in incisors, which are chisel-like teeth used for cutting; canines, which are used for gripping and tearing; and molars, which are used for crushing and breaking (see Figure 10-2, *A*).

There are three subclasses of mammals: monotremes, marsupials, and placental mammals. The only **monotremes** that exist are the duckbilled platypus and two genera of spiny anteaters (Figure 33-16). Monotremes lay eggs with leathery shells similar to those of reptiles. The platypus generally lays one egg and incubates it in a nest. The spiny anteater generally lays two eggs and incubates them in a pouch. When the young hatch, they feed on milk produced by specialized sweat glands of the mother. **Marsupials** are mammals in which the young are born early in their development and are retained in a pouch. After birth, the embryos crawl to the pouch and nurse there until they mature. The kangaroo and koala are familiar examples of marsupials (Figure 33-17). In **placental mammals,** the young develop to maturity within the mother. They are named for the first organ to form during the course of their embryonic development, the placenta. (The development of the human as an example of the development of a placental mammal is described in detail in Chapter 21.)

Placental mammals are extraordinarily diverse (Figure 33-18). There are 14 orders of mammals (see Figure 27-12), of which one is the primates, the order that includes monkeys, apes, and humans. The bats in the opening photograph (as well as all bats) belong to the order Chiroptera, a name that means the forelimbs are modified to form wings (*chiro* meaning "hand," *pteron* meaning "wing"). So although humans and bats are quite different, they have remarkable similarities in their circulatory, thermoregulatory, osmoregulatory, and reproductive patterns.

> Mammals are endothermic vertebrates that have hair and whose females secrete milk from mammary glands to feed their young. There are three subclasses of mammals: monotremes, marsupials, and placentals.

FIGURE 33-16 Monotremes (class Mammalia).
A Duckbilled platypus at the edge of a stream in Australia.
B Echidna.

FIGURE 33-17 Marsupials (class Mammalia).
A Kangaroo with young in its pouch.
B Koala in a eucalyptus tree.

FIGURE 33-18 Placental mammals (class Mammalia).
A Snow leopard, a cat.
B Starnosed mole, a burrowing insectivore.
C White-tailed deer, abundant in eastern and central temperate North America.
D Greater horseshoe bat in flight.
E Orca, a carnivorous whale.

Summary

1. The chordates are characterized by three main features: (1) a single, hollow nerve cord located along the back; (2) a rod-shaped notochord, which forms between the nerve cord and the gut (stomach and intestines) during development; and (3) pharyngeal (gill) arches, which are located at the throat (pharynx).

2. There are three subphyla of chordates: the tunicates (subphylum Urochordata), the lancelets (subphylum Cephalochordata), and the vertebrates (subphylum Vertebrata). The tunicates are sessile, sac-like, marine organisms that filter food from the surrounding water. Although the adults have gill slits, only their larvae have notochords and nerve cords. The lancelets are tiny, scaleless, fish-like marine chordates that are just a few centimeters long.

3. Vertebrates are a subphylum of chordates characterized by a vertebral column surrounding a dorsal nerve cord. Vertebrates have distinct heads with skulls that encase their brains, closed circulatory systems, and a heart to pump the blood. Most vertebrates also have livers, kidneys, and endocrine glands. In spite of these similarities, the vertebrates are an extremely diverse subphylum of organisms. Three classes of vertebrates are fishes; four are tetrapods—animals with four limbs.

4. The jawless fishes—lampreys and hagfishes—are long, tube-like animals that live in the sea or brackish water. They lack fins and scales. Lampreys are parasites of bony fishes, and hagfishes are scavengers.

5. The cartilaginous fishes—sharks, skates, and rays—are all marine fishes and are covered with tooth-like scales. They have sophisticated and diverse sensory systems. They fertilize their eggs internally, and most are ovoviparous, rataining fertilized eggs within their oviducts until the young hatch.

6. The vast majority of fishes are the bony fishes. Along with having bony internal skeletons, these fishes have thin, bony, plate-like scales. Both the bony fishes and the cartilaginous fishes have two-chambered hearts that pump blood to the gills. From there, the blood moves sluggishly around the body.

7. The amphibians live both in water and on land. Because their eggs have no shells or membranes to keep them from drying out, amphibians lay their eggs directly in water or moist places. The young of frogs and toads undergo change from larval to adult forms during development. The adults of certain salamanders live permanently in the water and retain gills and other larval features as adults. Although amphibians have a three-chambered heart, oxygenated and deoxygenated blood mix in the heart.

8. Reptiles are better adapted to life on land than the amphibians due to their dry, scaly skin that retards water loss and their shelled (amniotic) egg. The amniotic egg retains a watery environment within the egg while protecting and nourishing the developing embryo.

9. Fishes, amphibians, and reptiles are ectothermic; that is, they regulate body temperature by taking in heat from the environment. Ectothermic animals protect themselves from the cold and high heat in behavioral ways.

10. Birds are winged vertebrates that are covered with feathers and adapted to flight. They lay amniotic eggs like the reptiles but have a four-chambered heart like the mammals. Birds, like mammals and unlike reptiles, amphibians, and fishes, are endothermic; that is, they regulate body temperature internally.

11. Mammals are endothermic vertebrates that have hair and whose females secrete milk from mammary glands to feed their young.

REVIEW QUESTIONS

1. Fill in the blanks: The three major features that are characteristic of all vertebrates are: a(n) _____ _____, a(n) _____ _____, and _____ (_____) _____.
2. Into what structure(s) does each of the features in question 1 develop?

KEY TERMS

amniotic egg 589
amphibian 588
atrium 586
beak 591
bony fishes 586
cartilaginous fishes 584
cloaca 586
conus arteriosus 586
denticle 585
ectothermic 589
endothermic 592
external fertilization 584
gizzard 591
jawless fishes 584
lancelet 583
lateral line system 585
mammal 593
marsupial 593
monotreme 593
nerve cord 581
notochord 581
operculum 586
osmoregulation 585
oviparous 586
ovoviviparous 586
pharyngeal (gill) arches 581
placental mammal 593
reptile 588
sinus venosus 586
spawn 584
tetrapod 584
tunic 581
tunicate 581
ventricle 586
vertebrate 583
viviparous 586

3. Match each term to the most appropriate description:
   Terms
   (1) Jawless fishes
   (2) Tunicates
   (3) Lancelets
   Descriptions
   a. Scaleless, fish-like marine chordates that retain a notochord throughout life
   b. Tubular, scaleless, finless organisms; most members of this class have been extinct for millions of years
   c. Sac-like, sessile, marine filter-feeders
4. The "Jaws" movies starred members of which two classes of vertebrates?

THOUGHT QUESTIONS

1. How are the characteristics of the three subphyla of chordates related to the way of life of each group?
2. What limits the ability of amphibians to occupy the full range of terrestrial habitats and allows other terrestrial vertebrates to live in them successfully?
3. When you were an embryo, you had a notochord, just as a lancelet does. What do you think happened to it?

FOR FURTHER READING

Ehrlich, P. R., Dobkin, D. S., & Wheye, D. (1988). *The birder's handbook*. Simon and Schuster, Inc., New York.
A field guide to the natural history of North American birds, this book will greatly deepen your insight into the lives of these familiar animals.

Martin, J. (1990, June). The engaging habits of chameleons suggest mirth more than menace. *Smithsonian*, pp. 44-53.
The wonderful variety in a single group of tropical lizards is shown in this article.

5. What is a lateral line system? Which animals have one, and why is it important?
6. What is osmoregulation, and what is its significance?
7. Distinguish among ovoviviparous, oviparous, and viviparous.
8. What does the term *amphibian* mean?
9. Distinguish between ectothermic and endothermic. Give an example of an ectotherm and an endotherm.
10. Describe two important adaptations of reptiles to life on land.
12. What are feathers, and what functions do they perform for birds?
13. What characteristics differentiate mammals from other vertebrates?

Rismiller, P. O., & Seymour, R. S. (1991, February). The echidna. *Scientific American*.
This article outlines the natural history of this unusual and fascinating monotreme.

Waldvogel, J. A. (1990, July/August). The bird's eye view. *American Scientist*.
Scientists are trying to reconstruct the bird's excellent visual acuity in the laboratory.

PART EIGHT

HOW LIVING THINGS INTERACT WITH EACH OTHER AND WITH THEIR ENVIRONMENT

34 Innate behavior and learning in animals

HIGHLIGHTS

Ethology, the study of animal behavior in the natural environment, examines the biological basis of the patterns of movement, sounds, and body positions of animals.

All behaviors depend on nerve impulses, hormones, and other physiological mechanisms; behaviors are therefore directed in part by the genes that control the development of these systems and mechanisms.

Certain animal behaviors are unchangeable and are performed correctly the first time they are attempted; these innate, or instinctive, behaviors are directed by nerve pathways developed before birth.

Animals that can change their behaviors based on experience are capable of learning, a process that helps animals adapt to changes in the environment.

OUTLINE

Ethology: The biology of behavior
The link between genetics and behavior
Innate (instinctive) behaviors
  Kineses
  Taxes
  Reflexes
  Fixed action patterns
Learning
  Imprinting
  Habituation
  Trial-and-error learning (operant conditioning)
  Classical conditioning
  Insight

A case of mistaken identity? It would certainly seem to be the case. Scientists, however, would describe this "identity crisis" as a case of misfiring. Misfiring refers to a behavior that is normally appropriate being exhibited in an inappropriate circumstance.

Female birds like this cardinal are stimulated to feed their young when signaled by their tiny, gaping mouths. Such a signal is called a *releaser*. Releasers are structures or behavior patterns exhibited by one member of a species that trigger a response—a series of precise physical movements—in another member of that species. The behaviors triggered by releasers are also influenced by hormones or other internal stimuli. This cardinal recently lost her nest of young but is still hormonally "driven" to feed them. Therefore the tiny, wide-open mouth triggered a response in the cardinal—a response that will be quite a surprise for the fish!

Ethology: The biology of behavior

Animals (including humans) continually exhibit a range of behaviors. **Behaviors** include the patterns of movement, sounds (vocalizations), and body positions (postures) exhibited by an animal. In addition, behaviors include any type of change in an animal, such as a change in coloration or the releasing of a scent, that can trigger certain behaviors in another animal.

Some behaviors are simple, automatic responses to environmental stimuli. A bacterium "behaves" when it moves toward higher concentrations of sugar. You behave when you slam on the brakes to avoid a car accident. Other types of behaviors, such as the eating behavior of the sea otter, are quite complex (Figure 34-1). Complex behaviors are limited to multicellular animals with a neural network that can sense stimuli, process these stimuli in a central nervous system (brain), and send out appropriate motor impulses (Figure 34-2). You exhibit complex behavior when you walk and talk, drive a car, interact with your family, and do your job.

The study of the behavior of animals in their natural environments is called **ethology.** Ethologists observe and interpret the behavior of animals from a physiological rather than psychological perspective. They attempt to uncover the biological significance of animal behavior—the importance

FIGURE 34-2 Complex behaviors are characteristic of multicellular animals with a central nervous system and a neural network. These two penguins are exploring their interpersonal space. Most vertebrates exhibit complex behaviors.

of a behavior for a particular species in its natural environment. Ethologists bring order to their observations, breaking down behavior patterns into recognizable units, naming these patterns, and categorizing them.

One of the most famous ethologists is the Austrian scientist Konrad Z. Lorenz (see Figure 34-8), referred to by some as the father of modern ethology. In 1973, Lorenz, Dutch ethologist Nikolaas Tinbergen, and Austrian zoologist Karl von Frisch won the Nobel Prize for their contributions to the study of animal behavior. Lorenz based his work on the premise that animal behaviors are evolutionary adaptations, as are physical adaptations. He referred to behaviors as being part of an animal's "equipment for survival." Today, this concept is a fundamental premise in the study of animal behavior.

FIGURE 34-1 Complex behavior in a sea otter. This sea otter is having dinner while swimming on its back. It is using the rock as a hard surface against which to hit the clam and break it open. Often a sea otter will keep a favorite rock for a long time, suggesting that it has a clear idea of what it is going to use the rock for. The sea otter may learn this pattern of eating behavior from others while young, but the capacity to use tools and consciously foresee their future use certainly depends on inherited abilities.

> Scientists who study the behaviors of animals in their natural environments are ethologists. Behaviors include the patterns of movement, sounds, and body positions exhibited by animals.

INNATE BEHAVIOR AND LEARNING IN ANIMALS

The link between genetics and behavior

All behaviors depend on nerve impulses, hormones, and other physiological mechanisms such as sensory receptors. Therefore genes play a role in the development of behaviors because they direct the development of the nervous system. In addition, automatic responses depend on specific nerve pathways within the central nervous system of an organism. These pathways are neural programs and are genetically determined. Neural programs are part of the nervous system at the time of birth or develop at an appropriate point in maturation, resulting in behaviors that are called **instinctive, innate,** or **inborn.** Interestingly, innate behaviors are performed in a reasonably complete form the first time they are exhibited. A newborn baby, for example, will turn to suckle when touched on the cheek near the mouth. The baby performs this innate behavior without **learning,** or altering its behavior based on experience. Innate behaviors are important to the survival of an animal because they help an animal stay alive in certain situations and provide adaptive advantages that contribute to its fitness, or ability to achieve reproductive success.

> **The development of the nervous system and certain automatic nervous responses that are "preprogrammed" within the nervous system are directed by genes.**

The neural programs of innate behavior patterns can be likened to the "hard wiring" of the circuitry within a calculator. Certain functions within a calculator are preprogrammed and unchangeable; the range of hard-wired functions depends on the calculator. Basic functions such as addition, subtraction, multiplication, and division are always hard wired, meaning that the program a calculator needs to follow is permanently set within its electronic circuitry. Likewise, some calculators are hard wired for other functions such as determining square roots. In a similar manner, innate behaviors are preprogrammed into the nerve circuitry of an animal. These "nerve programs" are passed from parent to offspring by means of genetic (hereditary) material. In general, such hard-wired behaviors cannot be changed, but scientists have observed that some can be gradually perfected during the course of an animal's life.

In animals, some nerve circuits are not completely set as they are in the neural programs that govern innate behaviors. These neural programs can accept the input of growth and experience, resulting in learned behavior. Learned behavior can become automatic; you learned as a small child, for example, how to drink from a cup. This behavior quickly becomes set in your neural circuitry resulting in behavior that does not require conscious thought. Learned behavior is the area in which the effect of environment can be most dramatically seen on the inherited characteristics of the neural circuitry.

Innate (instinctive) behaviors

Ethologists group innate behaviors into four categories: kineses, taxes, reflexes, and fixed action patterns.

Kineses

A **kinesis** is the change in the speed of the random movements of an animal with respect to changes in certain environmental stimuli. Put simply, movement slows down in an environment favorable to the animal's survival and speeds up in an unfavorable one. Have you ever picked up a rotting log or a clump of damp leaves in a wooded area? The pillbugs that you may have seen living under the leaves or the log (Figure 34-3) stay in this favorable environment because of their low levels of activity and then scurry when the log is rolled over.

Taxes

Kineses are *nondirected* types of movements. In other words, an animal is not attracted by a favorable environment; it tends to "blunder" there as it moves about quickly and then stays there as it moves about slowly. A **taxis,** however, is a *directed* movement toward or away from a stimulus, such as light, chemicals, or heat. Animals having preprogrammed taxes also have receptors that sense the particular stimuli to which the animal can orient. Female mosquitos and ticks, for example, have sensory receptors that detect warmth, moisture, and certain chemicals emitted by mammals. Sensing these stimuli helps the insects orient to their victims. In

FIGURE 34-3 Kinesis.
These pillbugs scurry when the log they are living under is moved. This change in the level of activity that occurs with a change in the environmental stimulus is called *kinesis.*

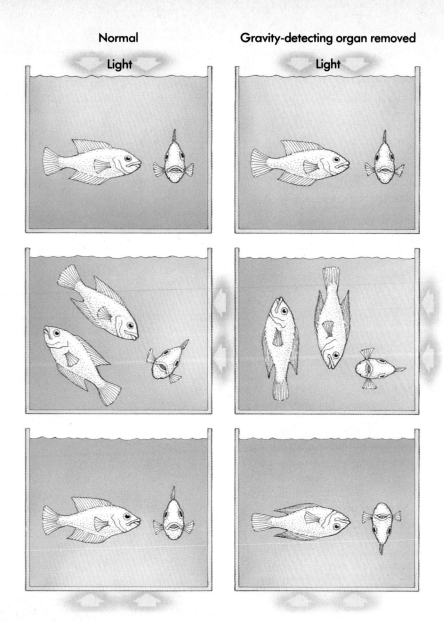

FIGURE 34-4 Phototaxis and gravitaxis in fish.
The normal fish in the left column orient to both light and gravity. The fish in the right column, which have had their gravity-detecting organ removed, orient to light only and have difficulty staying upright. The lower left diagram points out that orientation to gravity overrides orientation to light.

fact, some mosquito repellents work by "blocking" the insect's receptors so that it cannot sense and then locate its victim. The crowding of flying insects about outdoor lights is another familiar example of a taxis called *phototaxis*. Other insects, like the common cockroach, avoid light (are negatively phototactic). Fish swim upright by orienting to gravity and light. If the gravity-detecting organ in the inner ear is removed or if a light is shone into the sides or bottom of a fish tank, the fish become disoriented and do not swim upright (Figure 34-4). Certain species of fish such as trout and salmon also automatically orient against a current and therefore face and swim upstream (Figure 34-5). Even organisms without a nervous system such as protozoans and bacteria exhibit taxes. These organisms have cellular organelles or inclusions that act as receptors or that react to specific environmental stimuli.

INNATE BEHAVIOR AND LEARNING IN ANIMALS

FIGURE 34-5 Taxis in a salmon.
Salmon automatically orient themselves against a current and therefore always swim upstream.

Reflexes

A **reflex** is an automatic response to nerve stimulation. The knee jerk is one of the simplest types of reflexes in the human body. If the tendon just below the kneecap is struck lightly, the lower leg automatically "kicks" or extends. This behavior occurs as the tendon is stretched and pulls on the muscles of the upper leg that are attached to it. Stretch receptors in these muscles immediately send an impulse along sensory nerve fibers to the spinal cord. At the spinal cord these fibers synapse directly with motor neurons that extend back to upper leg muscles. The upper leg muscles are stimulated to contract and jerk the leg upward while opposing muscles simultaneously relax (see Figure 16-12). These reflexes play an important role in maintaining posture. More complex reflexes involve a relay of information from a sensory neuron through one or more interneurons to a motor neuron. The outcomes of these reflexes are modulated by the interneurons. In complex organisms, reflexes play a role in survival, such as when you jerk your hand away from a hot stove before you consciously realize that your hand hurts. In animals with extremely simple nervous systems such as the cnidarians (hydra, jellyfish, sea anemones, and corals), most behaviors are the result of reflexes, although some simple learning can take place (see p. 567). In these animals a stimulus is detected by sensory neurons, and the impulse is passed on to other neurons in the animal's nerve net, eventually reaching the body muscles and causing them to contract. There is no associative activity in which other neurons can influence the outcome, no control of complex actions, and little or no coordination. The nerve net of the cnidarians possesses only the barest essentials of nervous reaction.

Fixed action patterns

More complex than other innate behaviors are **fixed action patterns.** Fixed action patterns are sequences of innate behaviors in which the actions follow an unchanging order of muscular movements, such as the mother cardinal popping

FIGURE 34-6 Fixed action pattern in a graylag goose.
The egg retrieval movement in the graylag goose is an entirely programmed behavior, completely instinctive.

an insect into the mouth of her young. This sequence is a recognizable "unit" of behavior. When a releaser triggers the behavior, the sequence of activity begins and is carried through to completion. Fixed action patterns are often seen in body maintenance behaviors (such as a cat washing its face), courtship behavior, nest building, and attainment of food.

The classic example used to illustrate a fixed action pattern behavior is the retrieval of an egg that rolls out of the nest by the graylag goose (Figure 34-6). If a goose notices an egg has been knocked out of the nest, it will extend its neck toward the egg, reach up, and roll the egg back into the nest with its bill. Because this behavior seems so logical, it is tempting to believe that the goose saw the problem and figured out what to do. But, in fact, the entire behavior is totally instinctive. During experimentation, ethologists have discovered that any rounded object, regardless of size or color, acts as a releaser and triggers the response. Beer bottles, for example, are an effective releaser. And, as a fixed action pattern behavior, the goose completes the action even if the egg rolls away from its retrieving bill. Instead of stopping in the middle of the behavior, the goose will continually repeat the entire behavior until the egg is brought back to the nest.

> Innate behaviors are those that are preprogrammed within the nervous system before birth and are demonstrated in a somewhat complete form the first time they are exhibited. Four categories of innate behaviors are kineses, taxes, reflexes, and fixed action patterns.

Learning

Innate behaviors are certainly important to the survival of an animal. An animal with protective coloration, for example, does not have time to learn to "freeze" when it detects a predator; its survival depends on its instinct to do so (Figure 34-7). Many social behaviors, such as certain mating or food-attainment behaviors, also depend on each individual's performance of instinctive behaviors. In addition, organisms with simple nervous systems, such as the cnidarians, have an extremely limited capacity for learning and must rely primarily on hard-wired innate behavior patterns for their survival.

Although important in many respects to the survival and fitness of animals, innate behaviors can become a liability if environmental conditions change, and an animal's behavior cannot change to adapt to new conditions. For this reason, behavior patterns that can change in response to experience have adaptive advantages over the firmly set programs of instinct. In most animals, only some behaviors are innate and immutable; many behaviors can be changed or modified by an individual's experiences during the process of learning. Learned behaviors can help an animal become better suited to a particular environment or set of conditions.

> Behaviors based on experience are learned behaviors. Learning helps an animal change its behavior to adapt to changes in environmental conditions.

Ethologists distinguish between types of learning, grouping them into five categories: imprinting, habituation, trial-and-error learning, classical conditioning, and insight.

Imprinting

Innate behavior and learning interact closely in a time-dependent form of learning called **imprinting**. Imprinting is a rapid and irreversible type of learning that takes place during an early developmental stage of some animals. Various types of imprinting exist. One type is object imprinting and has been observed in birds such as ducks, geese, and chickens. During a short time early in the bird's life (optimally 13 to 16 hours after hatching in mallard ducks, for example), the young animal forms a learned attachment to a moving object. Usually this object is its mother. However, animals can imprint on various objects regardless of size or color, such as balloons, clocks, and people.

Konrad Lorenz performed classic experiments regarding object imprinting during the 1930s. In one of his most famous experiments, Lorenz divided a clutch of graylag goose eggs in half. He left one half of the eggs with the mother goose and put the other half in an incubator. The half that hatched with their mother displayed normal behavior, following her as she moved. As adults, these geese also exhibited normal behavior and mated with other graylag geese. The other half of the clutch, however, hatched in the incubator and then spent time with Lorenz. These goslings did not behave in the same manner as their siblings; they followed Lorenz around as if he were their mother (Figure 34-8). After their initial time with him, Lorenz introduced the geese to their mother, but they still preferred to follow Lorenz. As adults, these geese tried to "court" adult humans!

Other types of imprinting also exist. Many species of birds hone their singing skills by motoric imprinting. A young North American white-crowned sparrow, for example, listens to its father sing long before it has matured and is ready to do the same. The father's song forms a "mental template" for the young bird, which it uses to compare its song later in life (Figure 34-9).

Many migrating birds and fishes learn to recognize their birthplace by locality imprinting. Pacific salmon, for example, are imprinted with the odor of the stream or lake in which they were born. Amazingly, 2 to 5 years later when they return from the sea to spawn, they are able to find their birthplace by its odor. Such long-range two-way movements are called **migrations**. The physiological and behavioral mechanisms that interact to help migrating animals navigate have interested and puzzled biologists for centuries.

FIGURE 34-7 Protection against predation is instinctive. This tiny tropical frog has markings that resemble a bird dropping. The effectiveness of this protective coloration depends on the frog remaining completely still. This stillness is *not* learned but is an instinctive behavior on which the animal depends for survival.

FIGURE 34-8 Imprinting is a time-dependent form of learning.
The eager goslings following Konrad Lorenz think he is their mother. He is the first animal they saw when they hatched, and they have become imprinted with his image. The goslings will always recognize Lorenz as their parent.

FIGURE 34-9 Song learning in birds.
The singing of a sparrow, which sounds so spontaneous and free, is actually a carefully orchestrated performance that the immature bird learns from its father. The bird perfects its song by listening to others sing and by months of practice. A bird raised in isolation lacks singing partners and thus sings only a rough approximation of its song.

In many animals, migrations occur once a year and result from interactions among various environmental factors (such as day length) with animals' physiological and hormonal changes. Ducks and geese, for example, migrate down flyways from Canada across the United States each fall and return each spring. Caribou migrate across the top of North America each year (Figure 34-10). Monarch butterflies migrate from the eastern United States to Mexico and back, a journey of over 3000 kilometers, which takes from two to five generations of butterflies to complete. Perhaps the longest migration is that of the golden plover, a bird that flies from Arctic breeding grounds to wintering areas in southeastern South America, a distance of approximately 13,000 kilometers.

Migrating animals have the ability to orient themselves in relation to an environmental cue, such as the sun, the stars, or the Earth's magnetic field. In addition, they can navigate, or set a course and follow it. Day migrators such as some birds, ants, and bees use the sun's position to chart a course. Night migrators, a group that includes birds such as the indigo bunting, use the stars to chart a course. All these animals also have a biological clock, some sort of internal "timepiece" (of which scientists have limited understanding) that interacts with information from the environment to help migrating animals find their way. It helps animals compensate for the daily movements of the sun.

> Imprinting is a rapid and irreversible type of learning that takes place during an early developmental stage of some animals. Imprinting helps animals recognize kin, hone singing skills, or recognize their "home" or birthplace. The last type, locality imprinting, plays an important role in the migrations of some animals.

FIGURE 34-10 Caribou migration.
In caribou the seasonal migration across northern Canada may cover 1500 kilometers, although hunting and development are disrupting the long-used migration routes.

FIGURE 34-11 Habituation in gull chicks.
At first, a chick will crouch at any object that flies overhead. Gradually, the chick stops crouching at the familiar objects that fly overhead—the chick has gotten used to them. However, the chick will crouch when an unfamiliar flying object poses a potential threat.

Habituation

Habituation is the ability of animals to "get used to" certain types of stimuli. People get used to stimuli all the time. A single shotgun blast would startle most people, but this response would fade at the end of a day in a firing range. Not only would you get used to the noise level (and wear ear protection), but you would perceive the gunshots as nonthreatening. Likewise, you quickly get used to the feel of your clothing after dressing in the morning. Animals also stop responding to stimuli that they learn is neither harmful nor helpful. Learning to ignore unimportant stimuli is a critical ability in an animal confronting a barrage of stimuli in a complex environment and can help an animal conserve its energy. Hydra, for example, stop contracting if they are disturbed too often by water currents. Sea anemones stop withdrawing if they are touched repeatedly. Young black-headed gull chicks stop crouching as familiar-shaped birds fly overhead. However, they still crouch for the occasional unfamiliar shape that may pose a threat (Figure 34-11).

Trial-and-error learning (operant conditioning)

More complex than imprinting or habituation, **operant conditioning** is a form of learning in which an animal associates something that it does with a reward or punishment. (The word *operant* means "having the power to produce an effect" and is a form of the verb "to operate.") In operant conditioning an animal must make the proper association between its response (such as pressing a lever) and a reward (the appearance of a food pellet) before it receives this reinforcing stimulus. It may also learn to avoid a behavior when the stimulus is negative. Animal trainers use the techniques of operant conditioning.

The American psychologist B.F. Skinner studied such conditioning in rats by placing each in a specially designed box (today called a *Skinner box*) fitted with levers and other experimental devices (Figure 34-12). Once inside, the rat would explore the box feverishly, running this way and that. Occasionally, it would accidentally press a lever, and a pellet of food would appear. At first, a rat would ignore the lever and continue to move about, but soon it learned to press the lever to obtain food.

This sort of trial-and-error learning is of major importance to most vertebrates in nature. The toad in Figure 34-13 learned quickly, for example, not to eat a bumblebee.

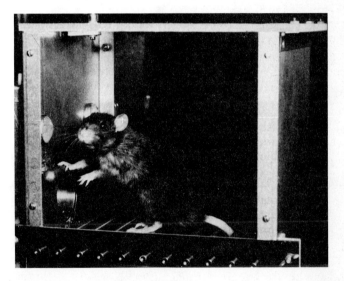

FIGURE 34-12 Operant conditioning.
This rat in a Skinner box rapidly learns that pressing the lever results in the appearance of a food pellet. This kind of learning is based on trial and error with a reward for success.

FIGURE 34-13 A toad is taught a lesson through trial and error. A toad gobbles a bumblebee, but as it does, it gets a violent sting on its tongue and spits out the bee.

The frog did not use reasoning to determine that the bumblebee was not good to eat—it merely became conditioned by experience to avoid a response that caused it pain.

Behavioral psychologists used to think that animals could be conditioned in the laboratory to perform *any* learnable behavior in response to *any* stimulus by operant conditioning. However, through experimentation, researchers have discovered that animals only tend to learn in ways that are compatible with their neural programs. Put simply, operant conditioning only works for stimuli and responses that have meaning for animals in nature. Rats can be conditioned, for example, to press levers with their paws to obtain food because in nature they obtain food with their paws. They cannot, however, be conditioned to obtain food by jumping, an unnatural food-attaining behavior for a rat.

> Some animals can learn by trial-and-error; by associating a specific behavior with a reward or punishment, they can learn to apply (or avoid) certain behaviors to affect outcomes in particular situations. Using the trial-and-error method, scientists can condition animals to perform specific behaviors in the laboratory. This type of learning is operant conditioning.

Classical conditioning

Classical conditioning is a form of learning in which an animal is taught to associate a new stimulus with a natural stimulus that normally evokes a response in the animal. Repeatedly presenting an animal with the new stimulus in association with the natural stimulus can cause the animal's brain to form an association between the two stimuli. Eventually, the animal will respond to the new stimulus alone; it will act as a substitute for the natural stimulus. The connection between the new stimulus and the natural stimulus is the result of a learning process but must be reinforced periodically with the presence of the natural stimulus or the animal will stop responding to the substitute.

In his famous study of classical conditioning, the Russian psychologist Ivan Pavlov worked with dogs to condition them to salivate in response to a stimulus normally unrelated to salivation. A natural stimulus to trigger the response of salivation in a dog is meat. In his experiments, Pavlov also presented the dog with a second, unrelated stimulus. He shone a light on the dog at the same time that meat powder was blown into its mouth. As expected, the dog salivated. After repeated trials, the dog eventually salivated in response to the light alone. The dog had learned to associate the unrelated light stimulus with the meat stimulus (Figure 34-14).

The natural stimulus-response connection in an animal is an inborn reflex, or an *unconditioned* response. Such innate responses are important to animals; many of them are protective, such as blinking, sneezing, vomiting, and coughing. Even the knee-jerk reflex mentioned earlier is part of the mechanism by which humans (and other vertebrates) maintain their posture. The work of researchers such as Pavlov contributed to the understanding that animal behavior depends on innate neural circuitry but that these neural programs can most often be modified and directed by the processes of learning.

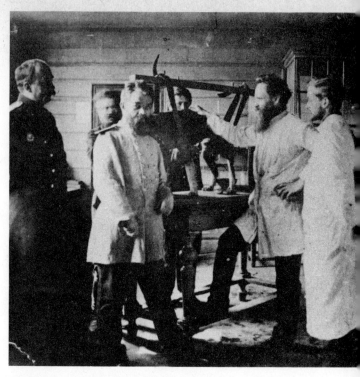

FIGURE 34-14 Classical conditioning.
The Russian psychologist Ivan Pavlov (second from right) in his laboratory. In the background a dog is suspended in a harness that Pavlov used to condition salivation in response to light.

Insight

Best developed in primates such as chimpanzees and humans, **insight,** or **reasoning,** is the most complex form of learning. An animal capable of insight can recognize a problem and solve it mentally before ever trying out a solution. Therefore the animal is able to perform a correct or appropriate behavior the first time it tries, without having been exposed to the specific situation.

German psychologist Wolfgang Kohler was the first to describe learning by insight, performing extensive experiments on chimpanzees in the 1920s. Kohler showed that animals must perceive relationships and manipulate concepts in its mind to solve a problem on the first try. In his classic experiments, he placed chimpanzees in a room with a few crates, poles that could be joined together, and a banana hung high above the grasp of the animals. The chimpanzees were able to stack the crates and join the poles appropriately to retrieve the banana (Figure 34-15, *A*). Unlike chimpanzees, most animals are unable to use insight to solve problems. Your dog or cat, for example, must use trial-and-error to "try out" solutions to problems. Your dog has no insight into the problem that having its leash wrapped around a tree is keeping it from reaching its food (Figure 34-15, *B*). The dog can merely continue walking in various directions and ultimately free itself by chance. This experience, however, may help the dog perform the appropriate behavior in the future, having learned by trial-and-error.

FIGURE 34-15 Insight is the most complex form of learning.
A A chimpanzee is able to see a problem and develop a solution, even before taking any action. To reach the bananas, the chimpanzee sees a solution in stacking the boxes one on top of the other.
B A dog, on the other hand, lacks insight and cannot develop a solution. A dog will eventually learn by trial and error to go around the post to reach the food.

BIOLOGY & you

Yawn!

Everyone yawns. Most people yawn within 5 minutes of birth and keep doing it at odd moments every day as long as they live. Did you ever wonder why?

First, there is nothing particularly human about yawning. Most carnivores yawn. Few herbivores seem to yawn, although there are exceptions—rhinoceroses have enormous yawns. Frogs yawn. Even fish appear to yawn. Whatever is going on, it must be a basic behavior, like eating or sleeping, that is exhibited in many animals.

It used to be felt that a yawn was a silent scream for oxygen that usually occurs when people are tired or bored, a type of deep breath to increase oxygen in the blood or get rid of excess carbon dioxide. Not so. When a group of freshman psychology students at the University of Maryland inhaled air containing different mixtures of oxygen and carbon dioxide and counted their yawns, only breathing rates went up or down to compensate for changing levels of oxygen and carbon dioxide. Yawning rates did not change.

Some researchers believe that yawning is the body's way of promoting arousal in situations where you have to stay awake. Most humans yawn when stimulation is lacking. When a team of yawn counters observed people engaged in various activities, they found that people riding subway cars yawned far more often when the cars were empty. Students yawned a lot studying in the library. But the highest yawn rates they recorded were in a college calculus class, where, on the average, each student racked up an astonishing 24.6 yawns per hour!

FIGURE 34-A Yawn!

The arousal hypothesis helps explain why people driving late at night on the highway yawn a lot and why very few people yawn when they are actually in bed—they don't need to because it's okay to go to sleep. Whether this hypothesis is plausible or not, one thing is sure: yawning is highly contagious. Seeing another person yawn releases a powerful urge for you to yawn yourself. This trait seems to be restricted to humans; no other animal responds in this way.

Yawning is not a hotbed of research, but studies are ongoing in the laboratories of Robert Provine at the University of Maryland and Ronald Baenninger at Temple University, where the work described here was carried out. Provine points out the research may be more important than people realize. "I think a lot of times yawning is trivialized," he says, "because people think it's a little behavior in a compartment off to the side of all other human experiences and actions." To the contrary, yawning seems associated with many diseases in ways not yet understood—brain lesions and epilepsy often lead to excessive yawning, but schizophrenics yawn very little. There is clearly much to learn.

Summary

1. The study of animal behavior encompasses the patterns of movement, sounds, and body positions of animals. Ethologists study the behavior of animals in their natural environments, observing and interpreting their behavior in the context of physiology while attempting to uncover the biological significance of the behavior.

2. Some patterns of behavior are inborn, or innate, resulting from neural pathways developed before birth. Such instinctive behaviors are important to animals, providing them with fixed patterns of survival, such as those used to elude predators, to mate, and to care for young.

3. Ethologists group innate behaviors into four categories: kineses, taxes, reflexes, and fixed action patterns. Certain animals slow down in suitable environments and speed up in unsuitable ones by means of kineses, move toward or away from certain stimuli by means of taxes, automatically respond to nerve stimulation by means of reflexes, and perform sequences of movements by means of fixed action patterns. These behaviors can be modified only slightly or not at all.

4. Animals can learn, or change their behaviors based on experience. Behaviors are changed within certain limits by means of various types of learning: imprinting, habituation, trial-and-error, and conditioning.

5. Certain animals learn to recognize their kin or birthplace by means of imprinting, a rapid and irreversible type of learning that takes place during an early developmental stage. Habituation helps animals adapt to certain types of stimuli. Some animals learn to associate a behavior with a reward or punishment by means of trial-and-error learning. During classical conditioning, animals learn to associate a new stimulus with a natural stimulus that normally evokes a response in the animal.

6. Insight, or reasoning, is the most complex form of learning. An animal capable of insight can recognize a problem and solve it mentally by using experience, thereby performing an appropriate behavior the first time it tries. Insight is best developed in primates such as chimpanzees and humans.

REVIEW QUESTIONS

1. What are behaviors?
2. Explain this statement: Behaviors are part of an animal's equipment for survival.
3. Fill in the blanks: Behaviors that are performed in a fairly complete form the first time they are exhibited are _____ _____. Behaviors that are based on experience are _____ _____.
4. Fill in the blanks: Ethologists group innate behaviors into four categories: _____, _____, _____, and _____ _____.
5. Define and give an example of each of the categories you listed for question 4.
6. Explain the importance of innate and learned behaviors. How do they complement each other?
7. Fill in the blanks: Ethologists distinguish between five types of learning: _____, _____, _____ _____, _____ _____, and _____.
8. Identify the type of behavior involved in each of the following situations:
   a. Fireflies are attracted to the flashing luminescence of other fireflies.
   b. While studying your biology text, you hear a faucet dripping. For a while it bothers you, but then you no longer notice it.
   c. You praise your puppy every time it sits down when you say "sit." Soon you've trained it to sit on command.
9. Distinguish between operant conditioning and classical conditioning.
10. What is insight? Explain its significance.
11. Identify the category of behavior involved in each of the following situations:
    a. Every year, purple martins fly to Brazil and return to the United States around April.
    b. When a bright light flashes near your eyes, you automatically blink.
    c. Your cat frequently grooms itself by licking its fur and rubbing its paws over its face.

KEY TERMS

behavior 599
classical conditioning 606
ethology 599
fixed action pattern 602
habituation 605
imprinting 603
insight (reasoning) 606
instinctive (innate, inborn) 600
kinesis 600
learning 600
migration 603
operant conditioning 605
reflex 602
taxis 600

THOUGHT QUESTIONS

1. Rats can be taught to press levers to get food but can never be taught to obtain food by jumping. Why not? Can you think of any behaviors dogs cannot be taught to do?
2. Monarch butterfly migrations take from two to five generations to complete, each way. How do you imagine a butterfly born midway along the journey knows which way to go?
3. Pacific salmon are born at the headwaters of rivers, then swim downstream hundreds of miles to the sea where they spend their adult lives. Years later, when it is time to spawn (that is, to lay and fertilize the eggs that will be the next generation), the adults swim up the same rivers to the precise location where they were born. Swimming up the river, how do they know which way to go when they come to a fork in the river?

FOR FURTHER READING

Dawkins, R. (1989). *The selfish gene*. ed. 2. New York: Oxford University Press.
This is an entertaining account of the sociobiologist's view of behavior.

Gould, J., & Marler, P. (1987, January). Learning by instinct. *Scientific American*, pp. 74-85.
This is a clear and interesting account of the relative roles of instinct and learning in behavior. The authors argue that learning is often limited or controlled by instinct.

Huber F., & Thorson, J. (1985, December). Cricket auditory communication. *Scientific American*, pp. 60-68.
This gives an unusually clear example of how nervous system activity underlies animal behavior.

35 Social behavior in animals

These kissing fish only appear to be kissing. In fact, their behavior is one of aggression, not affection. Engaged in a territorial war, each wants to win the rights to its own "space" within the water. Neither fish will die during this bloodless battle, but one will back down, allowing the other to lay claim to the win.

Their fight begins as they encounter one another at a territorial boundary. First, each tries to intimidate the other, puffing up its body with air to appear as big and threatening as possible. Then they swim side-by-side, pushing currents of water past one another with their fins. These currents provide clues to each contender regarding the size of the opponent. If neither retreats, the tug-of-war begins, mouths locked in battle. Soon, one fish emerges the victor. The loser signals submission, and the tournament is over.

OUTLINE

Types of social behavior
  Competitive behaviors
    Threat displays
    Submissive behavior
    Territorial behavior
  Reproductive behaviors
  Parenting behaviors
  Group behaviors
    Insect societies
    Rank order in vertebrate groups
Human behavior

HIGHLIGHTS

▼

Sociobiology, the biology of social behavior, applies the knowledge of evolution to the study of animal behavior.

▼

Social behaviors, those that help members of the same species communicate and interact with one another, are valuable mechanisms that ultimately aid the reproductive fitness of a species.

▼

Many species of animals live in social groups; some insects, in fact, live in societies having a division of labor.

▼

Many researchers question the scientific validity of applying sociobiology to human behavior and propose that the application may lead to racism, sexism, or other types of group stereotyping.

Types of social behavior

The aggressive behaviors exhibited by the kissing gouramis are types of **social behaviors,** which help members of the same species communicate and interact with one another. The advantages of social behaviors are numerous, and these advantages differ from species to species. In general, however, all species that reproduce sexually exhibit interactive patterns of behavior for the purposes of reproduction, the care of offspring, and the defense of a territory. In addition, some animals use social behaviors to hunt for and share food and to warn and defend against predators.

The biology of social behavior is called **sociobiology.** This science applies the knowledge of evolutionary biology to social behavior. Its purpose is to develop general laws of the biology and evolution of social behavior.

> Social behaviors are interactive patterns of animal behavior that help members of the same species communicate and interact with one another. The biology of social behavior is sociobiology.

Competitive behaviors

When two or more individuals strive to obtain the same needed resource, such as food, water, nesting sites, or mates, they are exhibiting **competitive behavior.** This type of behavior occurs when resources are scarce.

Threat displays

How do animals compete? Many first engage in **threat,** or **intimidation, displays,** a form of aggressive behavior. The purpose of these displays is to do as the name suggests: scare other animals away or cause them to "back down" before fighting takes place. Exact forms of this behavior vary widely among species but usually involve such behaviors as showing fangs or claws, making noises such as growls or roars, changing body color to one that is a releaser of aggression in an opponent, and making the body appear larger by standing upright, making the fur or hair stand on end, or inflating a throat sac (Figure 35-1). Threat displays are important social signals and communicate the intent to fight. Interestingly, some of the movements and body postures (body language) of threat displays that repel competitors also attract members of the opposite sex. Biologically, this makes sense because the competition may be over a mate.

Submissive behavior

Animals use other behaviors to avoid fighting. One type of behavior is called **submissive behavior** and is usually a behavior opposite to a threat display. The behavior might include making the body appear smaller, "putting away weapons" of fangs or claws, turning a vulnerable part of the body to an opponent, or removing body colors that are releasers of aggression. As Figure 35-2 shows, an animal may, in fact, possess an array of color "signals." If an animal is losing a fight, it might also display submissive behaviors to stop the fight. Contrary to popular belief, animals of the same species rarely fight to the death. One reason is that most animals do not have the means to do so—they do not have sufficiently dangerous fangs, claws, or horns, for example. Species with dangerous weapons usually have defenses against those weapons, too, such as a strong hide, long hair, or a thick layer of body fat. Scientists theorize that aggression within a species is meant to chase off rather than kill the rival. Such behaviors are adaptive for the species as a whole because they tend to disperse the population among the available resources rather than decrease the size of the population and conserve the energy of all the members of the species.

FIGURE 35-1 Defense by deception.
This common toad, confronted by a snake, inflates its body and sways from side to side in an attempt to make itself appear larger.

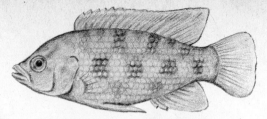
Frightened but well hidden

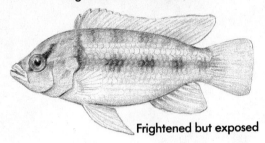
Frightened but exposed

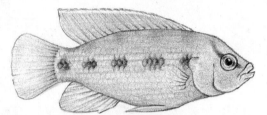
Neutral mood

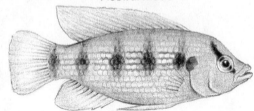
Mild territorial aggression

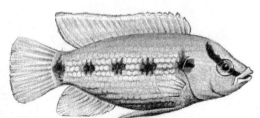
Rising aggression

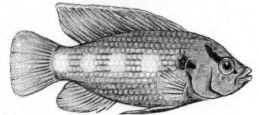
Full aggression

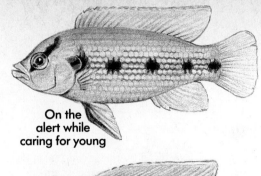
On the alert while caring for young

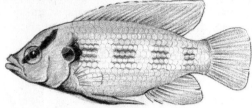
Spawning male

FIGURE 35-2 The technicolor fish.
This freshwater fish, a member of the cichlid family, changes its color patterns in different situations. It does this by expanding and contracting special red and black color cells in its skin.

Territorial behavior

Another means by which animals limit competition and aggression within their species is by the use of **territorial behavior.** A territory is an area that an animal marks off as its own, defending it against the same sex members of its species. Members of the opposite sex are always allowed into the territory; in fact, the territory is their meeting and mating place.

Territoriality is very common in all classes of vertebrates and is even found in some invertebrates such as crickets, wasps, and praying mantises. Some animals such as dragonflies mark their territories by conspicuously patrolling their borders, resting and then moving from prominent landmarks. As an animal demonstrates the borders of its territory (Figure 35-3), it may exhibit specific movements or body postures as signals. Some animals, such as song birds, monkeys, apes, frogs, and lizards, mark their territory by producing sounds that announce their ownership. Animals with a well-developed sense of smell, such as wolves, hippopotamuses, rhinoceroses, some rodents, and even domestic dogs, mark their territories with substances that have an odor. Many species, in fact, possess special scent glands that secrete substances just for this purpose. The odor marking helps keep out members of the same sex and helps owners of large territories orient themselves to its borders.

Territorial behavior has several adaptively important consequences. In many mammalian societies, territoriality often has an important influence on the establishment of new populations, with the dominant male defeating the lesser males and leaving them to found their own populations. Individuals surviving in the surrounding marginal areas repopulate any vacant territories. Territoriality is also a means to ration resources among members of a species.

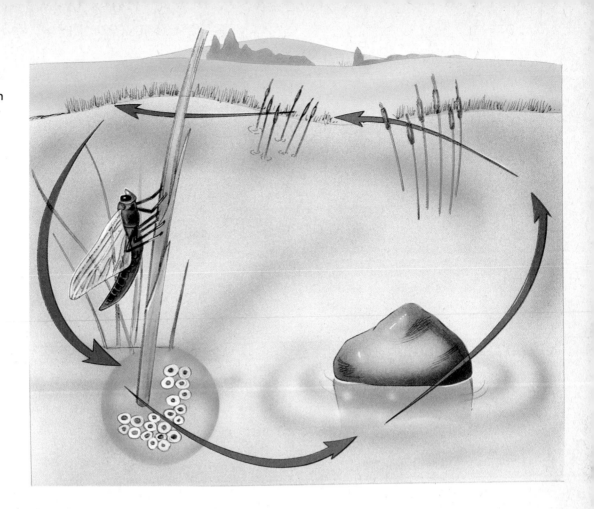

FIGURE 35-3 Territory of a male Demoiselle dragonfly. The male dragonfly sits on a perch above the egg deposit site. He periodically flies around his territory *(red arrows)* or circles within the territory.

When resources such as nesting sites and food are limited (Figure 35-4), each member of a species is in danger of the resources being spread too thin, so that no member gets an adequate amount of what it needs. In addition, breeding pairs of animals will not reproduce unless they have adequate resources; territorial behavior ensures that at least some members of the species will have their own space, food, and nesting sites so that they will survive and reproduce. Territorial behaviors also enhance reproductive capabilities by placing the peak time of competition and aggressive behavior at the time of the marking of the territory, *before* the time of reproduction and raising of young.

> Individuals within the same species compete with each other for various resources. Competition among animals includes a repertoire of behaviors such as threat displays, submissive behavior, and territorial behavior. Such competitive behaviors disperse the population among the available resources and enhance the reproductive capabilities of the species.

FIGURE 35-4 The emperor penguins in Antarctica nest in rookery sites.
There are only a limited number of such sites available, which may reflect the limited amount of resources available to the breeding penguins.

BIOLOGY IN FOCUS

Elephant talk

How do male and female elephants find each other to reproduce? Consider the obstacles: females spend 2 years in gestation (the time from fertilization and implantation of the embryo until birth) and another 2 years nursing their young. During these periods, female elephants will not mate. The period of estrus, in which the female elephant is receptive to mating, occurs only once every 4 years—and lasts only 2 days! Male elephants have their work cut out for them. For half the year, male elephants are solitary creatures and travel long distances in a search for willing female partners. It seems a miracle that elephants reproduce at all.

But is it a miracle that elephants find each other? The answer is not at all. In 1989, researchers investigated the possibility that a female elephant in estrus sends out a signal that informs males for miles around of her condition. Using sophisticated microphones, the researchers recorded a sequence of intense, low-frequency calls emitted from an elephant in estrus. The estrus call is the same in all female elephants and is a song. The low frequency of the song places it in the infrasound frequency range—the same range that includes thunderstorms and earthquakes. Humans can barely detect infrasound—you "feel" more than hear the low, dull throbbing associated with these low-frequency sounds. The female elephant sings her song for only a half an hour, and within a day she is surrounded by male elephants, some of whom have heard the song from several miles away.

Infrasound is used by elephants for purposes other then mating as well. Warnings are issued by infrasound, and the whereabouts of groups of elephants are transmitted to other groups by infrasound. Observers have noted for years that entire groups of elephants will suddenly freeze, raising and spreading their ears. Researchers surmise that these elephants are listening to faint, distant calls and are remaining as silent as possible to do so. Moments later, the elephants will come out of their trance and rush off together to follow the signal.

The elephant's "talk" is a factor that enhances elephants' survival by communicating useful information and facilitating the reproductive process. Researchers are currently looking into the possibility that elephants use their infrasound calls to maintain their elaborate hierarchical society over distances of several miles.

FIGURE 35-A A bull elephant can locate a female elephant in estrus by listening for her infrasound mating call.

FIGURE 35-5 Asexual reproduction. Offspring of sea anemones can be produced by budding. A bud grows from a few cells of a single parent and, when developed, breaks off and becomes completely separate individuals. Budding results in genetically identical parents and offspring. Buds can be seen emerging at the sea anemone's midsection.

Reproductive behaviors

Sexual reproduction, the union of a male sex cell (sperm) and a female sex cell (egg) to form the first cell of a new individual (zygote), takes place in most animals. Even some animals that can reproduce asexually, such as sponges that can bud off bits of themselves that grow into new individuals, reproduce sexually at certain times. Sexual reproduction is biologically significant because it provides genetic variability among organisms of the same species. Put simply, the combining of genes from two parents produces offspring that are similar but different from their parents. In addition, because the genes of each parent become assorted independently during the meiotic process that produces sex cells (see Chapters 5 and 22), the offspring are different not only from their parents, but from one another. Differences among organisms within a species are the "raw materials" needed for adaptation to environmental change. The ability of some organisms to survive and reproduce in an environment in which others of their species cannot is, in turn, the basis of natural selection.

Genetic variability does not exist in organisms that reproduce asexually; these organisms are clones of one parent. That is, organisms produced by means of asexual reproduction develop from one or a few cells of a single parent (Figure 35-5) and are therefore genetically identical to that parent. Because organisms within a population of clones do not vary from one another, the entire population can be at risk if environmental conditions change.

Behaviors that promote successful sexual reproduction, then, are highly adaptive behaviors. Organisms that reproduce sexually must have patterns of male-female interactions that lead to fertilization—the penetration of an egg by a sperm. Fertilization can be preceded by copulation, the joining of male and female reproductive organs. However, some organisms such as many fish and frogs do not copulate; instead the male releases sperm over the eggs after they have been deposited by the female. The term **mating** refers to male-female behaviors that result in fertilization, regardless of whether copulation occurs. The term **courtship** refers to the behavior patterns that lead to mating.

FIGURE 35-6 A male peacock.
Males of certain species, such as the peacock, use fantastic displays of color or markings to attract females.

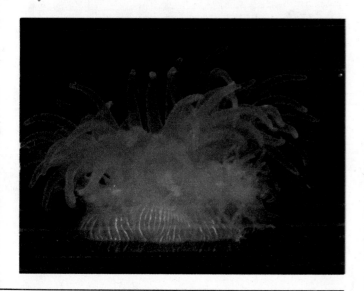

For courtship to take place, males and females must first find each other. In some species, males and females live together in pairs or groups, but many animal species do not. Most often the job of attracting a mate goes to the male of the species. Notable exceptions, however, are often found within insect populations.

Males often attract a mate by marking a territory and defending it against intrusion by other males. As mentioned previously, aggressive behaviors against other males are often the same behaviors that attract a female. In addition, males can attract females by displaying body colors or markings like the male peacock shown in Figure 35-6. Many birds sing songs to attract a mate, and certain frogs and many insects use other sounds or mating calls for this purpose. Many mammals and insects produce odors that are attractive to females. In some species, the males congregate and perform various dances or songs as a group to attract their mates (Figure 35-7).

SOCIAL BEHAVIOR IN ANIMALS

FIGURE 35-7 Courtship dance in male prairie chickens.
Male prairie chickens perform a courtship dance to attract females. They produce sounds as part of their behaviors, puffing up the orange air sacs on the sides of their necks.

After the formation of a male-female pair, a **courtship ritual,** unique to each species, leads to mating. Courtship behaviors usually consist of a series of fixed action patterns of movement (see Chapter 34), each triggered by some action of the partner that acts as a releaser and triggers in turn a preprogrammed movement pattern by the potential mate (Figure 35-8). For example, cranes dance around one another, peacocks strut with their tail feathers extended in an elaborate fan, and hummingbirds participate in a courtship flight (Figure 35-9). In addition, some species, such as various types of geese, exhibit postcopulatory behaviors. Both birds rise out of the water, wings extended, facing one another (Figure 35-10). Such behaviors appear to strengthen the pair bond, a behavior that is necessary when both parents care for the young.

> Reproductive behaviors within a sexually reproducing species serve to bring males and females together in a manner that leads to the fertilization of eggs by sperm.

FIGURE 35-8 Stickleback courtship.
Each set of movements in the courtship ritual is a fixed action pattern that triggers the next set of movements by the partner.

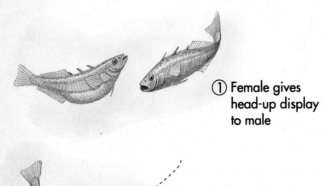

① Female gives head-up display to male

② Male swims zigzag to female and then leads her to nest

③ Male shows female entrance to nest

④ Female enters nest and spawns while male stimulates tail

⑤ Male enters nest and fertilizes eggs

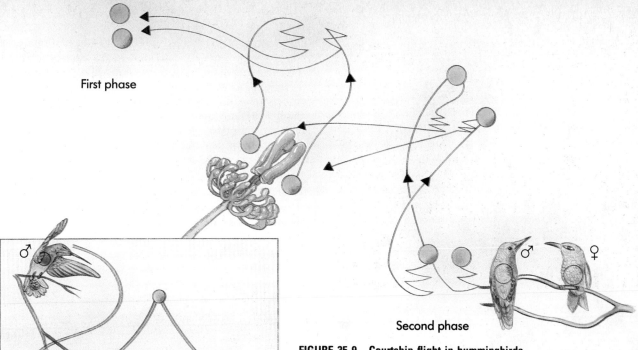

FIGURE 35-9 Courtship flight in hummingbirds.
In the first phase of the courtship flight, the male flies around the female in an undulating pattern. In the second phase, both partners fly around each other and may mirror their partner's movements.

FIGURE 35-10 Geese exhibiting postcopulatory behavior.
Here, a goose rises out of the water, wings extended. These postcopulatory behaviors seem to strengthen the pair bond between the mating geese.

SOCIAL BEHAVIOR IN ANIMALS

FIGURE 35-11 A marsupial with young.
The kangaroo protects its young in its pouch. The young are nourished by mammary glands located in the pouch.

Some species exhibit quite interesting and unique parenting behaviors. Very early in his studies of behavior, Tinbergen noted that certain birds remove the shell fragments of broken eggs from their nests. Any shell fragment that Tinbergen added was removed quickly. The adaptive advantage of this parenting behavior, Tinbergen discovered, was that the odor from shell fragments attracted predators and their prompt removal helped protect the other eggs in the nest.

> **Parenting behaviors lead to the reproductive fitness of a species by helping offspring survive and grow to reproductive age.**

Group behaviors

Individuals within many species of animals form pair bonds for the purposes of reproduction and parenting. In addition, the individuals of many species form temporary or permanent associations with many other members of their species, forming social groups. Animals within social groups work together for common purposes. Individuals within a group, for example, have more protection against predators than single individuals or pairs. They can warn each other of danger by using calls or chemical signals or cluster together when a predator appears, forcing the predator to either separate individuals from the group or fight the entire group. In addition, some animals such as musk-oxen and bison place their young within the center of the group (Figure 35-12), forming a circle around them much like the early American settlers did with their covered wagons.

Parenting behaviors

For the offspring to survive, grow, and eventually reproduce, parents must either make preparations for the care of their young or care for them themselves. Parenting behaviors are found in almost all animal groups, but the participation of the parents varies. In many species of birds and fishes, both parents care for the young. In these species, a division of labor usually exists, with each parent carrying out a specific role. Because female mammals produce milk to nourish the young, most often the mother assumes the parenting role in this class of animals. In only a few species, such as certain fishes, sea horses, and birds, does the male alone take care of the offspring.

To prepare for young, animals build protective structures such as nests or cocoons. Usually a food supply is gathered, and the eggs are deposited in areas that are protected and that may also lie near a ready food source. Marsupials, mammals in which the young are born developmentally early, keep their young safe and nourish them as they complete their development by storing them in a pouch, conveniently situated at the mammary glands (Figure 35-11). Some animals simply carry their young to a safe place when they sense danger.

FIGURE 35-12 Defensive formation of a herd of musk-oxen.
Like the pioneers who circled their wagons against danger, these musk-oxen form a defensive ring with horns pointing outward. The young musk-oxen are inside the ring.

FIGURE 35-13 Hunting behaviors of pelicans.
Eastern white pelicans hunt in groups **(A)** and then circle their prey **(B)**.

Many animals living within social groups search for food together and, scientists have observed, are more successful as group hunters than as single individuals. Food-gathering strategies become group strategies; often animals such as those that prey on fish will drive the prey to one location, encircle them, and feast. The eastern white pelicans shown in Figure 35-13 hunt for fish in this way.

In groups, animals can also ration the "work load" so that each animal does not have to perform an array of daily tasks but may focus on one task (Figure 35-14). Many species of insects, particularly bees and ants, are organized into social groups having a division of labor. These social groups are called **insect societies.**

FIGURE 35-14 Ant society.
This worker ant, a member of the genus *Polyergus,* is responsible for capturing slaves for the nest. The slave, a member of the genus *Formica,* is in its pupa stage. When it emerges as an adult, it will be a nonreproductive worker slave.

Insect societies

In insect societies, individuals are organized into highly integrated groups in which each member of the society performs one special task or a series of tasks that contributes to the survival of the group. The role an individual plays depends on its body structure which, in turn depends on its sex or age. In some animal societies the same animal performs different roles during the course of its life. In addition, a common feature among social insects is a queen that outlives other members of the society. The role of the queen is to produce offspring and to promote cooperation among members of the group.

A honeybee colony, for example, is made up of three different types of bees called **castes:** the queen, the workers, and the drones. The queen is the focus of the colony; she lays the eggs from which all the other bees develop (Figure 35-15). The drones are the male bees that fertilize the eggs of the queen. Interestingly, they develop from unfertilized eggs and so have no male parent themselves—and only one set of chromosomes! The workers are females that develop from fertilized eggs. Genetically, a queen and a worker are no different; any fertilized egg can develop into either a queen or a worker depending on how the egg is housed and fed.

In one hive, there may be up to 50,000 workers, 5000 drones, and of course 1 queen. The workers perform all the tasks of the hive except for mating and egg laying. In the winter months, the average length of the worker bee's life is 6 months. In the summer, workers live only 38 days! During the first 20 days of a worker's life, she performs various duties within the hive sequentially as she matures: feeding larvae, producing wax, building the honeycomb, passing out food, and guarding the hive. Although a worker may go on short play flights around the hive during this time, she does not venture very far. It is only after the first 20 days of her life that the worker takes long flights away from the hive to forage for food.

FIGURE 35-15 A bee colony.
The queen bee has a red spot on her back.

An interesting means of communication among the workers has evolved that enables them to communicate the location of flowers at which they can collect pollen and nectar. This communication is a form of body language called the **waggle dance.** The manner in which a worker shakes, or waggles, her abdomen communicates the distance to the flowers. The angle at which the worker dances indicates the direction her co-workers must fly (Figure 35-16).

The drones play almost no role in the hive other than to mate with the queen. They help keep the larvae warm by congregating near them in the morning while it is still cool. Drones also help distribute food. At the time of mating, many drones follow the queen out of the hive, following her pheromone trail, and mate in flight. After mating, the drones die. Drones that do not mate have no better fate, since the workers pinch them, sting them, and then throw them out of the hive as the winter approaches.

The queen bee controls the worker bees with a chemical she produces known as *queen substance*. This substance is a **pheromone,** a chemical produced by one individual that alters the physiology or behavior of other individuals of the same species. The bees ingest the queen substance by licking the queen and then pass the substance around from one to the other as they pass food. Queen substance renders the worker bees sterile and also inhibits the workers from making queen cells, compartments in which fertilized eggs can develop into queens. If the hive becomes too congested, however, the queen may begin producing less queen substance, resulting in the removal of the inhibition to produce queen cells. Workers make a half-dozen or more new queen cells in which replacement queens begin to develop. The old queen and a swarm of females and male drones leave to establish a new hive (Figure 35-17). The first new queen to emerge may kill the other candidate queens and assume rule or may create another swarm and leave to establish another hive.

The life-styles of social insects can be so unusual as to be bizarre, none more so than the little reddish ants—the leafcutters (Figure 35-18). Leafcutters live in the tropics, organized into colonies of up to several million individuals. These ants are farmers, growing crops of fungi beneath the ground. Their mound-like nests look like tiny underground cities covering more than a hundred square yards, with hundreds of entrances and chambers as deep as 16 feet underground. Long lines of leafcutters march daily from the mound to a tree or bush, hack its leaves into small pieces, and carry the pieces back to the mound. Each ant finds its way by following a trail of secretions left by those who came before it. Although these ants are nearly blind, they follow the scent by holding their antennae close to the ground. At the underground site, worker ants chew the leaf fragments into a mulch that they spread like a carpet in underground chambers. They wet the leaf mulch with their saliva and fertilize it with their feces. Soon a luxuriant lawn of fungi is growing, which serves as the sole food for all the ants,

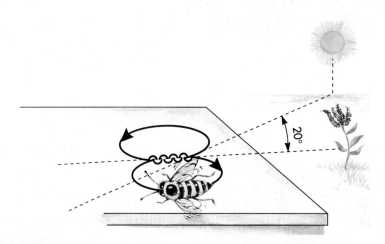

FIGURE 35-16 The waggle dance.
This dance is performed by a forager bee who, on spotting a food supply (a flower), goes back to the hive and communicates to the other bees through dance the exact location of the flower. The dance tells how far away the flower is and also the angle of the flower to the sun. The part of the dance that yields both kinds of information is the straight run, *(zigzag line between two circles).* The forager shakes its abdomen back and forth and buzzes with its wings while moving in the straight run. Then it makes a semicircle and returns to the starting point of the straight run. After completing this run, it circles in the opposite direction to land again at the beginning of the straight run and so on. How fast the bee performs the straight run determines how far away the flower is—a distance of 500 meters equals a straight run duration of about 1 second. The direction of the flower is indicated by the angle that the straight run deviates from the vertical. This angle equals the angle of the flower to the sun.

no matter what their age. Nurse ants even feed this fungus to the larvae, carrying them around to browse on choice spots.

The complex caste systems of bees and ants apparently evolved a long time ago. Ant fossils 80 million years old exhibit three castes (males, queens, and workers), indicating that their complex social system had already evolved.

> Many animals not only interact with other members of their species but form social groups and aid one another in various ways. Many species of insects, particularly bees and ants, form highly organized social groups exhibiting a division of labor.

Rank order in vertebrate groups

A social hierarchy, or **rank order,** exists in groups of fishes, reptiles, birds, and mammals. In chickens, the rank order is often called a **peck order** because these animals establish the order by pecking at one another. In many species of animals, the rank order is linear, with the highest-ranking individual dominant over all the others and the lowest-ranking individual submissive to all others. Some rank orders show certain complexities within the ranking, such as a rank order in which animal "A," for example, may be dominant over animal "B" and "B" dominant over "C," but "C" may be dominant over "A." In some rankings, the sex of the animal may play no role. In other rankings, males and females are ranked separately, and in still other vertebrate groups, the female takes on the rank of the male. Such rankings help reduce aggression and fighting within social groups, focusing it on the time that the ranking is developed. This behavior is only one of a repertoire of behaviors, including territoriality and submissive behaviors, that help contribute to the stability of the social relationships among animals and within groups.

Human behavior

Humans are unique animals. Because of this uniqueness, many scientists disagree with applying the principles of sociobiology to human behavior. Many scientists question the scientific validity of sociobiological research on human activity. Others question the social and political implications of human sociobiology, proposing that such endeavors might lead to yet another form of racism, sexism, or other types of group stereotyping. In addition, the study of human sociobiology has inherent limitations: scientists cannot experimentally manipulate the genes or vary the environments of humans for scientific studies. It is therefore extremely difficult to learn whether observed behaviors are a product of heredity or a person's environment. This "nature or nurture" dichotomy continually raises questions that scientists have a difficult time researching. Interesting observations have been made, however, in the study of identical twins raised by different families in different environments. From these studies and others, most biologists hold that much of the behavioral *capacity* of humans is genetically determined but that the neural circuits specified by genes can be shaped and molded—within certain limits—by learning.

FIGURE 35-17 A swarm resting on a horse chestnut branch.
These bees are in the process of establishing a new hive.

FIGURE 35-18 Leafcutter ants.
These leafcutter ants are carrying their day's harvest back to the nest. The ants do not live on the leaf material—they use it as a substrate on which to grow the fungi that they cultivate in underground farms.

Summary

1. Social behaviors help members of the same species communicate and interact with one another. Scientists who study the social behaviors of animals are sociobiologists. They seek to determine general laws of the biology and evolution of behavior.

2. Animals of the same species naturally compete for resources they need to survive. Within their repertoire of behaviors are aggressive behaviors that threaten and intimidate but do not kill rivals. Other behaviors also avoid fights to the death, such as submissive behaviors—signals of "backing down."

3. Animals also use territorial behavior to limit aggression within their species. Territoriality is very common in all classes of vertebrates and in some invertebrates. A territory is an area that an animal marks off as its own, defending it against same sex members of its species. This behavior helps disperse a population among limited resources and provides a breeding ground.

4. Reproductive behaviors include patterns of male-female interactions that lead to successful fertilization—the penetration of an egg by a sperm. These behaviors include methods by which mates attract each other and then interact with one another in types of courtships that lead to mating and fertilization.

5. Parenting behaviors include preparations for the care of young and their direct care to aid their survival and ultimate reproduction.

6. Individuals within species form temporary or permanent associations within social groups. Animals within social groups work together for a variety of common purposes. Some species of insects, particularly various species of bees and ants, are organized into social groups having a division of labor.

7. A social hierarchy, or rank order, exists in groups of fishes, reptiles, birds, and mammals. Behaviors relating to rank order help contribute to the stability of the social relationships among animals and within groups.

8. Some scientists question the scientific validity of applying sociobiological principles to human behavior.

REVIEW QUESTIONS

1. Distinguish between behaviors and social behaviors.
2. Explain the functions of social behaviors.
3. What is competitive behavior? Explain its significance.
4. Identify the type of behavior involved in the following situation:
   a. The hair on a dog's back stands on end as the dog growls at an approaching stranger.
   b. When you scold the dog for growling, it rolls over on its back and exposes its belly.
5. Define *territorial behavior*. Why is it important?
6. What is sexual reproduction? Explain the advantages that it can offer to a species.
7. Distinguish between mating and courtship.
8. Identify the type of behavior shown in each of the following situations:
   a. A robin builds its nest.
   b. A male peacock extends its tail feathers into a colorful fan.
   c. A cat chases another cat away from its food bowl.
9. Why do some animals form social groups?
10. Summarize the social structure of a honeybee colony.
11. Explain *rank order*. What is its significance, and in what animal(s) does it appear?
12. Do you feel the principles of sociobiology can be applied to human behavior? Explain your answer.

KEY TERMS

caste 619
competitive behavior 611
courtship 615
courtship ritual 616
insect societies 619
mating 615
peck order 621
pheromone 620
rank order 621
social behavior 611
sociobiology 611
submissive behavior 611
territorial behavior 612
threat (intimidation) displays 611
waggle dance 620

THOUGHT QUESTION

1. Very few mammals exhibit the complex societies seen among bees and ants—but a few do. Naked mole rats of the Middle East, for example, maintain large colonies with queens and special worker castes, organized remarkably like the colonies of bees. Why do you think such societies are so much rarer among mammals?

FOR FURTHER READING

Borgia, G. (1986, June). Sexual selection in bowers birds. *Scientific American*, pp. 92-100. *Behavior among birds is often bizarre, but none is more so than that of the Australasian bower birds, in which the females choose their mates depending on how well they adorn their bowers or ritualized nests.*

Bull, J. J., et al. (1992, April 3). Selfish genes. *Science*, pp. 65-66. *An analysis of a new class of selfish genes—genes that are successful at propagating themselves while being detrimental to the organisms that carry them—is presented in this article. The authors offer a lucid explanation for the paradox of selfish genes, implicating the more controversial aspects of sociobiology.*

Emlen, S. T., & Wrege, P. (1992, March 26). Parent-offspring conflict and recruitment of helpers among bee-eaters. *Nature*, pp. 331-332. *This short article describes the conflict between parent and offspring bee-eaters, in which fathers disrupt their sons' attempts to breed so that the sons can help raise their father's other offspring. Surprisingly, the sons do not resist their father's harassment.*

Heinrich, B. (1989). *Ravens in winter: A zoological detective story.* New York: Summit Books. *What do ravens do in winter, and how are their social systems organized? Vermont zoologist Bernd Heinrich talks about his outstanding field investigations of these intelligent animals.*

36 Population ecology

HIGHLIGHTS

Under ideal conditions, populations grow at an exponential rate, leveling off at the carrying capacity of the environment.

Very small populations are less able to survive than large populations; in addition, inbreeding in small populations leads to a loss of genetic diversity and increases the probability of the extinction of that species.

Large populations face threats to survival such as competition for resources such as food, light, and shelter and predation by other organisms.

In 1992, the global human population of more than 5.3 billion people was growing at approximately 1.8% per year; at that rate, the human population will reach well over 10 billion people by 2029.

OUTLINE

An introduction to ecology
Population growth
   Exponential growth
   Carrying capacity
Population size
Population density and dispersion
Regulation of population size
   Density-independent limiting factors
   Density-dependent limiting factors
Mortality and survivorship
Demography
The human population explosion

Looking like homes for oversized mud wasps, these cliff swallow nests hang precariously from the face of a rock outcropping. Any rough, vertical surface will serve as a nesting site for these birds. In fact, cliff swallows have recently discovered that the sides of bridges work well for this purpose . . . and come complete with protective overhangs! As a result, cliff swallows, once found mainly in the western part of the United States, can now be found inhabiting the prairie, nesting on the sides of bridges that span the major prairie rivers.

The cliff swallows that inhabit these nests are a population of organisms. A **population** consists of the individuals of a given species that occur together at one place and at one time. This flexible definition allows the use of this term in many contexts, such as the world's human population, the population of protozoans in the gut of an individual termite, or the population of blood-sucking swallow bugs living in the feathers of a cliff swallow.

An introduction to ecology

Population ecologists study how populations grow and interact. The science of ecology is much more broad, however, and includes the study of interactions between organisms and the environment. Scientists usually classify the study of ecological interactions into four levels: populations, communities, ecosystems, and the biosphere. This chapter discusses the first level of this hierarchy: populations.

Population growth

Most populations will grow rapidly if the ideal conditions for growth and reproduction of its individuals exist. But why, then, is the Earth not completely covered in bacteria, cockroaches, or houseflies? Why do some populations change from season to season or year to year?

To answer these questions, you need to understand how the size of a population is determined. The size of a population at any given time is the result of additions to the population from births and from **immigration,** the movement of organisms into a population, and deletions from the population from deaths and **emigration,** movement of organisms out of a population. Put simply, (births + immigrants) - (deaths + emigrants) = population change.

These statistics (births and deaths) are often expressed as a *rate:* numbers of individuals per thousand per year. For example, the population of the United States at the beginning of 1989 was approximately 250 million people. During 1989 there were the following:

- 3,975,000 live births: The birth rate was 3,975,000 per 250,000,000 people, or 15.9 births per 1000.
- 2,175,000 deaths: The death rate was 2,175,000 per 250,000,000 people, or 8.7 deaths per 1000.
- 570,009 immigrations: The immigration rate was 570,009 per 250,000,000 people, or 2.4 immigrants per 1000.
- 177,600 emigrations: The emigration rate was 177,600 per 250,000,000 people, or 0.7 emigrants per 1000.

The population change in the United States in 1989 can be calculated as follows: (15.9 births/1000/year + 2.4 immigrants/1000/year) - (8.7 deaths/1000/year + 0.7 emigrants/1000/year) = 8.9 people/1000/year. This figure can also be expressed as a population change of 0.89%—an increase of nearly 1%.

> The size of a population is the result of additions to the base population due to births and immigrations, and deletions from the population due to deaths and emigrations.

In natural populations of plants and animals, immigration and emigration are often minimal. Therefore, a determination of the **growth rate** of a population does not include these two factors. Growth rate (r) is determined by subtracting the death rate (d) from the birth rate (b):

$$r = b - d$$

Using the figures from our previous example:

$$r = 15.9 \text{ births}/1000/\text{year} - 8.7 \text{ deaths}/1000/\text{year}$$
$$r = 7.2/1000 \text{ or } 0.0072$$

To figure out the number of individuals added to a population of a specific size (N) in a given time *without* regard to immigration and emigration, r is multiplied by N:

$$\text{population growth} = rN$$

Therefore the population growth in the United States in 1989 solely from births and deaths was 0.0072 x 250,000,000 people = 1,800,000 people

> The determination of the growth rate of a population does not include the factors of immigration and emigration. It is determined by subtracting the death rate from the birth rate.

Exponential growth

Although the *rate* of increase in population size *may stay the same*, the *actual increase* in the number of individuals *grows*. This sort of growth pattern is similar to the growth pattern of money in the bank as interest is earned and compounded. If you put $1000 in the bank at 8% per year, the first year you will earn $80. The second year you will earn 8% interest on $1080, or $86.40. Although your interest rate has stayed the same, the amount of money you earn grows as your money grows. Actually, in a bank your earnings would grow even more quickly because interest would be posted and compounded more often than once a year.

POPULATION ECOLOGY

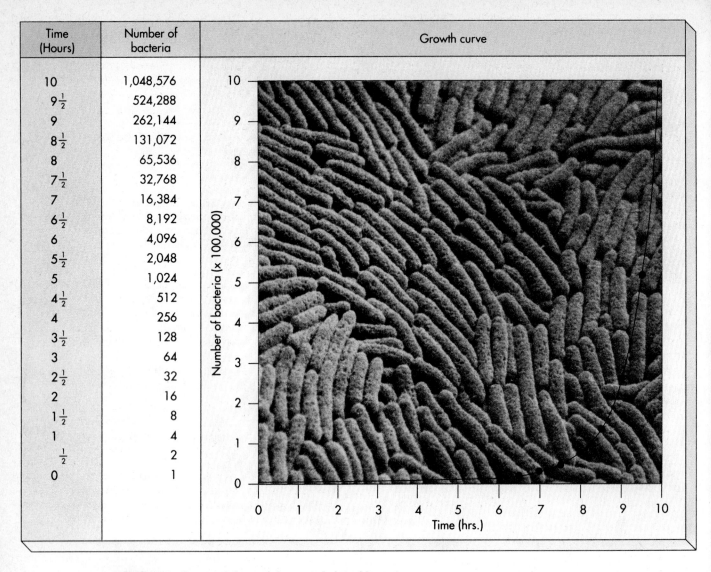

FIGURE 36-1 Exponential growth in a population of bacteria.
This period of rapid exponential growth can only be sustained as long as resources are abundant.

Figure 36-1 illustrates this principle with a population of bacteria in which each individual divides into two every half hour. The rate of increase remains constant, but the actual increase in the number of individuals accelerates rapidly as the size of the population grows. This type of mathematical progression found in the growth pattern of bacteria is termed **exponential growth.** (For example, 2 cells split to form 2^2 or 4, 4 become 2^3 or 8, and so on. The number, or power, to which 2 is raised is called an *exponent*.) Exponential growth refers to the rapid growth in numbers of a population of any species of organism, even though most organisms reproduce sexually and one organism does not split into two "new" organisms.

A period of exponential growth can occur only as long as the conditions for growth are ideal. In nature, exponential growth often takes place when an organism begins to grow in a new location having abundant resources. Such a situation occurred when the prickly pear cactus was introduced into Australia from Latin America. The species flourished, overrunning the ranges. In fact, the cactus became so abundant that cattle were unable to graze (Figure 36-2, *A*). Scientists regulated the population by introducing a cactus-eating moth to the area. The larvae of the moth fed on the pads of the cactus and rapidly destroyed the plants. Within relatively few years, the moth had reduced the population; the prickly pear cactus became rare in many regions where it was formerly abundant (Figure 36-2, *B*).

FIGURE 36-2 A cactus takes over Australia. After an initial period in which prickly pear cacti, introduced from Latin America, choked many of the pastures of Australia with their rampant growth, they were controlled by the introduction of a cactus-feeding moth from the areas where the cacti were native. **A** An infestation of prickly pear cacti in scrub in Queensland, Australia, in October 1926. **B** The same view in October 1929, after the introduction of the cactus-feeding moth.

Carrying capacity

No matter how rapidly a population may grow under ideal conditions, however, it cannot grow at an exponential rate indefinitely. As a population grows, each individual takes up space, uses resources such as food and water, and produces wastes. Eventually, shortages of important growth factors will limit the size of the population. In some populations such as bacteria, a buildup of toxic wastes may also limit population growth. Ultimately, a population stabilizes at a certain size, called the **carrying capacity** of the particular place where it lives. The carrying capacity is the number of individuals within a population that can be supported within a particular environment for an indefinite period. A population actually rises and falls in numbers at the level of the carrying capacity but tends to be maintained at an average number of individuals (Figure 36-3). The exponential growth of a population and its subsequent stabilization at the level of the carrying capacity is represented by an S-shaped **sigmoid growth curve** (after the Greek letter *sigma*).

> Under ideal conditions, populations grow at an exponential rate and show some stability in size at the carrying capacity of that place for that species.

FIGURE 36-3 The sigmoid growth curve. The sigmoid growth curve, characteristic of biological populations, begins with a slow period of growth quickly followed by a period of exponential growth. When the population approaches its environmental limits, the growth begins to slow down, and it finally stabilizes, fluctuating around the maximum number of individuals that the environment will hold. This maximum number is the carrying capacity.

Population size

The size of a population has a direct bearing on its ability to survive. Very small populations are less able to survive than large populations and are more likely to become extinct. Random events or natural disturbances can wipe out a small population, whereas a large population—simply due to its larger numbers—is more likely to have survivors. Inbreeding, reproduction between closely related individuals, is also a negative factor in the survival of small populations. Inbreeding tends to produce many homozygous offspring (see Chapter 22), which results in the expression of many recessive deleterious traits that are usually masked by dominant genes. In addition, inbreeding reduces the level of variability in the gene pool (the genes of all breeding individuals) of the population, detracting from the population's ability to adjust to changing conditions. Loss of genetic diversity therefore increases the probability of extinction of that species.

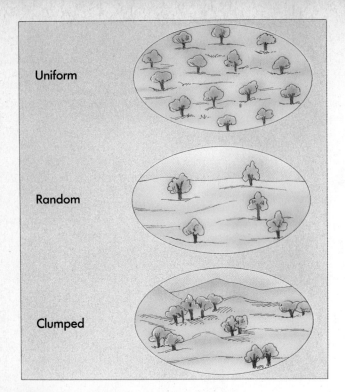

FIGURE 36-4 Distribution patterns in populations.
Uniform distribution patterns are rare in nature but are sometimes caused by allelopathy, in which plants secrete toxic chemicals that discourage the growth of other plants near them. Random patterns in plants are often caused by random seed dispersal, such as by wind. Clumped patterns, the most common type, result in an environment in which resources are unevenly distributed.

Population density and dispersion

In addition to a population's size, its **density**—the number of organisms per unit of area—influences its survival. For example, if the individuals of a population are spaced far from one another, they may rarely come into contact. Sexually reproducing animals cannot produce offspring if they do not mate. Therefore the future of such a population may be limited even if the absolute numbers of individuals over a wide area are relatively high.

A factor related to population density is **dispersion,** the way in which the individuals of a population are arranged. In nature, organisms within a population may be distributed in one of three different patterns: uniform, random, and clumped (Figure 36-4). Each of these patterns reflects the interactions between a given population and its environment, including the other species that are present.

Uniform, or evenly spaced, distributions are rare in nature. Populations of plants exhibiting allelopathy, the secretion of toxic chemicals that harm other plants, often show a uniform distribution. For example, the creosote bush, often the dominant vegetation covering wide areas of the deserts of Mexico and the southwestern United States, grows well spaced and evenly dispersed. This uniform pattern of distribution is probably due to chemicals secreted by the bush that retard the establishment of other individuals near established ones.

Random distributions occur if individuals within a population do not influence each other's growth and if environmental conditions are uniform—that is, if the resources necessary for growth are distributed equally throughout the area. Random distributions are often seen in plants as the result of certain types of seed dispersal, such as scattering by the wind.

Clumped distributions are by far the most frequent in nature. Organisms that show a clumped distribution are close to one another but far from others within the population. Clumping occurs as a result of the interactions among animals, plants, microorganisms, and unevenly distributed resources in an environment. Organisms are found grouped in areas of the environment that have the resources they need. Furthermore, animals often congregate for a variety of other reasons, such as for hunting, mating, or caring for their young; territorial behavior (see Chapter 35) is a means by which some animals group themselves for such purposes.

Regulation of population size

As a population grows and its density increases, competition among organisms for resources such as food, shelter, light, and mating sites increases and toxic waste products accumulate. Factors resulting from the growth of a population *regulate its subsequent growth,* and are "nature's way" of keeping the population size of every species in check. Such factors increase in effectiveness as population density increases and are appropriately termed **density-dependent limiting factors.** Other factors such as the weather, availability of soil nutrients, and physical disruptions of an area (such as volcanoes or earthquakes) can also limit the growth of a population. Because these factors operate regardless of the density of a population, they are called **density-independent limiting factors.**

> Density-dependent limiting factors come into play when a population size increases in a given area; density-independent limiting factors operate regardless of population size.

Density-independent limiting factors

A variety of environmental conditions can limit populations. For example, freak snowstorms in the Rocky Mountains of Colorado in the summer can kill butterfly populations there. The size of insect populations that feed on pollen and flower tissues vary seasonally with the blooming of flowering plants. Humans, too, can affect the sizes of populations. Poachers have killed so many African elephants for their ivory, for example, that the species may become extinct.

Density-dependent limiting factors

Individuals within a species and individuals of differing species **compete** for the same limited resources. (Chapter 37 discusses interspecific competition in detail and points out that competition among different species of organisms is

The urban explosion

Cities are a relatively new phenomenon. Villages and towns were first organized only about 5000 years ago, long after the development of agriculture. The first great cities of Mesopotamia date soon after that, but relative to a modern metropolis, the cities of ancient times were small. Only in the last thousand years have people lived in cities containing hundreds of thousands of people. Until the Industrial Revolution, only one person in five lived in a town of over 10,000 people. In the past hundred years, however, there has been a mass influx of people into cities; 14% of the world's population was urbanized in 1900. Starting in the industrial north, this mass migration has since spread to the less developed nations of the tropics. The cities of the Third World have swollen to hold one third of its total population. In 1920 the world's urban population was 360 million. By the year 2000, it will be near 3 billion—over 50% of the world's population will be urbanized. Two thirds of all people will be in cities of 100,000 or more, and over 20% in cities with more than a million people.

This massive movement of people is alarming because it is difficult to manage so many people crowded together. Mexico City, the largest city in the world today, is plagued by smog, traffic, waste disposal, and other problems, all worsened by the incredible congestion of over 20 million inhabitants (Figure 36-A). The prospects of supplying adequate food, water, and sanitation to these people, in a city whose population will increase to over 30 million by the end of the century, are almost unimaginable.

Nor is Mexico City's astonishing urban growth unique. Only 7 cities had a population larger than 5 million in 1950; by 2000 there will be 57 such megacities, 42 of them in the Third World. New York City has almost 16 million inhabitants, and Tokyo and São Paulo are bigger than New York! These four cities alone contain over 70 million inhabitants—half as many as the entire human population in the year 1 AD. In 1992 there were 16 cities with populations greater than 10 million. By 2000 there will be 25 such cities—and all but 3 of these will be in the northern hemisphere.

Most major Third World cities are really two cities: Shanty towns ring the outskirts of most megacities—in Calcutta they house 67% of the city's population. What draws the people to the city? Urban incomes in the northern hemisphere average three times rural incomes, and modern services such as doctors, teachers, sanitation, clean water, and electricity are at least within reach. Appalling though conditions are, they can be even worse in the countryside. As population numbers swell and resources become even more stretched, the difference between urban and rural becomes greater and the mass movement to the cities intensifies even more.

Management of cities will be a priority for governments in the twenty-first century. Environmentally sound ways of disposing of waste, provision of livable housing, and the supply of safe food and water to the multitudes living in large urban areas are just some of the concerns that will be at the forefront of city management.

FIGURE 36-A Mexico City, the world's largest city.

---

often greatest between those that obtain their food in similar ways.) In fact, if two species are competing with one another for the same limited resource in a specific location, the species able to use that resource most efficiently will eventually eliminate the other species in that location. This concept is called the principle of **competitive exclusion.** In fact, competition among organisms of the same and differing species was described by Charles Darwin as resulting in natural selection and survival of the fittest or most well-adapted organisms. Competition, therefore, not only limits the sizes of populations but is one of the driving forces of evolutionary change.

Predation is another factor that limits the size of populations and works most effectively as the density of a population increases. Predators are organisms of one species that kill and eat organisms of another—their prey. The intricate interactions between predators and prey are an essential factor in the maintenance of diverse species living in the same area. By controlling the levels of some species, the predators make the continued existence of other species in that same community possible. In other words, by keeping the numbers of individuals of some of the competing species low, the predators prevent or greatly reduce competitive exclusion. In fact, a given predator may very often feed on two or more different kinds of plants or animals, switching from one to the other as their relative abundance changes. Similarly, a given prey species may be a primary source of food for increasing species of predators as it becomes more abundant, a factor that will limit the size of its population automatically.

POPULATION ECOLOGY

Parasitism also limits the size of populations by weakening or killing host organisms. Parasites live on or in larger species of organisms and derive nourishment from them. As a population increases in density, parasites such as bacteria, viruses, and a variety of invertebrates can more easily move from one organism to another, infecting an increasing proportion of a population. Once again, this limiting factor to population size acts in negative feedback fashion, becoming more effective as the density of the population increases.

> Three density-dependent limiting factors are competition, predation, and parasitism.

Mortality and survivorship

A population's growth rate depends not only on the availability of needed resources and on the ability of its individuals to survive and compete effectively for those resources but also on the ages of the organisms in it. Interestingly, when a population lives in a constant environment for a few generations, its **age distribution**—the proportion of individuals in the different age categories—becomes stable. This distribution, however, differs greatly from species to species and even to some extent within a given species from place to place.

Scientists express the age distribution characteristics of a population by means of a survivorship curve. **Survivorship** refers to the proportion of an original population that survives to a certain age. The curve is developed by graphing the number of individuals within a population that survive through various stages of the life span. Samples of survivorship curves that represent certain types of populations are shown in Figure 36-5.

In the hydra, individuals are equally likely to die at any age, as indicated by the straight survivorship curve (type II). This type of survivorship curve is characteristic of organisms that reproduce asexually, such as hydra, bacteria, and asexually reproducing protists. Oysters, on the other hand, produce vast numbers of offspring, but few of these offspring live to reproduce. The rate of death, or **mortality**, of organisms that reach reproductive age is extremely low (type III survivorship curve). This type of survivorship curve is characteristic of organisms producing offspring that must survive on their own and therefore die in large numbers when young because of predation or their inability to acquire the resources they need. The survivorship curve for humans and other large vertebrates is much different from that for hydra and oysters. Humans, for example, produce few offspring but protect and nurture them; therefore most humans (except in areas of great poverty, hunger, and disease) survive past their reproductive years (type I survivorship curve).

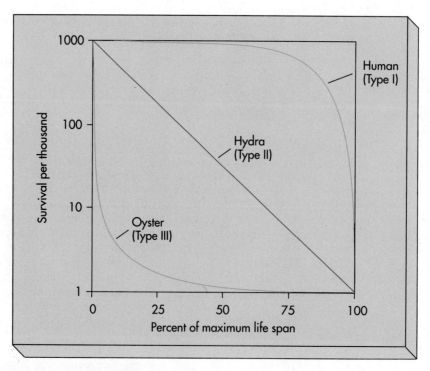

FIGURE 36-5 Survivorship curves.
The shapes of the respective curves are determined by the percentages of individuals in populations that die at different ages.

Many animal and protist populations in nature have survivorship curves that lie somewhere between those characteristic of type II and type III. Many plant populations, with high mortality at the seed and seedling stages, have survivorship curves close to type III. Humans have probably approached type I more and more closely through the years, with the life span being extended because of better health care and new medical technology.

▶ When a population lives in a constant environment for a few generations, the proportion of individuals in various age categories becomes stable. Characteristic age distributions occur among species and are closely linked in animal populations to parental care for offspring. ◀

Demography

Demography is the statistical study of populations. The term comes from two Greek words: *demos,* "the people" (the same root in the word *democracy*), and *graphos,* "to write." It therefore means the description of peoples and the characteristics of populations. Demographers predict the ways in which the sizes of populations will alter in the future, taking into account the age distribution of the population and its changing size through time.

A population whose size remains the same through time is called a **stable population.** In such a population, births plus immigration exactly balances deaths plus emigration. In addition, the number of females of each age group within the population is similar. If this were not the case, the population would not remain stable. For example, if there were many more females entering their reproductive years than older females leaving the population, the population would grow.

The age distribution of males and females in human populations of Kenya, the United States, and Austria is shown in Figure 36-6. A **population pyramid** is a bar graph that shows the composition of a population by age and sex. Males are conventionally enumerated to the left of the vertical age axis and females to the right. By using population pyramids, scientists can predict the future size of a population. First, the number of females in each age group is multiplied by the average number of female babies women in that age group bear. These numbers are added for each age group to see whether the new number will exceed, equal, or be less than the number of females in the population being studied. By such means the future growth trends of the human population as a whole and of individual countries and regions can be determined.

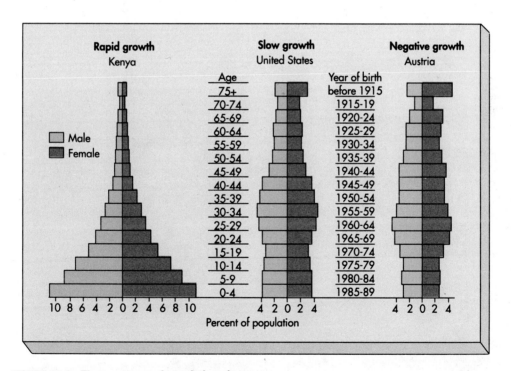

FIGURE 36-6 Three patterns of population change.
Kenya shows a rapid growth pattern—its pyramid has a broad base because of high fertility. The United States and Austria both show declining fertility and mortality, but the United States' "baby boom" has resulted in higher fertility rates than Austria, whose population is declining.

POPULATION ECOLOGY

The population pyramids in Figure 36-6 show the differences in the pattern of a rapidly growing population (Kenya), a slowly growing population (the United States), and a country experiencing negative growth (Austria). The population pyramid of Kenya is characteristic of **developing countries,** those that have not yet become industrialized, such as countries in Africa, Asia, and Latin America. Each of these countries has a population pyramid with a broad base, reflecting the large numbers of individuals yet to enter their reproductive years. In Kenya, for example, 50% of the population is under the age of 15. These children will reach reproductive age in the near future, leading to a population explosion in Kenya over the next decade.

In the United States, birth rates are higher than death rates at present, producing a growth rate of approximately 0.7%. The high birth rate is not due to couples having large families but to the large size of the "baby boom" generation who are at the peak of their reproductive years. Individuals in this age group were born within the 20 years or so following World War II. The large number of women in this group causes the births to still outnumber the deaths. Austria and the United States are both experiencing a decline in fertility and mortality. However, Austria's population does not include as high a percentage of women in their childbearing years as the United States, so that deaths are outnumbering births.

> Developing countries, those that have not yet become industrialized, have proportionately young populations and are experiencing rapid growth. Developed countries have populations with similar proportions of their populations in each age group, and are growing very slowly or not at all.

The human population explosion

Although some countries have populations that are no longer growing, such as Denmark, West Germany, Hungary, and Italy, and some countries such as Austria are declining in numbers, the population of the world as a whole is growing at the rate of 1.8% a year. This growth rate may sound low, but with the world population numbering over 5.3 billion, it adds over 95 *million* people to the population each year. Scientists estimate that the world population will double between 1990 and 2029!

How did the human population reach its present-day size? With the development of agriculture 11,000 years ago (Figure 36-7), human populations began to grow steadily (Figure 36-8). Villages and towns were first organized about 5000 years ago, and human effects on the environment began to intensify. In these centers of civilization, however, the specialization of professions such as metallurgy became possible; technology advanced. By 1650, the world population totaled approximately 500 million people. The Renaissance in Europe, with its renewed interest in science, ultimately led to the establishment of industry in the seventeenth century and to the Industrial Revolution of the late eighteenth and early nineteenth centuries. By the mid-nineteenth century, Louis Pasteur put forth the germ theory of disease, the understanding that microbes caused infection. With this understanding came new medical technology and discoveries. In the 1920s, Alexander Fleming accidentally discovered penicillin and opened the door to antibiotic therapy—medicine's "magic bullets" against bacterial infection. These medical advancements decreased the death rate by increasing the number of individuals surviving infection.

FIGURE 36-7 The development of agriculture was a key step in the growth of human populations. By producing abundant supplies of food, agriculture made the growth of cities and the future development of human culture possible. These workers are using a primitive method to thresh sorghum in southern India.

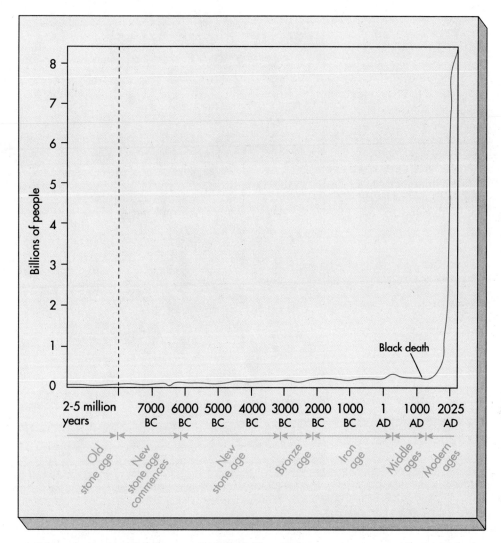

FIGURE 36-8 World population growth through history.
Except for the blip caused by the Black Death (25 to 35 million people died in Europe alone), the human population has grown steadily since the Stone Age.

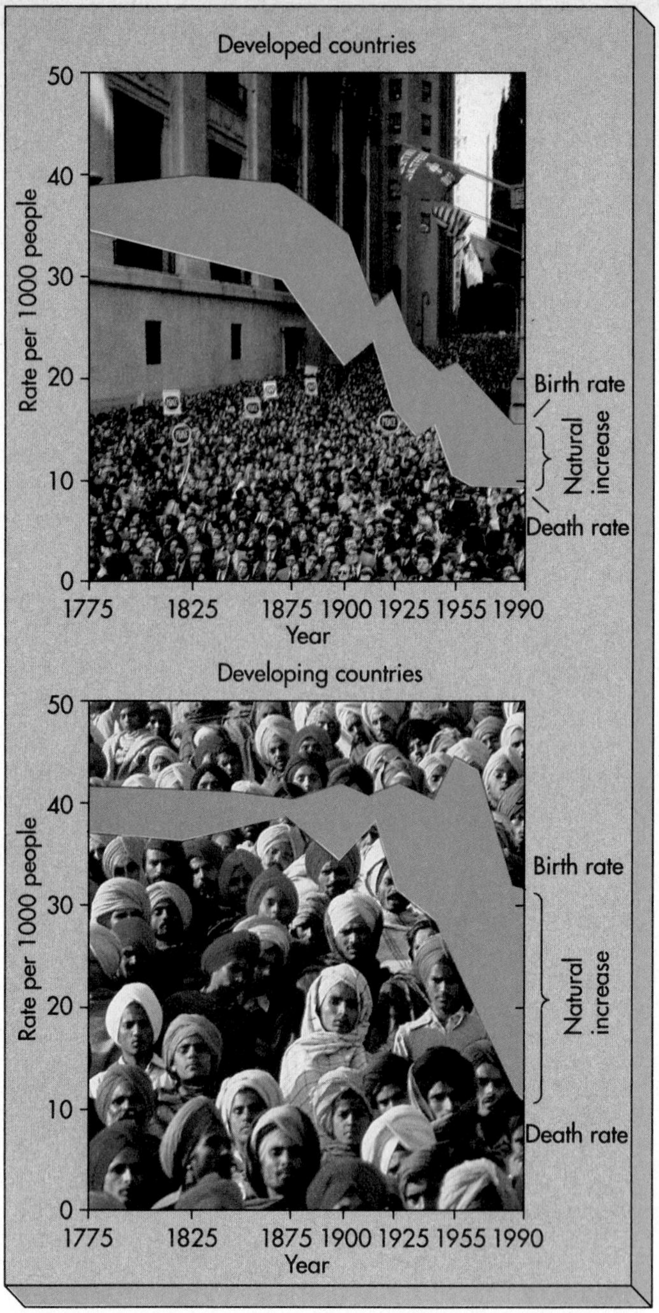

FIGURE 36-9 Population growth from 1775 to 1990. In developed countries, the Industrial Revolution caused mortality rates to drop and birth rates to remain unchanged **(A)**. In developing countries, mortality rates began to decline after World War II, but birth rates remained high **(B)**. Taken together, these changes caused human population growth rates to soar, especially in developing countries.

The advent of the Industrial Revolution also heralded new farming and transportation technology, which helped provide better nutrition for many people, especially those in industrialized countries. With better nutrition and increased medical understanding and technology, the death rate fell steadily and dramatically from the mid-nineteenth century on (Figure 36-9, *A*). In developing countries, international foreign aid imported this new technology along with food aid after World War II. The mortality rate plunged in a matter of years (Figure 36-9, *B*). However, birth rates remained largely unchanged, and as a result, the world's population growth rate soared. In fact, at their present rates of growth, the population of tropical South America will double in 31 years and that of tropical Africa will double in only 24 years. By the year 2000, about 60% of the people in the world will be living in countries that are at least partly tropical or subtropical, 20% will be living in China, and the remaining 20% will be in the developed countries of Europe, the Commonwealth of Independent States (formerly the Soviet Union), Japan, the United States, Canada, Australia, and New Zealand together (Figure 36-10). Putting it another way, for every person living in an industrialized country like the United States in 1950, there were two people living elsewhere; by 2020, just 70 years later, there will be five.

> The world population rose sharply and dramatically after the Industrial Revolution because of new technology in agriculture, transportation, industry, and medicine.

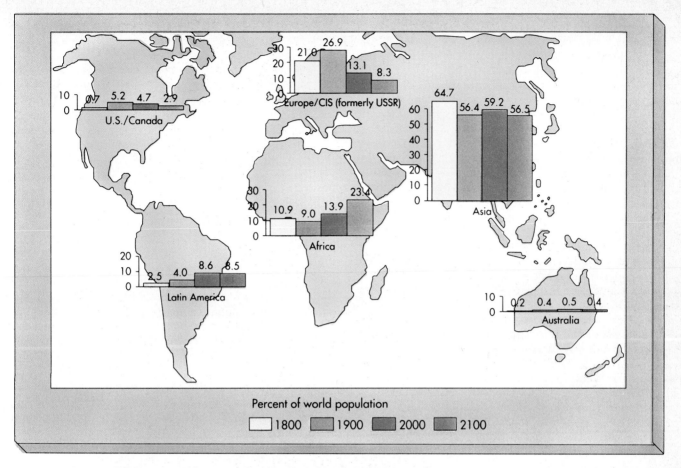

FIGURE 36-10 World population distribution by region, 1880 to 2100.
If current trends continue, Asia will have 57% of the total world population in 2100, Africa nearly a quarter, and Europe's share will drop to less than 10%.

Of the estimated 2.8 billion people living in the tropics in the late 1980s, the World Bank estimated that about 1.2 billion people were living in absolute poverty. (The World Bank is an international bank that provides loans and technical assistance for economic development projects in developing member countries.) These people cannot reasonably expect to be able to consistently provide adequate food for themselves and their children. Even though experts estimate that enough food is produced in the world to provide an adequate diet for everyone in it, the distribution is so unequal that large numbers of people live in hunger. The United Nations Children's Fund (UNICEF) estimates that in the developing world, about 14 million children under the age of 5—40,000 per day—starve to death each year, mainly of malnutrition and the complications associated with it (Figure 36-11).

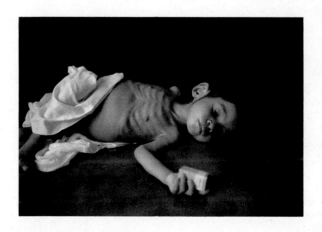

FIGURE 36-11 The face of starvation.
This Colombian child suffers from malnutrition. UNICEF estimates that 14 million children die each year as a consequence of malnutrition.

The size of human populations, like those of other organisms, is or will be controlled by the environment. Early in its history, human populations were regulated by both density-dependent and density-independent limiting factors, including food supply, disease, and predators; there was also ample room on Earth for migration to new areas to relieve overcrowding in specific regions. In the past century, however, humans have been able to expand the carrying capacity of the Earth because of their ability to develop technological innovations. Gradually, changes in technology have given humans more control over their food supply and enabled them to develop superior weapons to ward off predators, as well as the means to cure diseases. Improvements in transportation and housing have increased the efficiency of migration. At the same time, improvements in shelter and storage capabilities have made humans less vulnerable to climatic uncertainties.

As a result of the ability to manipulate these factors, the human population has been able to grow explosively to its present level of more than 5.3 billion people. Both the current human population level and the projected rate of growth have potential consequences for the future that are extremely grave. Pressures are placed on the land, water, forests, and other natural resources. Industrialization, although raising the standard of living by increasing the availability of goods and services, adds to air and water pollution. In the developed countries, nations of consumers have developed into "throwaway" societies, adding billions of tons of solid waste to landfills every year.

The most effective means of dealing with the population explosion has been the support of governments to encourage small families, the establishment of family planning clinics, improvement in education, and socioeconomic development. The developed countries of Western Europe and North America, Japan, and Australia have very low birth rates at this time. Mexico, Indonesia, Thailand, South Korea, Hong Kong, and Singapore have had considerable success with lowering their birth rates. Countries such as China and India have had some success in reducing their birth rates but are still striving toward this goal. The countries in sub-Sahara Africa have the highest birth rate of any countries in the world—some as high as 4.1%. Family planning programs are now being implemented across the continent.

Summary

1. Populations consist of the individuals of a given species that occur together at one place and at one time. They may be dispersed in an evenly spaced, clumped, or random manner. Clumped patterns are the most frequent.

2. The rate of growth of any population is the difference between the birth rate and the death rate per individual per unit of time. The actual rate is affected by emigration from the population and immigration into it.

3. Most populations exhibit a sigmoid growth curve, which implies a relatively slow start in growth, a rapid increase, and then a leveling off when the carrying capacity of the species' environment is reached.

4. Survivorship curves are used to describe the characteristics of growth in different kinds of populations. Type I populations are those in which a large proportion of the individuals approach their physiologically determined limits of age. Type II populations have a constant mortality throughout their lives. Type III populations have very high mortality in their early stages of growth, but an individual surviving beyond that point is likely to live a very long time.

5. Each population grows in size until it eventually reaches the limits of its environment to support it; resources are always limiting. Some of the limits to the growth of a population are related to the density of that population, but others are not. Competition within the populations and between populations of any two species limits their coexistence. Predation and other forms of interaction among populations also play an important role in limiting population size.

6. Agriculture was developed in several centers about 11,000 years ago when about 5 million people lived throughout the world. By 1990, the global population was more than 5.3 billion people and was growing at a rate of 1.8% per year, a rate that will double the population in approximately 39 years.

7. In the 1980s, the World Bank estimated that about 1.2 billion of the more than 2.75 billion people living in the tropics and subtropics existed in absolute poverty. Some 4 million children under 5 years of age were starving to death each year. With the populations of tropical countries growing from 1.8% to 4.1% per year, the task of feeding these people will be extraordinarily difficult.

REVIEW QUESTIONS

1. What is ecology?
2. Fill in the blanks, and explain what this equation means: _____ _____ = (births + immigrants) − (deaths + emigrants)
3. Fill in the blanks: Under ideal conditions, a population grows rapidly (_____ _____) until it stabilizes at a certain size called the _____ _____ of its particular environment.
4. Which is more likely to survive, a small population or a large one? Why?
5. Distinguish between density and dispersion. How does each affect a population's chances for survival?
6. Identify the three patterns of population dispersion found in nature. Into which pattern do human populations fall?
7. Distinguish between density-dependent and density-independent limiting factors. Give an example of each.
8. What do predation and parasitism have in common? How do they differ?
9. Draw type I, type II, and type III survivorship curves. Summarize the types of organisms that are characteristic of each curve, and give an example of each.
10. Compare the typical population pyramids of a developing country and industrialized country.
11. Fill in the blanks: The population of the world as a whole is growing at an annual rate of _____%, adding over _____ people to the population each year. Scientists estimate the world population will _____ between 1990 and 2029.
12. Both the current human population level and the projected rate of growth worldwide have potential consequences for the future that are extremely grave. Explain why.
13. Suppose that you were given the political power to deal with the world's population explosion. What steps would you take?

KEY TERMS

age distribution 630
carrying capacity 627
competitive exclusion 629
demography 631
density 628
density-dependent limiting factors 628
density-independent limiting factors 628
developing countries 632
dispersion 628
emigration 625
exponential growth 626
growth rate 625
immigration 625
mortality 630
parasitism 630
population ecologist 625
population pyramid 631
predation 629
sigmoid growth curve 627
stable population 631
survivorship 630

THOUGHT QUESTIONS

1. No other primate species has undergone a population explosion like humans. Why do you think humans alone have not controlled their population growth?
2. If female elephants have a single baby every 5 years and live on the average 60 years, how long would it take under ideal conditions for an elephant population to grow from 1000 individuals to 1 million individuals? Why does this not happen?
3. Much of the world's future seems to depend on bringing the world's human population growth rate under control. What do you think ought to be done?

FOR FURTHER READING

Ehrlich, P. R., & Roughgarden, J. (1987). *The science of ecology.* New York: Macmillan Publishing Company.
An excellent account of the entire field of ecology, this book is especially useful in defining the dynamics of populations.

Keyfitz, N. (1989, September). The growing human population. *Scientific American,* pp. 119-126.
This article discusses the general issues surrounding the burgeoning human population.

Population Reference Bureau. (Annual). *World population data sheet.* Washington, DC: Population Reference Bureau.
This annual data sheet provides the latest population statistics from around the world.

37 > Interactions within communities of organisms

HIGHLIGHTS

▼

A community is made up of populations of species that interact with one another.

▼

Each organism plays a special role within a community; this role is its niche.

▼

Members of the various species within a community interact by competing with one another for resources, killing one another for food, and living in close associations that both benefit and harm one another; these interactions affect the niche that a species can occupy in that community.

▼

Communities, like species, change over time in the dynamic process of succession.

OUTLINE

Ecosystems

Communities

Types of interactions within communities

  Competition

    The study of competition in the laboratory

    The study of competition in nature

  Predation

    Plant-herbivore coevolution

    Protective coloration

  Symbiosis

    Commensalism

    Mutualism

    Parasitism

Changes in communities over time: Succession

Although this photograph may look like an aerial view of a volcano-riddled landscape, these "volcanoes" will never compete with Krakatoa or Mount St. Helens. In fact, they are not volcanoes but tiny marine organisms living on a seashore rock. Each mound is an individual whose cone-like shell is toughened by lime and whose underside is firmly cemented to the rock. Together they form a population of acorn barnacles—organisms that are relatives of lobsters, crayfish, crabs, and shrimp. The acorn barnacles compete for food and space with a population of colonial invertebrates commonly called *moss animals*. These low-growing organisms look like dimpled areas near some of the barnacles.

Ecosystems

Scientists have long known that the nonliving or **abiotic** factors within the environment—such as air, water, and even rocks—affect an organism's survival, as do the living or **biotic** factors—such as surrounding plants, animals, and microorganisms. All of the biotic and abiotic factors together within a certain area are called an **ecosystem**. Therefore the rock is as important to the makeup of the ecosystem as the barnacles.

Within an ecosystem, each living thing has a home, an actual area in which it resides. This space, including the factors within it, is an organism's **habitat**. Organisms not only reside in their habitats, they interact with the biotic and abiotic factors within it and use them to survive. Each organism also plays a special role within an ecosystem; this role is called a **niche**. A niche may be described in terms of space, food, temperature, appropriate conditions for mating, and requirements for moisture. A full portrait of an organism's niche also includes the organism's behavior and the ways in which this behavior changes at different seasons and at different times of the day. These concepts are important to the understanding of the concept of communities.

Communities

The interactions among organisms within ecosystems are varied. Individuals of the same species make up a **population** of organisms. (The individuals within a population interact with one another in ways described in Chapters 35 and 36.) Populations also interact with one another, forming **communities**. Thus, an ecosystem can be thought of as a community of organisms, along with the abiotic factors with which the community interacts.

The magnificent redwood forest that extends along the coast of central and northern California and into the southwestern corner of Oregon is an example of a community. Within it, the most obvious organisms are redwood trees (Figure 37-1). However, populations of other organisms, such as the sword fern and the ground beetle in Figure 37-2, make up this community too. The coexistence of these various populations is made possible in part because of the special conditions that are created by the redwoods: shade, water (dripping from the branches), and relatively cool temperatures. For this reason and because the redwoods visually dominate the area, this distinctive group of populations is the redwood community.

> Individuals of a species compose a population; many populations make up a community. These populations of living things interact with one another in many complex ways.

FIGURE 37-1 The redwood community.
The redwood forests in coastal California and southwestern Oregon are dominated by the redwoods themselves.

FIGURE 37-2 Plants that occur regularly in the redwood community.
A The redwood sorrel.
B A ground beetle is feeding on a slug on a leaf of sword fern.

Types of interactions within communities

Interactions within communities can be grouped into the following three categories:
1. **Competition:** Organisms of different species that live near one another strive to obtain the same limited resources.
2. **Predation:** An organism of one species kills and eats an organism of another. (Animals that kill and eat members of their own species are cannibals.)
3. **Symbiosis:** Two or more organisms of different species live together in close association over a prolonged time (Figure 37-3).

Scientists study the interactions among organisms and between organisms and their environments in the laboratory and in nature. This specialized field of biology is called **ecology**; the scientists who work within this field are ecologists. Ecologists study the physical and biological variables governing the distribution and growth of living things. Ecologists also study the theoretical bases of these interactions, and some use computers to develop mathematical models of ecological systems. The knowledge gained by ecologists is essential to the basic understanding of the world and provides a foundation to find solutions to the many environmental problems created by humans.

Competition

When two kinds of organisms use the same limited resource for survival, they compete for that resource. Complex animals such as vertebrates compete by using innate behaviors such as threat displays and territorial behavior (see Chapter 35). Lower forms of animals and plants do not exhibit complex behaviors; they compete with one another simply by their adaptive fitness—by reproducing more and better-adapted offspring that crowd out opponents.

Competition among different species of organisms is often greatest between organisms that obtain their food in similar ways. Thus green plants compete mainly with other green plants, meat-eating animals with other meat-eating animals, and so forth. Competition within a genus or between individuals of the same species occurs as well.

> Organisms of different species compete with one another when they need the same limited resource for survival. This competition is greatest among organisms that obtain their food in similar ways.

Competition among organisms has been observed by scientists for a long time. Over 50 years ago, the Russian scientist G.F. Gause formulated the principle of **competitive**

FIGURE 37-3 Living—and dying—together.
Organisms that live within a community interact in a number of ways.
A Competition. The kudzu vine in the southern United States competes with other plants in the area for resources. However, the kudzu vine seems to be winning out in many locations.
B Predation. These lions have successfully ambushed their dinner.
C Symbiosis. This moray eel shelters a smaller fish in its mouth. The eel benefits from a thorough teeth cleaning, and the smaller fish receives an undisturbed meal.

exclusion, which was based on his experimental work. This principle states that if two species are competing with one another for the same limited resource in a specific location, the species able to use that resource most efficiently will eventually eliminate the other species in that location.

The study of competition in the laboratory

Scientists sometimes study competition between species in the laboratory so that they can control the environmental conditions. John Harper and his colleagues at the University College of North Wales, Australia, for example, performed competition experiments with two species of clover: white clover and strawberry clover. Each species was sown with the other at one of two densities: 36 or 64 plants per square foot. Various plots of the two species were planted, using all the possible combinations of the two densities of plants. One plot contained 36 white clover plants and 36 strawberry clover plants per square foot. Another contained 64 white and 36 strawberry per square foot, and so forth. The white clover initially formed a dense canopy of leaves in each experimental plot. However, the slower-growing strawberry clover, whose leaf stalks are taller, eventually produced leaves that grew above the white clover leaves. In competing more effectively for light, the strawberry clover overcame the white clover, causing it to die out. The outcome was the same, regardless of the initial densities at which the seeds of the plants were sown.

> In a competitive situation and under controlled laboratory conditions, the species able to use a particular resource more efficiently will be the species to survive.

The study of competition in nature

As well as competing for sunlight, plants compete for soil nutrients. The roots of one species, for example, may outcompete another species by using up minerals in the soil essential to both species. In addition, one species may secrete poisonous substances that depress the growth of other species. Sage plants, for example, inhibit the establishment of other plant species nearby, producing bare zones around populations of these plants (Figure 37-4). These interactions are poorly understood, but experimental studies are beginning to resolve their complexity.

Another interesting view of competition in nature is demonstrated by the acorn barnacles. Highly adapted to their environment, acorn barnacles are typically found in the intertidal zone of rocky shores—the narrow strip of land exposed during low tide and covered during high tide. When submerged, an acorn barnacle feeds by extending appendages from the hole in its shell. Spread out, these appendages act like a net, sweeping the water and collecting food that it then brings into its shell and eats. When exposed to the air, an acorn barnacle pulls in its feeding appendages and shuts down, actually using much less oxygen than when underwater. Interestingly, barnacles of the genus *Balanus* have been kept out of the water as long as 6 weeks without detectable ill effects. However, a relative of *Balanus* organisms, barnacles of the genus *Chthamalus* (pronounced with the first two letters silent), have been kept out of water for

FIGURE 37-4 Competition among plants.
The bare zones around these colonies of sage are plainly visible in this aerial photograph taken in the mountains above Santa Barbara, California. The plants secrete chemicals that suppress the growth of other plants nearby.

3 years, being submerged only 1 or 2 days a month—and they survived! Although both organisms have adaptations that make them well suited to the intertidal environment, their differences play an important role in determining where each genus lives. Of the two, *Chthamalus* barnacles live in shallower water, where they are often exposed to air as the tide rolls in and out. *Balanus* barnacles live deeper in the intertidal zone and are covered by water most of the time (Figure 37-5).

In studying these two genera of barnacles, J.H. Connell of the University of California, Santa Barbara, found that in this deeper zone, *Balanus* barnacles could always outcom-

FIGURE 37-5 *Chthamalus* and *Balanus* barnacles growing together on a rock.
Competing species belonging to these two genera were studied by J.H. Connell along the coast of Scotland. *Balanus* are the larger organisms of the two genera.

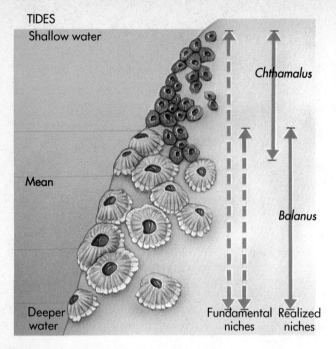

FIGURE 37-6 Competition can limit niche use.
The distribution of *Chthamalus* and *Balanus* barnacles with respect to different water levels is shown. In theory, *Chthamalus* barnacles can live in both deep and shallow zones, whereas *Balanus* organisms can live only in the deeper zone. These are the barnacles' fundamental niches. In reality, however, *Balanus* organisms outcompete *Chthamalus* barnacles in the deeper zone because *Balanus* barnacles use the resources of the deeper zone more efficiently. These are the barnacles' realized niches.

pete *Chthamalus* barnacles. *Balanus* organisms would crowd *Chthamalus* barnacles off the rocks, replacing it even where it had begun to grow. When Connell removed *Balanus* barnacles from the area, however, *Chthamalus* organisms were easily able to occupy the deeper zone, indicating that no physiological or other general obstacles prevented it from becoming established there. *Balanus* barnacles, however, must use the resources of the deeper zone more efficiently than *Chthamalus* organisms do, even though *Chthamalus* barnacles are able to survive there in the absence of its competitor. In contrast, *Balanus* barnacles cannot survive in the shallow-water where *Chthamalus* organisms normally occur. It evidently does not have the special physiological and morphological adaptations that allow *Chthamalus* barnacles to occupy this zone.

Along with illustrating the principle of competitive exclusion, these experiments with *Balanus* and *Chthamalus* barnacles illustrate that the role an organism plays in an ecosystem—its niche—can vary depending on the biotic and abiotic factors in the ecosystem. In this example, the niche occupied by *Chthamalus* barnacles is its **realized niche**—the role it *actually plays* in the ecosystem. It is distinguished from its **fundamental niche**—the niche that it *might* occupy if competitors were not present. Thus the fundamental niche of the barnacle *Chthamalus* in Connell's experiments included that of *Balanus* barnacles, but its realized niche was much narrower because *Chthamalus* organisms were outcompeted by *Balanus* organisms. However, the realized and fundamental niches of *Balanus* barnacles are the same (Figure 37-6).

Gause's principle of competitive exclusion can be restated in terms of niches as follows: no two species can occupy exactly the same niche indefinitely. Certainly, species can and do coexist while competing for the same resources. Nevertheless, Gause's theory predicts that when two species do coexist on a long-term basis, one or more features of their niches will always differ; otherwise the extinction of one species will inevitably result. The factors that are important in defining a niche are often difficult to determine, however, so Gause's theory can sometimes be difficult to apply or investigate.

> The role an organism plays in an ecosystem can vary depending on the biotic and abiotic factors in the ecosystem.

Predation

A **predator** is an organism that kills and eats other organisms, or **prey**. Predation includes one kind of animal capturing and eating another, an animal feeding on plants, and even a plant, such as the Venus flytrap shown in Figure 37-7, capturing and eating insects! In a broad sense, parasitism is also considered a form of predation. Parasitism is a close association between two organisms in which the parasite (predator) is much smaller than the prey but feeds on the prey, harming it and benefiting the parasite.

How do predator populations affect prey populations? When experimental populations are set up under very simple conditions in the laboratory, the predator often exterminates its prey and then becomes extinct itself because it has nothing to eat. This fact was illustrated nicely in experiments performed by Gause. In his experiments, Gause used populations of the two protozoans shown in Figure 37-8: *Didinium*, and its prey *Paramecium*. As shown in Figure 37-9, *A*, when *Didinium* protozoans are introduced into a growing population of *Paramecium* protozoans, the population instantly begins to decline and quickly dies out. The *Didinium* population lives on for a short while, then dies out itself.

If refuges are provided for the prey, however, its population can be driven to low levels but can recover. In another of Gause's experiments, he provided sediment in the bottom of the test tubes in which he was growing *Didinium* and *Paramecium* protozoans. Interestingly, as *Didinium* began to prey on *Paramecium*, only those organisms in the clear fluid of the test tubes were killed. Those in the sediment were not eaten. Eventually, *Didinium* protozoans died from lack of food; meanwhile, the *Paramecium* prey multiplied (Figure 37-9, *B*) and overtook the culture!

In another series of experiments, Gause discovered that when he introduced new prey at successive intervals (Figure 37-9, *C*), the decline and rise in the numbers of the predator-prey populations followed a cyclical pattern. As the number of prey increased, the number of predators in-

FIGURE 37-7 Different forms of predation.
A A cheetah eats its kill, an impala, while vultures wait patiently in the background.
B A gazelle feeds on the grasses of the African plain.
C This predatory plant, the Venus flytrap, inhabits the low, boggy ground of North and South Carolina. The Venus flytrap traps insects within its snapping leaves.

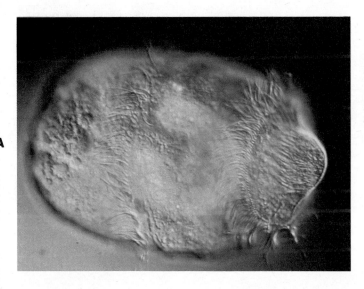
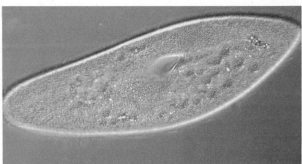

FIGURE 37-8 Gause's experimental subjects.
A *Didinium*.
B *Didinium*'s prey, *Paramecium*.

INTERACTIONS WITHIN COMMUNITIES OF ORGANISMS

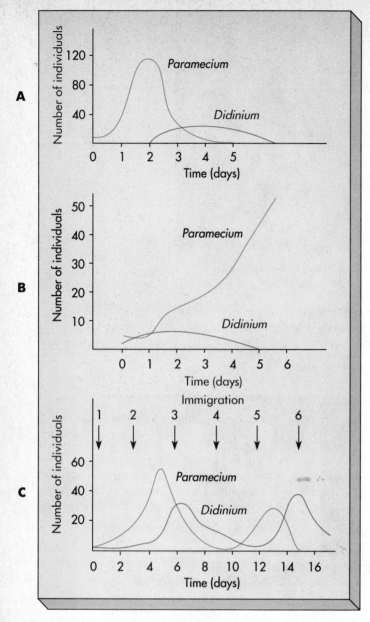

FIGURE 37-9 Outcome of Gause's experiments with *Paramecium* and *Didinium*.
A Results obtained when *Didinium* is introduced into a population of *Paramecium* protozoans.
B Results of the same experiment, only this time sediment was added to the test tubes, providing a place for *Paramecium* protozoans to hide.
C Results from third experiment in which new prey were introduced at intervals. As the number of prey increased, so did the number of predators.

creased. As the numbers of the prey were lowered by predation, the large predator population did not have enough food to eat; some died and the predator population declined. As this decline occurred, the prey recovered (aided by the addition of new organisms) and again became abundant, starting the cycle once again.

At one time, scientists thought that predator-prey populations always cycled in this manner. However, they have come to realize that in nature, conditions for survival are complex and do not always lead to such a cycling of populations. From experiments such as those described, scientists know that predators cannot survive when the prey population is low. Immigration of prey (movement of new prey into the community) may be necessary to sustain the predator population. From other experiments, scientists have come to learn that changes in predator-prey populations also depend on how the prey are dispersed in an area and the manner in which the predator searches for the prey. Factors other than the relationship between a single predator population and a single prey population also influence the survival and abundance of both predator and prey. For example, adverse weather conditions may result in the death of the predator and/or prey species, the predator may eat more than one type of prey and/or the prey may be eaten by more than one predator, and fluctuations may occur in the food source of the prey, limiting the survival of this population.

> The effect of predator populations on prey populations is difficult to predict because their complex interactions depend on their interactions with other organisms in the community, movement of new organisms into and out of the community, and the abiotic factors that influence their survival.

The intricate interactions between predators and prey often affect the populations of other organisms in a community. By controlling the levels of some species, for example, predators help species survive that may compete with their prey. In other words, predators sometimes prevent or greatly reduce competitive exclusion by limiting the population of one of the competing species. Such interactions among organisms involving predator-prey relationships are key factors in determining the balance among populations of organisms in natural communities.

Plant-herbivore coevolution

Plants, animals, protists, fungi, and bacteria that live together in communities have changed and adjusted to one another continually over millions of years. Such interactions, which involve the long-term, mutual evolutionary adjustment of the characteristics of the members of biological communities in relation to one another, are examples of **coevolution**.

Plants and plant-eating predators called **herbivores** are a group of organisms that change and adjust to one another over time. Natural selection favors plants that have developed some means of protection against herbivores. In the dynamic equation of coevolution, however, natural selection also favors adaptations that enable animals to prey on plants in spite of their protective mechanisms. To avoid being eaten, for example, some plants have developed hard parts that are difficult to eat or unpalatable. In fact, certain grasses defend themselves by incorporating silica (a component of glass) in their structure (Figure 37-10). If enough silica is present, the plants may simply be too tough to eat. Some groups of herbivores, however, have developed strong, grinding teeth and powerful jaws. In addition, herbivores such as cattle have developed adaptations of their digestive systems. One such adaptation allows them to store the grass they have eaten in a digestive pouch called a *rumen*. Bacteria that live in the rumen attack the grass chemically, aiding in

FIGURE 37-10 Plants can defend themselves.
These zebras grazing on the East African savanna dislike eating grasses with leaves reinforced with silica. The more silica in the leaf cells, the less likely zebras and other grazing animals are to eat that particular kind of grass.

the digestive process. This stored food is then regurgitated and rechewed at a later time, providing a better breakdown of the cell walls within the grass.

Some plants have developed chemical defenses against herbivores. The best known plant groups with toxic effects are the poison ivy, poison oak, and poison sumac plants. All contain the contact poison urushiol. Castor bean seeds are also toxic to a wide variety of animals, producing a protein that attaches to ribosomes, blocking protein synthesis. Other plants produce toxins that inhibit the growth of bacteria, fungi, and roundworms. Still others produce chemicals having odors that act as a warning or as a repellent to a predator. Today, using the techniques of genetic engineering, scientists have been able to grow plants that chemically repel certain predators, thereby reducing the need for artificial pesticides (Figure 37-11).

However, associated with each family or other group of plants naturally protected by a particular kind of chemical compound are certain groups of herbivores that are adapted to feed on these plants, often as their exclusive food source. For example, the larvae of cabbage butterflies feed almost exclusively on plants of the mustard and caper families, which are characterized by the presence of protective chemicals—the mustard oils. Although these plants are protected against most potential herbivores, the cabbage butterfly caterpillars have developed the ability to break down the mustard oils, rendering them harmless. In a similar example of coevolution, the larvae of monarch butterflies are able to harmlessly feed on the toxic plants of the milkweed and dogbane families (Figure 37-12).

> Over time, plants have developed various morphological and chemical adaptations that help protect them against plant eaters, or herbivores. In turn, however, herbivores have changed and adjusted to the plants and their adaptations. Such interactions, involving long-term mutual evolutionary adjustment, are coevolution.

FIGURE 37-11 Defense through genetic engineering.
Recently, scientists have developed plants that through genetic engineering, chemically repel predators. The tobacco plant on the *right* is a nonengineered plant and shows the effects of insect predation. The tobacco plant on the *left*, however, has been engineered to produce an insect toxin that deters insects and protects the plant from insects.

FIGURE 37-12 Monarch butterflies make themselves poisonous.
All stages of the life cycle of the monarch butterfly are protected from predators by the poisonous chemicals that occur in the milkweeds and dogbanes on which they feed as larvae. Both caterpillars and adult butterflies advertise their poisonous nature with warning coloration.

INTERACTIONS WITHIN COMMUNITIES OF ORGANISMS

BIOLOGY IN FOCUS

Competition and killer bees

In the fall of 1956 a prominent Brazilian bee geneticist named Warwick Kerr traveled to Africa and brought back, at his government's request, a small number of queen bees. These bees were reported to be high honey producers, and the government was hoping to improve local bee varieties by breeding this desirable trait into them. On their return, the bees were placed in quarantine at a research station 200 miles inland from Kerr's university and crossed to local bees. Although most of the African queens were commercial bees obtained by Kerr in South Africa, one that was used much more than the others in crosses was a wild bee from Tanzania. Workers quickly confirmed that the African bees produced a great deal of honey and were very fast breeders. They also noticed that they were unusually sensitive to disturbance, becoming quite aggressive (Figure 37-A).

In the fall of 1957 a visting beekeeper, not knowing of the quarantine, removed the queen-excluder doors (holes too small for the more massive queen to pass through) from the hives, normal practice in a nonquarantine colony if a queen has started laying. That day 26 queens left their hives, each with her swarm of 1000 to 2000 bees.

The swarms traveled long distances and quickly supplanted the populations of wild bees in the surrounding countryside. In these new surroundings, they were freed from the competition and predation that had kept them in check in their African home and armed with evolutionary advantages: aggressiveness, fast reproduction, far traveling ability, and frequent founding of new hives.

FIGURE 37-A A killer bee.

Within months, reports began to come in of farm animals killed, followed by the first human fatalities. The African bees had evolved in the highly competitive environment of their homeland an effective defense against attack—at the slightest hint of provocation, they attacked first, in overwhelming numbers. A human not sensitive to bee stings can survive some 200 to 300 stings but not the number these attacks inflicted. A University of Miami graduate student killed in Costa Rica in 1986 by a swarm of African bees had 8000 stings.

By 1970 the African bees had overrun all of Brazil. Local geneticists suspected that they were breeding with local bees, but molecular studies by American geneticists suggest that they simply replaced them. By 1979 the bees had reached Central America. They first appeared in Mexico in 1986. The first African bee was detected in Texas in October of 1990. In 35 years the bees have conquered over 5 million square miles, killing hundreds of thousands of farm animals and over 1000 people. When they eventually reach their climatic limits in the next 20 years, where winters are too cold for them to survive, they will extend in an arc from mid-California to North Carolina.

FIGURE 37-13 Warning coloration.
The coloring of all these animals is meant to warn other animals to stay away.
A The skunk ejects a foul-smelling liquid when threatened.
B The poisonous gila monster is a member of the only genus of poisonous lizards in the world.
C The red-and-black African grasshopper feeds on highly poisonous *Euphorbia* plants.
D The tiny tropical tree frog is so poisonous that Indians in western Colombia use their venom to poison their blow darts.

Protective coloration

Some groups of animals that feed on toxic plants receive an extra benefit—one of great ecological importance. When the caterpillars of monarch butterflies feed on plants of the milkweed family, for example, they do not break down the chemicals that protect these plants from most herbivores. Instead, they store them in fat within their bodies. As a result, the caterpillars and all developmental stages of the monarch butterfly are protected against predators by this "plant" poison. A bird that eats a monarch butterfly quickly regurgitates it. Although this is no help to the eaten insect, the bird will soon learn not to eat another butterfly with the bright orange and black pattern that characterizes the adult monarch. Such conspicuous coloration, which "advertises" an insect's toxicity, is called **warning coloration.** Warning coloration is characteristic of animals that have effective defense systems, such as poisons, stings, or bites. Other examples of animals that exhibit warning coloration are shown in Figure 37-13.

During the course of their evolution, many unprotected species have come to resemble distasteful ones that exhibit warning coloration. Provided that the unprotected animals are present in low numbers relative to the species they resemble, they too will be avoided by predators. If the unprotected animals are too frequent, of course, many of them will be eaten by predators that have not yet learned to avoid individuals with a particular set of characteristics. Such a pattern of resemblance is called **batesian mimicry,** after the British naturalist H.W. Bates, who first described this concept in the 1860s. Many of the best-known examples of batesian mimicry occur among butterflies and moths (Figure 37-14, *A* and *B*).

INTERACTIONS WITHIN COMMUNITIES OF ORGANISMS

FIGURE 37-14 Mimicry.
A The viceroy butterfly mimics the poisonous monarch (**B**). Larvae of the viceroy feed on willows or other nontoxic plants.
C The red-spotted purple mimics another poisonous butterfly, the pipevine swallowtail, which obtains its toxic chemicals from plants of the pipevine family on which its larvae feed.

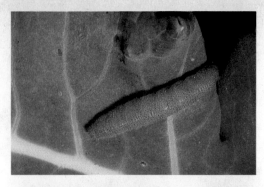

FIGURE 37-15 Insect herbivores are well suited to their hosts. The green caterpillars of the cabbage butterfly are camouflaged on the leaves of cabbage and other plants on which they feed. These caterpillars are able to break down the toxic mustard oils that prevent most insects from eating cabbage.

Another kind of mimicry, **müllerian mimicry,** was named for the German biologist Fritz Müller, a contemporary of Bates. Interestingly, in müllerian mimicry, the protective colorations of different animal species come to resemble one another as in batesian mimicry. However, unlike batesian mimicry, the organism and its mimic *do* possess similar defenses.

Some organisms are colored so as to blend in with their surroundings—a protective coloration called **camouflage.** Both cabbage caterpillars and cabbage butterflies have evolved a green coloration, allowing them to hide while feeding (Figure 37-15). Unlike monarchs, these insects do not store the toxic chemical in the plants they eat for use against predators. Instead, the cabbage butterflies break down the toxin. Similarly, insects that eat plants lacking specific chemical defenses are seldom brightly colored. Figure 37-16 shows examples of other animals who are camouflaged from their predators.

> Some organisms within communities exhibit conspicuous coloration, or warning coloration, that advertises their ability to poison, sting, or bite. Organisms without specific defenses are often colored, or camouflaged, so as to blend in with their surroundings.

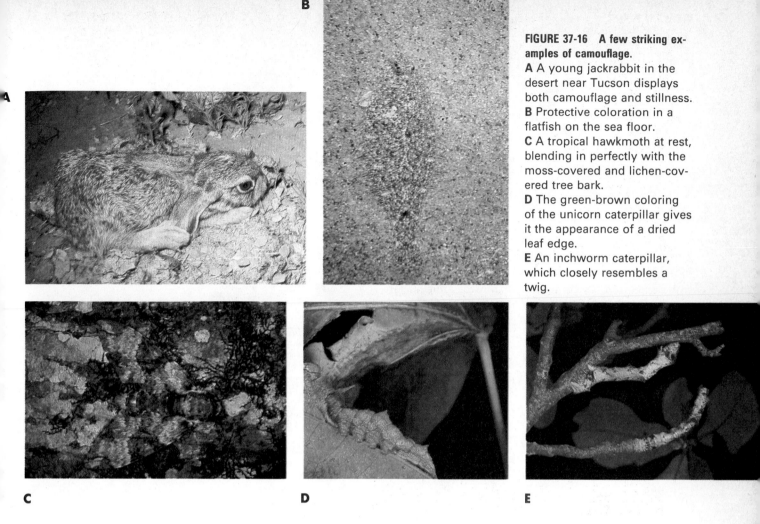

FIGURE 37-16 A few striking examples of camouflage.
A A young jackrabbit in the desert near Tucson displays both camouflage and stillness.
B Protective coloration in a flatfish on the sea floor.
C A tropical hawkmoth at rest, blending in perfectly with the moss-covered and lichen-covered tree bark.
D The green-brown coloring of the unicorn caterpillar gives it the appearance of a dried leaf edge.
E An inchworm caterpillar, which closely resembles a twig.

Symbiosis

Symbiotic relationships are those in which two different species of organisms live together in close association. All symbiotic relationships provide the potential for coevolution between the organisms involved, and in many instances the results of this coevolution are fascinating. The major kinds of symbiotic relationships include the following:

1. **Commensalism:** One species benefits and the other species neither benefits nor is harmed.
2. **Mutualism:** Both participating species benefit.
3. **Parasitism:** One species benefits but the other (the host) is harmed.

Parasitism is sometimes considered a form of predation. However, unlike a true predator, the successful parasite does not kill its host.

Commensalism

Many examples of the "one-sided" relationship of commensalism exist in nature. Often, the individuals deriving benefit are physically attached to the other species in the relationship. For example, plants called *epiphytes* grow on the branches of other plants. The epiphytes derive their nourishment from the air and the rain—not from the plants to which they attach for support (Figure 37-17). Similarly, various marine animals such as barnacles, grow on other,

FIGURE 37-17 Commensalism: epiphytes and trees.
The brownish growths attached to the tree limb are epiphytes. These plants derive their nourishment from the rain and air and use the tree limb for support only.

INTERACTIONS WITHIN COMMUNITIES OF ORGANISMS

FIGURE 37-18 Commensalism: barnacles and whales.
A This breaching whale displays the barnacles growing on its skin.
B A close-up of a gray whale's skin reveals hitchhikers—lice and barnacles. The lice are actually parasites, whereas the barnacles cause no harm to the whale.

often actively moving, sea animals (Figure 37-18). These "hitchhikers" gain more protection from predation than if they were fixed in one place, and they continually reach new sources of food. They do not, however, harm the organisms to which they are attached.

Possibly one of the best-known examples of commensalism involves the relationship between certain small tropical fish—the clownfish—and sea anemones, marine animals that have stinging tentacles. The fish have developed an adaptation that allows them to live among the deadly tentacles of the anemones (Figure 37-19). These tentacles quickly paralyze other species of fish, protecting the clownfish against predators.

Mutualism

When both species benefit in a close relationship, it becomes one of mutualism. A particularly striking example of mutualism involves one genus of stinging ants and a Latin American plant of the genus *Acacia*. The modified leaves of acacia plants appear as paired, hollowed thorns. These thorns provide a home for the ants, protecting them and their larvae. In addition, the ants eat nectar the plants produce. In turn, the ants attack any herbivore that lands on the branches or leaves of an acacia and clear away vegetation that comes in contact with their host shrub, increasing the plant's ability to survive.

FIGURE 37-19 Commensalism: clownfish and sea anemones.
Two clownfish peer out from the tentacles of a large red sea anemone off the coast of Australia.

Many other interesting examples of mutualism exist in nature. Certain birds, for example, spend most of their time clinging to grazing animals (such as cattle), picking insects from their hides. In fact, the birds carry out their entire life cycles in close association with the cattle. The birds are provided with food and the cattle benefit by having their parasites removed. In another similar mutualistic relationship, ants use the tiny insects aphids, or greenflies, as a provider of food. The aphids suck fluids from the phloem of plants, extracting a certain amount of sucrose and other nutrients. However, many of these nutrients are not absorbed within the digestive tract of the aphid. A substantial portion runs out (somewhat altered) through the anus. The ants use this nutritional excrement as a food source and, in turn, actually carry the aphids to new plants so that they can continue eating!

Parasitism

Parasites include viruses, many bacteria, fungi, and an array of invertebrates. A different species of organism, larger than the parasite itself, is "home" for a parasite. During the intimate relationship between parasite and host, the parasite derives nourishment and the host is harmed.

Many instances of parasitism are well known. Intestinal hookworms, for example, are parasites (Figure 37-20). A person is infected when walking barefoot in soil containing hookworm larvae. These larvae are able to penetrate the skin, entering the bloodstream. The blood carries the larvae to the lungs. From there, they are able to migrate up the windpipe to the esophagus. The larvae are then swallowed and reach the intestines. After growing into adult worms, they attach to the inner lining of the intestines. They remain attached here, feeding on the blood of the host.

Some parasites do not live within an organism as hookworms do but attach to the outer surface of a plant or animal. The attachment may be fleeting, as with the bite of a mosquito, or may take place over a longer period, such as the burrowing of mites.

FIGURE 37-21 Parasitism: dodder and plants.
Dodder has lost its chlorophyll, along with its leaves, in the course of evolution. Since it is unable to make its own food, it obtains food from host plants on which it grows.

Many fungi and some flowering plants are parasitic. The dodder plant, for example, has lost its chlorophyll and leaves in the course of its evolution and is unable to manufacture food. Instead, it obtains food from the host plants on which it grows (Figure 37-21).

The more closely the life of a parasite is linked with that of its host, the more its morphology and behavior are likely to have been modified during the course of its evolution. The human flea, for example, is flattened from side to side and slips easily through hair. The ancestors of this species of flea were brightly colored, large, winged insects. The structural and behavioral modifications of the human flea have come about in relation to a parasitic way of life.

Changes in communities over time: Succession

Communities, like species, change over time. Even when the climate of a given area remains stable year after year, the composition of the species making up a community, as well as the interactions within the community, shows a dynamic process of change known as **succession.** During the process of succession, a sequence of communities replaces one another in an orderly and predictable way. This process is familiar to anyone who has seen a vacant lot or cleared woods slowly become occupied with plants and animals or has seen a pond become filled with vegetation (Figure 37-22).

Primary succession takes place in areas not previously changed by organisms such as humans. Primary succession occurs in lakes left behind after the retreat of glaciers, for example, or on volcanic islands that may rise above the sea.

FIGURE 37-20 Parasitism: hookworms and humans.
These parasites of humans live in the intestine, feeding on the blood of the host.

FIGURE 37-22 Succession of a pond.
This pond in central Maine is beginning to fill with aquatic vegetation **(A)**. As the process of succession proceeds, the pond is slowly filled **(B)**. Gradually, the area where the pond once existed will become indistinguishable from the surrounding vegetation **(C)**.

In the latter case, succession may begin as lichens take hold on bare rock. Lichens are made up of an alga and a fungus living together in a symbiotic relationship. As lichens grow, they produce acids that can break down rock, forming small pockets of soil. When enough soil has accumulated, mosses may begin to grow. This first community, a pioneer community, consists of plants that are able to grow under harsh conditions. The pioneer community paves the way for the growth and development of vegetation native to that climate. Over many thousands of years or even longer, the rocks may be completely broken down, and vegetation may cover a once-rocky area. As plants take hold, the area becomes able to sustain other forms of life. And as the plant community changes, so too do the other living things. Eventually, the mix of plants and animals becomes somewhat stable, forming what is termed a **climax community** (Figure 37-23). However, with an increasing realization that (1) climates may change, (2) the process of succession is often

FIGURE 37-23 Climax community.
A stage in succession toward a climax community on the west side of the San Francisco Peaks in northern Arizona. Here, Ponderosa pine *(Pinus ponderosa)* is replacing aspen *(Populus tremuloides).*

very slow, and (3) the nature of a region's vegetation is determined to a great extent by human activities, ecologists do not consider the concept of a climax community as useful as they once did.

Succession occurs not only within terrestrial communities but in aquatic communities as well. A lake poor in nutrients, for example, may gradually become rich in nutrients as organic materials accumulate (see Figure 37-22). Plants growing along the edges of the lake, such as cattails and rushes, and those growing submerged, such as pondweeds, may contribute to the formation of a rich organic soil as they die and are decomposed by bacteria. As this process of soil formation continues, the pond may become filled in with terrestrial vegetation. Eventually, the area where the pond once stood may become an indistinguishable part of the surrounding vegetation.

Secondary succession occurs in areas that have been disturbed and that were originally occupied by living organisms. Humans are often responsible for initiating secondary succession throughout portions of the world that they inhabit. Abandoned farm fields, for example, undergo secondary succession as they revert to forest. Secondary succession may also take place after natural disasters, such as a forest fire or eruption of a volcano (Figure 37-24).

> **Communities change over time by means of the dynamic process of succession. During this process, a sequence of communities replaces one another in an orderly and predictable way.**

FIGURE 37-24 Secondary succession.
Mount St. Helens in the state of Washington erupted violently on May 18, 1980. The lateral blast devastated more than 600 square kilometers of forest and recreation lands within 15 minutes. **A** shows an area near Lang Ridge 4 months after the blast; 4 years later **(B)**, succession was underway at the same spot, with shrubs, blueberries, and dogwoods following the first plants that became established immediately after the blast. In 1989, more plants have become established and stable communities have formed **(C)**.

INTERACTIONS WITHIN COMMUNITIES OF ORGANISMS

Summary

1. All of the living (biotic) factors and nonliving (abiotic) factors together within a certain area are an ecosystem. Within an ecosystem, each living thing has a habitat, an area in which it lives. In addition, each organism plays a special role within an ecosystem, which is its niche.

2. Of the biotic factors within an ecosystem, groups of organisms of the same species are populations. Various populations of organisms make up a community. Thus an ecosystem can be thought of as a community of organisms, along with the abiotic factors with which the community interacts.

3. The interactions within communities fall into three categories: competition, predation, and symbiosis. Competition involves organisms of different species striving to obtain the same needed resource. During predation, one species (the prey) becomes a resource, being killed and eaten by another species (the predator). During a symbiotic relationship, two or more organisms of different species live together in close association over a prolonged time.

4. Symbiotic relationships also fall into three categories: commensalism, mutualism, and parasitism. Commensalism is a one-sided relationship: one species benefits whereas the other species neither benefits nor is harmed. In mutualism, both species benefit. In a parasitic relationship, one species (the parasite) benefits but the other (the host) is harmed.

5. Organisms living together in communities continually change and adjust to one another. Such interactions, which involve the long-term, mutual evolutionary adjustment of the characteristics of the members of biological communities in relation to one another, are forms of coevolution.

6. Communities, like species, change over time. This dynamic process of change, during which a sequence of communities replaces one another in an orderly and predictable way, is succession. Primary succession takes place in areas that are originally bare, like rocks or open water. Secondary succession takes place in areas where the communities of organisms that existed initially have been disturbed.

REVIEW QUESTIONS

1. Distinguish among population, community, and ecosystem.
2. Fill in the blanks: Within an ecosystem, each organism has a(n) _____, an area in which it resides. Each organism plays a role within the ecosystem, which is its _____.
3. Interactions within communities can be grouped into three categories. Name and describe each category.
4. True or false? (If it's false, change the statement so that it's correct): Competition among different species is often greatest between organisms that obtain their food in different ways.
5. State the principle of competitive exclusion in your own words.
6. Explain the terms *realized niche* and *fundamental niche*.
7. Fill in the blanks: A(n) _____ is an organism that kills and eats another organism called its _____. A(n) _____ also feeds on another organism but does not necessarily kill it.
8. Summarize the pattern of the predator-prey relationship as shown by Gause's experiments.
9. What other factors, in addition to those revealed in Gause's work, can affect the balance between predators and prey?
10. What is coevolution? Give an example of coevolution involving a herbivore.
11. What type of adaptation is shown in each of the following:
    a. The bright yellow and black stripes of a bee's body.
    b. The unobtrusive color of some lizards that make them difficult to see against surrounding rocks.
12. Distinguish among commensalism, mutualism, and parasitism. What do they have in common? Give an example of each.
13. What is primary succession? Summarize the process.
14. Fill in the blanks: Forests that have been cut down for logging can often no longer support the same animals and birds that lived in that ecosystem previously. Over time, the changes that take place in the communities of organisms in this logged area are examples of _____ _____.

KEY TERMS

abiotic 639
batesian mimicry 647
biotic 639
camouflage 648
climax community 652
coevolution 644
commensalism 649
community 639
competition 640
competitive exclusion 640
ecology 640
ecosystem 639
fundamental niche 642
habitat 639
herbivore 644
müllerian mimicry 648
mutualism 649
niche 639
parasitism 649
population 639
predation 640
predator 642
prey 642
primary succession 651
realized niche 642
secondary succession 653
succession 651
symbiosis 640
warning coloration 647

THOUGHT QUESTIONS

1. Can you think of a case of symbiosis that involves arthropods and humans? What type of symbiosis is involved? How did you make this determination?
2. How could you determine whether a resemblance between two butterfly species is mimicry or not?

FOR FURTHER READING

Breitwisch, R. (1992, March). Tickling for ticks. *Natural History,* pp. 57-63.
Red-billed oxpeckers spend most of their time preening mammals for parasites; the mammals seem to actively accommodate the birds' activities.

Creeping through the crinoids. (1992, May). *International Wildlife,* pp. 4-51.
This spectacular photo essay documents the symbiosis between the crinoids (relatives of seastars) and a variety of camouflaged animals that live among the crinoids' feathery arms.

Robinson, M. H. (1992, April). An ancient arms race shows no signs of letting up. *Smithsonian,* pp. 75-82.
In this engaging article, the author compares human military tactics to defensive tactics used by plants and animals.

38 Ecosystems

HIGHLIGHTS

▼

Ecosystems are communities of organisms and the nonliving environment with which they interact.

▼

Energy flows through ecosystems, initially captured from the sun by green plants and then passed from organism to organism as they feed on one another.

▼

The elements essential to life, primarily hydrogen, carbon, oxygen, and nitrogen, are cycled from the atmosphere, through living things and back to the atmosphere once again.

▼

Other elements such as phosphorus, potassium, calcium, and sodium are held in the soil in small amounts, incorporated in the tissues of living things, and cycled to the soil once again.

OUTLINE

Populations, communities, and ecosystems

The flow of energy through ecosystems
 Food chains and webs
 Food pyramids

The cycling of chemicals within ecosystems
 The water cycle
 The carbon cycle
 The nitrogen cycle
 The phosphorus cycle

A whole, raw mouse might not be your idea of a gourmet meal, but to this bullfrog it is a five-star dinner. Located in a rural village in South Africa, this "frog restaurant" is part of a food chain, a series of organisms that feed on one another. The bullfrog crushed the mouse with its powerful jaws. Its lower jaw has bony projections that help pull the prey into the frog's cavernous mouth and will soon help push it down into the stomach. Minutes before this picture was taken, it was the mouse that was feasting—on the seeds of nearby plants. And, unknown to the frog, a snake lies in wait behind the next branch, ready to add yet another link to this chain of gourmet relationships among the plants and animals of this community.

Populations, communities, and ecosystems

Individuals of a species, such as bullfrogs and mice, are each part of an individual population of organisms. Together, interacting populations are communities (see Chapters 36 and 37). The living organisms in a community interact not only with each other but with the nonliving substances in their environment, such as the soil, water, and air, to form an ecological system, or ecosystem.

An **ecosystem** is a community consisting of plants, animals, and microorganisms that interact with one another and with their environments and are interdependent on one another for survival. The living, or **biotic,** components of an ecosystem are made up of two types of organisms: those that can make their own food, or **producers,** and those that eat other organisms for food, or **consumers.** Many consumers kill and eat their food. A special group of consumers, called **decomposers,** obtains nourishment from dead matter such as fallen leaves or the bodies of dead animals. You know the decomposers as bacteria and fungi (although some species of bacteria can manufacture their own food and are therefore producers).

Figure 38-1 is a diagram of an ecosystem. Notice that the **abiotic,** or nonliving, components of the environment contribute substances needed for the ecosystem to function. In addition, notice that the exchanges of nutrients and other chemical substances among the organisms within the ecosystem form a cyclic pattern. Energy, however, does not cycle through the ecosystem but flows *through* the ecosystem, first captured from the sun by the producers, then used by herbivores that eat the producers, and ultimately used by all the consumers living in the ecosystem. Ecosystems are therefore systems in which there is a regulated transfer of energy and an orderly, controlled cycling of nutrients. The individual organisms and populations of organisms in an ecosystem act as parts of an integrated whole, adjust over time to their roles in the ecosystem, and relate to one another in complex ways that are only partly understood.

> An ecosystem is made up of communities of organisms living within a defined area and the nonliving environmental factors with which they interact. The organisms of an ecosystem—the producers, consumers, and decomposers—each play a specific role within it, contributing to the flow of energy and the cycling of nutrients.

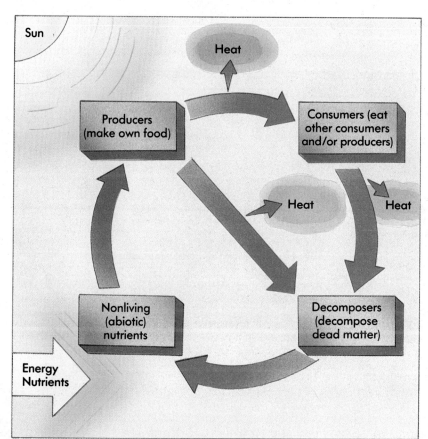

FIGURE 38-1 An ecosystem.
Nutrients flow through an ecosystem in a cyclic pattern, whereas energy flows through in one direction only.

FIGURE 38-2 Two distinct ecosystems.
In the coast ranges of California, the boundary between the evergreen shrub ecosystem known as chaparral and the grassland ecosystem is often sharp, as shown in this photograph taken along the western edge of the Santa Clara Valley near Morgan Hill. These two ecosystems each include a characteristic set of nonliving factors.

Where does one ecosystem begin and another end? Some ecosystems have clearly recognizable boundaries, such as that of a pond or a puddle or those found within the coastal ranges of California (Figure 38-2). Sometimes humans produce artifical ecosystems with human-made boundaries, such as the glass walls of an aquarium or the fencing surrounding a cultivated field. But the boundaries of many natural ecosystems blend with one another, sometimes almost imperceptibly. Ecosystems also change over time and with climate changes, slowly becoming modified into new ecosystems whose characteristics differ increasingly from those that preceded them.

The flow of energy through ecosystems

The energy that flows through an ecosystem comes from the sun. Green plants, the primary producers of terrestrial ecosystems, are able to capture some of the sun's radiant energy that falls on their leaves and convert it to chemical energy during the process of photosynthesis (see Chapter 8). Producers, then, are the key to life on Earth, since no other organisms can capture this energy for use in living systems. **Primary consumers,** or herbivores, feed directly on the green plants, incorporating some of this energy into molecules that make up their bodies and using the rest to perform the activities of life. **Secondary consumers** are meat-eaters, or carnivores, that feed in turn on the herbivores. And so the chain continues, with one living thing feeding on another, passing energy along that was once captured from the sun.

FIGURE 38-3 A decomposer doing its job.
Fungi are breaking down this hardwood stump, converting the organic materials contained within it into nutrients that can be reused by plants.

The refuse or waste material of an ecosystem is known as **detritus.** Organisms that are decomposers break down the organic materials of detritus into inorganic nutrients that can be reused by plants (Figure 38-3). Some of the energy still held in the tissues of once-living things is used by the decomposers, but they are the last link in this transfer of energy among organisms.

> The producers of an ecosystem capture energy from the sun and convert it to chemical energy usable by themselves and consumers. Consumers feed on the producers and other consumers in the ecosystem, passing energy along that was once captured from the sun. Decomposers break down the organic molecules of dead organisms, serving as the last link in the flow of energy through an ecosystem and contributing to the recycling of nutrients to the environment.

Food chains and webs

All of the feeding levels previously described and additional levels such as tertiary consumers are represented in any fairly complicated ecosystem. These feeding levels are called **trophic levels,** from the Greek word *trophos,* which means "feeder." (In fact, this Greek word is the root for the words **heterotroph,** or "other feeder"—another word for consumer, and for **autotroph,** or "self-feeder"—another word for producer.) Organisms from each of these levels, feeding on one another, make up a series of organisms called a **food chain.** An example of a food chain can be seen in a pond ecosystem in which water fleas (primary consumers) feed on green algae (producers). Sunfish (secondary consumers) eat the water fleas but in turn are eaten by green heron (tertiary consumers) (Figure 38-4). The length and complexity of food chains vary greatly.

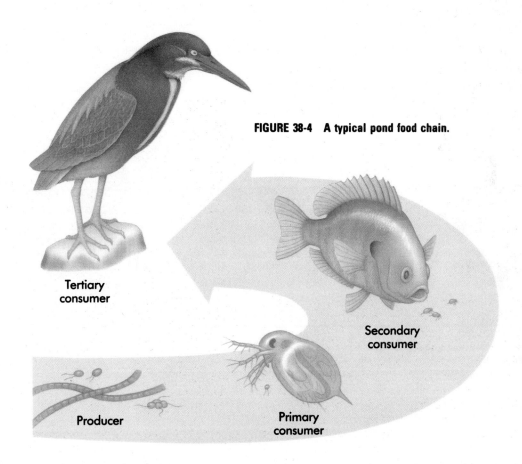

FIGURE 38-4 A typical pond food chain.

ECOSYSTEMS

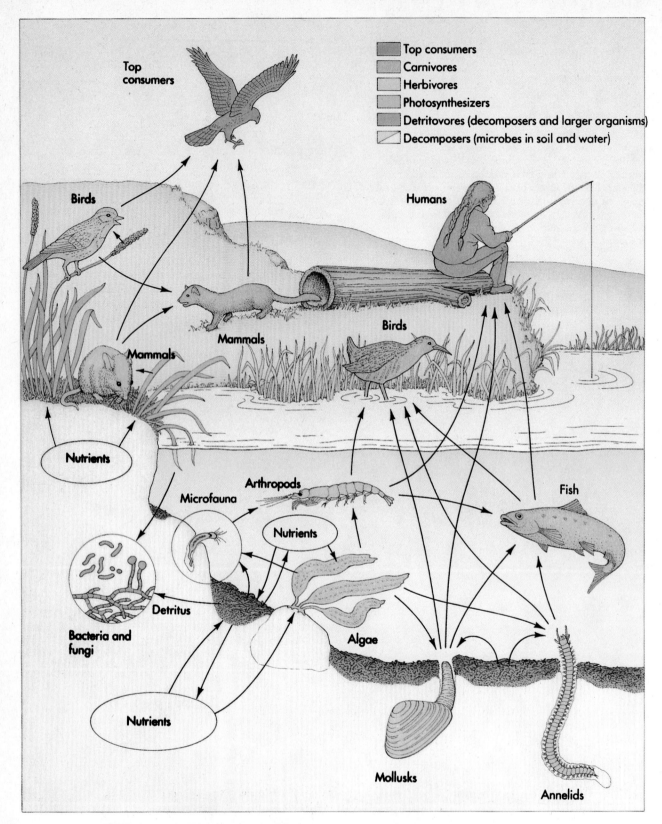

FIGURE 30-5 The food web in a salt marsh.
Each level of rectangles represents a trophic level. Each trophic level feeds on, or gains energy from, the layer below.

In reality, it is rare for any species of organism to feed on only one other species. Organisms feed on many different species and types of organisms and are, in turn, food for two or more other kinds. These relationships appear as a series of branching and overlapping lines rather than as one straight line. The organisms in an ecosystem that have such interconnected and interwoven feeding relationships make up a **food web.** Figure 38-5 shows a food web in a salt marsh ecosystem.

> A linear relationship among organisms that feed on one another is a food chain. Food chains that interweave are food webs.

Food pyramids

In any ecosystem the number of organisms and the amount of energy making up each successive trophic level are less than the level that preceded. Lamont Cole of Cornell University illustrated this concept in his study of the flow of energy in a fresh water ecosystem in Cayuga Lake, New York. He calculated that approximately 150 calories of each 1000 calories of energy "fixed" by producers during photosynthesis was transferred into the bodies of small heterotrophs that feed on these plants and bacteria (Figure 38-6). Smelt, which are tiny fish, eat the heterotrophs; these secondary consumers obtain about 30 calories of each original 1000. If human beings eat the smelt, they gain about 6 calories from each 1000 calories that originally entered the system. If trout eat the smelt and humans eat the trout, humans gain only about 1.2 calories from each original 1000.

These types of calculations show that, on average, only 10% of plants' accumulated energy is actually converted into the bodies of the organisms that consume them. What happens to the rest? A certain amount of the energy that is ingested by organisms goes toward heat production. A great deal of energy is used for digestion and work, and usually 40% or less goes toward growth and reproduction. An invertebrate, for example, typically uses about a quarter of this 40% for growth. In other words, about 10% of the food that an invertebrate eats is turned into new body tissue. This figure varies from approximately 5% for carnivores to nearly

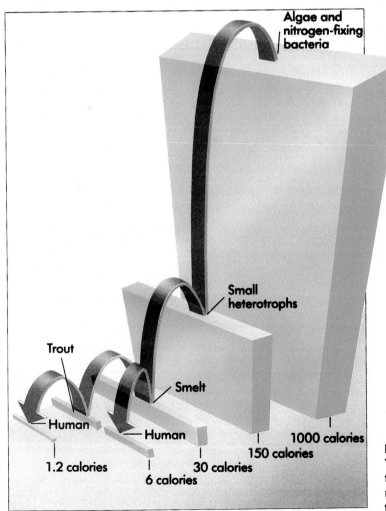

FIGURE 38-6 The flow of energy in Lake Cayuga.
The experiments in Lake Cayuga demonstrated that the number of organisms and the amount of energy making up each successive trophic level is smaller than the preceding level.

ECOSYSTEMS 661

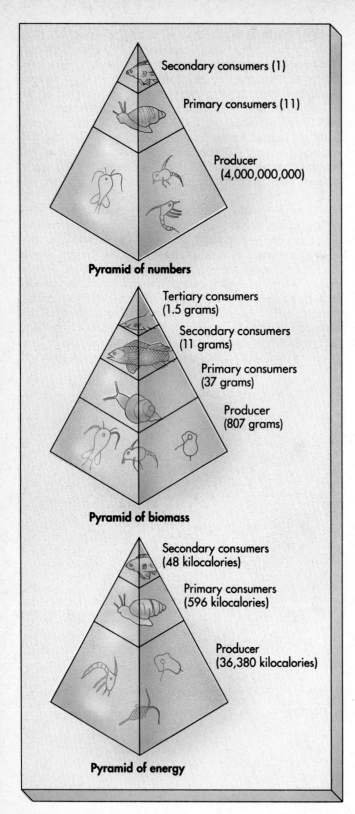

FIGURE 38-7 Pyramids of numbers, biomass, and energy for an aquatic ecosystem.
The pyramid of numbers shows the number of organisms at each trophic level in the ecosystem. The pyramid of biomass depicts the total weight of organisms supported at each level. The pyramid of energy depicts the amounts of energy available at each level.

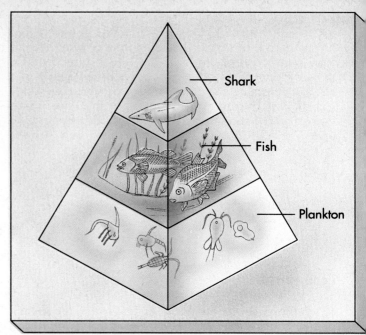

FIGURE 38-8 Food pyramid.
All food pyramids are basically the same—each successive trophic level is smaller than the preceding level (see Figure 38-7).

20% for herbivores, but 10% is an average value for the amount of energy (or organic matter) that organisms incorporate into their bodies from the energy available in the previous trophic level.

> A plant captures some of the sun's energy that falls on its green parts. The successive members of a food chain process about 10% of the energy available in the organisms on which they feed into their own bodies. The rest is released as heat (and is lost from the food chain) or is used during various metabolic activities.

If shown diagrammatically, the relationships between trophic levels appear as pyramids. Diagrams that depict the energy flow through an ecosystem are called *pyramids of energy* (Figure 38-7). Those that depict the total weight of organisms supported at each trophic level in an ecosystem are referred to as *pyramids of biomass* (Figure 38-7). *Pyramids of number* depict, as the name suggests, the total number of organisms at each feeding level (Figure 38-7). Figure 38-8 shows a food pyramid that incorporates all three concepts.

The relationships between organisms at the various trophic levels of a food pyramid make it clear that herbivores have more food available to them than carnivores. In other words, the lower a population eats in the food chain, the higher the number of individuals that can be fed. Such considerations are increasingly important as humans work to maximize the food available for a hungry and increasingly overcrowded world.

HOW LIVING THINGS INTERACT WITH EACH OTHER AND WITH THEIR ENVIRONMENT

Hubbard Brook and the cycling of nutrients in ecosystems

In 1963, scientists at the Yale School of Forestry and Environmental Studies began an interesting experiment designed to show the effects of the cycling of nutrients within ecosystems. Their studies have yielded much of the information that is now known about nutrient cycles, and their ingenious experimental design has provided the basis for the development of the experimental methods used in the study of other ecosystems.

Hubbard Brook is the central stream located in a temperate deciduous forest in New Hampshire. For measurement of the flow of water and nutrients within the Hubbard Brook ecosystem, concrete weirs with V-shaped notches were built across six tributary streams that were selected for study. All of the water that flowed out of the six valleys (the areas surrounding the six streams) had to pass through the notch. The precipitation that fell in the six streams was measured, and the amounts of nutrients present in the water flowing in the six streams were also determined. By these methods the scientists demonstrated that the undisturbed forests in this area were very efficient in retaining nutrients. The small amounts of nutrients that fell from the atmosphere with the rain and snow were approximately equal to the amounts of nutrients that ran out of the six valleys. There was a net loss of calcium—about 0.3% of the total calcium in the system per year—and small net gains of nitrogen and potassium.

Then the researchers disturbed the ecosystem. In 1965, the investigators felled all of the trees and shrubs in one of the six valleys and then prevented their regrowth by spraying the area with herbicides. The effects of these activities were dramatic. The amount of water running out of the valley was increased by 40%, indicating that water normally would have been taken up by the trees and shrubs and then evaporated into the atmosphere from their leaves but was now running off. The amounts of nutrients running out of the system also increased. The loss of calcium was 10 times higher than it had been previously. The change in the status of nitrogen was particularly striking. The undisturbed ecosystem in this valley had been accumulating nitrogen at a rate of 2 kilograms per hectare per year, but the cut-down ecosystem lost it at a rate of about 120 kilograms per hectare per year! The nitrate level of the water rapidly increased to a level exceeding that judged safe for human consumption, and the stream that drained the area generated massive blooms of cyanobacteria and algae.

The Hubbard Brook experiment demonstrated that nutrient cycling depends on the vegetation present in the ecosystem among other things. The fertility of the deforested valley decreased rapidly, and at the same time the danger of flooding greatly increased. The Hubbard Brook experiment is particularly instructive in the 1990s, since large areas of tropical rain forest are being destroyed to make way for cropland.

FIGURE 38-A Experimental weir at Hubbard Brook. Water is forced over the concrete, and samples of it are representative of the flow from the valley where the stream is located.

The cycling of chemicals within ecosystems

Although energy flows through ecosystems and most is lost at each successive level in food pyramids, the matter making up the organisms at each level is not lost. All of these substances are recycled and are only used temporarily by living things. Hydrogen, carbon, nitrogen, and oxygen—the principal elements that make up all living things—are primarily held in the atmosphere in molecules of water, carbon dioxide, nitrogen gas, and oxygen gas. Other recycled substances necessary for life such as phosphorus, potassium, sulfur, magnesium, calcium, sodium, iron, and cobalt are held in rocks and, after weathering, enter the soil. The atmosphere and rocks are therefore referred to as the **reservoirs** of inorganic substances that cycle within ecosystems.

The cycling of materials in ecosystems is usually described as beginning at the reservoirs. Living things incorporate substances into their bodies from their reservoirs or from other living things, passing these materials along the food chain. Ultimately these substances, with the help of decomposers, move from the living world back to the nonliving world, becoming part of the soil or the atmosphere once again.

▶ **Within ecosystems, matter cycles from its reservoir in the environment, to the bodies of living organisms, and back to the environment.** ◀

The water cycle

Heated by the sun, water evaporates into the atmosphere from the surfaces of oceans, lakes, and streams (Figure 38-9). In terrestrial (land-based) ecosystems, as much as 90% of the water that reaches the atmosphere comes from plants as they release water vapor into the air during the process of transpiration (see Chapter 31). But because oceans cover three fourths of the Earth's surface, these bodies contribute most of the water to the atmosphere worldwide.

Atmospheric water condenses in clouds and eventually falls back to the Earth as precipitation. Most of it falls directly into the oceans. But some falls onto the land, flowing into surface bodies of water or trickling through layers of soil and rock to form subsurface bodies of fresh water called **ground water.** Plants take up water as it trickles through the soil, almost in a continuous stream. Crop plants, for example, use about 1000 kilograms of water just to produce 1 kilogram of biomass. Animals obtain water directly from surface water or from the plants or other animals they eat.

FIGURE 38-9 The water cycle.
In terrestrial ecosystems, 90% of the water that reaches the atmosphere comes from plant transpiration. In general, however, oceans contribute the most water to the Earth's atmosphere.

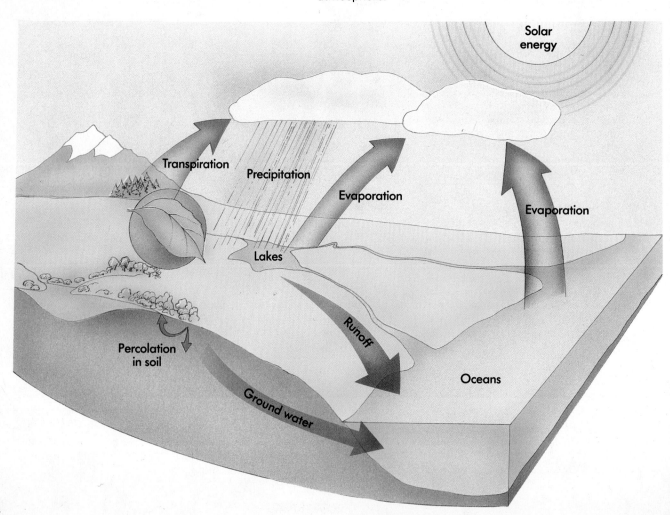

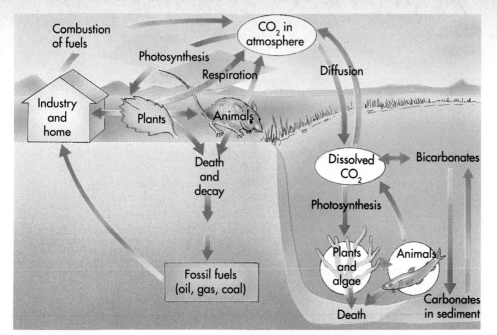

FIGURE 38-10 The carbon Carbon dioxide is found in oceans and in the atmosphere. Producers and some consumers incorporate this carbon into substances necessary for life.

In the United States, ground water provides about a quarter of the water used by humans for all purposes and provides about half of the population with drinking water.

> Water in the atmosphere condenses in clouds and falls back to the Earth as precipitation. Plants take up water from the soil, and animals obtain water from surface water or from the plants or other animals they eat. Water returns to the atmosphere through the evaporation of surface water and transpiration by plants.

Unfortunately, about 2% of the ground water in the United States is polluted, and the situation is worsening. Pesticides are one source, being carried to **aquifers,** underground reservoirs in which the ground water lies within porous rock as rain washes the chemicals from the surfaces of leaves and the topsoil. Chemical wastes, stored in surface pits, ponds, and lagoons, are another key source of groundwater pollution. Scientists have no technology that will remove pollutants from underground aquifers.

The carbon cycle

The carbon cycle is based on carbon dioxide (CO_2), which makes up about 0.03% of the atmosphere and is found dissolved in the oceans (Figure 38-10). Terrestrial as well as marine producers use CO_2—along with energy from the sun—to build carbon compounds such as glucose during the process of photosynthesis (see Chapter 8). The producers and the consumers that eat them break down these carbon compounds during cellular respiration and use the energy locked in their chemical bonds to carry on the metabolic processes of life. Consumers use some of the carbon atoms and compounds from the food they eat to produce needed substances. However, most of this carbon is waste and is released to the atmosphere (or to the oceans) as CO_2.

Intimately linked to the cycling of carbon is the cycling of oxygen (O_2). As plants use CO_2 during the process of photosynthesis, they produce O_2 as a waste product. This O_2 is released into the atmosphere and becomes available to organisms for the process of cellular respiration (Figure 38-11).

FIGURE 38-11 How carbon dioxide is linked to oxygen cycling. Plants use carbon dioxide in photosynthesis and give off oxygen. This oxygen is available to animals for the process of cellular respiration. Animals breathe out carbon dioxide; some carbon dioxide is liberated from the decomposition of dead organisms.

...rganisms such as mollusks use the CO_2 ...nd combine it with calcium to form their ...($CaCO_3$) shells. When these organisms ...ollect on the sea floor. Years of exposure ...dissolves the $CaCO_2$ releasing the CO_2 and ...n once again available to aquatic producers ...rocess of photosynthesis.

...ganisms die, decomposers break down the carbon compounds making up their bodies. Certain carbon-containing compounds, such as the cellulose found in the cell walls of plants, are more resistant to breakdown than others, but certain bacteria, fungi, and protozoans are able to accomplish this feat. Some cellulose, however, accumulates as undecomposed organic matter. Over time and with heat and pressure, this undecomposed matter results in the formation of fossil fuels such as oil and coal. When these fuels are burned, as when wood is burned, the CO_2 is returned to the atmosphere. The release of this carbon as CO_2, a process that is proceeding rapidly as a result of human activities, may change global climates.

Carbon is held in the atmosphere in the form of carbon dioxide. This gas is taken in by photosynthetic organisms and is used to build the carbon compounds that plants manufacture during the process of photosynthesis. Both producers and consumers use these compounds as an energy source, metabolizing them during cellular respiration and releasing carbon dioxide to the atmosphere.

The nitrogen cycle

Although nitrogen gas (N_2) makes up 78% of the Earth's atmosphere, only a minute amount is incorporated into chemical compounds in the soil, oceans, and bodies of organisms. However, this N_2 is an essential part of the proteins within living things. Relatively few kinds of organisms—only a few genera of bacteria—can convert N_2 into a form that can be used for biological processes, playing a crucial role in the cycling of nitrogen (Figure 38-12). These bacteria are called <u>nitrogen-fixing bacteria</u>, and convert N_2 to ammonia (NH_3). Living things depend on this process of nitrogen fixation. Without it, they would ultimately be unable to continue to synthesize proteins, nucleic acids, and other necessary nitrogen-containing compounds.

Certain of the nitrogen-fixing bacteria are free living in the soil. Others form mutualistic relationships (see Chapter 37) with plants by living within swellings, or nodules, of plant roots. Some of these plants are legumes—plants such as soybeans, alfalfa, and clover. Plants having mutualistic associations with nitrogen-fixing bacteria can grow in soils having such low amounts of available nitrogen that they are unsuitable for most other plants. Growth of a leguminous crop can enrich the nitrate level of poor soil enough to benefit the next year's nonleguminous crop. This is the basis for crop rotation in which, for example, a field may be planted with soybeans (a legume) and corn (a nonlegume) in alternating years.

The roots of a few other kinds of plants form associations with nitrogen-fixing bacteria of the group actinomy-

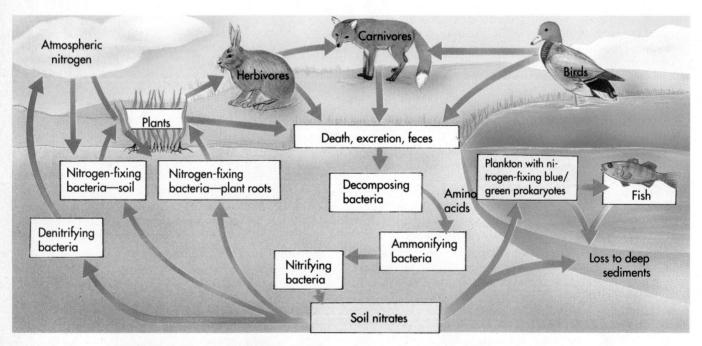

FIGURE 38-12 The nitrogen cycle.
Bacteria that can convert nitrogen into a usable form (nitrogen-fixing bacteria) are extremely important in the nitrogen cycle.

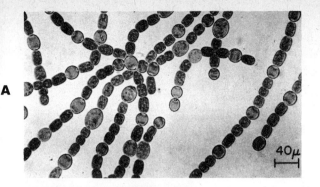

FIGURE 38-13 The use of nitrogen-fixing bacteria in agriculture.
The nitrogen-fixing bacteria *Anabaena azollae* (**A**) lives in the spaces between the leaves of the floating water fern *Azolla* (**B**), which is deliberately introduced into the rice paddies of the warmer parts of Asia. Rice, here cultivated in Sri Lanka (**C**), is the major food for well over one fourth of the human race.

cetes. Some of the plants involved are alders and mountain lilac. In addition, nitrogen-fixing bacteria of the genus *Anabaena* contribute large amounts of nitrogen to the rice paddies of China and Southeast Asia (Figure 38-13).

Other bacteria—certain decomposers—play another key role in the nitrogen cycle, producing NH_3 from the amino acids that make up the proteins and wastes of dead organisms. Still other bacteria convert NH_3 to nitrates (NO_3^-). NO_3^- is also produced by lightning, which causes N_2 to react with O_2 in the atmosphere. Humans add NO_3^- to the soil by spreading chemical fertilizers. Plants are able to use the nitrogen within molecules of NH_3 and NO_3^- to build their own proteins, nucleic acids, and vitamins. The nitrogen cycle comes full circle as nitrogen is continuously returned to the environment by bacteria that break down NO_3^- liberating N_2 to the atmosphere.

> Although nitrogen gas constitutes about 78% of the Earth's atmosphere, it becomes available to organisms only through the metabolic activities of a few genera of bacteria, some of which are free living and others of which live symbiotically on the roots of legumes and some other plants. This bound nitrogen is released back to the atmosphere by other microorganisms capable of breaking down certain nitrogen compounds.

The phosphorus cycle

The reservoir of the nutrients phosphorus, potassium, sulfur, magnesium, calcium, sodium, iron, and cobalt is in rocks and minerals rather than in the atmosphere.

Phosphorus, more than any of the other required plant nutrients except nitrogen, is apt to be so scarce that it limits plant growth. In the soil, phosphorus is found as relatively insoluble phosphate (PO_4^{-3}) compounds, present only in certain kinds of rocks. For this reason, phosphates exist in the soil only in small amounts. Therefore humans add millions of tons of phosphates to agricultural lands every year. Plants take up the phosphates from the soil, and animals obtain the phosphorus they need by eating plants or other plant-eating animals. When these organisms die, decomposers release the phosphorus incorporated in their tissues, making it again available for plant use (Figure 38-14).

> Phosphorus is held in the soil in the form of relatively insoluble phosphate compounds, found in most soils in small amounts. Therefore humans add phosphate compounds to the soil for plants to take up, and animals obtain the phosphates they need by eating plants or plant-eating animals. When organisms die, decomposers release phosphorus back to the soil.

ECOSYSTEMS

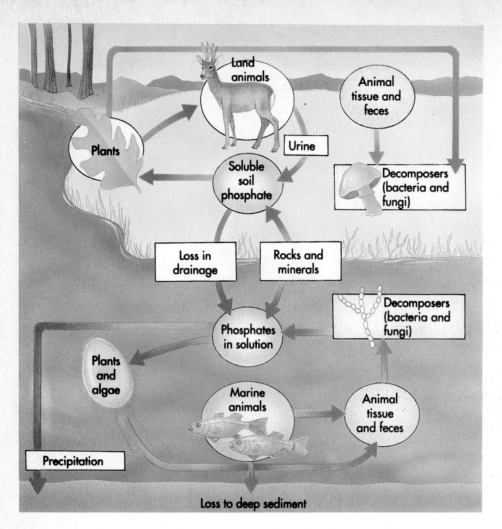

FIGURE 38-14 The phosphorus cycle. Phosphorus is found in only very small amounts in the soil; to make up for this lack, humans add phosphates to the soil. Plants take up phosphate from the soil, and animals obtain the necessary levels of phosphate by eating plants.

Humans unnecessarily alter the phosphorus cycle by adding up to four times as much phosphate as a crop requires each year. Much of this phosphate runs off the land or is eroded into rivers, streams, and the oceans. In addition, phosphates in sewage and waste water from homes and industry find their way into surface water. Rivers and streams become **eutrophic,** or "well fed," causing algae and other aquatic plants to overgrow. As they die, the decomposers feed on them, using up much of the oxygen in the water and choking out other forms of life.

Rivers and streams carry phosphates to the oceans. Some of these phosphates become incorporated into the bodies of fishes and other marine animals. The sea birds that eat these animals deposit enormous amounts of guano (feces) rich in phosphorus along certain coasts (these deposits have traditionally been used for fertilizer) (Figure 38-15). Phosphates not incorporated into the bodies of animals precipitate out of the water and become part of the bottom sediment. These phosphates become available again only if the sea floor rises up during eras of climatic change, such as along the Pacific coast of North and South America.

FIGURE 38-15 A guano coast. Marine animals incorporate phosphates from runoff into their bodies. The seabirds that feast on these marine animals deposit feces rich in phosphorus on certain coasts.

Summary

1. Ecosystems are communities of organisms and the nonliving factors of their environments through which energy flows and nutrients cycle.
2. Through photosynthesis, plants growing under favorable circumstances capture and lock up some of the sun's energy that falls on their green parts. They may then be eaten by herbivores (primary consumers), which in turn may be eaten by secondary consumers (carnivores). The dead remains of all organisms are broken down by the decomposers. This sequence of organisms, one feeding on another, constitutes a food chain.
3. Each link in a food chain is a trophic, or feeding, level. Each trophic level includes organisms that can transfer about 10% of the energy that exists at each level to the next level. The rest is lost as heat or is used for various metabolic activities.
4. Carbon dioxide, nitrogen gas, oxygen gas, and water are the atmospheric reservoirs of the carbon, nitrogen, oxygen, and hydrogen used in biological processes. All of the other elements that organisms incorporate into their bodies come from the Earth's rocks.
5. Water in the atmosphere condenses in clouds and falls to the Earth as precipitation. Plants take up water from the soil, and animals obtain water from surface water or from the plants or other animals they eat. Water returns to the atmosphere through the evaporation of surface water and transpiration by plants.
6. Carbon is held in the atmosphere in the form of carbon dioxide. This gas is taken in by photosynthetic organisms and is used to build the carbon compounds that plants manufacture during the process of photosynthesis. Both producers and consumers use these compounds as an energy source, metabolizing them during cellular respiration and releasing carbon dioxide to the atmosphere.
7. Atmospheric nitrogen is converted to ammonia by several genera of symbiotic and free-living bacteria. The ammonia, in turn, is assimilated into amino groups in proteins of cells or is converted to nitrites and then to nitrates by other bacteria. Nitrates are incorporated into the bodies of plants and are converted back into ammonium ions, which are used in the manufacture of many kinds of molecules in the bodies of living organisms. The breakdown of these molecules either converts them to recyclable forms or results in the release of atmospheric nitrogen.
8. Phosphorus is a key component of many biological molecules; it weathers out of soils and is transported to the world's oceans where it tends to be lost. Phosphorus is relatively scarce in rocks; this scarcity often limits or excludes the growth of certain kinds of plants. Therefore humans add phosphate compounds to the soil for plants to take up, and animals obtain the phosphates they need by eating plants or plant-eating animals. When organisms die, decomposers release phosphorus back to the soil.

REVIEW QUESTIONS

1. What is an ecosystem? Briefly describe three of the living groups it contains with respect to mode of nutrition.
2. To which of the groups in question 1 do you belong?
3. Draw a diagram that illustrates the relationships between the biotic and abiotic components of a generalized ecosystem.
4. Fill in the blanks: _____ _____ feed directly on green plants. _____ _____ feed on herbivores. Other organisms known as _____ live on the refuse or waste material of an ecosystem, called _____.
5. Distinguish among trophic level, food chain, and food web.
6. Summarize what happens to the sun's energy as it travels through a food chain.
7. Explain how chemicals are cycled through an ecosystem.
8. Describe how water is cycled within ecosystems.
9. What is ground water, and how can it become polluted? Explain why this is a serious problem.
10. Summarize the carbon cycle.
11. Why can crop rotation make soil more fertile? To what chemical cycle does this relate?
12. Many farmers fertilize their crops heavily with phosphates, believing that this will improve the soil. Explain how this practice can affect an ecosystem.

KEY TERMS

abiotic 657
aquifer 665
autotroph 659
biotic 657
consumer 657
decomposer 657
detritus 659
ecosystem 657
eutrophic 668
food chain 659
food web 661
ground water 664
heterotroph 659
primary consumer 658
producer 657
reservoir 664
secondary consumer 658
trophic level 659

THOUGHT QUESTIONS

1. Imagine you had a hollow glass sphere the size of a basketball and you wanted to create within it a self-sustaining stable ecosystem that needed only sunlight and moderate temperature to persist indefinitely. What would you put in this ecosphere?
2. Most terrestrial food chains have only three or rarely four links. Why do you imagine this is so? Why not forest food chains with ten links?
3. Extensive cutting and burning of tropical rain forests often results in a drastic and permanent lowering of rainfall in the cleared area (an effect noted by Alexander von Humboldt over 100 years ago). Why?

FOR FURTHER READING

Colinvaux, P. A. (1989, May). The past and future Amazon. *Scientific American*, pp. 102-108.
This is a very interesting article about historical changes in the Amazon rain forest ecosystem, based on research.

Houghton, R. A., & Woodwell, G. M. (1989, April). Global climatic change. *Scientific American*, pp. 36-44.
This is a good summary of research on global warming.

Biomes and life zones of the world

39

Does your backyard look like the photo? Probably not, unless you happen to live near the equator. In this steamy climate, the vegetation is lush and the trees are tall, "topping out" at about 160 feet! Very little light penetrates this screen of foliage and finds its way to the forest floor beneath. Vines and other plants cling high up on the trunks of trees, competing for the light that filters through their leaves. Animal and plant species abound; so many species inhabit the tropical rain forest, in fact, that thousands remain unknown, unclassified, and unstudied.

Interestingly, not all regions of the Earth lying near the equator are tropical rain forests. Latitude is only one factor having an impact on the growth and distribution of living things. This chapter will describe other factors, explaining why organisms live where they do . . . and why your backyard is probably not a tropical rain forest.

HIGHLIGHTS

▼
Biomes are distinctive, broad, terrestrial ecosystems of plants and animals that occur over wide areas within specific climatic regions.

▼
Seven categories of biomes, listed by their distance from the equator, are tropical rain forests, savannas, deserts, temperate grasslands, temperate deciduous forests, taiga, and tundra.

▼
The fresh water of ponds, lakes, streams, and rivers also provides various habitats for organisms to live; estuaries, places where fresh water meets salt water as rivers open into oceans, provide nutrient-rich environments intermediate between fresh and salt water.

▼
Marine organisms occur most plentifully in the intertidal zone and the photic zone of the neritic and open sea; here, they have abundant light and oxygen.

OUTLINE

Biomes and climate
    The sun and its effects on climate
    Atmospheric circulation and its effects on climate

Life on land: The biomes of the world
    Tropical rain forests
    Savannas
    Deserts
    Temperate grasslands
    Temperate deciduous forests
    Taiga
    Tundra

Life in fresh water

Estuaries: Life between rivers and oceans

Life in the oceans
    The intertidal zone
    The neritic zone
    The open-sea zone

Biomes and climate

Biomes are ecosystems of plants and animals that occur over wide areas of land within specific climatic regions and are easily recognized by their overall appearance. Each biome is similar in its structure and appearance wherever it occurs on Earth and differs significantly from other biomes. Biomes are sometimes named by the climax vegetation (stable plant communities) of the region (see Chapter 37), such as the tropical rain forest.

The characteristics of biomes are a direct result of their temperature and rainfall patterns. These patterns result from the interaction of the features of the Earth itself (such as the presence of mountains and valleys) with two physical factors:

1. The amounts of *heat from the sun* that reach different parts of the Earth and the seasonal variations in that heat
2. *Global atmospheric circulation* and the resulting patterns of oceanic circulation

Together these factors determine the local climate, including the amounts and distribution of precipitation.

> Biomes are broad, distinctive climatic regions containing specific types of plants and animals. Temperature and rainfall patterns are the primary determinants of the characteristics of these huge ecosystems.

The sun and its effects on climate

Because the Earth is a sphere, some parts receive more energy from the sun than others. The tropics are warmer than the temperate regions because the sun's rays arrive almost perpendicular to regions near the equator, whereas near the poles the rays' angle of incidence spreads them over a much greater area (Figure 39-1). Therefore the greater the latitude (distance from the equator), the colder the climate.

The northern and southern hemispheres also experience a change of seasons as well as the gradations in temperature that vary with latitude. Seasons occur because the Earth is tilted on its axis, so the northern and southern hemispheres receive unequal amounts of sunlight at various times during the Earth's year-long journey around the sun. One of the poles is closer to the sun than the other, except during the spring and autumn equinoxes (Figure 39-2).

Atmospheric circulation and its effects on climate

Near the equator, warm air rises and flows toward the poles. If the Earth were a stationary sphere, this air would reach the poles, cool, fall to the ground, and flow back toward the equator. The Earth is not stationary, however; as it spins, its rotary movement breaks the air into six "coils" of rising and falling air that surround the Earth (Figure 39-3, *A*). With increasing latitude, each air mass is cooler than the one before but warmer than the next. Therefore the air in each mass rises at the region of its lowest latitude, moves toward the poles, sinks to the ground at the region of its highest latitude, and flows back toward its lowest latitude. As it warms, it rises again, completing the cycle.

These moving masses of air that encircle the earth set up the prevailing wind patterns (Figure 39-3, *B*) and also have an effect on precipitation. The moisture-holding capacity of air increases when it is warmed and decreases when it is cooled. Therefore precipitation is relatively high near 60 degrees north and south latitude, where it is rising and being cooled. Conversely, precipitation is generally low near 30 degrees north and south latitude, where air is falling and being warmed (Figure 39-3, *A*).

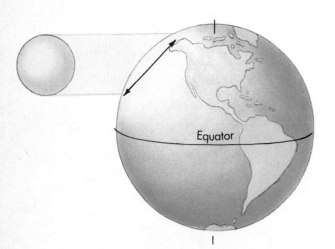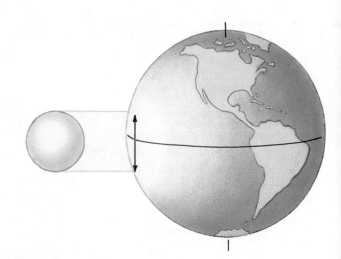

FIGURE 39-1 How the angle of solar energy striking the Earth affects climate.
The angle of a beam of solar energy striking the Earth at the equator is perpendicular to the surface of the Earth. The angle of this same beam striking an area at the middle latitude is larger, and the solar energy is more spread out over a larger area. In general, the greater the latitude, the colder the climate.

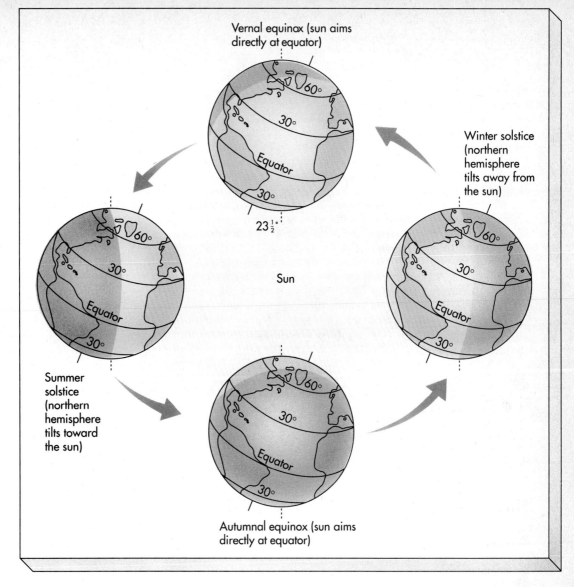

FIGURE 39-2 The rotation of the Earth around the sun has a profound effect on climate. In the northern and southern hemispheres, temperatures change in an annual cycle because the Earth is slightly tilted on its axis in relation to its pathway around the sun.

FIGURE 39-3 How the Earth's atmospheric circulation influences climate.
A Air surrounding the rotating Earth is broken into six coils of rising and falling air. These masses of air are associated with certain climates where moist air rises.
B The wind patterns around the Earth are set up by the six moving air masses.

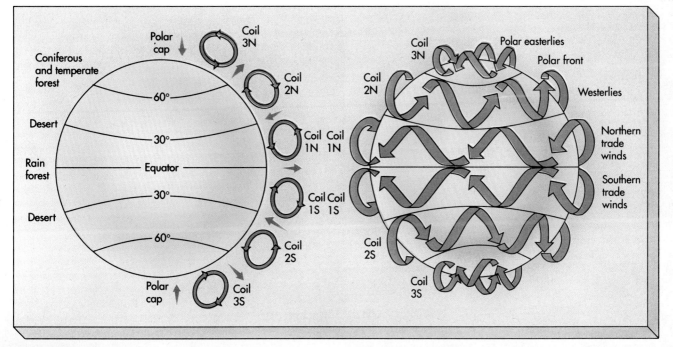

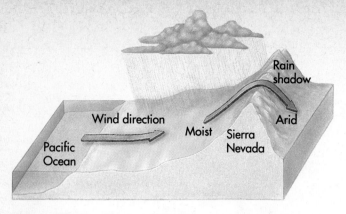

FIGURE 39-4 The rain shadow effect.
Moisture-laden winds from the Pacific Ocean rise and are cooled when they encounter the Sierra Nevada mountain range. As their moisture-laden capacity decreases, precipitation occurs, making the middle elevation of the range one of the snowiest regions on Earth. As the air descends on the east side of the range, its moisture-laden capacity increases again, and the air picks up moisture from its surroundings rather than releasing it. As a result, desert conditions prevail on the east side of the mountains.

Partly as a result of these factors, all the great deserts of the world lie near 30 degrees north or 30 degrees south latitude, and some of the great temperate forests are near 60 degrees north or south latitude. But other factors also come into play. Some major deserts are formed in the interiors of the large continents; these areas have limited precipitation because of their distance from the sea, the ultimate source of most precipitation. Other deserts sometimes occur because mountain ranges intercept the moisture-laden winds from the sea (Figure 39-4). As the air travels up a mountain (its windward side), it is cooled, and precipitation forms. As the air descends the other side of the mountain (its leeward side), it is warmed; its moisture-holding capacity increases. For these reasons, the windward sides of mountains are much wetter than the leeward sides, and the vegetation is often very different. This phenomenon is the rain shadow effect. Seattle, Washington, for example, lies on the windward side of the Cascade Mountain range in the northwestern United States. This city receives 99 centimeters (39 inches) of rainfall per year. Yakima, Washington, slightly south from Seattle on the leeward side of this mountain range, receives only 20 centimeters (8 inches) of rain per year.

> The climate of a region is determined primarily by its latitude and wind patterns. These factors, interacting with surface features of the Earth such as mountains and distance from the ocean, result in the particular rainfall patterns. The temperature, rainfall, and altitude of an area, in turn, provide conditions that result in the growth of vegetation characteristic of that area.

The patterns of atmospheric circulation influence patterns of circulation in the ocean, modified by the location of the land masses around and against which the ocean currents must flow. Oceanic circulation is dominated by huge surface gyrals (Figure 39-5), which move around the subtropical zones of high pressure between approximately 30 degrees north and 30 degrees south latitude. These gyrals move clockwise in the northern hemisphere and counterclockwise in the southern hemisphere. They profoundly affect life not only in the oceans but also on coastal lands

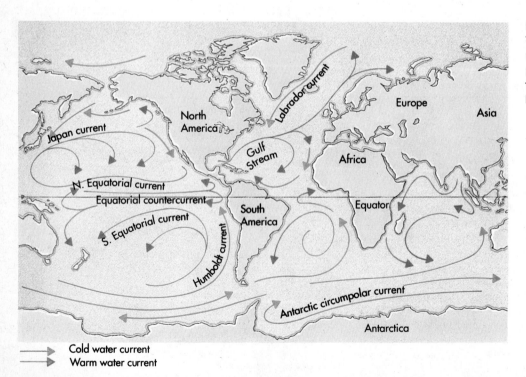

FIGURE 39-5 Ocean circulation.
The circulation in the oceans moves in great surface spiral patterns, or gyrals, and affects the climate on adjacent lands.

HOW LIVING THINGS INTERACT WITH EACH OTHER AND WITH THEIR ENVIRONMENT

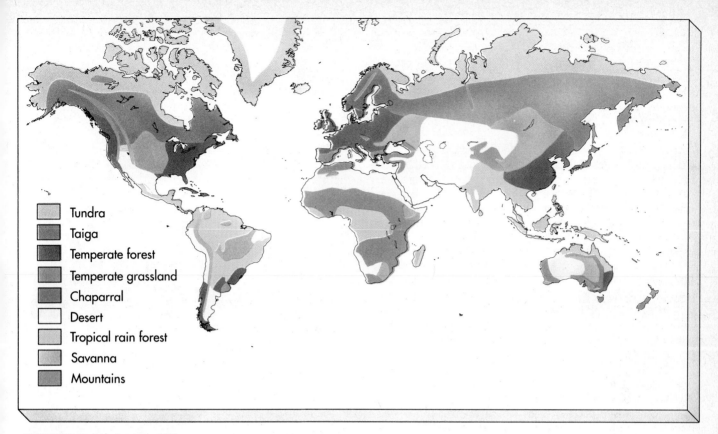

FIGURE 39-6 Distribution of biomes.
Mountain ranges and chaparrals are also shown on this map. Chaparrals are temperate scrublands that border grasslands and deserts in certain parts of the world.

because they redistribute heat. For example, the Gulf Stream in the North Atlantic swings away from North America near Cape Hatteras, North Carolina, and reaches Europe near the southern British Isles. Because of the Gulf Stream, western Europe is much warmer and thus more temperate than eastern North America at similar latitudes. As a general principle, the western sides of continents in the temperate zones of the northern hemisphere are warmer than their eastern sides; the opposite is true in the southern hemisphere.

Life on land: The biomes of the world

Biomes are often classified in seven categories: (1) tropical rain forests, (2) savannas, (3) deserts, (4) temperate grasslands, (5) temperate deciduous forests, (6) taiga, and (7) tundra (Figure 39-6). This list is arranged by distance from the equator, but the biomes do not encircle the Earth in neat bands. Their distribution is greatly affected by the climatic effects caused by the presence of mountains, the irregular outlines of the continents, and the temperature of the surrounding sea. In addition, the climate (and vegetation) changes with the elevation of land similar to changes with increasing latitude (Figure 39-7). Thus tundra-like vegetation occurs near the top of a tall mountain in the tropics, as well as near the North and South Pole.

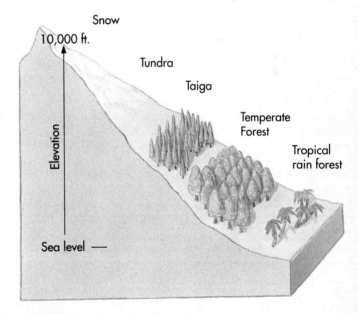

FIGURE 39-7 Elevation and biomes.
Biomes that normally occur far north and far south of the equator at sea level occur also in the tropics but at high mountain elevations. Thus on a tall mountain in southern Mexico or Guatemala, you might see a sequence of biomes like the one illustrated here.

BIOMES AND LIFE ZONES OF THE WORLD

Tropical rain forests

Tropical rain forests occur in Central America; in parts of South America, particularly in and around the Amazon Basin; in Africa, particularly in central and west Africa; and in southeast Asia. As the name suggests, the tropical rain forests occur in regions of high temperature and rainfall, generally 200 to 450 centimeters (80 to 175 inches) per year, with little difference in its distribution from season to season. The temperature averages 25° C (77° F). As a comparison, Houston, Texas—one of the hottest and wettest cities in the United States—receives an average of 121 centimeters (48 inches) of rainfall per year; the average temperature is 20.5° C (69° F). New Orleans, Louisiana—another hot, wet city—receives an average of 145 centimeters (57 inches) of rainfall per year; the average temperature is 20° C (68° F). These cities almost seem dry and cool in comparison to the tropical forest!

You may have the idea that the tropical rain forest is thick with lush vegetation and creeping vines, creating a network too dense to penetrate without a machete. These forests *are* thick and lush—but not at the forest floor. Little can grow on the ground far beneath the canopy of trees whose branches and leaves form an overlapping roof to the forest. In fact, only 2% of the light shining on the forest canopy reaches its floor! Plants that do grow there have large, dark green leaves adapted to conducting photosynthesis at low light levels. Other types of vegetation have interesting adaptations that enable them to compete successfully with the large trees for sunlight. Vines, for example, have their roots anchored in the soil but climb up the trees with their long stems. They reach the canopy where leaves grow to capture sunlight. Other interesting plants are the epiphytes, or "air plants." **Epiphytes** grow on the trees or other plants for support but draw their nourishment mostly from rain water. Some epiphytes catch moisture with modified leaves or flower parts; others have roots that hang free in the air and absorb water (with its dissolved minerals) from the rain (Figure 39-8).

The giant trees of the tropical forest support a rich and diverse community of animals on their branches. Figure 39-9 provides examples of just a few of these interesting organisms. The roots of the trees are interesting also, spreading out from thickened trunks into a thin layer of soil, often no more than a few centimeters deep. These roots transfer the nutrients from fallen leaves and other organic debris quickly and efficiently back to the trees after bacteria and fungi break it down. Very few nutrients remain in the soil. Therefore when humans cut down and then burn these trees to clear the land for agriculture, they are "burning away" the nutrients held in the trees as well as breaking down organic matter to carbon dioxide. The small amount of ash that is left provides few nutrients for the crops farmers try to grow. In 2 to 3 years of farming, these few remaining nutrients are depleted from the soil, and the land remains barren (Figure 39-10).

> The tropical rain forest, a biome that occurs in regions of high temperature and rainfall, is characterized by tall trees that support a variety of plant and animal life on their branches.

FIGURE 39-8 Tropical rain forests: lush equatorial forests.
A Epiphytes in the Monteverde Cloud Forest Reserve in Costa Rica. **B** Blooming only for 24 hours, the flower of a passion vine is a specialist in rain forest competition. Sweet spots reward vigilant guard ants, and poison leaves deter unwanted intruders.

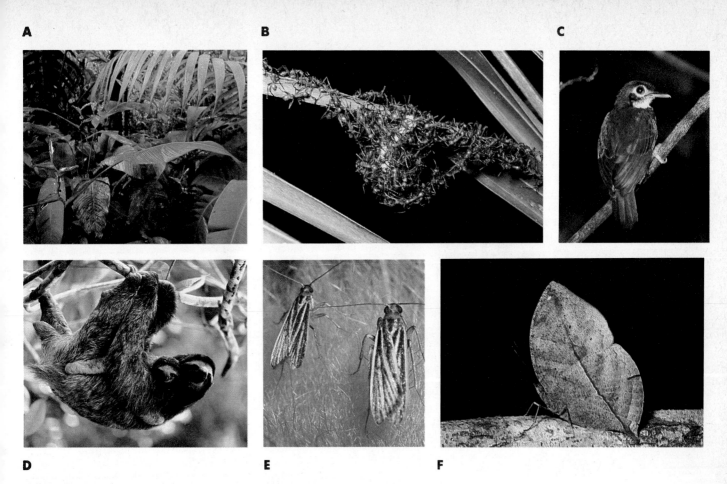

FIGURE 39-9 Ecological specializations in the tropical rain forest.
A Orange and red flower clusters of *Heliconia irrasa* stand out like beacons in the Costa Rican rain forest. They are visited and pollinated by the hummingbirds that they attract.
B Army ants live as huge, mobile, foraging communities in the tropical rain forest. Here a small column is transporting a wasp larva to be used as food.
C Some species of birds, such as the bicolored antbird, fly above the columns of army ants and feed on the insects that they flush from the foliage.
D A mother and baby three-toed sloth hang from a tree limb. Sloths reside in the canopy of the tropical rain forest, as shown here in Panama.
E Sloth moths carry out their entire life cycle in the fur of the three-toed sloth, where their larvae feed on the green algae that grow luxuriantly there.
F A dead leaf butterfly in the forests of Sumatra is superbly camouflaged.

FIGURE 39-10 The destruction of the rain forest.
To clear land for agriculture, farmers first cut down the trees and then burn the stumps in the technique of slash-and-burn agriculture. Here, a rain forest is being destroyed in the Amazon basin of Peru.

BIOMES AND LIFE ZONES OF THE WORLD

BIOLOGY IN FOCUS: The El Niño southern oscillation

The west coast of South America is normally a highly productive area, cooled by the Humboldt current, which sweeps up from the south and makes it possible for penguins (cold-adapted birds) to live on the Galapagos Islands near the equator. The cool Humboldt current is also responsible for bringing the cool, nutrient-rich waters up from the depths and replacing the depleted waters at the surface. Every year, usually in December or January, this cool current begins to warm as the trade winds slack off and warm water flows down the coast to southern Peru and northern Chile. Local fishermen are familiar with this event, since it signals the end of the fishing season, and have named it *El Niño* for "the Christ Child" because the warm current occurs around the Christmas season.

Every few years, however, the El Niño current is much warmer than usual. Accompanying the warmer waters, scientists have also noted changes in atmospheric pressure systems around the world. The term *southern oscillation* is used to describe these atmospheric results of a severe El Niño southern oscillation (ENSO) event such as the one that occurred in 1972 can be devastating on fisheries. Commercial fish stocks disappeared from the waters of Peru and northern Chile. The anchovy fishery was especially hard hit—the commercially valuable anchovy fisheries of Peru were destroyed.

Other populations of animals are affected as well by a severe ENSO. In the ENSO of 1982 to 1983, many birds starved to death on islands throughout the central Pacific, and a huge number of bird colonies simply gave up breeding. Christmas Island, located about 2000 kilometers (1243 miles) south of Honolulu, is an important breeding site for many species of birds. During the spring and summer of 1983, 95% of the 14 million birds that normally nest on the island, representing 18 species, simply left, abandoning their eggs; by June 1983, there were only 150,000 birds on the island.

The effects of an ENSO on the weather are more well known. The severe ENSO of 1986 to 1988 correlated with unusual drought in the states of Washington and Oregon in the summer of 1987. The ENSO event of 1991 to 1992 brought heavy flooding to southern California and parts of Texas.

Not all the effects of an ENSO are negative and catastrophic. The moist, rainy conditions that accompany an ENSO are favorable for crops on land. The Peruvian rice crop, for example, benefits from the ENSO rains. Land birds, such as Darwin's finch on the Galapagos Islands, breed abundantly during an ENSO, and their population sizes increased remarkably during the ENSO of 1982 to 1983.

Scientists still do not know what causes an ENSO event; they do not know, for instance, whether the warm current causes the southern oscillation, or vice versa. Because the cause is still unknown, an ENSO cannot be predicted. Nevertheless, scientists have used the knowledge they have gained to recognize ENSO events in their early stages. Further study of the ENSO may hold the key to understanding the complex interplay of atmospheric and water currents around the globe.

Savannas

Not all areas near the equator are wet; some areas experience prolonged dry seasons or have a lower annual rainfall than the tropical forests (generally, about 90 to 150 centimeters per year, or 35 to 60 inches). The heat, periodic dryness, and poor soils cannot support a forest but have led to the evolution of savannas: open grasslands with scattered shrubs and trees. These areas, situated between the tropical forests and deserts, cover much of central and southern Africa, western India, northern Australia, large areas of northern and east-central South America, and some of Malaysia.

The vegetation of the savanna supports large grazing herbivores such as buffalo and zebra (Figure 39-11); these animals, in turn, are food for carnivores such as lions. The savanna also supports a large number of plant-eating invertebrates, such as mites, grasshoppers, ants, beetles, and termites. The termites, in fact, are one of the most important soil organisms, breaking down dried twigs, leaves, and grass to usable nutrients. In addition, their huge, complex mounds (Figure 39-12) provide passageways for rain water to deeply penetrate the ground rather than just running off or evaporating from the surface.

> Like tropical forests, savannas are found near the equator but in areas having less annual rainfall. This climate supports grasslands with only scattered trees and shrubs.

FIGURE 39-11 Savannas: dry, tropical grasslands.
These zebras are grazing on a savanna in Tanzania.

FIGURE 39-12 Termite mound in southern Africa.
Termite mounds allow rainwater to penetrate the ground, thwarting run-off and evaporation.

BIOMES AND LIFE ZONES OF THE WORLD

FIGURE 39-13 Deserts: arid and hot.
The semidesert in **A** is in South Mountain Park in Phoenix, Arizona. The true desert in **B** is the Great Basin desert near Baker, Nevada. The extreme desert in **C** is the Namib desert in Namibia.

Deserts

Deserts are biomes that have 25 centimeters (10 inches) or less of precipitation annually. For this reason, the vegetation in deserts is characteristically sparse. The higher the annual rainfall a desert has, however, the greater the amount of vegetation it will be able to support. In fact, ecologists classify deserts based on their annual rainfall: semideserts receive about 25 centimeters (10 inches) per year (Phoenix, Arizona, and San Diego, California, for example), true deserts receive less than 12 centimeters (4.7 inches) per year (Las Vegas, Nevada), and extreme deserts average below 7 centimeters (2.8 inches) per year (Namib Desert in southwestern Africa). The photos of these three types of deserts shown in Figure 39-13 point out the differences in their patterns of vegetation.

Major deserts occur around 20 to 30 degrees north and south latitude, where the warm air that rose from the equator falls. As previously mentioned, the air at the equator rises, cooling and releasing its moisture, which falls on the tropical forests. The dry air then falls over desert regions, resulting in little precipitation. Deserts also occur in the interiors of continents far from the moist sea air, especially in Africa (the Sahara Desert), Eurasia, and Australia. Some deserts, such as the Baja region of California, are near the ocean yet dry; the winds blow from the north, carrying little moisture because they are cool. High pressure areas off the West Coast of the United States also deflect storms moving down from the north. In addition, some deserts form on the leeward side of mountain ranges, such as in the Great Basin of Nevada and Utah in the United States.

Because desert vegetation is sparse and the skies are usually clear, deserts radiate heat rapidly at night. This situation results in substantial daily changes in temperature, sometimes more than 30 degrees between day and night. Although both hot deserts (the Sahara, for example) and cool deserts (the Great Basin of North America) exist, summer daytime temperatures in all deserts are extremely high, frequently exceeding 40° C (104° F). In fact, temperatures of 58° C (136.4° F) have been recorded both in Libya and in San Luis Potosi, Mexico—the highest that have been recorded on Earth.

Plants have developed a wide variety of adaptations to survive in this difficult environment. Annual plants are often abundant in deserts and simply bypass the unfavorable dry season in the form of seeds. After sufficient rainfall, many germinate and grow rapidly, sometimes forming spectacular natural displays. Characteristics of deserts, of course, are the many species of succulent plants, those with tissues adapted to store water, such as cacti. The trees and shrubs that live in deserts often have deep roots that reach sources of water far below the surface of the ground. The woody plants that grow in deserts may be either deciduous, losing their leaves during the hot, dry seasons of the year, or evergreen, with hard, reduced leaves. The creosote bush of the deserts of North and South America is an example of an evergreen desert shrub. Near the coasts in areas where there are cold waters offshore, deserts may be foggy, and the water that the plants obtain from the fog may allow them to grow quite luxuriantly.

FIGURE 39-14 How two desert animals conserve water.
A Spadefoot toads, which live in the deserts of North America, can burrow nearly a meter below the surface and remain there for as much as 9 months of each year. Under such circumstances, their metabolic rate is greatly reduced, and they depend largely on their fat reserves. When moist, cool conditions return to the desert, they emerge and breed rapidly. The young toads mature quickly and burrow back underground, using the horny projections on their feet that gave them their name.
B On the very dry sand dunes of the Namib desert in southwestern Africa, the beetle *Onymacris unguicularis* collects fog water by holding up its abdomen at the crest of a dune, thus gathering condensed water on its body.

FIGURE 39-15 Temperate grasslands: seas of grass.
Among the other names for these grasslands are prairie, steppe, veld, and pampa.

Desert animals, too, have fascinating adaptations that enable them to cope with the limited water of the deserts (Figure 39-14). Many limit their activity to a relatively short period of the year when water is available or even plentiful; they resemble annual plants in this respect. Many desert vertebrates live in deep, cool, and sometimes moist burrows. Organisms that are active for much of the year emerge from their burrows only at night, when temperatures are relatively cool. Other organisms, such as camels, can drink large quantities of water when it is available and store it for use when water is unavailable. A few animals simply migrate to or through the desert and exploit food that may be abundant seasonally; when the food disappears, the animals move on to more favorable areas.

Temperate grasslands

Temperate grasslands have various names in different parts of the world: the prairies of North America, steppes of Russia, pusztas of Hungary, veld of South Africa, and pampas of South America. All temperate grasslands have 25 to 75 centimeters (10 to 30 inches) of rainfall annually, much less than that of savannas but more than that of deserts. Temperate grasslands also occur at higher latitudes than savannas but are often found bordering deserts as savannas do.

Temperate grasslands are characterized by large quantities of perennial grasses; the rainfall is insufficient to support forests or shrublands. Grasslands are often populated by herds of grazing mammals, such as the North American bison (Figure 39-15). The soil in temperate grasslands is rich; in fact, much of the temperate grasslands are farmed for this reason. Grasslands are often highly productive when they are converted to agriculture, and many of the rich agricultural lands in the United States and southern Canada were originally occupied by prairies.

> Deserts occur around 20 to 30 degrees north and south latitude and in other areas that have 25 centimeters (10 inches) or less of precipitation annually. Desert life is somewhat sparse but exhibits fascinating adaptations to life in a dry environment.

> Temperate grasslands experience a greater amount of rainfall than deserts but a lesser amount than savannas. They occur at higher latitudes than savannas, but, like savannas, are characterized by perennial grasses and herds of grazing mammals.

FIGURE 39-16 Temperate deciduous forests: rich hardwood forests. The leaves of the trees in temperate deciduous forests often change color in the autumn before they fall from the trees as winter approaches.

Temperate deciduous forests

The climate in areas of the northern hemisphere such as the eastern United States and Canada and an extensive region in Eurasia supports the growth of trees that lose their leaves during the winter. Such trees are called *deciduous* (from a Latin word meaning "to fall"), dropping their leaves and remaining dormant throughout the winter. These vast areas of trees are therefore temperate deciduous forests (Figure 39-16) and thrive in climates where summers are warm, winters are cold, and the precipitation is moderate, generally from 75 to 150 centimeters (30 to 60 inches) annually. Precipitation is well distributed throughout the year, but water is generally unavailable during the winter because it is frozen.

The temperate forest differs from the tropical forest in that more vegetation grows near the forest floor than in the tropical forest. Although the temperate forest does have an upper canopy of dominant trees such as beech, oak, birch, and maple, there is a lower tree canopy and a layer of shrubs beneath. On the ground, herbs, ferns, and mosses abound (Figure 39-17). In addition, animal life in the temperate forest is abundant on the ground as well as in the trees; in the tropical forest, animal life is primarily arboreal, with the exception of large mammals, a few bird species, and soil invertebrates.

In areas having less than 75 centimeters (30 inches) of precipitation annually, temperate deciduous forests are replaced by grassland, as in the prairies of North America and the steppes of Eurasia. Where conditions are more limiting—restricted, for example, by intense cold—these forests may be replaced by coniferous forests.

> **Temperate deciduous forests occur in areas having warm summers, cold winters, and moderate amounts of precipitation. The trees of this forest lose their leaves and remain dormant throughout the winter.**

FIGURE 39-17 The floor of the temperate deciduous forest is rich in plant growth. Plants such as herbs, ferns, and mosses inhabit the forest floor.

Taiga

The northern coniferous forest is called **taiga.** The cone-bearing trees of this forest are primarily spruce, hemlock, and fir and extend across vast areas of Eurasia and North America (Figure 39-18). The taiga is characterized by long, cold winters with little precipitation; most of the precipitation falls in the summers. Because of the latitude where taiga occurs, the days are short in winter (as little as 6 hours) and correspondingly long in summer. The light, warmth, and rainfall of the summer allows plants to grow rapidly, and crops often attain a large size in a surprisingly short time.

The trees of the taiga occur in dense stands of one or a few species of cone-bearing trees. Alders, a common species, harbor nitrogen-fixing bacteria in nodules on their roots; for this reason, they are able to colonize the infertile soils of the taiga. Marshes, lakes, and ponds also characterize

FIGURE 39-18 Taiga: great evergreen forests of the north.
The taiga located in Alaska is dominated by spruces and alders.

the taiga; they are often fringed by willows or birches. Many large mammals can also be found there, such as the elk (Figure 39-19). Other herbivores, including moose and deer, are stalked by carnivores such as wolves, bear, lynx, and wolverines.

To the south, taiga grades into temperate forests or grasslands, depending on the amount of precipitation. Coniferous forests also occur in the mountains to the south, but these are often richer and more diverse in species than those of the taiga. Northward, the taiga gives way to open tundra.

> The taiga, or northern coniferous forest, consists of evergreen, cone-bearing trees. The climate of this biome is characterized by long, cold winters with little precipitation.

FIGURE 39-19 Bull elk crossing a mountain stream.
Elk, which are herbivores, are hunted by carnivores such as wolves, bears, lynx, and humans.

BIOMES AND LIFE ZONES OF THE WORLD

FIGURE 39-20 Tundra: cold boggy plains of the north. Mount McKinley National Park in Alaska.

Tundra

Farthest north in Eurasia, North America, and their associated islands, between the taiga and the permanent ice, is the open, often boggy community known as the **tundra** (Figure 39-20). Dotted with lakes and streams, this enormous biome encircles the top of the world, covering one fifth of the Earth's land surface. (A well-developed tundra does not occur in the Antarctic because there is no land at the right latitude.) The tundra is amazingly uniform in appearance, dominated by scattered patches of grasses and sedges (grass-like plants), heathers, and lichens. Some small trees do grow but are primarily confined to the margins of streams and lakes.

Annual precipitation in the tundra is very low, similar to desert-like precipitation of less than 25 centimeters (10 inches) annually. In addition, the precipitation that falls remains unavailable to plants for most of the year because it freezes. During the brief Arctic summers, some of the ice melts. The **permafrost,** or permanent ice, found about a meter down from the surface never melts, however, and is impenetrable to both water and roots. When the surface ice melts in the summer, it has nowhere to go and forms puddles on the land. In contrast, the alpine tundra found at high elevations in temperate or tropical regions does not have this layer of permafrost.

The tundra teems with life during its short summers. As in the taiga, perennial herbs grow rapidly then, along with various grasses and sedges. Large grazing mammals, including musk-oxen, caribou, and reindeer, migrate from the taiga. Many species of birds and waterfowl nest in the tundra in the summer and then return to warmer climates for the winter. Populations of lemmings, small rodents that breed throughout the year beneath the snow, rise rapidly and then crash on a long-term cycle, influencing the populations of the carnivores that prey on them, such as wolves, foxes, and lynx.

> The tundra encircles the top of the world. This biome is characterized by desert-like levels of precipitation, extremely long and cold winters, and short, warmer summers.

Life in fresh water

Only 2% of the Earth is covered by fresh water, found standing in lakes and ponds or moving in rivers and streams. Fresh water ecosystems lie near and are intertwined with terrestrial ecosystems. For example, some organisms such as amphibians may move from one ecosystem to another. In addition, organic and inorganic material continuously enters bodies of fresh water from terrestrial communities. Often, the wet, spongy land of marshes and swamps provides habitats intermediate between the two.

Ponds and lakes have three life zones, or regions, in which organisms live: the shore zone, the open-water zone, and the deep zone. The **shore zone** is the shallow water near edges of a lake or pond in which plants with roots, such as cattails and water lilies, may grow. Consumers such as frogs, snails, dragonflies, and tiny shrimp-like organisms live among these producers. The **open-water zone** is the main body of water through which light penetrates. Microscopic floating algae, or phytoplankton, grow here. Microscopic floating animals (zooplankton) feed on the phytoplankton in this aquatic food chain. They, in turn, are eaten by small fish, which are eaten by larger fish. The **deep zone,** the water into which light does not penetrate, is devoid of producers. This dark zone is inhabited mainly by decomposers and other organisms such as clams that feed on the organic material that filters down to them (Figure 39-21). Ponds differ from lakes in that they are smaller and shallower. Therefore light usually reaches to the bottom of all levels of a pond; it has no deep zone.

Rivers and streams differ from ponds and lakes primarily in that their water flows rather than remains stationary. The nature of this ecosystem is therefore different from that of a pond or lake. One difference is that the level of dissolved oxygen is usually much higher in a river or stream than in a standing body of water because moving water mixes with

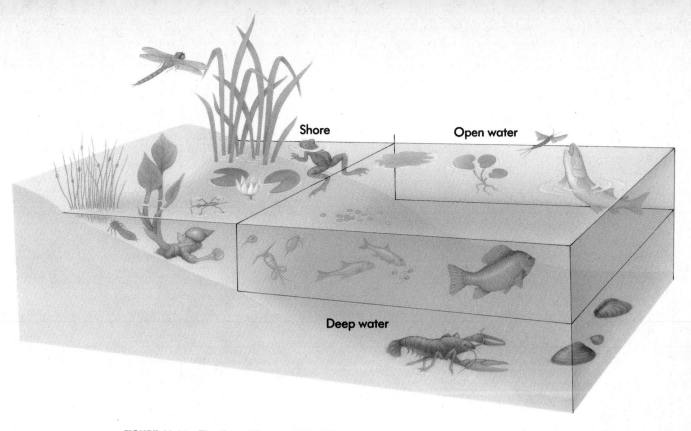

FIGURE 39-21 The three life zones of a lake.
Organisms live in the shore zone, the open-water zone, and the deep-water zone.

the air as it churns and bubbles along. A high level of dissolved oxygen allows an abundance of fish and invertebrates to survive. In addition, only a few types of producers inhabit rivers and streams: various species of algae that grow on rocks and a few types of rooted plants such as water moss.

A river or stream is characterized as an open ecosystem; that is, it derives most of its organic material from sources other than itself. Detritus (debris or decomposing material) flows from upstream or enters from the land. Leaves and woody material drop into the stream from vegetation bordering its banks. Rainwater washes organic material from overhanging leaves. In addition, water seeps into a river or stream from below the surface of adjoining land, carrying with it organic materials and, in some cases, fertilizers and other chemicals. These nutrients feed the producers and small consumers. As commonly occurs in pond and lake food chains, large fish feed on smaller fish that feed on tiny invertebrates. The river/stream ecosystem is largely heterotrophic and is strongly tied to terrestrial ecosystems that surround it.

> **Fresh water ecosystems lie near and are intertwined with terrestrial ecosystems. From them, organic and inorganic material continuously enters fresh water ecosystems.**

As the water from a stream or river flows into a lake or the sea, the velocity of the water decreases. As the water slows, sediment carried along by faster water now sinks to the bottom. These deposits form fan-shaped areas called **deltas.** Accumulated sediment breaks the path of the water into many small channels that course through the delta. Occasionally, the path of the water is actually stopped as sediment accumulates. For this reason a delta created by a large river, such as the Mississippi River delta, consists of a great deal of swampy, marshy land.

Estuaries: Life between rivers and oceans

As rivers and streams flow into the sea, an environment called an **estuary** is created where fresh water joins salt water. In the shallow water of the delta areas, rooted grasses often grow. Other producers of the estuary are various types of algae and phytoplankton. Consumers are primarily mollusks, crustaceans, and fish. All organisms inhabiting estuaries, however, have adaptations that allow them to survive in an area of moving water and changing salinity. (*Salinity* refers to the concentration of dissolved salts in the water.) Therefore many estuarine organisms are bottom-dwellers and attach themselves to bottom material or burrow in the mud. These organisms are found living in regions of optimum salinity for their survival.

Oysters are one of the most important bottom-dwellers in estuaries, providing a habitat for many other species of organisms. These mollusks either bury themselves in the mud, forming oyster beds, or cement themselves in clusters to the partially buried shells of dead oysters, forming oyster reefs. Many other invertebrates, such as sponges and bar-

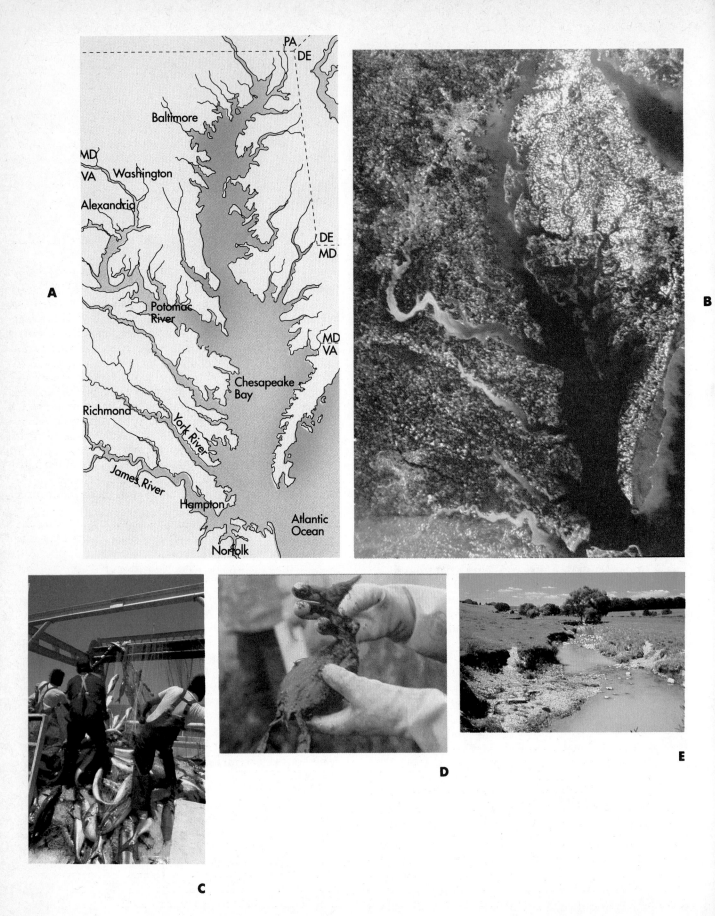

686 HOW LIVING THINGS INTERACT WITH EACH OTHER AND WITH THEIR ENVIRONMENT

FIGURE 39-22 Chesapeake Bay.
Chesapeake Bay has more than 11,300 kilometers (7020 miles) of shoreline and drains more than 166,000 square kilometers (64,100 square miles) in one of the most densely populated and heavily industrialized areas in North America. The body of open water is about 320 kilometers (199 miles) long and, at some points, nearly 50 kilometers (31 miles) wide.
A Large metropolitan areas and shipping facilities make the bay one of the busiest natural harbors anywhere.
B Aerial view of Chesapeake Bay.
C One of the most biologically productive bodies of water in the world, the bay yielded an annual average of about 275,000 kilograms (600,000 pounds) of fish in the 1960s but only a tenth as much in the 1980s. The human population grew 50% during the same period.
D Oil transport and commercial shipping is expected to double by the year 2020. This grebe is coated from a recent oil spill off the mouth of the Potomac River.
E Uncontrolled erosion from certain agricultural practices, pesticides, and increases in nutrients block the light needed for photosynthesis and upset the delicate ecological balance on which the productivity of the bay depends. The states that border the bay are cooperating, with the assistance of the Environmental Protection Agency, to try to bring the bay back to its former productivity.

nacles, attach themselves to the oysters and feed on algae. Other species such as crabs, snails, and worms live on, beneath, and between the oysters, feeding on the oysters themselves or on detritus trapped in the oyster reef. One researcher, in fact, has documented over 300 species of organisms living in association with a single oyster reef!

The motile organisms of the estuary are primarily crustaceans such as crabs, lobsters, and shrimp and various species of fish. Fish exhibit interesting reproductive adaptations to the varying salinities of the estuary. Some species, such as the striped bass, spawn upstream from the estuary where the salinity of the water is low. The larvae and young fish move downstream through increasing concentrations of salt as they develop, moving into the ocean in adulthood. In a similar manner, shad spawn upstream in fresh water, and the young spend their first summer in the estuary before swimming to the sea.

> Estuaries are places where the fresh water of rivers and streams meets the salt water of oceans. These ecosystems are made up of plentiful communities of organisms exhibiting behaviors and growth characteristics adapted to changing salinity.

Nutrients are more abundant in estuaries than in the open ocean because they are close to terrestrial ecosystems and derive much of their nutrients from them as rivers and streams do. Unfortunately, estuaries are also easily polluted from these sources. In Chesapeake Bay (Figure 39-22), for example, complex systems of rivers enter the Atlantic Ocean, forming one of the most biologically productive bodies of water in the world. In the 1960s, the bay yielded an annual average of about 275,000 kilograms (600,000 pounds) of fish. As the human population in this area increased, however, along with oil transport and commercial shipping, pollution also increased. More than 290 oil spills were reported in the bay in 1983 alone. In addition, uncontrolled erosion of the land into the water and leaching of pesticides has caused a 90% decrease in the yield of fish. Maryland and Virginia, the states that border the bay, along with the Environmental Protection Agency, are working to bring the Chesapeake Bay estuary back to its former productivity.

Life in the oceans

Although only 2% of the Earth is covered by fresh water, nearly three quarters of the Earth's surface is covered by ocean. These seas have an average depth of 4 kilometers (approximately 2½ miles), and they are, for the most part, cold and dark. The concentration of oxygen, as well as the availability of light and food, is a factor that limits life in the ocean. Although cold water is able to "hold" more oxygen than warm water, the warmer sea water near the surface of the ocean mixes with the oxygen in the atmosphere. Therefore oxygen is present in its highest concentrations in the upper 200 meters (650 feet) or so of the sea. Light is most abundant in the top 100 meters (325 feet).

The marine environment provides a variety of habitats, but it can be divided into three major life zones:
1. The **intertidal zone,** the area between the high tide mark and the low tide mark
2. The **neritic zone,** the area of shallow waters along the coasts of the continents, which extends from the low tide mark to waters up to 200 meters deep
3. The **open-sea zone,** composing the remainder of the ocean

The intertidal zone

The wind-swept shoreline is a harsh place for organisms to live. As the tide rolls in and out, environmental conditions change from hour to hour: wet to dry, sun protected to sun parched, and wave battered to calm. Nevertheless, life abounds in the intertidal zone, exhibiting interesting adaptations and characteristics necessary for survival.

Figure 39-23 shows the rocky shore at low tide at the Pacific Grove Monterey Bay in California. The exposed rocks are teeming with life, but these organisms vary along a continuum from the driest areas (those least often covered with water) to the wettest areas (those most often covered with water). Highest up on the rocks, in areas that the high tide sometimes does not reach, grow certain lichens and algae. Somewhat lower on the rocks grow barnacles. Then oysters, blue mussels, and limpets (mollusks that have conical shells) take over, followed by brown algae and red algae. "Forests" of large brown algae, or kelp, take over in areas that are exposed for only short periods of time. All of these organisms have adaptations such as hard shells (Figure 39-24) or

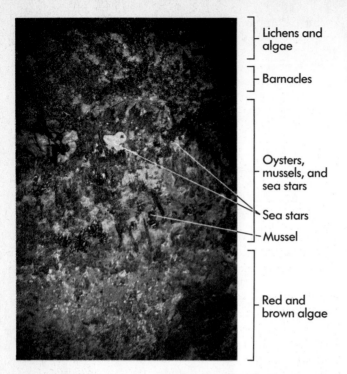

FIGURE 39-23 A rocky shore in Pacific Grove in Monterey Bay, California.
The different areas on the rock all receive different amounts of moisture, providing a variety of habitats for a variety of organisms.

gelatinous coverings that keep them from drying out and are either anchored within the sand or stick to the surfaces of rocks so that they will not be washed away.

In contrast to the rocky shore, the sandy shore (Figure 39-25) looks as though no life is present. However, life abounds beneath the sand and mud. Because organisms have

FIGURE 39-24 Rocky shore organisms.
A rocky shore in Maine shows a few of the organisms that make this shore their home. The pink sea star clings to the rocks, as do barnacles, which use special filaments to anchor themselves.

FIGURE 39-25 A sandy shore.
In contrast to the rocky shore, there seems to be no life present on this sandy shore in Hawaii. However, microscopic animals make their homes in the spaces between grains of sand, and larger animals burrow beneath the sand during low tide.

no large surfaces to which they can attach, they are adapted to burrowing under the sand during low tide. Copepods, tiny "micro" crustaceans, are predominant organisms. In addition, worms, crabs, and mollusks such as clams burrow to safety when the tides roll out. In areas near the low tide mark, sea anemones, sea urchins, and sea stars make their home.

The intertidal zone has plentiful light and is home to a variety of producers. Along with the algae, phytoplankton float in the water and are used for food by the zooplankton and many other consumers. In addition, the heterotrophs of the intertidal zone have the waves to thank for bringing fresh organic material to them and for washing away their wastes.

The neritic zone

Surrounding the continents of the world is a shelf of land that extends out from the intertidal zone, sloping to a depth of approximately 200 meters (approximately 650 feet) beneath the sea. This margin of land is called the **continental shelf.** The waters lying above it make up the neritic zone, which is derived from a Greek root referring to the sea. Because light reaches the waters of most of this zone, it supports an abundant array of plant and therefore animal life.

One outstanding community in the neritic zone is the **coral reef.** The term *reef* refers to a mass of rocks in the ocean lying at or near the surface of the water. Coral reefs are built by marine animals, or corals (phylum Cnidaria), that secrete carbonate, a hard, shell-like substance. With the help of algae that reside in their bodies, the coral build on already shallow portions of the continental shelf or on submerged volcanoes in the ocean. Complex and fascinating ecosystems, coral reefs provide habitats for a variety of invertebrates and fish (Figure 39-26). Along with the tropical rain forests, coral reefs are the most highly productive ecosystems in terms of biomass (see Chapter 38).

FIGURE 39-26 The coral reef: a delicate biome in danger.
The living corals in coral reefs depend on constant conditions for survival. Increasingly, the conditions in which coral reefs live are being disrupted by pollution and human carelessness. Preventative measures and a sensitivity to the precarious balance necessary for the coral reefs' survival will be instrumental in saving the reefs from destruction.
A Coral reef lagoon, Fiji.
B Coral reef in Fiji, showing hard and soft corals.
C Coral reef in Borneo, showing the brilliant colors of the different coral formations.
D Green turtle hovering over a soft and hard coral-encrusted reef off Sipadan Island.
E Scallop attached to a limestone reef in the Indo-Pacific.
F Diver destroying a reef in Key Largo, Florida. When exploring a coral reef, divers must never touch or stand on the corals.

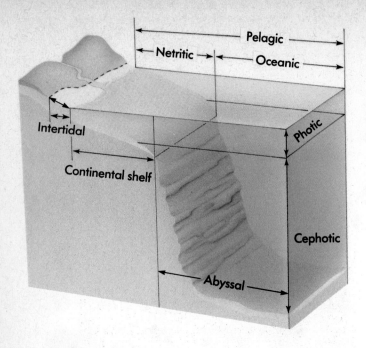

FIGURE 39-27 Zones of the ocean.
The pelagic zone encompasses the oceanic and neritic zones. Nearer to shore is the continental shelf and the intertidal zone. The photic zone and aphotic zones are measured by the amount of light that penetrates the water.

The open-sea zone

Beyond the continental shelf lies the great expanse of the open ocean. This open-sea zone is often referred to as the **pelagic zone,** a term derived from another Greek word meaning "ocean" (Figure 39-27). Within this huge ecosystem exists many diverse forms of life—some with which you are familiar, such as the floating plankton and various species of fishes. But other forms of life are unfamiliar—even bizarre—such as the anglerfish shown in Figure 39-28, *D*.

Organisms live in the vast expanse of the ocean based on the availability of light and food. Temperature, salinity, and water pressure also play roles in creating the various habitats of the ocean. Light is only available to organisms from the water's surface to an approximate depth of 200 meters (650 feet). In fact, this area of the open ocean is called the **photic zone** for this reason. Phytoplankton thrives in this well-lighted layer of the ocean (especially within the better-lighted upper 100 meters [325 feet]), drifting freely with the ocean currents and serving as the base of oceanic food webs. Zooplankton float with the phytoplankton and are first-order consumers in many photic food webs. Other typical heterotrophs of this zone are most air-breathing mammals such as whales, porpoises, dolphins, seals, and sea lions and fishes such as herring, tuna, and sharks. These fishes and mammals of the sea are called **nekton** and feed on the plankton and on one another. Together, the organisms that make up the plankton and the nekton provide all of the food for those that live below.

Little light penetrates the ocean from 200 to 1000 meters (650 to 3250 feet). Therefore no photosynthetic organisms live in this zone known as the **mesopelagic,** or middle ocean. At this depth, temperatures remain somewhat steady throughout the year but are cooler than the water above; water pressure increases steadily with depth. Under these conditions, many bizarre organisms have evolved, such as those that exhibit bioluminescence (see Figure 26-12), which they use to communicate with one another or attract prey. The organisms common to middle ocean life are fish with descriptive names such as swordfish, lanternfish, and hatchetfish; certain sharks and whales; and cephalopods such as octopi and squid.

Peculiar creatures also live in the ocean depths of 1000 or more meters—a mile or so (and more) beneath the surface. This region of the ocean is called the **abyssal zone,** meaning "bottomless." Although the ocean is not really bottomless, the water at this tremendous depth contains high concentrations of salt, is under immense pressure from the water above, and is very cold. The organisms living at this level cannot make trips into the photic zone to capture food as some mesopelagic organisms do, but must feed on material that settles from above.

Organisms collectively referred to as **benthos,** or bottom-dwellers, live on the ocean floor. Sea cucumbers and sea urchins crawl around eating detritus. Various species of clams and worms burrow in the mud, feeding on a similar array of decaying organic material. Bacteria are also rather common in the deeper layers of the sea, playing important roles as decomposers as they do on land and in fresh water habitats.

> Of the three main life zones in the ocean, the intertidal zone, the neritic zone and the photic area of the open-sea zone support the most life because of the presence of light and the availability of oxygen. Organisms that live in waters beneath the photic zone must feed on organic material and detritus that falls from above and be adapted to increased water pressure and salinity.

FIGURE 39-28   The open ocean.
A What mysteries lie beneath the ocean's deepest waters?
B Flying fish in the eastern Pacific ocean.
C School of yellowfin tuna swimming around a coral reef.
D A deep-sea anglerfish showing the lure used by the fish to attract prey. The lure is actually part of the fish, but to prey, the lure looks like a tempting bit of food. When the prey comes close to investigate, the anglerfish quickly eats the prey.
E Hydrothermal vent community located in the deep waters off the Galapagos Islands.

BIOMES AND LIFE ZONES OF THE WORLD

Summary

1. Biomes are wide-ranging, distinctive ecosystems of plants and animals that occur together within certain climatic regions and have definite characteristics. These characteristics are a direct result of temperature and rainfall patterns.

2. The climate of a region is determined primarily by its latitude and wind patterns. These factors interact with the surface features of the Earth, resulting in particular rainfall patterns. The temperature, rainfall, and altitude of an area, in turn, provide conditions that result in the growth of vegetation characteristic of that area.

3. Seven categories of biomes, arranged by distance from the equator, are (1) tropical rain forests, (2) savannas, (3) deserts, (4) temperate grasslands, (5) temperate deciduous forests, (6) taiga, and (7) tundra.

4. The tropical rain forests and savannas lie nearest the equator. The tropical rain forest receives an enormous amount of rain year round, whereas the savanna experiences less rain and prolonged dry spells. These differences promote the growth of tall trees and other lush vegetation in the tropical rain forest and open grasslands with scattered trees and shrubs in the savanna.

5. Deserts are extremely dry biomes. Hot deserts are hot year round, whereas cool deserts are hot only in the summer. Deserts are of great biological interest due to the extreme behavioral, morphological, and physiological adaptations of the plants and animals that live there.

6. Temperate grasslands receive less rainfall than savannas but more than deserts. The soil in these grasslands is rich, so they are well suited to agriculture.

7. The temperate deciduous forests receive moderate precipitation that is well distributed throughout the year. The climate of these forests differs from tropical forests in that they receive less rainfall, are found at a higher and cooler latitude, and experience cold winters. The trees are therefore deciduous, losing their leaves and remaining dormant throughout the winter.

8. The taiga is the coniferous forest of the north. It consists primarily of cone-bearing evergreen trees, which are able to survive long, cold winters and low levels of precipitation. Even farther north is the tundra, which covers about 20% of the Earth's land surface and consists largely of open grassland, often boggy in summer, which lies over a layer of permafrost.

9. Fresh water ecosystems make up only about 2% of the Earth's surface; most of them are ponds and lakes. Ponds and lakes have a shore zone and an open-water zone. Lakes have a deep zone. Rivers and streams differ from these bodies of water because they contain moving water that mixes with the air to provide high levels of oxygen for its fish and invertebrate inhabitants.

10. Estuaries are places where fresh water meets salt water as rivers empty into the ocean. These ecosystems receive nutrients from the surrounding land and support a large number of living things.

11. The marine environment consists of three major life zones: the intertidal zone, between the high and low tide marks; the neritic zone, the water from the low tide mark that lies over the continental shelf, and the open-sea zone. Within these zones, the ocean supports the most life in areas that have light and sufficient quantities of dissolved oxygen.

REVIEW QUESTIONS

1. What is a biome?
2. January is a winter month in the United States but part of summer in Australia. Why?
3. Summarize how atmospheric circulation affects climates around the world.
4. What factors determine the climate of a region?
5. Identify the terrestrial biome associated with each of the following:
   a. Animals in these areas live mostly in trees.
   b. Dominant trees such as oak and beech allow enough light so that herbs, ferns, and mosses grow on the ground.
   c. Because of permafrost, when surface ice melts, the ground cannot absorb the water.
   d. Although located near the equator, these regions are mostly open grasslands.
   e. These coniferous forests receive most of their precipitation during the summer.
   f. Herds of North American bison once roamed these areas, grazing on large expanses of perennial grasses.
   g. These regions radiate heat rapidly at night, causing wide temperature differences between day and night.
6. About 100,000 square kilometers of tropical rain forest are being destroyed each year (more than 1.5 acres per second), often to make room for planting crops. Do you think these areas will make good farmland? Explain your answer.
7. What factors limit rainfall in a desert?
8. Distinguish between taiga and tundra.
9. Describe the three life zones found in fresh water. What types of organisms live in each?
10. What is an estuary? Summarize how its inhabitants are adapted to its conditions.
11. Why are estuaries and fresh water regions so vulnerable to human pollution?
12. Identify and briefly describe the three major life zones of the ocean.
13. The ocean is divided into three zones based on depth. Describe what types of organisms are found in each zone.

KEY TERMS

abyssal zone 690
benthos 690
biome 672
continental shelf 688
coral reef 688
deep zone 684
delta 685
epiphyte 676
estuary 685
intertidal zone 687
mesopelagic zone 690
nekton 690
neritic zone 687
open-sea zone 687
open-water zone 684
pelagic zone 690
permafrost 684
photic zone 690
shore zone 684
taiga 682
tundra 684

THOUGHT QUESTIONS

1. Why do you think there are so many more species in the tropical rain forests of Brazil than in the temperate forests of Michigan? How would you go about testing your idea?
2. Temperatures over the antarctic ice cap (South Pole) are markedly colder than over the arctic ice cap (North Pole). Why do you think this is so?
3. Much of the world's most productive agriculture is carried out on soil of temperate grasslands. Agriculture in the tropics is far less productive. Why do you think temperate grassland soil is so much richer than soil in a tropical rain forest?

FOR FURTHER READING

Attenborough, D. (1984). *The living planet: A portrait of the Earth*. London: William Collins, Sons & Co., Ltd., and British Broadcasting Corporation.
This is a beautifully written and illustrated account of the biomes.

Gore, R. (1990, February). Between Monterey tides. *National Geographic*, pp. 2-43.
This is a beautifully illustrated account of life along the California coast.

Ropelewski, C. F. (1992, April 9). Predicting El Niño events. *Nature*, vol. 38.
Using statistical correlations of patterns of circulation in the upper atmoshpere, it may be possible to predict the onset of an El Niño event.

Swan, L. (1992, April). The aeolian biome. *BioScience*, 42(4):262-270.
The aeolian biome is a region around and beyond the limits of flowering plants, such as a high altitude spot on Mt. Everest. The animals in these biomes are nourished by airborne nutrients, and studies of aeolian zones can reveal much about how life began on Earth billions of years ago.

Whitehead, J.A. (1989, February). Giant ocean cataracts. *Scientific American*, pp. 50-57.
Enormous undersea cataracts play a crucial role in determining the chemistry and climate of the deep ocean.

40 The biosphere
Today and tomorrow

HIGHLIGHTS

▼

The biosphere extends from the tops of the mountains to the depths of the seas; it is all the parts of the Earth where biological activity occurs.

▼

Humans need to work toward a sustainable society that uses nonfinite and renewable sources of fuel, recycles resources to the fullest extent, and manages forests to meet global needs while not compromising the ability of future generations to survive.

▼

Water is being polluted by multiple sources, causing physical or chemical changes that harm living and nonliving things; prevention of this contamination is essential to ensure a safe water supply.

▼

A primary cause of air pollution is the combustion of fossil fuels; energy conservation measures and new energy technology are needed.

OUTLINE

The biosphere
The land
  Diminishing natural resources
    Fossil fuels
    Mineral resources
    Deforestation
  Species extinction
  Solid waste
The water
  Surface water pollution
  Ground water pollution
  Acid rain
The atmosphere
  Air pollution
  Ozone depletion
Overpopulation and environmental problems

Just imagine that you could hold the world in your hands and make a difference in its future. Well, imagining isn't even necessary. All people do hold the world in their hands in a figurative way. Everyone's actions have a direct impact on the Earth and on the quality of life that you and others will experience for generations to come.

However, you must ask yourself some crucial questions. How *are* you affecting the Earth? What environmental problems do people face today? And what can you do to deal with those problems to be sure that the quality of life on Earth will be enhanced for yourself and your children? The answers to these questions are the focus of this chapter and will help you understand more about how you shape this fragile planet on which you live.

The biosphere

Life on Earth is confined to a region called the **biosphere,** the global ecosystem in which all others exist. The biosphere extends from approximately 9000 meters (30,000 feet) above sea level to about 11,000 meters (36,000 feet) below sea level. You can think of it as extending from the tops of the highest mountains (such as Mount Everest or some of the Himalayas) to the depths of the deepest oceans (such as the Mariana Trench of the Pacific Ocean)—the part of our Earth in which the land, air and water come together to help sustain life (Figure 40-1).

The biosphere is often spoken of as the **environment.** This general term refers to everything around you—not only the land, air, and water but other living things as well. You can speak, for example, of the environment of an ant, a water lily, or all the peoples of the world. Your particular environment can change during the day from a home environment, to a classroom environment, and then to an office environment. The environment can include a great deal—or very little—of the total biosphere and its living things.

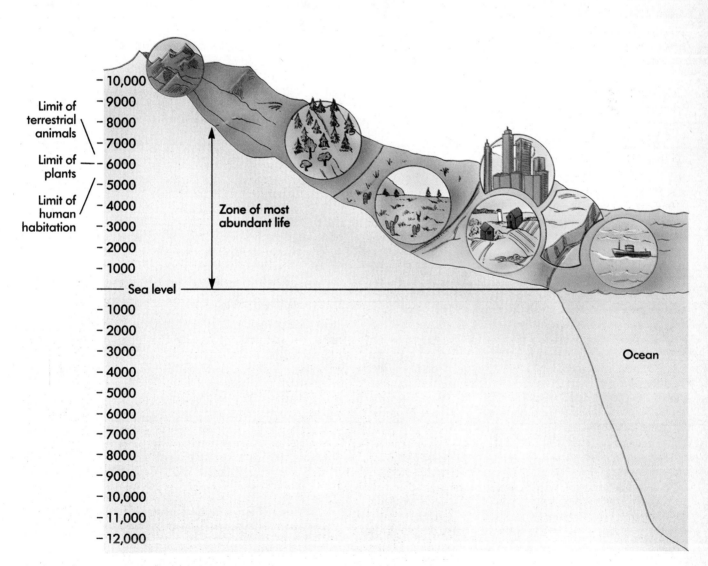

FIGURE 40-1 The biosphere.
The biosphere is that part of the Earth where life can be found, from the highest mountain peaks to the depths of the ocean. Most life is found in the area indicated by the arrows.

THE BIOSPHERE: TODAY AND TOMORROW

The land

Humans interact with the land and its inhabitants in a myriad of ways. Its natural resources are used. Wastes are produced that are filling and in some cases polluting the land. Forests are cut down, stripping many species of their habitats and therefore their life. And people continue to procreate at a rate that will result in the doubling of the world's population by the year 2030. What can people do to stop using the land in destructive ways? What can people do to help heal the Earth?

Diminishing natural resources

The land gives humans many things: fossil fuels, timber, food, and minerals. Water, too, is an important natural resource. Some of these resources—fossil fuels and minerals—are finite; that is, they cannot be replaced. Forests *can* be used and replaced *if managed wisely*. However, when whole sections of forest are cut and burned and their soil depleted, they cannot be replaced . . . nor can the wildlife that once lived there.

Fossil fuels

Many of the fuels used to heat homes and run cars are **fossil fuels.** These substances—coal, oil, and natural gas—are formed over time (acted on by heat and pressure) from the undecomposed carbon compounds of organisms that died millions of years ago. Environmental scientists suggest that instead of depending on these fuels for energy—fuels that are finite—people should work toward a **sustainable society** that uses nonfinite and renewable sources of fuel. In a sustainable society, the needs of the society are satisfied without compromising the ability of future generations to survive and without diminishing the available natural resources.

The primary renewable and nonfinite sources of power that are available are the sun, the wind, moving water, geothermal energy, and bioenergy. Technology is available now to harness these energies but are not yet in widespread use. **Nuclear power,** once considered a viable alternative energy source, will probably contribute only 6% to 8% of the world's energy by the year 2000. Although nuclear power is dependent on uranium, a finite but abundant natural resource, its problems lie primarily with the cost of building nuclear power plants, disposal of highly radioactive wastes, and the public fear regarding safety. The waste disposal problem would be virtually eliminated if nuclear fusion reactors can be developed to replace today's reactors, all of which use nuclear fission (splitting) reactions. Nuclear power could become a primary source for generating electricity.

Solar power is not a new technology; panels mounted on rooftops, for example, heat water that can be used for bathing, cooking, and space heating (Figure 40-2, *A*). In addition, homes and businesses can be built using the strategies of passive solar architecture (Figure 40-2, *B*). The technology is also available to use solar power to produce electricity by converting the sun's energy to electricity after it has been captured within a fluid such as oil or by the use of photovoltaic solar cells that can convert sunlight directly into electricity.

Water has been used as an energy source for decades (Figure 40-2, *C*) and is currently supplying 20% of the world's electricity. A new form of hydropower called **wave power** is now being tested. Wave power uses the vertical motion of sea waves to produce electricity. In addition, **tidal power,** the use of the movement of the tides of the oceans to generate electricity, is currently under study in France and England (Figure 40-2, *D*). The wind, too, has been used not for decades but for centuries as a source of energy. Today, windmills are being used in developing countries (those not yet industrialized) for pumping water to livestock and to irrigate the land. In developed countries, however, windmills are being used in an entirely new way—to generate electricity. This "new breed" of windmill has rigid blades fashioned from lightweight materials (Figure 40-2, *E*), and looks unlike the picturesque windmills of Holland and Cape Cod, Massachusetts. Currently, the United States and China are the two countries that lead in the use of **wind power** to generate electricity. In fact, some scientists predict that the United States will be generating 10% to 20% of its electricity from wind power by the year 2030.

Geothermal energy refers to the use of heat deep within the Earth. In some places, reservoirs of hot water exist that can be extracted from the Earth using drilling procedures much like those used to tap into the Earth's oil and natural gas reserves. Alternatively, dry, hot rock can be drilled and water flushed through it. After the water is heated within the Earth, it can be used directly for heating purposes or as part of a process to produce electricity. Currently, this technology is being used and further developed in the United States, England, Italy, New Zealand, and Japan.

Bioenergy refers to the use of living plants to produce energy. The most obvious type of bioenergy, the burning of wood, was first used by our ancestors approximately 1 million years ago. In fact, until the Industrial Revolution of the 1800s, wood, not coal, supplied most of the world's energy. Today, wood supplies 12% of the world's energy, primarily in Latin America, Asia, India, and Africa. Unfortunately, however, the world is experiencing a fuel wood crisis—demand is exceeding the supply. The reasons for this crisis are complex but include the high world population and therefore high demand for wood, the degradation of woodlands without proper reforestation (replanting) techniques, and the cutting and burning of huge areas of tropical rain forests. In developing countries, the use of **bio-gas machines** is helping ease the shortage of fuel wood. These stoves use microorganisms to decompose animal or human excrement or other organic wastes in a closed container. This process yields a methane-rich gas that can be used as to fuel stoves, light lamps, and produce electricity.

> **Environmental scientists suggest that humans use renewable and nonfinite energy sources such as solar, wind, and water power, as well as tap the Earth's geothermal energy. The use of such fuels rather than the extensive use of fossil fuels will help satisfy people's needs without diminishing the available natural resources.**

FIGURE 40-2 Alternative energy sources.
To decrease reliance on fossil fuels, many individuals and nations are experimenting with alternative energy sources.
A Solar power. These panels collect the energy from sunlight and use this energy to heat homes and water.
B A passive solar home, with solar panels mounted on the roof.
C Hoover Dam in Arizona. Water power has been used for many centuries.
D Tidal power plant in the Bay of Fundy. Using the ocean's tides to generate power is still in the experimental stages.
E Modern windmills in southern California.

THE BIOSPHERE: TODAY AND TOMORROW

A newer form of bioenergy is the use of plants such as corn and sugar cane to produce carbohydrates that are fermented, producing liquid fuels like ethanol. This technology, however, produces a product that appears to be an expensive alternative to fossil fuels. In addition, many cars do not function well on gasoline with ethanol additives.

Mineral resources

Minerals are inorganic substances that occur naturally within the Earth's crust. Zinc, lead, copper, aluminum, and iron are the minerals that humans use in the greatest quantities. However, other minerals are also mined, such as gold, silver, copper, and mercury. These minerals are present in the Earth in fixed amounts; once they are used, they are gone. Figure 40-3 shows the number of years the world's reserves of various minerals will last based on the rate of current consumption.

What can be done to avert a mineral shortage in years to come? First of all, people can cut down on their consumption. An American uses, on the average, 6 times more zinc, lead, and iron than a person living outside of the United States. In addition, an American uses 11 times more copper and 14 times more aluminum! Researchers agree that increased use of plastics and technology such as microelectronics will lower the demand for certain minerals. Plastics, however, are derived from petroleum products, so their manufacture increases consumption of fossil fuels.

Recycling is another way to help avert a mineral shortage in the coming years. Aluminum soda cans are produced from the clay-like mineral bauxite. Producing cans from recycled aluminum uses 95% less energy than producing them from bauxite, saves mineral resources, and saves waste disposal costs and problems. Plastics are not easily recyclable and are not "truly" recyclable. That is, plastics cannot be

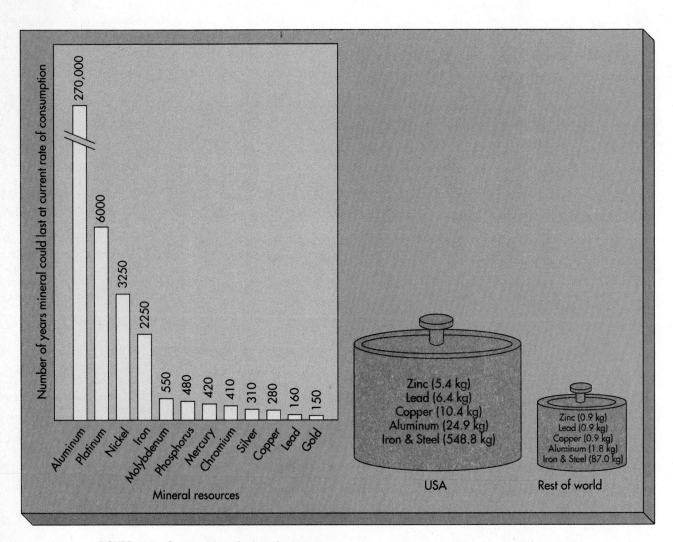

FIGURE 40-3 Consumption of mineral resources.
According to *The Earth Report,* modern industrialized societies require large amounts of minerals to sustain their high standard of living. The left shows a number of major minerals and the years they could last at the current rates of consumption. The right compares the consumption of major minerals by the United States with the rest of the world.

refashioned into the product from which they were claimed; plastic milk jugs cannot be made into plastic milk jugs but must be made into park benches or parking lot curbs. Grocery store plastic bags are the closest truly recycled plastic products, being made from used plastic bags and other additives. Therefore purchasing goods in recyclable containers such as paper, glass, and aluminum *and recycling those containers* is a wise choice. Scrap metal and old automobiles are also recyclable, as are automobile tires and batteries.

> Recycling paper, glass, and various metals will help preserve mineral resources as well as other natural resources.

Deforestation

Paper makes up 40% of the garbage in the typical American household. Recycling just your daily newspaper will save four trees every year. Buying products made from recycled paper will save even more. But recycling is not enough. The forests of the world are in severe crisis and will be lost if steps are not taken to stop their destruction.

The most severe crisis is that of **tropical rain forest deforestation.** Tropical rain forests (see Chapter 39) are located in Central and South America, tropical Asia, and central Africa, forming a belt around the equatorial "waist" of the Earth. Although this belt of forest covers only 2% of the Earth's surface, it is home to over *half* the world's species of plants, animals and insects. These organisms contribute 25% of medicines, along with fuel wood, charcoal, oils, and nuts. In addition, the tropical rain forests play an important role in the world climate.

Population and poverty are both high in rain forest countries. People with few resources move from towns and cities to the rain forest and cut the trees for sale as lumber or burn them to clear a patch of land to grow crops and raise cattle to sustain themselves and their families (Figure 40-4). Commercial ranchers also cut and burn the forests to make way for pastureland to feed beef-producing cattle. Unfortunately, the soil of the rain forests is poor, with few nutrients, and does not support crops. Before it is cut, the forest sustains itself because of symbiotic relationships between the trees and microorganisms that quickly decompose dead and dying material on the forest floor. These "processed" nutrients are quickly reabsorbed by the tree roots. Few nutrients stay in the soil; most of the nutrients are in the vegetation. Cutting these trees down and burning them releases the nutrients from the trees. Crops grow poorly on this land after it is stripped. After a year or two, crops will not grow at all. The people move on, cutting yet another portion of the forest.

Commercial logging also takes its toll on the tropical rain forests. Many of the trees are cut to supply fuel wood, paper, wood panels such as plywood, and charcoal and to supply furniture manufacturers with mahogany and other woods demanded by consumers around the world. At this time, the tropical forest is being slashed and burned at a devastating rate. By 1950, two thirds of the forests of Central America had been cleared. Madagascar, an island off the southwestern coast of Africa, had lost about half its rain forest by 1950, but as Figure 40-5 shows, has lost half again since then! The United Nations Food and Agriculture Organization estimates that 100,000 square kilometers (62,000 square miles) of rain forest are now being lost each year worldwide—about the size of New England. At the present rate, scientists estimate, nearly all the tropical rain forests will be gone—including their rich diversity of animal life—within the next 50 years.

In addition to losing a rich natural resource, many scientists agree the the burning of the tropical forests is adding tremendous quantities of carbon dioxide (CO_2) to the air.

FIGURE 40-4 Deforestation: a desperate effort to stave off poverty results in an environmental disaster.
These farmers live near the Andasibe reserve in Madagascar, an island where the per capita income is less than $250 a year. Clearing the rain forest allows these impoverished farmers to plant rice and graze cattle. But the environmental price may be too steep. In Madagascar alone, 80% of the rain forest has been destroyed.

THE BIOSPHERE: TODAY AND TOMORROW

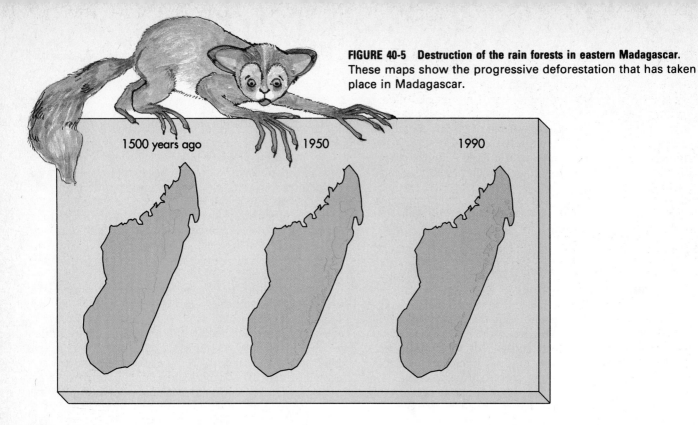

FIGURE 40-5 Destruction of the rain forests in eastern Madagascar. These maps show the progressive deforestation that has taken place in Madagascar.

CO_2 acts like the glass in a greenhouse (or the windows in your car), allowing heat to enter the Earth's atmosphere but preventing it from leaving. This selective energy absorption by CO_2 in the atmosphere is called the **greenhouse effect.** Rising levels of atmospheric CO_2 may therefore result in a worldwide temperature increase, a situation called **global warming.** Global warming could melt the arctic and antarctic ice packs, causing a rise in the sea level and the flooding of one fifth of the world's land area. In addition, a rise in global temperatures could affect rain patterns and agricultural lands.

FIGURE 40-6 A rain forest reserve.
In 1979, the World Wildlife Fund established a research program—the Biological Dynamics of Forest Fragments Project. The program isolates "islands" of rain forests out of the larger forest during conversion to pastureland. Scientists can study these islands to determine what their optimum size and shape should be if they are to provide a suitable habitat for various species.

To help stop rain forest destruction, individuals can join and support organizations involved in rain forest conservation, as well as avoid buying furniture constructed from tropical hardwoods. Research programs are currently being conducted that are exploring the concept of conserving forest "fragments" as reserves that will provide suitable habitats for species (Figure 40-6).

> The world's forests are being destroyed at an alarming rate, resulting in the loss not only of this precious resource but of the habitats of thousands of species of organisms. In addition, the burning of the forests adds tremendous quantities of carbon dioxide to the air, a situation that may result in a worldwide temperature increase, or global warming.

Species extinction

Although scientists have classified approximately 1,700,000 species of organisms, they estimate that approximately 40,000,000 exist. But many of these species are becoming extinct—dying out—and will never be seen again. An estimated 1000 extinctions are taking place each year, a number that translates into more than two species per day. Although such mass extinctions have occurred in the past, as with the dying out of the dinosaurs, these former extinctions were caused by climatic and geophysical factors. The extinctions today are caused by human activity.

One way in which humans destroy species is by destroying their habitats. Widespread habitat destruction is taking place as the tropical rain forests are cut down. This one factor alone will cause the extinction of one third of the world's species.

FIGURE 40-7 A germ plasm bank at the United States Department of Agriculture.
Researchers are in a cold storage room that contains seeds from around the world.

Humans are also destroying coral reefs at an alarming rate. Because the base of the coral reef food chain is algae, the reef ecosystem depends on sunlight for its existence. In areas where soil erosion muddies the water, the reef dies from insufficient sunlight. In some cases the water may become polluted with fertilizer runoff or with sewage, which causes the algae to overgrow, smothering the corals. In some cases, humans use dynamite to kill and harvest fish, a practice that obviously destroys the coral reef. In addition, coral is often harvested to sell to tourists.

Although habitat destruction and pollution is the most serious threat to the existence of certain species, exploitation of commercially valuable species threatens them as well. The Convention on International Trade in Endangered Species of Wild Fauna and Flora (CITES) regulates trade in live wildlife and products. However, illegal trade still takes place because of smuggling and the inability of law enforcement inspectors to check all shipments of goods. Certain types of alligator, crocodile, sea turtle, snake, and lizard trade are illegal and endanger the existence of various species. Buying other products such as coral and ivory endangers the existence of coral reefs and African elephants. Even certain plants such as cacti and succulents should be purchased only if cultivated in greenhouses; they should not be taken from the wild. Organizations such as the World Wildlife Fund are invaluable sources of information regarding which products should be avoided by consumers so that they can help stop trade that threatens certain species.

> Humans are destroying species by destroying their habitats, polluting their habitats, and illegally trading endangered species.

The extinction of many species leads to a reduction in **biological diversity,** a loss of richness of species. This loss is certainly regrettable; each species is not only the result of millions of years of evolution and can *never* be reproduced once it is gone but also is an intricate part of the interwoven relationships among organisms within the ecosystems of the world. The extinction of just one species can affect ecosystems in many unforeseen ways. In addition, many people feel that humans have no right to influence the world in such a devastating way. And looking at the issue from a utilitarian standpoint, humans rely on the other species of the world as sources of food, medicine, and other substances.

Recently, for example, the genetic diversity in crop plants has declined as scientists have used selective breeding techniques to produce plants with specific characteristics, such as resistance to particular diseases or pests, hardiness, and other characteristics considered important. However, when only a few species or varieties of a species are cultivated or survive, the genetic diversity of the organism declines. Populations of species that have little diversity are more vulnerable to being wiped out by new diseases or climatic changes.

In the 1960s, the United Nations Food and Agricultural Organization (FAO) made recommendations that have led to the establishment of "banks" that store plant seeds and genetic material, or germ plasm (Figure 40-7). Zoos around the world are making similar efforts to preserve genetic diversity among animals by establishing and carrying out sophisticated breeding programs to increase genetic diversity among endangered species (Figure 40-8).

FIGURE 40-8 Zoos are now responsible for preserving genetic diversity in endangered species.
These workers at the San Diego Zoo are helping this baby elephant to nurse. The efforts of zoos around the world will be instrumental in saving some endangered species from extinction.

THE BIOSPHERE: TODAY AND TOMORROW

BIOLOGY TECHNOLOGY & society

Ten things you can do to change the world

1. Make up your mind that the conditions on the planet are serious enough to merit your attention and commitment. Human beings have had a tremendously negative effect on the environment but can have a positive impact. Knowledge and awareness precede action.
2. Water: Install low-flow systems on all faucets. Use less. Take short showers. Clean water is a finite resource. Landscape for low water use.
3. Trash: How many things do you really need? Buy products in minimal packaging. Reuse what you can. Recycle everything else you can.
4. Life-style: Choose carefully where you live. Can you walk, bike, or take public transportation to work, to play, and to buy necessities? How much space do you need to live?
5. Energy: Use less. Turn out lights. Install compact fluorescents where possible. Encourage energy conservation. Learn about how conservation has an impact.
6. Transportation: If you need a car, buy for high mileage. Support mass-transit legislation. Walk, ride a bike, or take the bus.
7. Population: Learn about population as a problem and what has worked to control population growth in other parts of the world.
8. Work: Find a job that supports positive environmental change whether it is actually working directly for change, or teaching, or bringing positive changes to your workplace.
9. Attitude: Learn about what sustainable living means. How you manage resources today will affect how the next generations can live.
10. Political action: Be informed about legislation in process at state and national levels. Write and call your representatives. Find out how candidates have voted on environmental issues. Vote for the ones who are concerned.

Solid waste

The average American produces about 19 pounds of garbage and trash—referred to as **solid waste**—per week. If that does not sound like very much, multiply that number by 52, and you will realize that each person produces approximately *half ton* of solid waste per year. In 1992, the population of the United States was over 250 million—a population that produced approximately 125 million tons of waste per year, not including the waste produced by schools, stores, and manufacturing. Americans, often described as having a "disposable society," are beginning to realize much of the wastes produced should be viewed as resources that must be reclaimed rather than as substances to throw away and clog the landscapes.

What happens to your trash when you put it out for collection? The burial site for your throwaways is the **sanitary landfill,** an enormous depression in the ground where trash and garbage is dumped, compacted, and then covered with dirt (Figure 40-9). In 1983, the Federal Resource Conservation and Recovery Act forced the closing of all **open dumps** or required them to be converted to landfill sites. At an open dump, solid waste is heaped on the ground, periodically burned, and left uncovered. Landfills are considered superior to dumps because landfill wastes are covered, reducing the number of flying insects and rodents that are attracted to the site and reducing the odor produced by open, rotting organic material. In addition, wastes are not burned at landfills, decreasing the problem of air pollution. Additionally, when the capacity of a landfill site is reached, it may be used as a building site or recreational area. Examples of landfill reuse are Mount Trashmore recreational complex in Evanston, Illinois, and Mile High Stadium in Denver, Colorado.

FIGURE 40-9 Sanitary landfill, New York City.
Sanitary landfills around the country are filling up rapidly. Communities must quickly find new ways to dispose of their trash. One of the most popular and efficient methods is recycling.

Problems do exist with landfills, however. First of all, space is running out! In just one year, the population of New York City alone will produce enough trash to cover over 700 acres of land 10 feet deep. Another problem with landfills is that liquid wastes can trickle down through a landfill, reaching and contaminating ground water below. Liquids leaching from landfills can also pollute nearby streams, lakes, and wells (therefore *never* put batteries, paint solvents, drain cleaners, and pesticides in with your trash). In addition, as the organic material compacted in landfills is decomposed in the absence of oxygen, methane gas is produced. This highly explosive gas rises from landfills and can seep into buildings constructed on or near reclaimed sites.

To reduce solid waste and landfill problems, everyone should recycle paper, glass, aluminum cans, and even clothing. An important part of recycling is buying recycled goods so that a market is maintained for them. In addition, people should purchase products in recyclable containers, avoid purchasing "overpackaged" products, and compost yard waste. Composting involves piling (and periodically turning to aerate) grass clippings, wood shavings, and similar yard wastes. Bacteria will degrade these substances, and they can be used to fertilize flower beds and vegetable gardens. By composting, you will be eliminating the second largest waste by volume in landfills.

In the future, careful planning will be necessary to ensure that landfill sites are located away from streams, lakes, and wells and that they have proper drainage and venting systems. Most likely, landfills will be only a portion of a solid waste disposal system that incorporates recycling, safe incineration, and waste-to-energy reclamation.

> Solid waste is disposed of in sanitary landfills, depressions in the ground where trash and garbage is dumped, compacted, and covered with dirt. Solid waste management could be improved with increased and better-organized recycling efforts, safe incineration, better-engineered landfill sites, and waste-to-energy reclamation.

The water

The water, or hydrosphere, of this planet lies mainly in the oceans but is also found in fresh water lakes and ponds, in the atmosphere as water vapor, and as subsurface reservoirs called **ground water.** Heated by the sun, water continually cycles from the land to the air, condenses, and falls back to the Earth (see Chapter 38). As this cycling of water repeats continually, contaminants may mix with the water, **polluting** it—causing physical or chemical changes in the water that harm living and nonliving things.

Surface water pollution

Surface water can be polluted by factories, power plants, and sewage treatment plants that dump waste chemicals, heated water, or human sewage into a lake, stream, or river. These sources of pollutants are called **point sources** because they enter the water at one or a few distinct places. Other types of pollutants may enter surface water at a variety of places and are called **nonpoint sources.** Examples of nonpoint sources of pollution are (1) sediments in land runoff caused by erosion from poor agricultural practices (a major type of water pollution), (2) metals and acids draining from mines, (3) poisons leaching from hazardous waste dumps

FIGURE 40-10 A toxic chemical dump in northern New Jersey. Toxic chemical dumps are serious threats to ground and surface water. Pollution occurs when the drums rust through and release their contents, which then enter the surface water and may eventually percolate down to the ground water.

(Figure 40-10), and (4) pesticides, herbicides, and fertilizers washing into surface waters after a rain. These pollutants affect aquatic organisms and also affect terrestrial organisms that drink the water. The manner in which a pollutant affects living things depends on its type: nutrient, infectious agent, toxin (poison), sediment, or thermal pollutant.

Organic nutrients are sometimes discharged into rivers or streams by sewage treatment plants, paper mills, and meat-packing plants. These "organics" are food for bacteria. If high amounts of organic nutrients are available to bacteria, their populations will grow exponentially. As they grow and reproduce, they use oxygen—oxygen that fish need. Therefore as the bacterial populations rise, the only organisms that survive are those that can live on little oxygen. So-called "trash fish" such as carp can out survive other species such as trout and bass, but if oxygen levels become extremely low, all the fish die, survived only by various worms and insects.

The accumulation of **inorganic nutrients** in a lake is called **eutrophication,** meaning "good feeding." Certain inorganics such as nitrogen and phosphorus, which come from croplands or laundry detergents, stimulate plant growth. Although heavy plant growth makes swimming, fishing, or boating difficult, it does not cause most of its problems until the autumn (in most regions of the United States), when the plants die. At that time, bacteria decompose the dead plant material, and problems similar to those of organic nutrient pollution arise. In addition, the decomposed materials begin to fill the bottom of the lake. Eventually, the lake may become transformed into a marsh and then into a terrestrial community by an accelerated process of succession (see Chapter 37). **Sediment** that flows into lakes from erosion of the land caused by certain agricultural practices, mining, and road construction also fills in lakes and hastens

THE BIOSPHERE: TODAY AND TOMORROW

al succession. An example of eutrophication that began
 to the successional "death" of a lake occurred in
Lake Washington near Seattle. The problem stemmed from local communities dumping their wastes into the lake. In 1968 the dumping was halted, and Lake Washington is now fully recovered.

Surface waters are rarely polluted with **infectious agents,** or disease-causing microbes, in the United States. However, in areas such as Africa, Asia, and Latin America, waterborne diseases are common. Surface waters become polluted from untreated human wastes and from animal wastes, causing diseases such as hepatitis, polio, amebic dysentery, and cholera.

An array of **toxic substances** pollutes surface waters worldwide. Toxic substances include both organic compounds such as PCBs (polychlorinated biphenyls) and phenols and inorganic substances such as metals, acids, and salts. These toxic, or poisonous, substances come from a diverse array of sources such as industrial discharge, mining, air pollution, soil erosion, old lead pipes, and many natural sources. The effects on humans from drinking these substances in water range from numbness, deafness, vision problems, digestive problems, and the development of cancers.

Unfortunately, most toxic pollutants do not **degrade,** or break down, and are therefore present in bottom sediments of surface waters for decades or more. In fact, some organisms accumulate certain chemicals (often deadly ones) within their bodies, a process called **biological concentration.** Oysters, for example, accumulate heavy metals such as mercury, so these organisms might be highly toxic when living in waters with relatively low concentrations of this metal. Also, as organisms higher on the food chain eat organisms lower on the food chain, toxins accumulate in the predators in a high concentration, a concept called **biological magnification.** As discussed in Chapter 38, the relationships between trophic levels of food chains are pyramidal, so a smaller number of organisms eat a larger number of organisms as you progress "up" the chain. (Figure 40-11 illustrates this concept). Scientists estimate that the concentration of a toxin in polluted water may be magnified from 75,000 to 150,000 times in humans consuming tainted fish.

The electric power industry and various other industries such as steel mills, refineries, and paper mills use river water for cooling purposes, discharging the subsequently heated water back into the river (Figure 40-12). Small levels of **thermal pollution** do not cause serious problems in aquatic

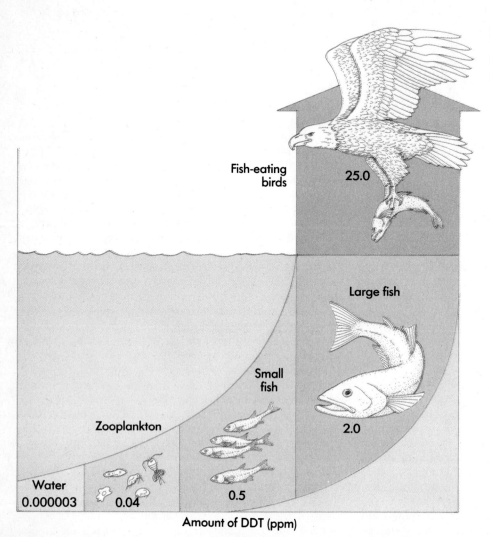

FIGURE 40-11 Biological magnification. The amount of DDT in an organism increases as you go up the food chain.

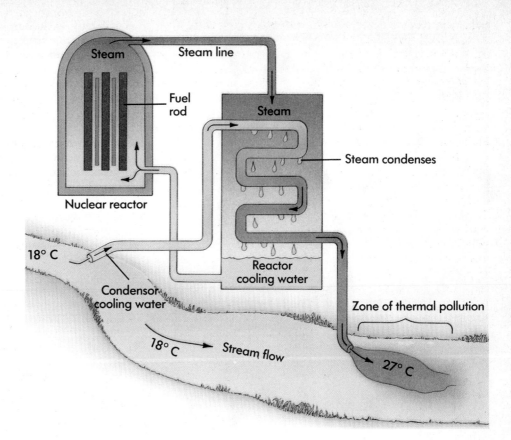

FIGURE 40-12 Thermal pollution. Water from streams or rivers is used to cool nuclear power plants and is used to condense steam to water. As it cools the reactor, the water is heated forming steam, which is then cooled again in the condenser. The cool stream water used in the condenser is discharged into the stream after it has absorbed heat from the steam.

ecosystems, but sudden, large temperature changes kill heat-intolerant plants and animals. Interestingly, ecosystems adjust to artificially heated waters and are damaged if the heat source is shut down, as when a power plant closes.

> Surface water can be polluted by waste chemicals, heated water, or human sewage put into the water by factories, power plants, and sewage treatment plants or by silt and chemicals that can leach into the water from surrounding sites. These substances all affect the aquatic ecosystem in different ways.

Ground water pollution

The surface water and rainwater that trickles through the soil to underground reservoirs is ground water. This water can be contaminated with some of the same substances as surface water. The primary ground water contaminants are toxic chemicals that seep into the ground from hazardous waste dump sites and chemicals such as pesticides used in agriculture. Although ground water does get filtered and "cleansed" of some substances as it trickles through the soil, toxic chemicals consist of molecules too small to be filtered in this manner. At this time, prevention of contamination is the cheapest and most feasible way to end ground water contamination. Pumping contaminated water from underground sources to the surface, purifying it, and returning it to the ground is extremely costly.

Acid rain

In recent years, the water in the atmosphere has become polluted with sulfur dioxide and nitrogen dioxide, two chemical compounds that form acids when combined with water. As this water vapor condenses and falls to the ground, it is commonly referred to as **acid rain**, although acid precipitation also falls as snow or as dry "micro" particles, mixing with water when it reaches surfaces on the ground. "Normal" rain has a pH of approximately 5.7, primarily because of dissolved carbon dioxide (see Chapter 2 for a discussion of pH). The pH of acid rain is lower than 5.7 and usually falls between 3.5 and 5.5. However, rainfall samples taken in the eastern United States have measured as low as 1.5—a pH lower than that of lemon juice and approaching that of battery acid!

Acid rain results in many devastating effects on the environment. As it mixes with surface water, it acidifies lakes and streams, killing fish and other aquatic life. It seeps into ground water, causing heavy metals to leach out of the soil. The result is that these heavy metals enter the ground and surface water, posing health problems for humans as well as fish. Acid rain also eats away stone buildings and monuments as well as metal and painted surfaces (Figure 40-13).

The effect of acid rain falling on plant life has been hypothesized but undocumented until recently. Botanists realized that acid rain not only leached many of the minerals essential to plant growth from the soil but also liberated toxic minerals such as aluminum. Recent research suggests that acid rain also kills or damages microorganisms that live in symbiotic associations with forest trees, helping them extract water and needed minerals from the soil. Without these organisms, the trees die (Figure 40-14).

Although sulfur and nitrous oxides are produced naturally during volcanic eruptions and forest fires, humans produce more than half these chemicals from the burning of coal by electricity generating plants, industrial boilers, and large smelters that obtain metals from ores. In addition, nitrogen oxides are emitted by cars and trucks. The situation also becomes complicated because countries pollute their own air as well as the air of other countries. Emissions produced in the Midwest and eastern portions of the United States not only affect those areas, for example, but are carried by the wind into Canada. Emissions produced in England move into the Scandinavian countries.

> **Sulfur dioxide and nitrogen dioxide, gases emitted primarily by coal-fired power plants, combine with water in the atmosphere to produce acid rain, a precipitation that harms both living and nonliving things.**

Solving the problem of acid rain and curtailing these emissions is not easy or inexpensive. One technique used to reduce sulfur dioxide is to put **scrubbers** on coal-burning power plants. This technology can remove up to 95% of the sulfur dioxide emissions produced. Unfortunately, the United States lags behind most sulfur-emitting countries of the world in implementing this technology. A solution to this problem depends on international agreements to reduce emissions, energy conservation measures, and the increased use of public transportation.

FIGURE 40-13 Acid rain damage.
This statue has been eroded by acid precipitation.

FIGURE 40-14 Effects of acid rain.
These balsam fir trees in North Carolina have been killed by acid rain. The chemicals in the acid rain, which are carried on prevailing winds, come from as far away as the Midwest.

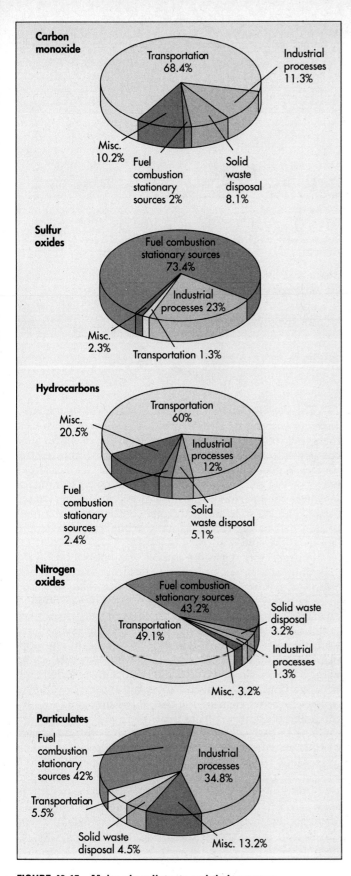

FIGURE 40-15 Major air pollutants and their sources. Transportation and fuel combustion at stationary sources are the main contributors to pollution.

The atmosphere

The Earth's atmosphere actually extends much higher than the portion within the biosphere, a part of the atmosphere more technically called the **troposphere.** The troposphere extends approximately 11 kilometers (36,000 feet) into the atmosphere but slopes downward toward the poles and upward toward the equator. The word *troposphere* literally means "turning over," a name extremely descriptive of the atmosphere. As the Earth is heated by the sun, air rises from its surface, cooling as it ascends. Cooler air then falls, resulting in a constant turnover of the air, aided by the prevailing winds.

Air pollution

About 99% of the clouds, dust, and other substances in the atmosphere are located in the troposphere. Some of these "other substances" are nitrogen and sulfur oxide pollutants. Other major air pollutants are carbon monoxide, hydrocarbons, and tiny particles, or particulates. As Figure 40-15 shows, sulfur oxides and particulates are produced primarily by the burning of coal in electricity-generating plants. Carbon monoxide and hydrocarbons are emitted primarily by cars, buses, and trucks; all these sources spew out nitrogen oxides.

The type of air pollution in a city depends not only on which of the pollutants are in the air, but on the climate of the city. **Gray-air cities,** such as New York, Philadelphia, and Pittsburgh, are located in relatively cold but moist climates and have an abundance of sulfur oxides and particulates in the air. The haze, or **smog,** that can be seen in the air is the result of the the burning of fossil fuels in power plants, other industries, and homes. **Brown-air cities,** such as Los Angeles, Denver, and Albuquerque, have an abundance of hydrocarbons and nitrogen oxides in the air. In these sun-drenched cities the hydrocarbons and nitrogen oxides undergo photochemical reactions that produce "new" pollutants called **secondary pollutants.** The principal secondary pollutant formed is **ozone** (O_3), a chemical that is extremely irritating to the eyes and the upper respiratory tract. Smog caused by pollutants reacting in the presence of sunlight is called **photochemical smog** (Figure 40-16).

The solutions to cutting down on smog are the same solutions to problems of acid rain formation and fossil fuel depletion: the use of coal-fired power plant scrubbers, energy conservation, and the recycling of resources leading to a reduction in energy use for manufacturing.

> The primary type of air pollution in the first 11 kilometers of the atmosphere consists of smoke and fog, or smog. Smog is caused by sulfur oxides, nitrogen oxides, hydrocarbons, and particulates in the air as a result of the combustion of fossil fuels.

THE BIOSPHERE: TODAY AND TOMORROW

FIGURE 40-16 Air pollution.
A "Brown air" in Los Angeles. Sunlight heats the chemicals spewed from automobiles to create ozone, which in the upper atmosphere protects you from the sun's harmful rays but in the lower atmosphere is poisonous.
B "Gray air" in New York City is caused by fossil fuel pollutants from power plants.

Ozone depletion

Ironically, humans are producing ozone in the troposphere that is polluting the environment but destroying it in the **stratosphere** where it is needed. The stratosphere is the "layer" of the atmosphere directly above the biosphere. It contains a layer of ozone that is formed when sunlight reacts with oxygen. Although ozone is harmful in air that is breathed, it is helpful in the stratosphere, acting as a shield against the sun's powerful ultraviolet (UV) rays. Excess exposure to UV rays can cause serious burns and skin cancers and can harm or kill bacteria and plants.

Scientists have measured "holes" in the ozone layer where it is thinnest, over the polar ice caps. The chemical chlorofluorocarbon, or CFC, is the main culprit. The spray-can propellant and refrigerant used in air conditioners, freon, is a CFC. In addition, CFCs are used in the manufacture of styrofoam and foam insulation. To protect the ozone layer, people should have their air conditioners serviced by persons who use equipment that does not allow the escape of freon. In addition, they should not use aerosol products containing CFC and should avoid the use of styrofoam products.

Overpopulation and environmental problems

Some scientists hold that the enormous world population is the key to the problems of pollution and diminishing natural resources. Others state that **technology**, the application of science to industrial use, is the culprit. Most, however, would agree that neither factor alone is the cause of these complex problems. Many factors interact to affect population size and the "health" of the environment.

The size of the human population influences the environment because it puts a demand on resources. However, the social, economic, and technological development of a country affects the demand its population places on resources. A person living in a rural area of Kenya, for example, does not use the same amount and kind of environmental resources as a person living in a large city in the United States. The life-style and per capita consumption of a population make a big difference on the impact of that population on the environment.

Other factors regarding impact of a population on the environment is how a population uses its resources and the effects of that use. Using resources and disposing of wastes in certain ways pollute the environment. The pollution of one aspect of the environment (such as the air) can lead to pollution of other aspects of the environment (such as the water through acid rain). Pollution, in turn, can limit population size by increasing the death rate. However, living with pollution can lead to attitudinal changes among members of a population, which may result in the development of laws to better manage the use of resources and to curb the pollution resulting from their use.

The impact of the human population on the environment is extremely complex and has multiple causes and effects. Many factors, such as population size, per capita consumption, technology, and politics, interact in complex ways, resulting not only in the problems faced today but in solutions for the future. Through research, environmental scientists will help everyone understand how each person can live on this planet without harming it and everyone's future existence.

Summary

1. All life on Earth exists in the biosphere. The biosphere is the interface of the land, air, and water, extending from approximately 9 kilometers (30,000 feet) above sea level to 11 kilometers (36,000 feet) below.
2. Many of the natural resources of the Earth are finite or nonrenewable; that is, they cannot be replaced. Coal, oil, and natural gas are fossil fuels, nonrenewable resources formed over time from the remains of organisms that lived long ago. To curb the use of these resources, scientists are testing and refining alternative renewable energy sources, such as solar, wind, water, and geothermal power.
3. The most severe crisis of natural resource destruction is occurring today in the tropical rain forests. Approximately 100,000 square kilometers (62,000 square miles) of rain forest are being lost each year because of slashing and burning for agriculture and logging and to create pastures for cattle. Scientists estimate that at the present rate, nearly all the tropical rain forest will be gone within the next 50 years.
4. The atmospheric rise in carbon dioxide levels, which is caused by the burning of the tropical forests and the combustion of fossil fuels, is acting as a barrier against the escape of heat from the surface of the Earth. This greenhouse effect could affect global temperatures, rainfall patterns, and agricultural lands.
5. Species are dying out at a rate of 1000 extinctions per year. A major cause of species extinction is habitat loss, which occurs when the tropical rain forests are cut. The extinction of many species leads to a reduction in biological diversity, a loss of species richness. In addition, the loss of species affects ecosystems and diminishes future sources of food, medicine, and other substances.
6. The populations of the world are literally drowning in their trash and garbage, running out of room in which to put it. Solutions to this problem lie in recycling, lowering consumption, composting yard waste, and implementing landfill technology that incorporates recycling, safe incineration, and energy reclamation.
7. Surface water is contaminated by factories, power plants, and sewage treatment plants that dump waste chemicals, heated water, and human sewage into lakes, streams, and rivers. Other sources of surface water pollution are sediment runoff during erosion and the leaching of chemicals from mines, hazardous waste dumps, and croplands. These pollutants affect aquatic life in different ways and can become concentrated in their bodies.
8. Underground reservoirs of water can be contaminated by the same sources as surface water when the contaminants trickle through the earth; however, the primary ground water contaminants are toxic chemicals. At this time, prevention of contamination is the cheapest and most feasible way to end ground water contamination.
9. The atmosphere has become polluted with sulfur and nitrogen oxides, carbon monoxide, hydrocarbons, and particulates. The primary source of these pollutants is the combustion of fossil fuels in automobiles and electricity-generating plants. As the sulfur and nitrogen oxides mix with water vapor in the atmosphere, they form acid rain, which harms both living and nonliving things. As the nitrogen oxides and hydrocarbons react with sunlight, they form photochemical smog, an upper respiratory irritant that is a health hazard.
10. The environmental problems of this and the next century are the result of many interacting factors. Through research, environmental scientists will help people understand how they can live on this planet without harming it and their future existence.

REVIEW QUESTIONS

1. What is the biosphere?
2. Many scientists feel people should work toward creating a "sustainable society." What does this mean?
3. Discuss several energy sources that could be used to help create a sustainable society.
4. What can be done to help prevent a mineral shortage in the future? Why are preventative measures important?
5. Why is the destruction of tropical rain forests a serious problem for everyone—not just the people living in tropical countries?
6. What is meant by global warming? Why is it dangerous?
7. Define *biological diversity*. How are humans affecting the biological diversity of the world's species? Why is this serious?
8. How can the amount of solid waste thrown away be reduced and waste management improved?
9. Distinguish between point and nonpoint sources of surface water pollution. Give an example of each.
10. Describe the different effects of pollution by organic nutrients, inorganic nutrients, infectious agents, and toxic substances in surface water.
11. How is acid rain produced? What are its effects?
12. Explain the terms *gray-air cities* and *brown-air cities*. What pollutants are involved?
13. Ozone is a dangerous pollutant in air that is breathed. Why should people be concerned that the ozone layer in the stratosphere is being depleted?
14. Suppose the President of the United States asked you to put together a plan of action for addressing environmental problems. What steps would you recommend?

KEY TERMS

acid rain 705
bioenergy 696
bio-gas machines 696
biological concentration 704
biological diversity 701
biological magnification 704
biosphere 695
brown-air cities 707
degrade 704
environment 695
eutrophication 703
fossil fuels 696
geothermal energy 696
gray-air cities 707
greenhouse effect 700
ground water 703
infectious agent 704
inorganic nutrient 703
nonpoint source 703
nuclear power 696
open dump 702
organic nutrient 703
ozone 707
photochemical smog 707
point source 703
sanitary landfill 702
scrubber 706
secondary pollutant 707
sediment 703
smog 707
solar power 696
solid waste 702
stratosphere 708
sustainable society 696
thermal pollution 704
tidal power 696
toxic substance 704
tropical rain forest deforestation 699
troposphere 707
wave power 696
wind power 696

THOUGHT QUESTIONS

1. The most tragic of environmental injuries are those that needlessly consume or destroy irreplaceable resources. In the United States, two of the most important resources that are being seriously compromised today are ground water and topsoil. List potential sources of damage to these resources. What might be done to limit the damage?
2. Many environmentalists believe that the economic system of the United States subsidizes pollution of the environment. Consumption is kept higher than it should be, they claim, because the price a consumer pays for a manufactured item does not include the entire cost—the costs of injury to the environment are instead passed on to a third party (the future). As a result, consumers buy more of the item than they would if it were priced at its true cost. What might realistically be done to address this problem?
3. Imagine for a moment that no serious effective program is mounted in the next decades to address the Earth's environmental woes. Outline a scenario of the likely consequences—what would such a world be like in 20 years?

FOR FURTHER READING

Brown, L., editor. (1990). *State of the world, 1990*, New York: W. W. Norton & Co.
This is a highly recommended, easily read summary of the ecological problems faced by an overcrowded and hungry world, with an excellent chapter on suggested solutions. A new edition appears every year.

Matthews, S., & Sugar, J. (1990, October). Is our world warming? *National Geographic*, pp. 66-99.
This is a vivid, well-illustrated discussion of the greenhouse effect, its causes, and the possible consequences of global warming.

National Wildlife. (1990). What on Earth are we doing? *28*(2).
An excellent issue of this outstanding magazine is entirely devoted to the state of the environment and presents many actions that you can take to help solve the problems discussed in this chapter.

National Wildlife. (1992, April-May). Endangered species: preserving pieces of the puzzle. *30*(5).
This entire issue is devoted to the fight to protect plants and animals from extinction. Indepth features include the endangered species in the tropical rain forest, the effects of poaching on certain species, and the success stories of bald eagles, gray whales, and alligators.

Answers to review questions

APPENDIX A

CHAPTER 1

1. The study of biology includes a wide variety of topics, for example, studying hormonal changes in one organism, researching changes in populations over time, and learning how your body protects itself against disease.
2. Inductive reasoning develops generalizations from specific instances (for instance, the large number of dying fish in a particular river could be due to rising pollution from factories in the area). Deductive reasoning begins with a general statement and proceeds to a specific statement (pollution from factories could kill wildlife; we will measure the effects of pollution on fish in a specific river).
3. Your hypothesis could be the following: If the students walk 1 hour at a time, 4 days a week, then they will each lose at least 5 pounds during the semester, assuming that this exercise is in addition to any exercise they're currently getting. The independent variable is the walking that the students will do; the dependent variable refers to the amount of weight they may or may not lose.
4. The control is the standard against which the subjects who go through the experiment's treatment would be compared. In this case, it would be a group of students who do not take the hour-long walks that the other students take.
5. The scientific method is a process that scientists use to answer the questions they ask. Its steps are (1) posing a question, (2) forming a hypothesis, (3) testing the hypothesis with a controlled experiment, (4) recording and analyzing the data, (5) drawing a conclusion, and (6) repeating the experiment.
6. Scientists cannot actually prove hypotheses; they can only disprove them or support them with evidence. However, there is no guarantee that further evidence might not disprove a hypothesis.
7. A theory is a synthesis of hypotheses that are supported by so much evidence they are commonly accepted as "true." However, scientists cannot state with certainty that they are true, since future evidence could disprove any theory.
8. (1) Living things display both diversity and unity. (2) Living things are composed of cells and are hierarchically organized. (3) Living things interact with each other and with their environments. (4) Living things transform energy and maintain a steady internal environment. (5) Living things exhibit forms that fit their functions. (6) Living things reproduce and pass on biological information to their offspring. (7) Living things change over time, or evolve.
9. The levels are (1) the cell (microscopic mass of protoplasm), (2) tissue (group of similar cells that work together to perform a function), (3) organ (group of tissues forming a structural and functional unit), and (4) organ system (group of organs that function together to carry out the principle activities of the organism).
10. The four levels are (1) population (individuals of a given species that occur together at one place and time), (2) community (populations of different species that interact with each other), (3) ecosystem (community of plants, animals, and microorganisms that interact with each other and with their environment), and (4) biosphere (the part of the Earth where biological activity takes place).
11. Unlike all other organisms, people use many nonrenewable resources and fill the biosphere with wastes. In addition, people are overpopulating the land and destroying habitats, killing off various species, and using up resources that cannot be renewed.
12. Ultimately, all energy comes from the sun. Solar energy is captured by producers, which use photosynthesis to convert it into chemical energy. Other organisms (consumers) feed on the producers and on each other, passing the energy along. When you eat plants or animals, you are harvesting their stored energy.
13. DNA is the code of life that contains the instructions translated into a working organism. Using only four different nucleotides, DNA codes for all of the structural and functional components of an organism.
14. The theory of evolution states that organisms alive today are descendants of organisms that lived long ago and that organisms have changed and diverged from one another over time. Fossils are remains of earlier cells and organisms that have been preserved in rocks, they provide a record of how living things have changed over billions of years.
15. (1) Monera (bacteria), (2) Protista (algae), (3) Plantae (rosebushes), (4) Fungi (mushrooms), (5) Animalia (you).
16. Kingdom, phylum, class, order, family, genus, species.
17. a. True.
    b. False. (they might be closely enough related to belong to the same genus, but this cannot be assumed).
    c. False. (Binomial nomenclature only gives the genus and species of an organism. These organisms may not belong to the same genus.)

CHAPTER 2

1. a. See Figure 2-1. Because it has an atomic number of 3, your atom would have three protons and three electrons, although not necessarily three neutrons.
   b. The atom would still have one proton and one electron, but it would have a different number of neutrons.
2. Protons; electrons.
3. Atoms are the tiny parts that make up all matter. Molecules are combinations of tightly bonded atoms of the same element. Compounds are molecules that consist of atoms from two or more elements. For example, one atom is oxygen (O), a molecule of two oxygen atoms is atmospheric oxygen (O_2), and a compound that includes oxygen is water (H_2O).
4. The three chemical forces that influence how an atom interacts with other

A-1

atoms are (1) the tendency of electrons to occur in pairs, (2) the tendency of atoms to balance positive and negative charges, and (3) the tendency of the outer shell (energy level) of electrons to be full. The third point is known as the *octet rule*, since an atom with an unfilled outer shell tends to interact with other atoms to gain a complete set of eight electrons in this outer shell.

5. An ionic bond forms when atoms are attracted to each other by opposite electrical charges. A common example is table salt (NaCl). A covalent bond forms when two atoms share one or more pairs of electrons; an example is hydrogen gas (H—H).

6. Water is unusual because it is the only common molecule on Earth that exists as a liquid in the natural environment. This liquid enables other molecules to move and interact; life evolved as a result. A second important characteristic is its ability to form hydrogen bonds because of its polar molecule. Water molecules are strongly attracted to ions and other polar molecules, giving water another important trait: it is an excellent solvent. Chemical interactions readily take place in water because so many molecules are water soluble.

7. Oil is a nonpolar molecule, which means that is cannot form hydrogen bonds with water and therefore does not dissolve.

8. The pH value refers to the concentration of H^+ ions in a solution. The values on the pH scale are expressed as positive numbers, which are determined by taking the negative value of the exponent of the hydrogen ion concentration. Pure water has a pH of 7.0. A substance with a lower pH is an acid: one with a higher pH is a base.

9. Organic molecules tend to be large, contain carbon, and interact with each other via covalent bonding. Inorganic molecules tend to be small, do not usually contain carbon, and interact by means of ionic bonding. If you wanted to study living things—such as human beings—you would learn about organic molecules.

10. Polymers and composite molecules are types of macromolecules, large organic molecules with many functional groups. They differ in their structures. Polymers are built by forming covalent bonds between a long chain of similar components (such as complex carbodyrates, proteins, and nucleic acids). Composite molecules (such as lipids) have several different components.

11. Monosaccharides, disaccharides, and polysaccharides are all carbohydrates. Most organisms use carbohydrates as an important fuel. Monosaccharides are among the least complex carbohydrates. Many organisms link monosaccharides to form disaccharides that are less readily broken down as they are transported within the organism. To store the energy from carbohydrates, organisms convert monosaccharides and disaccharides into polysaccharides, long polymers of soluble sugars.

12. a. The formula for the monosaccharides glucose, fructose, and galactose. Glucose is the primary energy-storage molecule used by living things.
    b. A generalized formula for an amino acid. Proteins perform many important functions; proteins called *enzymes*, for example, speed up chemical reactions.

13. The three major types of lipids are (1) oils, fats, and waxes; (2) phospholipids; and (3) steroids. Lipids have various functions. Fats are important energy-storing molecules, and land plants and some animals use waxes as a waterproofing material. Phospholipids and steroids are important components of many cell membranes; male and female sex hormones are steroids.

14. Saturated fats are fatty acids that carry as many hydrogen atoms as possible because they have only single bonds between their component carbon atoms. Polyunsaturated fats are fatty acids that carry fewer hydrogen atoms because they have more than one double bond between their carbon atoms. This product contains lipids (fats.) (It is olive oil, to be exact.)

15. Proteins are the third major group of macromolecules that make up the bodies of organisms. They play diverse roles in living things. Some speed up chemical reactions; others are chemical messengers throughout the body. Still others are structural proteins, an important component of bones, cartilage, and tendons.

16. The primary structure of a protein is the sequence of amino acids that make up the polypeptide chain. The amino acids interact by forming hydrogen bonds that cause the chains to fold into sheets or wrap into coils (secondary structure). The secondary structure of a protein largely determines its tertiary structure (three-dimensional shape). Proteins made up mostly of sheets often form fibers, whereas those with regions forming coils frequently fold into globular shapes. (When two protein chains associate to form a subunit (functional unit), these subunits assemble to form the quaternary structure.

17. The two forms of nucleic acids are DNA (deoxyribonucleic acid) and RNA (ribonucleic acid.) DNA stores the information for making proteins; RNA directs the production of proteins.

CHAPTER 3

1. Cells are the smallest unit of life that can exist independently. They can take in food, break it down to release energy, and expel wastes. They can also reproduce, react to stimuli, and maintain an internal environment that is different from their surroundings.

2. The cell theory includes three basic principles: (1) all living things are made up of one or more cells, (2) the smallest living unit of structure and function of all organisms is the cell, and (3) all cells arise from preexisting cells.

3. Cells must maintain a large surface area-to-volume ratio because they are constantly working. A relatively large surface area allows them to move substances in and out fast enough to survive.

4. Large, complex cells that are active often have many nuclei. A single nucleus could not control the activities of such a large cell.

5. Almost all cells share these four traits: (1) a surrounding membrane, (2) protoplasm that is enlosed by this membrane, (3) organelles (in eukaryotes) that carry out certain metabolic functions, and (4) a control center (nucleus or nucleoid) that contains DNA.

6. a. Eukaryotes.
   b. Eukaryotes.
   c. Prokaryotes.
   d. Eukaryotes.

7. See Figures 3-5 and 3-6.

8. Plasma membrane.

9. The cytoplasm, a viscous fluid inside the cell's plasma membrane, contains a gel-like fluid, storage substances, a network of interconnected filaments and fibers, and organelles. The cytoskeleton is a network of filaments and fibers that helps support and shape the cell, anchor the organelles, and move substances within the cell.

10. See Table 3-1.

11. Rough ER (endoplasmic reticulum) has surfaces studded with ribosomes; the ribosomes on its surface are the sites of protein synthesis. It transports proteins destined to leave the cell. Smooth ER, which lacks ribosomes, helps build carbohydrates and lipids.

12. If the lysosomes stopped working, your cells would soon fill up with old cell parts and foreign substances such as bacteria. In short, they would die.

13. DNA and RNA are both nucleic acids found in a eukaryotic cell's nucleus. DNA is the hereditary material that

directs the synthesis of RNA. RNA directs the synthesis of proteins.
14. Chromosomes. (This takes place inside the nucleus of the cell.)
15. Both mitochrondria and chloroplasts may have originated as endosymbiotic bacteria. They differ in that mitochondria use oxygen to break down cell fuel and release energy; chloroplasts, found in plants and algae, use carbon dioxide, water, and light energy to produce the cell's fuel (the process of photosynthesis.)
16. Flagella.
17. (1) Prokaryotic cells are smaller. (2) They lack membrane-bound compartments. (3) They have a nucleoid instead of a nucleus.
18. See Table 3-2 for a summary of the differences. For example, the cell with chloroplasts would be the plant, the cell with mitochondria but no chloroplasts would be the animal, and the one with neither mitochondria nor chloroplasts would be the bacterium.

CHAPTER 4

1. A lipid bilayer is the basic foundation of biological membranes. It consists of phospholipid molecules, with the nonpolar tails pointed toward the interior of the cell and the polar heads pointed outward. It forms a fluid, flexible cell covering and separates the cell's watery contents from its watery environment.
2. Molecules can move into and out of cells because of proteins floating in the cell's membrane. These proteins form specialized channels that regulate the types and amounts of substances that enter and leave the cell. These varied proteins allow the cell to interact with its environment; without them, the cell would soon die.
3. The carbohydrates attached to the cell membrane serve as cell name tags that help cells to recognize one another.
4. A gradient is the difference in concentration, pressure, or electrical charge between two areas. Molecules undergo net movement in response to this difference.
5. Passive transport is molecular movement down a gradient and across a cell membrane from an area of higher concentration to one of lower concentration. It does not require energy. Active transport involves molecular movement against the concentration gradient; it requires the cell to expend energy.
6. a. Facilitated diffusion (passive).
   b. Osmosis (passive).
   c. Diffusion (passive).
   d. Active transport (active).
7. The solute is the instant coffee crystals, the solvent is water, and the solution is the liquid coffee.
8. Hypotonic; hypertonic; isotonic.
9. Diffusion is the net movement of molecules down the concentration gradient as a result of random molecular motion, until the molecules are uniformly distributed. Facilitated diffusion is the movement of selected molecules across the cell membrane and down the concentration gradient; it requires specific transport proteins.
10. The sodium-potassium pump, a type of active transport, uses enegy and a transport protein to move sodium ions out of the cell and potassium ions into the cell. Maintaining appropriate concentrations of each ion, both inside and outside the cell, is important to the functioning of muscle and nerve cells.
11. See Figure 4-10. White blood cells use endocytosis to engulf bacteria and other foreign substances.
12. See Figure 4-11. In plants, exocytosis is the main way that cells move the materials out of the cytoplasm that are needed to build the cell wall.

CHAPTER 5

1. Chromatin is a complex of DNA and proteins found in the cell nucleus. DNA is the hereditary material that contains the code of life.
2. A karyotype is the particular array of chromosomes that belongs to an individual. The sex chromosomes are those that determine the gender of an individual. In addition to the pair of sex chromosomes, humans have 22 pairs of chromosomes, called *autosomes*. Chromosomes carry the genes (pieces of hereditary information) that determine the characteristics of the individual.
3. Histones; supercoils; sister chromatids; centromere.
4. Eukaryotic cells divide for three reasons: growth, repair, and reproduction. Generally, they undergo mitosis for the first two reasons, and reproduce by meiosis. Some simple animals and many plants, however, reproduce by mitosis as well.
5. Mitosis produces two daughter cells from one parent cell; the daughter cells are identical with each other and with the parent. Meiosis produces four sex cells, each of which contains half the amount of the parent cell's hereditary material.
6. Interphase is the portion of the cell cycle during which the cell grows and produces an exact copy of its DNA as it prepares for cell division.
7. The correct sequence is c (prophase), a (metaphase), d (anaphase), b (telophase). this is the process of mitosis.
8. The correct sequence is c (prophase I), d (metaphase I), a (anaphase I), b (telophase I), e (interphase). This is the process of meiosis I.
9. Cytokinesis is the physical division of the cytoplasm of a eukaryotic cell into two daughter cells. In animal cells and in eukaryotes that lack cell walls, cytokinesis involves "pinching" the cell in two by forming a cleavage furrow around its circumference. Plant cells divide by forming a cell plate (partition) inside which cellulose forms, producing a new cell wall and separating the two cells.
10. Diploid; haploid.
11. Homologous; synapsis.
12. Meiosis II produces four haploid cells that may function as gametes (animals) or spores (plants).
13. Genetic recombination during meiosis allows the hereditary material to be recombined during sexual reproduction. It is the principal factor that has made the evolution of eukaryotic organisms possible.

CHAPTER 6

1. Free energy of activation; substrates; products.
2. Exergonic reactions break apart substrates to produce chemically smaller products; they release energy. Endergonic reactions bond substrates together to form chemically larger products; they store enegy. In coupled reactions, exergonic reactions supply the energy needed for endergonic reactions.
3. Kinetic energy; potential energy.
4. The first law of thermodynamics states that energy cannot be created or destroyed, it can only be changed from one form to another. The second law states that disorder in the universe constantly increases, energy spontaneously converts to less organized forms.
5. Entrophy is the energy lost to disorder. Your friend is restating the second law of thermodynamics, which states that disorder in the universe constantly increases.
6. The Earth constantly receives energy from the sun. Photosynthetic organisms change this energy to other forms of energy that drive life processes.
7. Metabolic pathways are the chains of reactions within your body that move, store, and free energy. By means of metabolic pathways, people obtain energy from food, repair damaged tissues, and in general avoid increasing entropy.
8. Enzymes are proteins that act as biological catalysts. They speed up chemical reactions within living organisms. This reduces the amount of activation

energy needed to perform vital chemical reactions; without enzymes, these reactions would occur too slowly or require excessive amounts of energy to support life.
9. Enzymes catalyze reactions by bringing substrates together so that they react more easily and by placing stress on bonds that must break before a reaction can proceed.
10. Most human enzymes work best within a fairly narrow temperature range. If your body temperature became too high, this could affect the bonds between the amino acids that are responsible for the enzymes' three-dimensional shapes. Eventually this could stop chemical reactions from occurring.
11. Cofactors are special nonprotein molecules that help enzymes catalyze chemical reactions; one example is a zinc ion that helps digestive enzymes break down proteins in food. Coenzymes are cofactors that are nonprotein organic molecules. Many vitamins are synthesized into coenzymes in your body.
12. Your body can store energy for later use by converting it to fat, glycogen, or ATP (adenosine triphosphate). This is important because otherwise you would expend your available energy too quickly.
13. ATP is a high-energy compound that can be broken down easily to release energy for use by cells. It is called "energy currency" because cells can save energy released in exergonic reactions by storing it as ATP; cells can also spend ATP to provide necessary energy for endergonic reactions.
14. ATP, APD, and phosphate are continually being recycled within living cells. ATP is split into ADP plus phosphate to drive endergonic reactions, and ATP is constantly being re-formed from ADP, phosphate, and energy from exergonic reactions. These interrelating reactions allow cells to store and release energy easily.
15. Most enzymes function best within a narrow range of pH. If an environment becomes too acidic, enzymes could be affected, causing chemical reactions to stop and ultimately affecting the metabolic pathways of organisms.

CHAPTER 7

1. ATP, created by chemical energy, is the energy currency of cells. The splitting of ATP is coupled to endergonic reactions that provide the energy necessary to keep cells alive and functioning.
2. Cellular respiration is the process by which living things make the ATP that powers cellular operations. During respiration, fuel molecules are broken down in the presence of oxygen to release energy. Fermentation is the process by which organisms that do not breathe oxygen make ATP.
3. This simplified formula summarizes the process of cellular respiration. It shows that the net effect is to break down one molecule of glucose in the presence of six oxygen molecules to yield six molecules of carbon dioxide and six molecules of water.
4. Glycolysis; Krebs cycle; electron transport chain.
5. Oxidized; increases; reduced; decreases.
6. This is an oxidation-reduction reaction, an essential part of the flow of energy in cellular respiration.
7. Your body metabolizes foods by breaking down complex molecules into simple ones. These simple molecules are then broken apart through metabolic pathways (the primary pathways are glycolysis, the Krebs cycle, and the electron transport chain) to release their stored energy and capture it as ATP.
8. Three changes occur during glycolysis: (1) glucose is converted to pyruvate, (2) ADP + P_i is converted to ATP, and (3) NAD^+ is converted to NADH.
9. Glycolysis is inefficient because it captures only about 2% of the available chemical energy of glucose. However, cells continue to do it, probably because it was one of the first biochemical processes to evolve. As cells evolved, they retained glycolysis as the first step in catabolic metabolism but added additional steps that were more efficient.
10. The Krebs cycle consists of nine reactions. The first three reactions are preparation reactions in which acetyl-CoA joins a four-carbon molecule from the previous cycle to form citric acid, and the chemical groups are rearranged. The last six reactions are energy extraction reactions. At every turn of the cycle, acetyl-CoA enters and is oxidized to CO_2 and H_2O, and the hydrogen ions and electrons are donated to electron carriers.
11. See Table 7-1.
12. The electron transport chain is a series of carrier proteins that pass along electrons harvested from glucose. These electrons drive protons out across the inner membrane; the return of the protons by diffusion generates ATP.
13. See below.

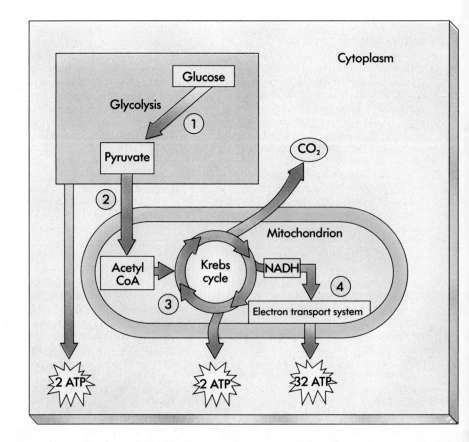

CHAPTER 8

1. Autotrophs are organisms that produce their own food either by photosynthesis or by chemical energy. Heterotrophs are organisms that cannot produce their own food; they depend on the autotrophs and the sun for survival. Human beings are heterotrophs.
2. The visible light in sunlight consists of many wavelengths. The prism separated the various wavelengths so that each color became visible as the light shone on the floor.
3. A pigment is a molecule that absorbs some wavelengths of light and reflects others. An object appears a particular color because it contains (or is covered with) a pigment that absorbs all the wavelenghts except that color. Those wavelengths bounce off the object and are reflected back to your eyes.
4. Carotenoids; chlorophylls.
5. This equation describes the process of photosynthesis: carbon dioxide and water, in the presence of light, produce glucose and oxygen. By means of photosynthesis, energy from the sun is captured and stored in molecules of food. This food ultimately builds and fuels the bodies of virtually all living things.
6. Photosynthesis involves three processes: (1) synthesis of ATP and NADPH, (2) the use of ATP and NADPH to produce glucose, and (3) the recycling of the photosynthetic pigment. The reactions that produce ATP and NADPH are called "light-dependent reactions" because they occur only in the presence of light. During the light-independent reactions (which do not need light to take place), ATP and NADPH are used to produce glucose from carbon dioxide.
7. A photocenter is an array of pigment molecules within the thylakoid membranes of a chloroplast. It channels photon energy to chlorophyll a, a special molecule of chlorophyll that will participate in photosynthesis.
8. In Photosystem I, the first to evolve in photosynthetic organisms, energy is transferred to a molecule of chlorophyll a called P700. In Photosystem II a different chlorophyll a molecule, P680, acts as a reaction center. Most bacteria use Photosystem I to produce ATP, whereas plants and algae use a two-stage photosystem. (Organisms that use a two-stage photosystem, however, can use Photosystem I to produce ATP without producing NADPH.)
9. During noncyclic electron flow, energized electrons ejected from Photosystem II are passed down an electron transport chain, triggering ATP production. Their places in Photosystem II are taken by electrons from the breakdown of water, a process that releases oxygen as a by-product. Energized electrons from Photosystem I, along with hydrogen ions, reduce $NADP^+$ to NADPH.
10. During the light-independent reactions (1) carbon is fixed with the attachment of CO_2 to an organic molecule, (2) glucose is produced from molecules of CO_2, and (3) the organic molecule that hooks to CO_2 re-forms to begin the cyclic again.
11. The metabolic pathways of photosynthesis and cellular respiration form a continuous cycle. The by-products of cellular respiration are carbon dioxide and water, the substrates of photosynthesis; the by-product of photosynthesis, oxygen gas, is essential to cellular respiration. Reducing one component of the cycle could harm animal life in the area; it could also affect the soil and the atmosphere.
12. Chloroplasts.
13. Such a dust cloud would interfere with photosynthesis, on which life on Earth ultimately depends. After a certain point, all living organisms would die.
14. Electromagnetic energy; radiation; waves; photons.

CHAPTER 9

1. A cell is the smallest structure in an organism that is capable of performing all the functions necessary for life. An example is a cardiac muscle cell. The cells of the body are organized into tissues (groups of similar cells that work together to perform a function, such as cardiac muscle tissue). Several different tissues group together to form a structural and functional unit—an organ (the heart.) An organ system is a group of organs that works together to carry out the body's principal activities (such as the circulatory system).
2. The four types of tissue in the body are (1) epithelial (skin cells), (2) connective (chondrocytes), (3) muscle (skeletal muscles such as leg muscles), and (4) nervous (neurons in your brain.)
3. Epithelial cells cover and line the surfaces of the body, both internal and external. They have six major functions: (1) protect tissues, (2) help or hinder absorption, (3) provide a sensory surface, (4) secrete substances, (5) excrete waste products, and (6) transport substances.
4. See Table 9-2. Squamous cells are found in the air sacs of the lungs, cuboidal cells line ducts in glands, and columnar cells line much of the digestive tract.
5. Connective tissue and its cells provide a framework for the body, join its tissues, help defend it from foreign invaders, and store specific substances.
6. Lymphocytes, erythrocytes, and fibroblasts are all cells found in connective tissue, but tney have different functions. Lymphocytes defend the body against infection, erythrocytes pick up and deliver gases in the blood, and fibroblasts produce fibers that are found in various types of connective tissue.
7. This is part of the body's inflammation response to injury. Mast cells produce histamine, which dilates blood vessels and increases blood flow to the area. This makes the area turn red and feel warm.
8. Fibroblasts; chondrocytes; osteocytes.
9. Chondrocytes; osteocytes; spongy; compact.
10. Platelets (e)
    Plasma (d)
    Erythrocytes (a)
    Leukocytes (b)
    Hemoglobin (c)
11. Yes, fat cells have their uses. In addition to storing fuel, fat tissue helps shape and pad the body and insulate it against heat loss.
12. The three kinds of muscle cells are (1) smooth muscle (intestinal walls), (2) skeletal muscle (arm muscles), and (3) cardiac muscle (the heart).
13. Neurons, which transmit nerve impulses; supporting cells, which nourish and protect the neurons.
14. See Table 9-1.
15. Homeostasis is the maintenance of a stable environment inside your body despite conditions in the external environment. The molecules, cells, tissues, organs, and organ systems in your body must all work together to maintain this internal equilibrium.

CHAPTER 10

1. The six classes of nutrients are carbohydrates, fats, proteins, vitamins, minerals, and water. The first three are organic compounds that your body uses as a source of energy; they are also used as building blocks for growth and repair and to produce other important substances. Vitamins, minerals, and water help body processes take place. Some minerals are also part of body structures.
2. Three types of enzymes help to digest the energy nutrients. Proteases break down proteins into peptides and amino acids. Amylases break down carbohydrates (starches and glycogen) to sugars. Lipases break down the tri-

glycerides in lipids to fatty acids and glycerol.
3. Saliva is the solution secreted by the salivary glands. It moistens food and begins the digestive process of breaking down food into component molecules.
4. Herbivore; carnivore; omnivore. Human beings are omnivores.
5. You would bite off the mouthful with your incisors, your canines would help you tear the food away, and your premolars and molars allow you to crush and grind the food thoroughly.
6. Pharynx; swallowing reflex; esophagus; peristalsis.
7. See Table 10-2.
8. Small intestine.
9. Two important accessory organs are the pancreas and liver. The pancreas secretes digestive enzymes. The liver, among other functions, produces bile, which helps digest lipids.
10. The duodenum is the first part of the small intestine; digestion is completed there. Starch and glycogen are broken down to disaccharides and then to monosaccharides. Proteins and polypeptides are broken down to shorter peptides and then to amino acids. Triglycerides are digested to fatty acids and glycerol.
11. These are hormones that help regulate digestion. Gastrin controls the release of hydrochloric acid in the stomach. Secretin stimulates the release of sodium bicarbonate to neutralize acid in the chyme and increases bile secretion in the liver. Cholecystokinin stimulates the gallbladder to release bile into the small intestine and stimulates the pancreas to release digestive enzymes.
12. Sodium bicarbonate neutralizes the acid in chyme. Thus it can help reduce the uncomfortable symptoms of acid indigestion.
13. Increasing the internal surface area of the small intestine helps absorb more nutrients from the same amount of food. Three features accomplish this: inner folds, projections (villi) on the folds, and additional projections on the projections (microvilli).
14. (1) Stomach (collects and partially digests food), (2) small intestine (finishes the digestive process), (3) large intestine (reabsorbs water and sodium and eliminates the rest as digestive wastes), and (4) anus (the opening through which feces are excreted.)
15. Glucose is broken down by glycolysis, the Krebs cycle, and the electron transport chain. Amino acids and triglycerides are converted to substances that can be metabolized by two of these pathways. Fatty acids are metabolized by another pathway.
16. The essential amino acids are eight amino acids that must be obtained in the diet. Humans cannot manufacture them.
17. See Figure 10-16. Answers will vary.

CHAPTER 11

1. Respiration is the uptake of oxygen and the release of carbon dioxide by the body. Cellular respiration is the process at the cellular level that uses the oxygen you breathe in and produces the carbon dioxide you breathe out. Internal respiration is the exchange of oxygen and carbon dioxide between the blood and tissue fluid; the exchange of the two gases between the blood and alveoli is external respiration.
2. Inspiration; expiration.
3. Your vocal cords produced the sound for speech as air rushed by and made them vibrate. Your lungs served as a power supply and volume control for you voice, and your lips and tongue formed the sounds into words.
4. Trachea; bronchi; respiratory bronchioles; alveoli.
5. Increases; decreases.
6. As the volume of the thoracic cavity increases, the air pressure within it decreases. Outside air is pulled in, equalizing the pressures inside and outside the thoracic cavity. As the volume of the thoracic cavity decreases, the air pressure inside it increases and forces air out, equalizing the pressures.
7. Tidal volume; residual air.
8. Deoxygenated blood has collected carbon dioxide during its passage through the body, so it contains more carbon dioxide than oxygen. Oxygenated blood contains more oxygen than carbon dioxide.
9. The circulatory system provides a transport system that distributes gases throughout the body. Without it you would not survive because it would take too long for oxygen to diffuse from the lungs to the rest of the body.
10. At the alveoli, carbon dioxide diffuses out of the blood into the alveoli because it moves down the pressure gradient (its pressure in the blood is greater than its pressure within the alveoli.) Meanwhile, oxygen diffuses from the alveoli into the blood because its pressure within the lungs is greater than in the blood.
11. At high altitudes the pressure of the oxygen molecules in the air is lower than at sea level. This means that the pressure gradient is lower at the alveoli and less oxygen diffuses into the blood. This can cause you to feel dizzy and short of breath.
12. Within the capillaries, blood pH decreases as carbon dioxide (in the form of bicarbonate ions) diffuses in and hydrogen ions accumulate. This more acid environment helps split oxygen from hemoglobin and enhances the diffusion of oxygen into the tissue fluid.
13. At the capillaries, oxygen moves down a pressure gradient to diffuse from the blood into the tissue fluid. Meanwhile, carbon dioxide moves down its own pressure gradient and diffuses from the tissue fluid into the blood.
14. Cigarette smoking paralyzes ciliated epithelial cells lining respiratory passageways, causing mucus to build up and plug passageways and the passageways to narrow. Smoking also impairs the action of the macrophages at the alveoli. The result is that smokers increase their risk of infection and their lungs and passageways become less elastic and efficient. The accumulated mucus, bacteria, viruses, and particles apparently contribute to the development of cancer.

CHAPTER 12

1. The circulatory system has four main functions: (1) nutrient and waste transport, (2) oxygen and carbon dioxide transport, (3) temperature maintenance, and (4) hormone circulation.
2. The blood carriers sugars, amino acids, and fatty acids to the liver, where some of the molecules are converted to glucose and released into the bloodstream. Excess energy molecules are stored in the liver for later use. Essential amino acids and vitamins pass through the liver into the bloodstream. The cells releases their metabolic waste products into the blood, which carries them to the kidneys.
3. A regulatory center in your brain constantly monitors your body temperature. If your temperature gets too high, signals from this center dilate your surface blood vessels and increase blood flow to the surface of your skin. This increases heat loss and lowers your body temperature.
4. The circulatory system has three components: the heart, the blood vessels, and the blood. The term *cardiovascular system* refers to the plumbing (heart and blood vessels) of the circulatory system.
5. Arteries; arterioles; capillaries; venules; veins.
6. When people are scared (or cold), the walls of the arterioles can contract and the blood flow decreases, routing more blood to other areas of the body. This can cause light-skinned people to "turn pale."

7. The entry to a capillary is guarded by a ring of muscle, which can constrict to block blood flow into that vessel. This is an important means of limiting heat loss. It also permits adjustments in blood flow, allowing more blood to go to areas where it is needed, such as to the intestines after a meal.
8. The walls of arteries have more elastic tissue and smooth muscle than those of veins; this helps arteries accommodate the pulses and high pressures of blood pumped from the heart. The lumen of veins is larger because blood flows through them at lower pressures; a larger diameter offers less resistance. The walls of capillaries are only one cell thick, which permits the exchange of substances between the cells and the blood.
9. See Figures 12-10 and 12-11.
10. Pulmonary circulation; systemic circulation.
11. Lower body.
12. The SA node is the pacemaker of the heart; this cluster of cells initiates the excitatory impulse that causes the atria to contract and the impulse to be passed along to other cardiac cells. The AV node conducts the impulse from the SA node to other cardiac cells, initiating the contraction of the ventricles.
13. Plasma is the liquid portion of blood in which the formed elements (cells and cell parts) are suspended. Plasma contains nutrients, hormones, respiratory gases, wastes, ions, salts, and blood proteins.
14. a. Erythrocytes are red blood cells; they carry oxygen.
    b. Leukocytes are white blood cells; they defend the body against microorganisms and other foreign substances.
    c. Platelets play an important role in blood clotting. These are all types of formed elements in blood.
15. Neutrophils (c)
    Basophils (a)
    Macrophages (d)
    Lymphocytes (b)
    All four cells are types of leukocytes.
16. The lymphatic system is a system of blind-ended vessels that collects the tissue fluid diffusing from blood plasma and returns it to the bloodstream. This counteracts the effects of the net movement of fluid out of the blood. The structures and organs of the lymphatic system also play a role in the immune system.
17. A high-fat diet can encourage the development of fatty deposits (plaque) on the inner walls of arteries, making the vessels narrower and less elastic. This can lead to atherosclerosis, heart attacks, and strokes.

CHAPTER 13

1. Nonspecific defenses act against any foreign invader, whereas specific defenses work against particular types of microbes. Nonspecific defenses are the first line of defense because they are the first barriers that a microorganism must bypass to enter your body.
2. The nonspecific defenses include (1) the skin (which if unbroken provides a barrier against invasion, (2) mucous membranes (mucus can trap particles and move then away from delicate areas), (3) chemicals that kill bacteria (many membranes are washed by fluids deadly to bacteria or have an acidic environment), (4) the inflammatory process (a series of events that remove the irritation, repair the damage, and protect the body from further damage).
3. The immune system is the body's specific defense; it protects you from particular microorganisms and other invaders.
4. The immune system differs from other body systems because it is not a system or organs and it lacks a single "controlling" organ. Instead, it consists of cells scattered throughout the body that have a common function: reacting to specific foreign molecules.
5. Antigens.
6. Phagocytes; T cells; B cells; NK (natural killer) cells; leukocytes.
7. Monocytes are precursors of macrophages; they develop into macrophages in response to infection. Macrophages are the phagocytes that play a key role in the immune response; they phagocytize foreign particles and cells and secrete proteins that activate the immune response. NK cells recognize and destroy body cells that have become cancerous or infected.
8. The four principal types of T cells are (1) helper T cells (which initiate the immune response), (2) cytotoxic T cells (which break apart infected and foreign cells), (3) inducer T cells (which oversee the development of T cells in the thymus), and (4) suppressor T cells (which limit the immune response).
9. During the cell-mediated immune response, cytotoxic T cells recognize and destroy infected body cells in addition to destroying transplanted and cancer cells. Helper T cells initiate the response, activating cytotoxic T cells, macrophages, inducer T cells, and finally, suppressor T cells.
10. When you had measles, B cell clones that did not become plasma cells became circulating lymphocytes—memory B cells. These cells give you an accelerated response during subsequent exposures to measles, promptly defending you against infection.
11. During the humoral response, B cells recognize foreign antigens and, if activated by helper T cells, produce antibodies. The antibodies bind to the antigens and mark them for destruction.
12. An antibody molecule recognizes a specific antigen because it has binding sites into which the antigen can fit. This allows your body to defend against specific microorganisms.
13. Injection with antigens causes an antibody (B cell) response, with the production of memory cells. A booster shot induces these memory cells to differentiate into antibody-producing cells and form still more memory cells.
14. Active.
15. The AIDS virus is dangerous because it destroys helper T cells, thus destroying the immune system's ability to mount a defense against any infection.
16. An allergic reaction occurs when the immune system mounts a defense against a harmless antigen. When class E antibodies are involved and bind to the antigens, they cause a strong inflammatory response that can dilate blood vessels and lead to symptoms ranging from uncomfortable to life threatening.

CHAPTER 14

1. Excretion is the removal of metabolic wastes and excess water from the body. Without excretion, your body would soon become poisoned by the buildup of metabolic waste products. The process also helps maintain the balance of water and ions that is necessary for life.
2. The main metabolic waste products are carbon dioxide, water, salts, and nitrogen-containing molecules.
3. Urea, urine, and uric acid are all forms of metabolic waste products (nitrogenous wastes). Urea consists of ammonia and carbon dioxide. Uric acid forms from the breakdown of nucleic acids you eat and from the metabolic replacement of your nucleic acids and ATP. Urine is the excretion product consisting of water, urea, creatinine, uric acid, and other substances.
4. Lungs; kidneys.
5. Urea (d).
6. The kidneys maintain water balance, retain substances the body needs, and eliminate metabolic wastes by filtering most substances out of the blood and then reabsorbing what is needed.
7. See Figures 14-4 and 14-5.
8. A nephron is a filtering system that is the workhorse of the kidney. It has three major functions: filtration, selec-

tive reabsorption, and tubular secretion.
9. Kidneys help conserve water by reabsorbing it from the filtrate. By changing the concentrations of salts and urea in the filtrate, the kidneys create an osmotic gradient that results in water moving from the filtrate into the surrounding tissue fluid.
10. Urine can reveal the presence of certain substances in the body because the kidney detoxifies the blood. When these substances, such as drugs, are removed from the blood, they become part of the filtrate.
11. There is no right or wrong answer. Be sure to back up your opinion with reasons, however.
12. Urine is formed when the blood that flows through the kidneys is filtered, removing most of its water and all but its largest molecules and cells. The kidneys then selectively reabsorb certain substances.
13. The urinary system is a set of interconnected organs that removes excess wastes, water, and ions from the blood and stores them as urine until they are excreted. Its major components are the kidneys, ureters, urinary bladder, and urethra.
14. Homeostasis is the maintenance of constant physiological conditions in the body. ADH and aldosterone are two hormones that help regulate the water and ion balance necessary for homeostasis. ADH regulates that amount of water reabsorbed at the collecting ducts; aldosterone promotes the retention of sodium (and therefore water), while promoting the excretion of potassium.
15. Two problems with kidney function are kidney stones and renal failure. Kidney stones are crystals of certain salts that can block urine flow. Renal failure is a reduction in the rate of filtration of blood in the glomerulus.

CHAPTER 15

1. Neurons and hormones are both forms of communication that integrate and coordinate body functions. They differ in that neurons transmit rapid signals that report information or initiate quick responses in specific tissues. Hormones are chemical messengers that trigger widespread prolonged responses, often in a variety of tissues.
2. Brain; neurons; nerve impulses.
3. Dendrites and axons are cellular projections that extend from the cell body of a neuron. Dendrites bring messages in to the cell body; axons carry messages away from the cell body.
4. Sodium; potassium.
5. d, c, a, e, b; this describes the development and conduction of a nerve impulse.
6. *Resting potential* refers to the difference in electrical charge along the membrane of the resting neuron. The action of sodium-potassium transmembrane pumps and ion-specific membrane channels separates positive and negative ions along the inside and outside of the membrane, which creates the resting potential. The difference in electrical charges is the basis for the transmission of nerve impulses.
7. Stimulus; receptors; depolarization.
8. A nerve impulse travels because nearby transmembrane proteins respond to the electrical changes that accompany depolarization of the nerve cell membrane. The adjacent section of membrane depolarizes, followed by another section, leading to a wave of depolarization.
9. Saltatory conduction occurs along myelinated neurons when impulses jump to unmyelinated areas and skip over myelinated portions. Impulses travel much more quickly by this method than by continuous waves of depolarization that occur along continuous portions of the membrane.
10. Mylinated; unmylinated.
11. A synapse is a narrow gap between an axon tip of a neuron and the membrane of a muscle cell or another neuron. At a chemical synapse the impulse crosses this gap by stimulating the release of neurotransmitters that diffuse to the other side of the gap, combining with receptors in the membrane of the postsynaptic cell.
12. Synapses between neurons and skeletal muscle cells are neuromuscular junctions. At a neuromuscular junction, acetylcholine released from an axon tip depolarizes the muscle cell membrane, releasing calcium ions that trigger muscle contraction.
13. The enzyme acetylcholinesterase breaks down acetylcholine. If the neurotransmitter were not broken down, it would continue to signal the muscles to contract.
14. The postsynaptic neuron integrates the information it receives from presynaptic neurons. The summed effect of excitatory and inhibitory signals either facilitates or inhibits depolarization.

CHAPTER 16

1. The nervous system is a complex network of neurons that gathers information about the internal and external environment and processes and responds to this information. The central nervous system (brain and spinal cord) is the site of information processing. The peripheral nervous system (nerves) brings messages to and from the central nervous system.
2. Sensory neurons; afferent neurons; motor neurons; efferent neurons.
3. The somatic nervous system consists of motor neurons that control voluntary responses (for example, moving your leg to take a step). The autonomic nervous system consists of motor neurons that control involuntary activities (for example, digesting a meal.)
4. See Figure 16-3.
5. Cerebral cortex; gray matter; association areas; motor area; sensory area.
6. The thalamus and hypothalamus are masses of gray matter at the base of the cerebrum. The thalamus receives sensory stimuli, interprets some of them, and sends the rest to appropriate locations in the cerebrum. The hypothalamus controls various body organs and the secretions of some hormones.
7. a. Bundles of nerve fibers in the peripheral nervous system; carry information to and from the central nervous system.
   b. Network of neurons in the hypothalamus; responsible for many drives and emotions.
   c. The part of the brain that coordinates unconscious movements of skeletal muscles.
   d. Fluid in the spinal cord and surrounding the brain; acts as shock absorber and carries dissolved gases.
8. The brainstem (midbrain, pons, medulla) contains tracts of nerve fibers that carry messages to and from the spinal cord. Nuclei located there control important body reflexes.
9. The spinal cord is the part of the central nervous system that runs down the neck and back. It receives information from sense organs, carries it to the brain, and sends information from the brain to the rest of the body. Without it the brain could not control the rest of the body.
10. Sensory receptors change stimuli into nerve impulses. The nerve fibers of sensory neurons carry this information to the central nervous system, where interneurons interpret them and direct a response (in this case, turning your head.) The axons of motor neurons conduct these impulses to the appropriate muscles.
11. How you perceive stimuli depends on which receptors are stimulated. You registered this stimulus as a sound because it stimulated your auditory nerve.
12. A reflex is an automatic response to

nerve stimulation. A familiar example is the knee jerk reaction when tapped just below the kneecap.
13. These systems act in opposition to each other to maintain homeostasis and help you respond to environmental changes. The sympathetic nervous system generally mobilizes the body for greater activity (faster heart rate, increased respiration), whereas the parasympathetic nervous system stimulates normal body functions such as digestion.

CHAPTER 17

1. Sensory receptors are cells that can change environmental stimuli into nerve impulses. They provide information about the body's internal environment, its position in space, and its external environment.
2. For you to sense a stimulus, it must be of sufficient magnitude to open ion channels within the membrane of the receptor cell. This depolarizes the membrane, creating a generator potential that leads to an action potential (nerve impulse) in the sensory neurons with which the receptor synapses. Nerve fibers conduct the impulse to the central nervous system.
3. Receptors that sense the body's internal environment monitor body temperature, levels of carbon dioxide, blood pressure, and other bodily functions. This continual monitoring allows your body to respond and maintain homeostasis.
4. Proprioceptors are receptors that sense the body's position in space. These receptors help you to keep your balance, control your muscular movements, and protect your muscles, tendons, and joints from excessive tension and pulling.
5. The general senses are touch, pressure, pain, and temperature; their receptors are relatively simple and are located within the skin. Receptors for the special senses—smell, taste, sight, hearing, and balance—are more specialized and complex. Their receptors are located within sense organs.
6. Olfactory receptors detect different smells because of specific binding of airborne gases with the receptor chemicals located in the cilia of the nasal epithelium.
7. On its way to the olfactory area of the cerebral cortex, the nerve impulse travels through the limbic system, the area of the brain that is responsible for many of your drives and emotions. Thus certain odors become linked in your memory with emotions and events.
8. Taste buds are taste receptors concentrated on the tongue. They detect chemicals in food and register an overall taste that consists of different combinations of sweet, salty, sour, and bitter. The sense of taste interacts with the sense of smell to produce a taste sensation.
9. Rods and cones are sensory receptors located in the eye that detect certain wavelengths of electromagnetic energy. The pigment inside them absorbs photons of light, which causes a series of events leading to hyperpolarization of the receptor cells and firing of adjacent neurons.
10. Color vision is caused by cone cells. There are three types of cones, each of which absorbs a specific wavelength of light. The color you perceive depends on how strongly each group of cones is stimulated by a light source.
11. You see objects because light enters the eye through the pupil and the lens focuses it on the fovea, which is rich in cone cells. These cells, along with surrounding rods and cones, initiate nervous impulses that travel to the cerebrum via the optic nerve.
12. See Figure 18-8.
13. You have depth perception because of the positioning of your two eyes on either side of the head, which sends slightly different images to the brain. Covering one eye makes it more difficult to judge distance.
14. See Figure 18-12.
15. Your outer ear funnels sound waves toward the eardrum, which changes these waves into mechanical energy. This energy is then transmitted to the bones of the middle ear, which increase its force. Receptor cells in the inner ear change the mechanical energy into nerve impulses.
16. You can detect movements of your head because of the otoliths embedded in the jelly-like layer that covers the cilia in your inner ear. When you move your head, the otoliths slide and bend the underlying cilia. This generates signals to the brain, which interprets the type and degree of movement.

CHAPTER 18

1. Your movement resulted from the contraction of your skeletal muscles, which are anchored to bones. The muscles use bones like levers to direct force against an object. When you raised your hand, your skin stretched to accommodate the change in position.
2. Skin.
3. Skin has several important functions: (1) it serves as a protective barrier, (2) it provides a sensory surface, (3) it compensates for body movement, and (4) it helps control your internal temperature.
4. Bone is a type of connective tissue consisting of living cells that secrete collagen fibers into the surrounding matrix. The bones of the skeletal system support the body and permit movement by serving as points of attachment and acting as levers against which muscles can pull. They also protect delicate internal structures, store important minerals, and produce red and white blood cells.
5. Compact bone runs the length of long bones and has no spaces within its structure visible to the naked eye. Spongy bone is found in the ends of long bones and within short, flat, and irregularly shaped bones. It is an open latticwork of thin plates of bone, and its spaces are filled with red bone marrow. Compact bone gives bones strength, whereas spongy bone provides some support and stores red bone marrow as well as lightens the bones.
6. The axial skeleton includes the central axis (skull, vertebral column, and rib cage). The appendicular skeleton consists of the appendages (arms and legs) and the bones that help attach the appendages to the axial skeleton.
7. A "slipped disk" refers to one of the intervertebral disks of fibrocartilage that separate the vertebrae from each other (except in the sacrum and coccyx). These disks act as shock absorbers, provide the means of attachment between vertebra, and allow the vertebral column to move.
8. Pectoral girdle; pelvic girdle; appendicular.
9. An articulation is a joint—a place where bones, or bones and cartilage, come together. Joints are classified according to the degree to which they permit movement: immovable joints (the sutures in your skull), slightly movable joints (joints between your vertebra), and synovial joints (wrists).
10. These are all types of synovial (freely movable) joints. They differ in the type of movement they allow. A hinge joint allows movement in one plane only, a ball-and-socket joint allows rotation, and a pivot joint permits side-to-side movement.
11. A muscle consists of specialized cells packed with intracellular fibers capable of shortening. The three types of muscle tissue are (1) smooth muscle tissue (interior of arteries), (2) cardiac tissue (heart), and (3) skeletal muscle tissue (calf muscles).
12. Tendons; origin; insertion; origin; insertion.
13. Actin is a protein that makes up the

thin myofilaments of muscle; the protein myosin makes up the thick myofilaments. Myofilaments are the microfilaments of muscle cells. Muscle cells contract when the actin and myosin filaments slide past each other. Changes in the shape of the ends of the myosin molecules (located between adjacent actin filaments) cause the myosin molecule to move along the actin, causing the myofilament to contract.
14. When depolarization from a nerve impulse reaches a neuromuscular junction, it triggers the release of the neurotransmitter acetylcholine. The acetylcholine crosses over to the muscle fiber membrane and opens the ion channels of that membrane, depolarizing it. This sets off a series of reactions that release calcium ions; the calcium ions initiate the chemical reactions of contraction.

CHAPTER 19

1. Hormones are chemical messages secreted by cells that affect other cells. They are an important method the body uses to integrate the functioning of various tissues, organs, and organ systems.
2. The endocrine system consists of 10 glands that together produce over 30 different hormones. The messages carried by the endocrine hormones fall into four categories: (1) regulating the body's internal environment, (2) helping it respond to changes, (3) regulating reproduction, and (4) aiding normal growth and development.
3. The nervous system sends messages to glands and muscles, regulating glandular secretion and muscular contraction. Endocrine hormones carry messages to virtually any type of cell in the body.
4. The two main classes of endocrine hormones are peptide and steroid hormones. Peptide hormones bind to receptors on the cell membrane of target cells, ultimately triggering enzymes that alter cell functioning. Steroid hormones bind to receptors within the cytoplasm of target cells and ultimately cause the cell's hereditary material to produce specific proteins.
5. A feedback loop controls hormone production by initially stimulating a gland to produce the hormone. After the hormone has exerted its effect on the target cell, the body feeds back information to the endocrine gland. In a positive feedback loop, the feedback causes the gland to produce more hormone; in a negative feedback loop, it causes the gland to slow down or stop hormone production.
6. The hypothalamus stimulates or inhibits the secretion of the pituitary gland's hormones by means of chemicals called *regulating factors*. In addition, the hypothalamus produces two hormones that it stores in the pituitary.
7. Tropic hormones stimulate other endocrine glands. The four are the following: (1) Follicle-stimulating hormone (FSH): In women, FSH triggers the maturation of eggs in the ovaries and stimulates the secretion of estrogens. In men, it triggers the production of sperm. (2) Luteinizing hormone (LH): In women, LH stimulates the release of an egg from the ovary and fosters the development of progesterone. In men, it stimulates the production of testosterone. (3) Adrenocorticotropic hormone (ACTH): ACTH stimulates the adrenal cortex to produce steroid hormones. (4) Thyroid-stimulating hormone (TSH): TSH triggers the thyroid gland to produce the thyroid hormones.
8. a. Produced by the anterior pituitary; helps control normal growth.
   b. Produced by the posterior pituitary; enhances uterine contractions.
   c. Produced by the adrenal cortex; stimulates the kidneys to reabsorb sodium ions and water.
   d. Local hormones produced by cells throughout the body; stimulate smooth muscle contraction and the dilation and constriction of blood vessels.
9. No. Alcohol suppresses the release of ADH (antidiuretic hormone), which means that it encourages more water to leave your body in your urine. If you are already hot and thirsty, this will only dehydrate you further.
10. The thyroid gland produces thyroxine (T_4), triiodothyronine (T_3), and calcitonin (CT). Thyroid hormones regulate the body's metabolism.
11. Parathyroid hormone (PTH) and calcitonin (CT) work antagonistically to maintain appropriate calcium levels. If the level becomes too low, PTH stimulates osteoclasts to liberate calcium from the bones and stimulates the kidneys and intestines to reabsorb more calcium. When levels grow too high, more CT is secreted, which inhibits the release of calcium from bones and speeds up its absorption.
12. Both are produced by the adrenal cortex. The mineralocorticoids regulate the levels of specific ions in body fluids. The glucocorticoids affect glucose metabolism.
13. Over a prolonged period of stress, the body reacts in three stages: (1) alarm reaction (quickened metabolism triggered by adrenaline and noradrenaline), (2) resistance (glucose production and blood pressure rise; hormones involved are ACTH, GH, TSH, mineralocorticoids and glucocorticoids, and (3) exhaustion (loss of potassium and glucose; organs become weak and may stop functioning).
14. Chronic stress leads to long-term stimulation by the autonomic nervous system, which heightens metabolism, raises blood pressure, and speeds up internal chemical reactions. Over time this can physically stress the body and lead to health problems.
15. The pancreatic islets of Langerhans secrete two hormones that act antagonistically to one another to regulate glucose levels. Glucagon raises the glucose level by stimulating the liver to convert glycogen and other nutrients into glucose, whereas insulin decreases glucose levels in the blood by helping cells transport it across their membranes.
16. a. Tiny gland embedded in the brain; possible site of the biological clock that regulates daily rhythms.
    b. Small gland located in the neck; produces hormones that trigger maturation of T lymphocytes.
    c. In women, glands that produce female sex cells and the hormones estrogen and progesterone.
    d. In men, produce male sex cells and testosterone.

CHAPTER 20

1. Sexual reproduction is the process in which a male and a female sex cell combine to form the first cell of a new individual.
2. Spermatozoa; testes; scrotum; spermatogonia; spermatids.
3. Follicle-stimulating hormone (FSH) triggers sperm production. Luteinizing hormone (LH) regulates the testes' secretion of testosterone, a hormone responsible for the development of male secondary sexual characteristics.
4. A spermatozoon has a head that contains the hereditary material. Located at its leading tip is an acrosome that contains enzymes helping the sperm penetrate the egg's membrane. The sperm also has a flagellum that propels it and mitochondria that produce the ATP from which sperm derive the energy to power its flagellum.
5. These are accessory glands that add fluid to the sperm to produce semen. The seminal vesicles supply a fluid containing fructose, which serves as a source of energy for the sperm. The

prostate gland adds an alkaline fluid that neutralizes the acidity of any urine in the urethra and the acidity of the female vagina. The bulbourethral glands also contribute an alkaline fluid.

6. At birth, females have all of the oocytes that they will ever produce. As the woman ages, so do her oocytes, and the odds of a harmful mutation increase appreciably after age 35.

7. Oocyte; uterine tube; uterus; sperm; zygote.

8. The reproductive cycle of females occurs roughly every 28 days. The primary oocyte matures, is released from the ovary during ovulation, and journeys through the uterine tube to the uterus. The endometrial lining of the uterus has thickened to prepare for implantation; if fertilization does not occur, it sloughs off during menstruation. The hormones FSH, LH, estrogen, and progesterone orchestrate these events.

9. Fertilization occurs in the uterine tube. After the egg is fertilized, it completes a second meiotic division, joins its hereditary material with that of the sperm, begins dividing by mitosis, and implants itself on the thickened uterine wall. The placenta begins to form and human chorionic gonadotropin (HCG) is secreted.

10. The sexual response includes four phases: (1) excitement (sexual activity that precedes intercourse), (2) plateau (intercourse), (3) orgasm (reflexive muscular contractions; ejaculation), and (4) resolution (return to the normal physiological state).

11. The only methods of birth control that may prevent the transmission of sexually transmitted diseases are condoms and abstinence.

12. Birth control pills contain estrogen and progesterone, which shut down the production of FSH and LH. By maintaining high levels of estrogen and progesterone, the pills cause the body to act as if ovulation has already occurred; the ovarian follicles do not mature and ovulation does not occur.

13. The most effective birth control methods are vasectomies and tubal ligation. Birth control pills are very effective. Condoms and diaphragms are effective when used correctly, but mistakes are common. Least reliable are the rhythm method and withdrawal, which have high failure rates.

CHAPTER 21

1. Prenatal development refers to the growth and development that occur between conception and birth (about 8½ months).

2. After fertilization, the oocyte's surface changes so that no other sperm can penetrate, and oocyte meiosis is completed. The sperm sheds its tail and the sperm and egg nuclei fuse.

3. Zygote; cleavage; uterine tube; uterus; morula.

4. During implantation the blastocyst attaches to the inner wall of the uterus. This anchors it during development, allows it to obatin nutrients from the mother, and keeps it from being swept out of the uterus.

5. The three primary germ layers are the (1) ectoderm (outer layer of skin, the nervous system, and portions of sense organs), (2) endoderm (lining of the digestive tract and digestive organs, respiratory tract and lungs, and urinary bladder and urethra), and (3) mesoderm (skeleton, muscles, blood, reproductive organs, connective tissues, and innermost layer of skin).

6. The extraembryonic membranes play a role in the life support of the pre-embryo/embryo/fetus. They are the (1) amnion (cushions the embryo in amniotic fluid and keeps the temperature constant), (2) chorion (facilitates the exchange of nutrients, gases, and wastes between the embryo and mother), (3) yolk sac (produces blood for the embryo before its liver is functional and becomes the lining of the digestive tract), and (4) allantois (responsible for formation of the embryo's blood cells and vessels).

7. During gastrulation, groups of inner mass cells differentiate into the three primary germ layers from which all the organs and tissues will develop.

8. During neurulation, cells lying over the notochord curl upward to form a tube that will develop into the central nervous system.

9. The notochord is a structure that forms the midline axis of an embryo. In humans and in most other vertebrates, it develops into the vertebral column.

10. The correct sequence is b. morula (a mass of cells formed about three days after fertilization), d. blastula (a hollow ball of cells formed by cell migration; begins about 4 days after fertilization), a. gastrula (inner mass cells migrate and differentiate into the three primary germ layers; begins at the end of the second week and ends midway through the third week), c. neurula (development of a hollow nerve cord that later becomes the central nervous system and related structures; begins at the third week and ends at about 4½ weeks). These refer to the four stages of development of the pre-embryo and embryo.

11. Between the fourth and eighth weeks, the embryo grows much longer, establishes a primitive circulation, and begins to exchange gases, nutrients, and wastes with the mother. The central nervous system and body form begin to develop.

12. The first trimester is primarily a time of development; by the ninth week, most body systems are functional. During the second trimester the fetus grows, ossification is well underway, and it has a heartbeat. The third trimester is primarily a period of growth; by birth the fetus is able to exist on its own.

13. During pregnancy the fetus obtains all of its nutrients from the mother. Poor nutrition in the mother can damage the child, possibly resulting in retardation and stunted growth.

14. At birth the circulation of the newborn changes as the lungs, rather than the placenta, become the organ of gas exchange.

CHAPTER 22

1. Offspring of sexual reproduction inherit characteristics from both parents; those of asexual reproduction are exact duplicates of one parent. Sexual reproduction introduces variation within a species; the asexual variety does not.

2. Self-fertilization occurs when the gametes of one plant fertilize each other so that the offspring derive from one organism. Cross-fertilization occurs when the male gametes of one plant fertilize the female gametes of another.

3. A gene is the unit of transmission of hereditary characteristics in an organism. An allele is each member of a "factor pair"; each human receives one allele for each gene from the mother's egg and the allele for that gene from the father's sperm.

4. Mendel's Law of Segregation states that each gamete receives only one of an organism's pair of alleles. Random segregation occurs during meiosis.

5. Homozygous; heterozygous.

6. Genotype refers to the genetic make-up of an organism. Its phenotype is the expression of those genes in its outward appearance.

7. All of the F_1 offspring would have an Ll genotype and the phenotype of long leaves.

	L	L
l	Ll	Ll
l	Ll	Ll

8. The F_2 genotype would be 1:2:1

(LL:Ll:ll); the phenotype would be 3:1 (long-leaved:short-leaved).

	L	l
L	LL	Ll
l	Ll	ll

9. Mendel used a testcross to determine whether a phenotypically dominant plant was homozygous or heterozygous for the dominant trait. He crossed the plant with a homozygous recessive plant. When the test plant was homozygous, the offspring were hybrids and phenotypically dominant. When it was heterozygous, half the progeny were heterozygous and looked like the test plant, and half were homozygous recessive and resembled the recessive parent.
10. Monohybrids; dihybrids.
11. Mendel's Law of Independent Assortment states that the distribution of alleles for one trait into the gametes does not affect the distribution of alleles for other traits, unless they are on the same chromosome.
12. The probability is ¼.
13. a. ¹⁄₁₆.
    b. ⁹⁄₁₆.
14. Sex-linked.

GENETICS PROBLEMS

1. The probability of getting two genes on the same chromosome is (½)²³.
2. Somewhere in your herd you have cows and bulls that are not homozygous for the dominant gene "polled." Because you have many cows and probably only one or some small number of bulls, it would make sense to concentrate on the bulls. If you have only homozygous "polled" bulls, you could never produce a horned offspring regardless of the genotype of the mother. The most efficient thing to do would be to keep track of the matings and the phenotype of the offspring resulting from these matings and prevent any bull found to produce horned offspring from mating again.
3. Albinism, *a*, is a recessive gene. If heterozygotes mated, you would have the following:

	A	a
A	AA	Aa
a	Aa	aa

One-fourth would be expected to be albinos.
4. The best thing to do would be to mate Dingleberry to several dames homozygous for the recessive gene that causes the brittle bones. Half of the offspring would be expected to have brittle bones if Dingleberry were a heterozygous carrier of the disease gene. Although you could never be 100% certain Dingleberry was not a carrier, you could reduce the probability to a reasonable level.
5. Your mating of *DDWw* and *Ddww* individuals would look like the following:

	Dw	Dw	dw	dw
DW	DDWw	DDWw	DdWw	DdWw
Dw	DDww	DDww	Ddww	Ddww
DW	DDWw	DDWw	DdWw	DdWw
Dw	DDww	DDww	Ddww	Ddww

Long-wing, red-eyed individuals would result from 8 of the possible 16 combinations, and dumpy, white-eyed individuals would never be produced.
6. Breed Oscar to Heidi. If half of the offspring are white eyed, Oscar is a heterozygote.
7. Both parents carry at least one of the recessive genes. Because it is recessive, the trait is not manifested until they produce an offspring who is homozygous.
8. To solve this problem, let's first look at the second cross, where the individuals were crossed with the homozygous-recessive sepia flies *se/se*. In one case, all the flies were red eyed:

Unknown genotype

		Se	Se
Sepia	Se	Se/Se	Se/Se
	se	Se/se	Se/se

The only way to have all red-eyed flies when bred to homozygous sepia flies is to mate the sepia fly with a homozygous red-eyed fly. In the other case, half of the offspring were black eyed and the other half red eyed.

Unknown genotype

		Se	se
Sepia	se	Se/se	se/se
	se	Se/se	se/se

The unknown genotype in this case must have been *Se/se*, since this is the only mating that will produce the proper ratio of sepia-eyed flies to red-eyed flies. Since the ratio of this unknown genotype and the one previously determined was 1:1, you must deduce the genotype of the original flies, which when mated, will produce a 1:1 ratio of *Se/se* to *Se/Se* flies.

Unknown original 1

		Se	Se
Original 2	Se	Se/Se	Se/Se
	se	Se/se	Se/se

You can see from this diagram that if one of the original flies was homozygous for red eyes and the other was a heterozygous individual, the proper ratio of heterozygous and a homozygous offspring would be obtained.

CHAPTER 23

1. Sex chromosomes; 22; autosomes.
2. Nondisjunction occurs if homologous chromosomes fail to separate after synapsis. This results in gametes with abnormal numbers of chromosomes.
3. Down syndrome (trisomy 21) is a genetic disorder caused by having three 21 chromosomes. Individuals who inherit this condition show delayed maturation of the skeletal system and are often mentally retarded.
4. People who inherit abnormal numbers of sex chromosomes often have abnormal features and may be mentally retarded. Examples are (1) triple X females (XXX zygote), underdeveloped females who may have lower-than-average intelligence; (2) Klinefelter syndrome (XXY zygote), sterile males with some female characteristics; (3) Turner syndrome (XO zygote), sterile females with immature sex organs; and (4) XYY males, fertile males of normal appearance.
5. X.
6. A chromosomal duplication involves the replication of a portion of the chromosome. A translocation occurs when a section of a chromosome breaks off and reattaches to another chromosome. An inversion occurs if the broken piece reattaches to the same chromosome but is reversed.
7. The three major sources of damage to chromosomes are high-energy radiation (such as x-rays), low-energy radiation (such as UV light), and chemicals (such as certain legal and illegal drugs).
8. Heavy use of drugs could affect ova and sperm by damaging the DNA. In general, chemicals add or delete molecules from the structure of DNA. This can result in chromosomal breaks or changes as the cell works to repair the damage.
9. A point (gene) mutation is a change in the genetic message of a chromosome due to alterations of molecules within the chromosomal DNA. This produces a new allele of a gene.
10. a. Sex-linked.
    b. Dominant.
    c. Recessive.
    d. Autosomal.
11. Most human genetic disorders are recessive because those genes are able to persist in the population among carriers; people carrying lethal dominant disorders are more likely to die before

reproducing. Recessive disorders include cystic fibrosis, sickle-cell anemia, and Tay-Sachs disease.
12. In incomplete dominance, alternative forms of an allele are neither dominant nor recessive; heterozygotes are phenotypic intermediates. In codominance, alternative forms of an allele are both dominant, so heterozygotes exhibit both phenotypes.
13. Their children's genotypes would be 1:2:1 (AA [type A], AB [type AB], BB [type B]).

	A	B
A	AA	AB
B	AB	BB

14. Couples who suspect they may be at risk for genetic disorders can undergo genetic couseling to determine the probability of this risk. When a pregnancy is diagnosed as high risk, a woman can undergo amniocentesis (analysis of a sample of amniotic fluid) to test for many common genetic disorders.

GENETICS PROBLEMS

1.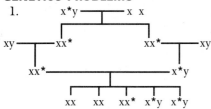

2. $I^AI^O \times I^BI^O \rightarrow I^OI^O$
3. 45 (44 autosomes + one X)
4. AO (blood type A)
5.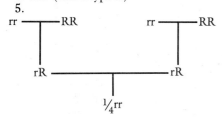

6. c—sex-linked recessive

CHAPTER 24

1. Mendel's work suggested that traits are inherited as discrete packets of information. Sutton theorized that these packets were located on chromosomes, and Morgan's experiments provided evidence that they were. Hammerling's experiments showed that this hereditary information is located in the nucleus of the cell.
2. The two types of nucleic acids are DNA (deoxyribonucleic acid) and RNA (ribonucleic acid). Both consist of nucleotides that are made up of three molecular parts: a sugar, a phosphate group, and a base. The sugar in RNA is ribose, whereas the sugar in DNA is deoxyribose. DNA contains the bases thymine and cytosine; RNA contains uracil instead of thymine.
3. Chargaff's experiments showed that the proportion of bases varied in the DNA of different types of organisms but that the amount of adenine always equalled the amount of thymine and guanine always equalled cytosine. This suggested that DNA was the hereditary material that encoded the information used by cells for growth, development, and repair.
4. *Double helix* refers to the DNA molecule, which is shaped like a double-stranded helical ladder. The bases of each strand together form rungs of uniform length, and alternating sugar-phosphate units form the ladder uprights.
5. DNA begins replicating by unwinding at intervals along its helix. Free nucleotides bind to the exposed bases, producing two new double-stranded DNA molecules that are identical to each other and to the original double strand.
6. Gene; nucleotides; amino acid.
7. The type of RNA found in ribosomes, along with the proteins also found there, is ribosomal RNA (rRNA). Transfer RNA (tRNA) is in the cytoplasm; during polypeptide synthesis, tRNA molecules transport amino acids to the ribosomes and position each amino acid at the correct place on the elongating polypeptide chain. Messenger RNA (mRNA) brings information from the DNA in the nucleus to the ribosomes in the cytoplasm to direct which polypeptide is assembled.
8. During transcription, the DNA message is copied onto a strand of mRNA. The mRNA is constructed from free RNA nucleotides by using a single strand of DNA as a template. The mRNA leaves the nucleus and travels to the ribosomes where polypeptide synthesis occurs.
9. c, a, d, b. This describes the process of translation (polypeptide synthesis).
10. Exons are the nucleotide sequences on a gene that encode the amino-acid sequence of the polypeptide. Introns are sequences of nucleotides that intervene between exons (introns may have evolutionary significance by allowing easier rearrangement of exons). Codons are the triplet-nucleotide code words on mRNA that reflect the sequence of the DNA code.
11. Regulatory genes are special sequences of nucleotides that control the transcription of structural genes (the genes that code for polypeptide synthesis). Regulatory genes are responsible for deciding which polypeptides are produced and when.
12. Repression; repressor protein; activation; activator protein.
13. In gene therapy, scientists attempt to cure genetic disorders by inserting genes with the correct genetic message into patients with incorrect genes.
14. Answers will vary. Be sure to back up your opinion with reasons.

CHAPTER 25

1. Darwin's observations led him to conclude that (1) organisms of the past and present are related, (2) factors other than or in addition to climate are involved in the development of plant and animal diversity, (3) members of the same species often change slightly in appearance after becoming geographically isolated from each other, and (4) organisms living on oceanic islands often resembled those living on a nearby mainland.
2. Three other factors that influenced Darwin were (1) evidence of geological layers and fossils that suggested the Earth was much older than traditionally thought, (2) the observed changes in species that occur with artificial selection, and (3) Malthus' writings on geometric population growth.
3. The two dogs look so different today because of artificial selection; people who bred them selected for desired characteristics, so over time the breeds changed. Artificial selection is based on the principle of natural variation exhibited by all organisms.
4. A geometric progression increases more rapidly because it involves multiplying a number by a constant factor; an arithmetic progression is one in which the elements increase by a constant difference. Malthus theorized that populations grow geometrically, but certain factors limit this growth. Darwin realized that nature acts to limit population numbers.
5. Adaptations are naturally occurring inheritable traits found within populations that confer a reproductive advantage on organisms that possess them. As adaptive (advantageous) traits are passed on from surviving individuals to their offspring, the individuals carrying those traits will increase, and the nature of the general population will gradually change.
6. *Survival of the fittest* refers to the fact that natural selection tends to favor those organisms that are most fit to survive to reproductive age in a particular environment and produce offspring.
7. Polar bears are white because they live

in the Arctic where there is snow year round. Thus their white color is an adaptation that camouflages them and has made them more fit to survive in this environment.
8. In adaptive radiation, the population of a species changes as it disperses into different habitats. Eventually some of these populations change so sufficiently that interbreeding is no longer possible. This is speciation, the formation of new species through the process of evolution.
9. Fossil; sedimentary; radioactive isotopes.
10. The carbon-14 method can help establish a date by estimating the relative amounts of the different isotopes of carbon present in a fossil. The half-life of carbon-14 is 5730 years (the amount of time for half of the ^{14}C to decay into nitrogen). Scientists can estimate the length of time that the carbon-14 has been decaying, which means the time that has elapsed since the organism died.
11. According to radioctive dating, the Earth is 4.6 billion years old. This is significant because it allows sufficient time for evolution to occur.
12. Vestigial organs are structures that are present in an organism of today but are no longer useful. A human example is the appendix.
13. Homologous structures share the same evolutionary origin but now differ in structure and function. Analogous structures have a similar form and function but different evolutionary origins.
14. Comparative studies of anatomy show that many organisms have groups of bones, nerves, muscles, and organs with the same anatomical plan but with different functions. These homologous structures imply evolutionary relatedness. Studies in comparative embryology show that many organisms have early developmental stages that are similar. This again implies evolutionary relatedness.
15. A phylogenetic tree diagrams the degree of relatedness among groups of organisms. Phylogenetic trees support evolution by showing the same evolutionary relationships revealed by anatomical studies.

CHAPTER 26

1. The primordial soup theory attempts to explain how life began on Earth. It states that life arose in the sea as elements and simple compounds in the atmosphere interacted to form simple organic molecules. Another theory holds that life might have arisen in hydrothermal vents in the oceans.
2. Dehydration synthesis is a type of chemical reaction that links molecules by the removal of water and the input of energy. These reactions build polymers (complex molecules). The synthesis of complex organic molecules would have been the first step toward the evolution of living organisms.
3. The formation of some sort of membrane would have allowed chemical reactions to occur in a closed environment. Enzymes could be organized to carry on life functions, chemicals could be selectively absorbed, wastes could be eliminated, and hereditary material could be passed on to future cells.
4. See Table 26-1.
5. Cyanobacteria are single-celled organisms that are similar to the first oxygen-producing bacteria. These evolved during the Archean Era and played an important evolutionary role by gradually oxygenating the atmosphere and oceans.
6. *Proterozoic* is actually inaccurate, since it comes from the Greek for "prior to life." During the Proterozoic Era the first eukaryotic organisms appeared.
7. The Cambrian Period was the oldest period in the Paleozoic Era (roughly 590 to 500 million years ago). All of the main phyla and divisions of organisms that exist today (except chordates and land plants) evolved by the end of the Cambrian Period.
8. The Devonian Period has been nicknamed "the age of fishes" because during this time fishes became abundant and diverse. Amphibians appeared by the end of this period.
9. During the Carboniferous Period, reptiles developed from amphibians and arthropods moved from the sea onto land. Insects, one arthropod group, evolved water-conserving characteristics such as a cuticle. Fungi evolved during the late Carboniferous Period. Much of the land was low and swampy with extensive forests; the worldwide climate was warm and moist.
10. To date, five major extinction events have occurred: (1) at the end of the Cambrian Period, roughly 500 million years ago; (2) about 440 million years ago; (3) about 360 million years ago; (4) during the Permian Period; and (5) at the end of the Mesozoic Era (65 million years ago).
11. Reptiles; mammals; birds; dinosaurs.
12. During this extinction event, many of the larger plankton disappeared, as did the ammonites and the dinosaurs. One theory suggests that a meteorite struck the Earth, creating a huge dust cloud that blocked photosynthesis, killing many organisms. Other theories are that the climate cooled, the sea level changed, or that the vegetation cover of the Earth changed.
13. The plate tectonics theory states that the Earth's outer shell consists of plates that drift over underlying rocks. This constant movement eventually moved land masses apart or closer together, changing habitats. Populations underwent adaptive radiation, speciation, and/or extinction in response to the changing pressures of natural selection.
14. The Cenozoic Era, which began 65 million years ago, saw the rapid evolution and growth of mammals, including primates.
15. Anthropoid primates; mammals.
16. Alas for Hollywood, no. Dinosaurs died out at the end of the Cretaceous Period (about 65 million years ago), long before the first primates appeared.

CHAPTER 27

1. Chordates are distinguished by three features: (1) a single, hollow nerve cord along the back, (2) a rod-shaped notochord located between the nerve cord and the developing gut, and (3) pharyngeal arches.
2. The vertebrates evolved from the chordates. Vertebrates differ from other chordates in that adult vertebrates have a vertebral column replacing the embryologic notochord. Most vertebrates also have a distinct head and bony skeleton.
3. Amphibians; reptiles.
4. c.
5. Warm; hair; milk; mammary glands.
6. Warm bloodedness is the ability to maintain a constant internal body temperature. This ability helps warm-blooded animals stay alive in environments that have extreme temperatures or widely varying temperatures.
7. These are all subclasses of mammals that are living today. Monotremes lay eggs with leathery shells and incubate them in a nest. Marsupials bear immature young and nurse them in a pouch until they are old enough to be on their own. In placental mammals, the young develop to maturity inside the mother.
8. Primates are the order of mammals with characteristics reflecting an arboreal life-style. They have developed two especially helpful characteristics: depth perception resulting from stereoscopic vision and flexible, grasping hands with opposable thumbs.
9. These are both suborders of the primates. Prosimians (lower primates such as lemurs) are small animals, usually nocturnal, with a well-developed

sense of smell. Anthropoids (higher primates such as apes and humans) have larger brains, flatter faces, eyes that are closer together, and relatively long front and hind limbs. Anthropoids are also diurnal and possess color vision.
10. (1) d, (2) c, (3) a, (4) b.
11. Statement (a) is correct; statement (b) is false. All hominids are a family within the hominoid superfamily.
12. *Australopithecus afarensis* is the oldest hominid fossil skeleton that has yet been found; it is not considered human. The first human fossils belong to the extinct species *Homo habilis,* which was more intelligent than its ancestors and able to make tools and clothing. The extinct species *Homo erectus* was close to modern human size, walked upright, and had a larger brain than its ancestors. Of the three, it is most like modern humans.
13. Using "neanderthal" in this way actually contradicts what scientists have learned about the Neanderthal subspecies of *Homo sapiens*. Neanderthals had larger brains than modern humans and made diverse tools. Neanderthal graves indicate that they cared for their injured and sick, believed in a life after death, and were capable of symbolic thought.
14. Early members of the species had smaller heads, brow ridges, teeth, jaws, and faces than earlier species did. They also fashioned sophisticated tools from stone and from other substances such as bone and ivory. Over time, groups of *Homo sapiens sapiens* moved away from the hunter-gatherer life-style to develop agriculture, complex social structures, and civilization.

CHAPTER 28

1. Viruses are infectious agents that lack a cellular structure, so they are not living. Instead, they are nonliving obligate parasites that must exist in association with and at the expense of other organisms.
2. See Figure 28-3.
3. Both cycles are patterns of viral replication. In the lytic cycle, viruses enter a cell, replicate, and cause the cell to burst and release new viruses. In the lysogenic cycle, viruses enter into a long-term relationship with the host cells, their nucleic acid replicating as the cells multiply.
4. A cold sore results from the lysogenic cycle of cell damage caused by the herpes simplex virus. The virus remains latent in nervous tissues until something triggers it, such as a cold.
5. Tumors are large masses of cells resulting from viral infections that caused these cells to grow faster than normal cells. A benign tumor is walled off from the rest of the body so that these cells cannot spread. A malignant (cancerous) tumor—the more dangerous of the two—consists of cells that invade and destroy other body tissues.
6. Bacteria are classified into their own kingdom because their cell structure is different from that of all other organisms. (They do not contain membrane-bounded organelles, they lack a nucleus, and the bacterial cell is bounded by a membrane encased within a cell wall.)
7. Bacteria perform many vital functions, such as decomposing organic materials and recycling inorganic compounds. They were also largely responsible for creating the properties of the atmosphere and soil that are present today.
8. Binary fission; exponential growth.
9. a. Chemoautotrophs (nitrifying bacteria).
   b. Heterotrophs (decomposers).
   c. Photoautotrophs.
   d. Archaebacteria.
10. These terms refer to the purposeful manipulation of genes within organisms.
11. These are all mechanisms by which genes move from one prokaryote to another. In transformation, free pieces of DNA move from a donor to a recipient cell. In transduction, DNA from a donor cell is carried to a recipient cell by a virus. During conjugation, DNA from the donor cell is injected into the recipient cell when the two cells make contact.
12. Researchers use restriction enzymes to cut desired pieces of human DNA from nuclei, open bacterial plasmid DNA, and insert the human fragment. These hybrid plasmids can be taken up by bacteria, which then express the human gene by producing a particular human protein. This technique can provide a relatively inexpensive and plentiful source of clinically useful substances; because the source is human, patients are less likely to suffer allergic reactions.
13. Gene therapy is the treatment of a genetic disorder by inserting normal genes into the patient's cells.
14. Genetic engineering offers many potential benefits, such as improving crops and forest trees, producing medically important substances, and treating some genetic disorders. It is important to establish appropriate safeguards for such experiments.

CHAPTER 29

1. Protists are single-celled eukaryotic organisms; this kingdom contains organisms that are animal-like (protozoa), plant-like (algae and diatoms), and fungus-like (slime molds).
2. An ameba is a protozoan that takes on changing shapes. It moves and obtains food by means of pseudopods (cytoplasmic extensions). Its pseudopods engulf organic matter, which the ameba then digests.
3. Both flagellates and ciliates are protozoans, and both have hair-like cellular processes that help them move and obtain food. However, flagellates have flagella (long hair-like processes), whereas ciliates have cilia (short hair-like extensions). Ciliates have a more complex internal organization then flagellates do.
4. Sporozoans are nonmotile protozoans that parasitize animals. They undergo complex life cycles in which they are passed from host to host by means of a vector.
5. Euglenoids, dinoflagellates, and golden algae are all phyla of plant-like protists. Euglenoids are flagellated protists with chloroplasts. Dinoflagellates have stiff outer coverings and their flagella beat in two grooves. The golden algae have gold-green photosynthetic pigments and store food as oil.
6. All three organisms belong to the phylum of golden algae.
7. Slime molds; fungus; ameba.
8. Saprophytic; parasite.
9. All fungi could be described as helpful, since as decomposers they are crucial to the cycling of materials in ecosystems. More specifically, helpful fungi include those that produce antibiotic drugs. Yeasts are fungi that are crucial in the production of bread, cheese, and beer. Other fungi, however, cause diseases in plants and animals, such as Dutch elm disease, wheat rust, and yeast infections.
10. Club fungi (c), zygote-forming fungi (d), sac fungi (a), water molds (e), imperfect fungi (b).
11. Lichens are symbiotic associations between fungi and photosynthetic partners. Lichens are able to tolerate harsh living conditions, so they are found in a wide range of habitats (such as deserts, extreme northern and southern latitudes, and mountaintops).
12. Radiolarians; foraminifera.

CHAPTER 30

1. Vascular; nonvascular.
2. Sporophytes; gametophytes; gametophytes; sporophytes.
3. The three classes are the mosses, liv-

erworts, and hornworts. All three share a fairly uniform life cycle involving two phases. The sporophyte generation grows out of or is embedded in the tissues of the gametophyte generation and depends on it for nutrition. Neither generation tends to dominate the life cycle.
4. Seeds are structures that protect the embryonic sporophyte plant from drying out or being eaten and contain stored food.
5. In seedless vascular plants, the sporophyte generation is dominant and lives separately from the gametophyte. The gametophytes produce motile sperm that swim to the eggs. The fertilized eggs produce sporophytes that at first grow within gametophyte tissues but eventually become free living.
6. Gymnosperms (c), angiosperms (d), bryophytes (b), ferns (a).
7. See Figure 30-7.
8. Sexual reproduction in gymnosperms and angiosperms produces seeds by the union of male and female gametes. Germinating pollen tubes convey the sperm to the eggs. The seeds protect and nourish the young sporophytes as they develop into new plants.
9. The relationship between flowering plants and their pollinators has influenced the evolution of both groups. Flowers have evolved colors and forms that attract certain pollinators, whereas insects, for example, have evolved special traits that are useful in obtaining food more efficiently from the flowers.
10. Fruits have evolved a range of shapes, textures, and tastes that help disperse their seeds more effectively. Fruits with brightly colored flesh are often eaten by animals and thus dispersed to different habitats. Other fruits have evolved hooked spines that stick to animals' fur; others have wings that are blown by the wind.
11. b, d, e, a, c.
12. Monocots and dicots are both angiosperms. They differ in the placement of the stored food in their seeds. In dicots, most of the stored food is in the cotyledons (seed leaves), whereas in monocots, most of it is stored in endosperm. In addition, monocots usually have parallel veins in their leaves, and dicots have net-like veins. Monocots also have flower parts in threes, whereas dicots have flower parts in fours and fives.
13. Vegetative propagation is an asexual reproductive process in which a new plant develops from a portion of a parent plant. An example is the white potato, a tuber which can grow new plants from its "eyes."
14. Hormones are chemical substances produced in one part of an organism and then transported to another part of the organism, where they bring about physiological responses. The five kinds of plant hormones are auxins, gibberellins, cytokinins (promote and regulate growth), abscisic acid (inhibits growth), and ethylene (affects the ripening of fruit).

CHAPTER 31

1. Vascular plants are plants having a system of vessels that transports water and nutrients.
2. Root; shoot; leaves; stem.
3. (1) c, (2) b, (3) a, (4) d.
4. Xylem and phloem are both types of vascular tissue, but they have different functions and contain different types of conducting cells. Xylem conducts water and dissolved minerals, whereas phloem conducts carbohydrates and other substances.
5. a. Cells in ground tissue that function in photosynthesis and storage.
   b. The most abundant type of cell in dermal tissue, often covered with a thick waxy layer that protects the plant and retards water loss.
   c. Openings that connect air spaces inside dicot leaves with the exterior; they are opened and closed by the change in shape of surrounding guard cells. Water vapor and gases move into and out of the leaves through these openings.
   d. Thimble-like mass of unorganized cells that covers and protects a root's apical meristem as it grows through soil.
6. Primary growth occurs mainly at the tips of roots and shoots, making plants taller; secondary growth occurs in the lateral meristem and makes plants larger in diameter.
7. Vegetables are a good source of minerals. Plants take in minerals from the soil to form proteins, fats, and vitamins.
8. See Figures 31-5 and 31-6.
9. A dandelion has a taproot, a single major root that can grow deep as a firm anchor. Some taproots also store food for the plant. An ivy plant has adventitious roots that develop from the lower part of the stem and help anchor the plant. Grass has fibrous roots, a root system without one major root; this type of system anchors the plant and absorbs nutrients over a large surface area.
10. Annual rings provide a clue to a tree's age by illustrating the annual pattern of rapid and slow growth in the cambium (lateral meristem tissue). This results in secondary growth.
11. Water rises beyond the point where it would be supported by air pressure because evaporation from the plant's leaves produces a force that pulls upward on the entire column of water. The forces invovled include air pressure, adhesion, cohesion, root pressure, and transpiration.
12. See Figure 31-19.
13. In general, nonvascular plants are shorter than vascular plants. This is because they lack the sophisticated transport systems of vascular plants.

CHAPTER 32

1. Both you and jellyfish are animals. An important difference is that jellyfish are invertebrates (they lack a backbone), whereas you are a vertebrate (with a backbone and dorsal nerve cord).
2. Eukaryotic organisms have a cellular structure different from that of bacteria. Animals are also multicellular (having more than one cell), and as heterotrophs, they are unable to make their own food. Therefore they must eat other organic matter for food. Beyond these basic similarities, however, animals are a diverse group.
3. Radial symmetry; bilateral symmetry.
4. Coelomates have lined, fluid-filled body cavities in which organs are suspended by thin sheets of connective tissue (such as in humans). Pseudocoelomates have body cavities, but the cavities are not lined and their organs are not suspended by connective tissue (such as in roundworms). Acoelomates have no body cavities (such as in flatworms).
5. Sponges are called the simplest animals because they lack tissues and organs.
6. Cnidarians exist as either polyps (cylindrically shaped animals that anchor to rocks) or as medusae (free-floating, umbrella-shaped animals).
7. The acoelomates are the most primitive bilaterally symmetrical animals. They have organs and some organ systems, including a primitive brain. Their organs are embedded within body tissues.
8. Roundworms can be dangerous if they parasitize humans, absorbing digested food from their hosts. Roundworms are cylindrical pseudocoelomates encased in a flexible outer cuticle; they have primitive excretory and nervous systems. They move by using muscles attached to the cuticle that push against the pseudocoelom.
9. In a closed circulatory system, blood is enclosed within vessels as it travels through the body. In an open system, blood flows through irregular chan-

nels (blood sinuses) in many parts of the body. Since you have blood vessels, you have a closed system.
10. Mollusks are coelomates with a muscular foot and soft body, usually covered by a shell. They use gills for respiration in water, whereas terrestrial species have adaptations to breathing on land. They show both open and closed circulatory systems and well-developed excretory systems.
11. Like many arthropods, flies have compound eyes that contain many independent visual units, each with its own lens. This feature helps them detect motion quickly.
12. Arthropods are a diverse phylum of organisms having jointed appendages and rigid exoskeletons.
13. Insects belong to a group of arthropods that have jaws (mandibles). Spiders belong to a different group of arthropods that lack mandibles.
14. Echinoderms are deuterostomes, whereas the other invertebrates are protostomes. These embryological differences make echinoderms more similar to the chordates, which implies that echinoderms and chordates evolved from a common ancestor.

CHAPTER 33
1. Nerve cord; notochord; pharyngeal (gill) slits.
2. The nerve cord develops into a brain and spinal cord; the notochord becomes a vertebral column enclosing the nerve cord; and the pharyngeal arches develop into gills (in fishes) or into ear, jaw, and throat structures (terrestrial vertebrates).
3. (1) b, (2) c, (3) a.
4. Vertebrates have (1) a vertebral column, (2) a distinct head with a skull enclosing their brain, (3) a closed circulatory system, and (4) a heart that pumps their blood.
5. The *Jaws* films starred sharks (class Chondrichthyes) and humans (class Mammalia).
6. A lateral line system (found in all fishes and amphibians) is a system of mechanoreceptors that detects sound, pressure, and movement.
7. Osmoregulation is the control of water movement into and out of organisms' bodies. Maintaining the proper balance of water and solutes is vital to life.
8. Ovoviviparous organisms (some fishes, reptiles; many insects) retain fertilized eggs within their oviducts until the young hatch; the young receive nourishment from the egg. Oviparous animals (birds) lay eggs and their young hatch outside the mother. Viviparous organisms (humans) bear their young alive, and the young are nourished by the mother, not the egg.
9. *Amphibian* means "two lives." This refers to the fact that amphibians live both aquatic and terrestrial existences.
10. Ectothermic animals (fish, amphibians, reptiles) regulate body temperature by taking in heat from the environment. Endothermic animals (birds, mammals) regulate their body temperatures internally.
11. Reptiles have dry skins covered with scales, retarding water loss. They also lay amniotic eggs, which contain nutrients and water for the embryos and protect the embryos from drying out.
12. Feathers are flexible, light, waterproof epidermal structures; birds are the only animals that have them. Feathers are important to flight; they also give birds waterproof coverings and insulate them against temperature changes.
13. Mammals are endothermic vertebrates that have hair and whose females secrete milk from mammary glands to feed their young.

CHAPTER 34
1. Behaviors include the patterns of movement, sounds, and body positions exhibited by animals. They also include any type of change in an animal that can trigger behaviors in other animals.
2. This statement expresses a fundamental idea in the study of animal behavior, that animal behaviors are evolutionary adaptations that make an organism more fit to survive in and adjust to its environment.
3. Innate (inborn) behaviors; learned behaviors.
4. Kineses; taxes; reflexes; fixed action patterns.
5. A kinesis is a change in the speed of an animals's random movements in response to environmental stimuli (for example, insects moving around underneath a fallen log). A taxis is a directed movement toward or away from a stimulus (for example, moths being attracted to a light at night). A reflex is an automatic response to nerve stimulation (for example, knee jerk reflex in humans). A fixed action pattern is a sequence of innate behaviors in which the actions follow an unchanging order of muscular movements (for example, a mother bird stuffing an insect into the gaping beak of her offspring).
6. Innate behaviors protect an animal from environmental hazards without it having learned to do so; many social behaviors also depend on innate behaviors. Organisms with simple nervous systems rely primarily on innate behaviors for survival. Learned behaviors, however, allow an animal to adjust its behavior based on its experiences, helping it to adapt better to its environment or to environmental changes.
7. Imprinting; habituation; trial-and-error learning; classical conditioning; insight.
8. a. Taxis (phototaxis).
   b. Habituation.
   c. Trial-and-error conditioning (operant conditioning).
9. Operant conditioning is a form of learning in which an animal associates something that it does with a reward or punishment. In classical conditioning, an animal learns to associate a new stimulus with a natural stimulus that normally evokes a particular response. Eventually the animal will respond to the new stimulus alone.
10. Insight (reasoning) is the most complex form of learning in which an animal recognizes a problem and solves it mentally before trying out the solution. Insight allows an animal to perform a correct behavior the first time it tries; it helps the animal adjust and respond to new situations and perhaps determine better ways to deal with its environment.
11. a. Locality imprinting (migration).
    b. Reflex.
    c. Fixed action pattern.

CHAPTER 35
1. Behaviors are the patterns of movement, vocalizations, and body positions exhibited by animals. Social behaviors are behaviors that help members of the same species communicate and interact.
2. Social behaviors help sexually reproducing species to reproduce, care for offspring, and defend territories. Other behaviors may help animals to hunt for and share food and to warn and defend against predators.
3. Competitive behavior results when two or more individuals are competing for the same resource (such as food, territories, or mates) and there is not enough of the resource to be shared. This competition encourages the population to disperse among the resources, conserving time and energy, rather than killing each other.
4. a. Threat display.
   b. Submissive behavior.
5. An animal exhibiting territorial behavior marks off an area as its own and defends it against same-sex members of its species. Territorial behavior limits competition and aggression within

a species and encourages the founding of new populations. It also rations resources among individuals, ensures that at least some individuals will have enough resources to survive and reproduce, and places the peak time of competition/aggression before the time of reproduction and parenting.
6. Sexual reproduction involves the union of male and female sex cells to form the first cells of new individuals. It provides genetic variability to a species, giving it the raw materials needed to adapt to environmental change.
7. Mating includes male-female behaviors that result in fertilization; courtship includes the behavior patterns that lead to mating.
8. a. Parenting behavior.
   b. Courtship ritual (reproductive behavior).
   c. Territorial behavior.
9. Animals form social groups to work together toward common goals: to protect against predators, hunt for food, and create a division of labor.
10. A honeybee colony consists of three castes: (1) the queen, which lays the eggs; (2) the drones, male bees that fertilize the queen's eggs; and (3) the workers, female bees that develop from fertilized eggs and do everything around the hive except lay eggs and fertilize them. (This includes feeding larvae, guarding the hive, and foraging for food).
11. A rank order is the social hierarchy that appears in groups of fishes, reptiles, birds, and mammals. This hierarchy helps reduce aggression and fighting within social groups, focusing these behaviors to a short time when the rank order develops.
12. Answers will vary; human behavior appears to result from a combination of hereditary and environmental factors.

CHAPTER 36

1. Ecology is the study of interactions between organisms and the environment.
2. Population change; this equation shows that the size of a population results from the number of births and of organisms moving into the population (immigrants), balanced against the number of deaths and organisms moving out of the population.
3. Exponential growth; carrying capacity.
4. A large population is more likely to survive because it is less vulnerable to random events and natural disasters and because its size promotes genetic diversity.
5. Density refers to the number of organisms per unit of area; dispersion refers to the way in which the individuals of a population are arranged. If individuals are too far apart, they may not be able to reproduce. If population becomes too dense, however, factors can arise that limit the population's size (disease, predation, starvation).
6. The three patterns are (1) uniform (evenly spaced), (2) random, and (3) clumped (organisms are grouped in areas of the environment that have the necessary resources). Human populations are distributed in the clumped pattern.
7. Density-dependent limiting factors (such as competition) come into play when a population size is large, whereas density-independent factors (such as weather) operate regardless of population size.
8. Both predation and parasitism are density-dependent limiting factors. Predators are organisms that kill and eat organisms of another species; parasites live on or in other species and derive nourishment from them but do not necessarily kill them.
9. See Figure 36-7. Type I is typical of organisms that tend to survive past their reproductive years (such as humans). Type II characterizes organisms that reproduce asexually and are equally likely to die at any age (such as the hydra). Type III is characteristic of organisms that produce offspring that must survive on their own and therefore die in large numbers when young (oysters).
10. Developing countries tend to have a population pyramid with a broad base, reflecting a rapidly growing population with large numbers of individuals entering their reproductive years. Industrialized countries tend to have populations with similar proportions of their populations in each age group and are growing slowly or not at all.
11. 1.8%; 95 million; double.
12. People are already putting pressure on our natural resources (land, water, forests, atmosphere). Increasing industrialization leads to rising levels of pollution and solid waste. Even though it is estimated that there is enough food produced to feed the world's population, many people live—and die—in poverty and hunger.
13. Answers will vary.

CHAPTER 37

1. A population consists of individuals of the same species. Populations that interact with each other form a community. An ecosystem is a community of organisms along with the abiotic factors with which the community interacts.
2. Habitat; niche.
3. Interactions can be grouped into (1) competition (organisms of different species living near each other strive to obtain the same limited resources), (2) predation (organisms of one species kill and eat organisms of another species), and (3) symbiosis (organisms of different species live together in close association over a period of time).
4. False. Competition is often greatest between organisms that obtain their food in similar ways.
5. The principle of competitive exclusion states that if two species are competing for the same limited resource in a location, the species that can use the resource most efficiently will eventually eliminate the other species in that area.
6. An organism's realized niche is the role it actually plays in the ecosystem; its fundamental niche is the role it might play if competitors were not present.
7. Predator; prey; parasite.
8. Gause's experiments showed that predator-prey relationships follow a cyclical pattern: as the number of prey increases, the number of predators increases. When the prey decrease because of predation, the predator population also declines until an upsurge in the prey population starts the cycle again.
9. The balance between predator and prey populations is complex because interactions depend on many factors: interactions with other organisms, movement of new organisms into or out of the community, and abiotic factors in the ecosystem.
10. Coevolution is the long-term, mutual evolutionary adjustment of characteristics of members of biological communities is relation to one another. Cattle are one example: their digestive systems have adapted (grinding teeth, strong jaws, a rumen) to allow them to digest silica in certain grasses.
11. a. Warning coloration.
    b. Camouflage.
12. These are all types of symbiotic relationships. In commensalism, one species benefits and the other is neither hurt nor helped (epiphytes). In mutualism, both participating species benefit (ants and aphids). In parasitism, one species benefits and the other is harmed (viruses).
13. Primary succession takes place in areas not previously changed by organisms. It could begin, for instance, with lichens growing on bare rock, producing acids that break down the rock to form soil. This gives rise to a pioneer community, which in turn leads to

more vegetation, eventually making the area able to support other forms of life. At some point the mix of plants and animals becomes somewhat stable, forming a climax community.
14. Secondary succession.

CHAPTER 38

1. An ecosystem is a community of plants, animals, and microorganisms that interact with one another and their environment and that are interdependent. The living (biotic) groups it contains include producers (organisms that can make their own food), consumers (organisms that eat other organisms for food), and decomposers (consumers that eat dead matter).
2. Human beings are consumers.
3. See Figure 38-1.
4. Primary consumers; secondary consumers; decomposers; detritus.
5. A trophic level is a feeding level within a food chain. A food chain is a linear relationship among organisms that feed one on another. Food webs are food chains that interweave with one another.
6. A green plant captures some of the sun's energy. The successive members of a food chain, in turn, incorporate into their own bodies about 10% of the energy in the organisms on which they feed. The rest is released as heat or is used to power metabolic activities.
7. Chemicals are stored in the atmosphere and in rocks (reservoirs). These substances enter the soil, are incorporated into the bodies of living organisms, and are passed along the food chain. Ultimately, the chemicals, with help from decomposers, return to the nonliving reservoirs.
8. Water in the atmosphere falls to the Earth as precipitation. Plants take up water from the soil, and animals obtain water from drinking it or from eating plants or other animals. Water returns to the atmosphere through the evaporation of surface water and transpiration by plants.
9. Ground water includes subsurface bodies of fresh water. It can be polluted by pesticides that are washed from plants and topsoil by the rain and by chemical wastes that are dumped in surface water. It is serious because at the present time there is no technology to remove these pollutants from ground water; already about 2% of ground water in the United States is polluted, and the situation is getting worse.
10. Carbon is held in the atmosphere as carbon dioxide. Photosynthetic organisms take in this gas and use it to build carbon compounds. Both producers and consumers use carbon compounds as an energy source, metabolizing them and releasing carbon dioxide to the atmosphere again.
11. Crop rotation can improve soil by alternating a nonleguminous crop (which consumes nitrogen from the soil) with a legumiunous crop (which can enrich the nitrate level of depleted soil by forming mutualistic associations with nitrogen-fixing bacteria). This relates to the nitrogen cycle.
12. Adding more phosphates than crops need causes the extra phosphates to wash off or erode into water. Bodies of water can become eutrophic; algae and other aquatic plants overgrow and decomposers feed on them, using much of the water's oxygen and choking out other organisms.

CHAPTER 39

1. A biome is an ecosystem of plants and animals that occurs over wide areas within specific climatic regions.
2. Australia is in the southern hemisphere, whereas the United States is in the northern hemisphere. Thus when the Earth tilts so that the United States is further from the sun, people in the United States experience winter but it is summer in Australia. When the Earth tilts so that the North Pole is closer to the sun than the South Pole, it is summer in America but winter in Australia.
3. Near the equator, warm air rises and flows toward the poles; the Earth's rotation breaks the air into six coils of rising and falling air that surround the Earth. The air in each coil rises at the region of its lowest latitude, moves toward the poles, sinks at the region of its highest latitude, and flows back to the equator. These air masses affect wind patterns and precipitation worldwide.
4. The climate of an area is determined by its latitude, wind patterns, and surface features of the Earth.
5. a. Tropical rainforests.
   b. Temperate deciduous forests.
   c. Arctic tundra.
   d. Savannahs.
   e. Taiga.
   f. Temperate grasslands.
   g. Desert.
6. Tropical rainforests make poor farmland, since most nutrients are found in the vegetation, not the soil. The soil itself is thin, often only a few centimeters deep. Cutting down and burning the trees removes the nutrients and breaks down the organic matter to carbon dioxide. In 2 to 3 years the land becomes barren.
7. Desert rainfall is limited by latitude because the air, which has released its moisture over the rain forests, is dry over desert latitudes. Deserts can also occur in the interior of continents, far from the ocean, and on the leeward side of mountains that block rainfall.
8. The taiga is the northern coniferous forest with long, cold winters and little precipitation. The tundra lies further north, and is open, often boggy, with desert-like levels of precipitation.
9. The three fresh water life zones are (1) shore zone (shallow water in which plants with roots may grow; some consumers live here), (2) open-water zone (the main body of water through which light penetrates; floating plants, microscopic floating animals, and fish live here), (3) deep zone (light does not penetrate, so there are no producers; inhabited mainly by decomposers or organisms that feed on organic material that filters down).
10. An estuary is an environment where fresh water and salt water meet. All organisms living here are adapted to living in an area of moving water and changing salt concentrations. Some are bottom-dwellers that attach to bottom material or burrow into the mud; others spawn in less salty water.
11. Estuaries are ecosystems where fresh water rivers and streams meet the ocean. They are vulnerable to pollutants that flow from upstream as well as to pesticides that leach from the land and soil.
12. The three major life zones of the ocean are (1) intertidal (area between high and low tides), (2) neritic (shallow waters along the coasts of the continents, and (3) open-sea (the rest of the ocean).
13. (1) Photic zone (well-lit layer of the ocean, which contains phytoplankton, zooplankton, fish, and air-breathing aquatic mammals (nekton), (2) mesopelagic zone (organisms exhibiting bioluminescence, some fish [swordfish, certain sharks and whales], and cephalopods), (3) abyssal zone (benthos or bottom-dwellers that eat detritus).

CHAPTER 40

1. The biosphere is the global ecosystem in which all other ecosystems on the Earth exist.
2. A sustainable society uses nonfinite, renewable sources of energy. Society's needs are satisfied without compromising future generations and natural resources.
3. People could use solar, wind, and water power, as well as the Earth's geothermal energy.

4. People can recycle paper, glass, and metal products to help preserve mineral resources. This is important because mineral supplies are finite; once used up, they are gone forever.
5. Tropical rainforests are home to over half the world's species of plants, animals, and insects from which people get many medicines as well as wood, charcoal, oils, and nuts. Burning the forests adds carbon dioxide to the air, which may raise temperatures worldwide. Destroying the forests also destroys the habitat for many species and leads to extinctions.
6. Global warming refers to a worldwide increase in temperature. Even a small increase in temperature could melt ice at the poles, raising the sea level, and flooding one fifth of the world's land area. Higher temperatures also affect rain patterns and agriculture.
7. Biological diversity refers to the richness of species. Humans are reducing the world's biological diversity through extinctions and selective breeding. Reducing diversity makes species more vulnerable to being wiped out by disease or environmental changes.
8. Humans can improve solid waste management by increasing and improving recycling, using safe incineration, engineering better landfill sites, and practicing waste-to-energy reclamation.
9. Point sources are sources of pollutants that enter the water at one or a few distinct places (industrial waste dumped by a specific factory). Nonpoint sources refer to pollutants that enter water at a variety of places (metals and acids draining from mines).
10. Organic nutrients are food for bacteria; as the bacteria multiply, they use up the oxygen in the water, killing the fish. Inorganic nutrients can stimulate plant growth in the water. Bacteria decompose the plants after they die; again, when the bacteria multiply, they consume the water's oxygen. The decomposed materials also begin to fill in the body of water, eventually turning it into a terrestrial community. Pollution with infectious agents can cause serious diseases. Toxic substances in water can poison the organisms that take in this water; sometimes organisms accumulate these substances in their bodies.
11. Acid rain results from gases such as sulfur dioxide and nitrogen dioxide that combine with water in the atmosphere. When this acidic rain falls, it kills aquatic life, pollutes groundwater, damages plants, and eats away at stone, metal, and painted surfaces.
12. Gray-air cities are those with relatively cold, moist climates. They develop a layer of smog because of the burning of fossil fuels and the abundance of sulfur oxides and particulates in the air. The air over brown-air cities contains hydrocarbons and nitrogen oxides that react with the sunny climate to produce secondary pollutants such as ozone.
13. Ozone in the stratosphere helps shield you from the sun's ultraviolet rays. Depletion of this layer can damage bacteria and plants and cause burns and skin cancers.
14. Answers will vary.

APPENDIX B: Classification of organisms

The purpose of the classification scheme presented here is for student reference. It was compiled from a variety of sources in microbiology, botany, and zoology, taking into account the most widely accepted schemes at this time. It is not intended to include all of the phyla of organisms currently identified by taxonomists, but it includes all of the phyla and divisions described and referred to in *Biology Today*. Because this textbook describes humans in its extensive physiology section and includes a chapter on chordate and human evolution, the classification of the chordates is listed in more detail than other phyla.

Kingdom Monera

Prokaryotic, single-celled organisms that have neither membrane-bounded nuclei nor membrane-bounded organelles.

Subkingdom Arcaebacteria: Methanogens (methane-producing bacteria), halophiles (bacteria that live in salt marshes), thermaocidophiles (bacteria that live in hot springs and deep-sea vents). A separate kingdom has been proposed for this subkingdom.

Subkingdom Eubacteria: All other bacteria, including nitrogen-fixing bacteria and cyanobacteria.

Kingdom Protista

A varied group of eukaryotic organisms. Many are single celled, although some phyla include multicellular or colonial forms. Some forms are photosynthetic.

Phylum Rhizopoda: Amebas
Phylum Zoomastigina: Flagellates
Phylum Ciliophora: Ciliates
Phylum Sporozoa: Sporozoans
Phylum Dinoflagellata: Dinoflagellates
Phylum Chrysophyta: Diatoms, yellow-green algae, golden-brown algae
Phylum Phaeophyta: Brown algae
Phylum Chlorophyta: Green algae
Phylum Rhodophyta: Red algae
Phylum Acrasiomycota: Cellular slime molds
Phylum Myxomycota: Plasmodial slime molds
Phylum Oomycota: Water molds, white rusts, downy mildews

Kingdom Fungi

Mostly multicellular eukaryotic organisms that are saprophytes, feeding on dead or decaying organic material. Some fungi are parasitic, feeding on living organisms.

Division Zygomycota: Zygote-forming fungi (black bread mold)
Division Ascomycota: Sac fungi (yeasts, cup fungi)
Division Basidiomycota: Club fungi (mushrooms, puffballs, shelf fungi)
Division Fungi Imperfect: Imperfect fungi (fungi that have no sexual stages of reproduction, such as *Penicillium* and *Aspergillus*)
Lichens: Associations between fungi and green algae and/or cyanobacteria

Kingdom Plantae

Multicellular, eukaryotic organisms that evolved on land and that perform photosynthesis, producing their own food by using energy from the sun and carbon dioxide from the atmosphere.

    Division Bryophyta: Mosses, hornworts, liverworts
    Division Psilophyta: Whisk ferns
    Division Lycophyta: Club mosses
    Division Sphenophyta: Horsetails
    Division Pterophyta: Ferns
    Division Coniferophyta: Conifers
    Division Cycadophyta: Cycads
    Division Ginkgophyta: Ginkgos
    Division Gnetophyta: Gnetae
    Division Anthophyta: Flowering plants (angiosperms)
        Class Monocotyledons: Grasses, irises
        Class Dicotyledons: Flowering trees, shrubs, roses

Kindgom Animalia

Multicellular, eukaryotic organisms that are heterotrophic, eating other organisms for food.

    Phylum Porifera: Sponges
    Phylum Cnidaria
        Class Hydrozoa: Hydra
        Class Scyphozoa: Jellyfish
        Class Anthozoa: Corals, sea anemones
    Phylum Ctenophora: Comb jellies, sea walnuts
    Phylum Platyhelminthes
        Class Turbellaria: Free-living flatworms
        Class Trematoda: Flukes
        Class Cestoda: Tapeworms
    Phylum Nematoda: Roundworms
    Phylum Mollusca
        Class Polyplacophora: Chitons
        Class Gastropoda: Snails and slugs
        Class Bivalvia: Bivalves
        Class Cephalopoda: Octopuses, squids, nautilus
    Phylum Annelida
        Class Polychaeta: Marine worms
        Class Oligochaeta: Earthworms and fresh water worms
        Class Hirundinea: Leeches
    Phylum Arthropoda
      Subphylum Chelicerata
        Class Arachnida: Spiders, mites, ticks
        Class Merostomata: Horseshoe crabs
        Class Pycnogonida: Sea spiders
      Subphylum Crustacea
        Class Crustacea: Lobsters, Crayfish, shrimps, crabs
      Subphylum Uniramia
        Class Chilopoda: Centipedes
        Class Diplopoda: Millipedes
        Class Insecta: Insects
    Phylum Echinodermata
        Class Crinoidea: Sea lilies
        Class Asteroidea: Sea stars (starfish)
        Class Ophiuroidea: Brittle stars
        Class Echinoidea: Sea urchins and sand dollars
        Class Holothuroidea: Sea cucumbers

Phylum Chordata: Chordates
    Subphylum Urochordata: Tunicates (sea squirts)
    Subphylum Cephalochordata: Lancelets
    Subphylum Vertebrata: Vertebrates
        Class Agnatha: Jawless fishes (lamprey eels, hagfishes)
        Class Chondrichthyes: Cartilaginous fishes (sharks, skates, rays)
        Class Osteichthyes: Bony fishes (perch, cod, and trout)
        Class Amphibia: Salamanders, frogs, toads
        Class Reptilia: Reptiles (lizards, snakes, turtles, crocodiles)
        Class Aves: birds
        Class Mammalia: Mammals
          Subclass Prototheria: Egg-laying mammals (duck-billed platypus, spiny anteater)
          Subclass Metatheria: Pouched mammals or marsupials (oppossums, kangaroos, wombats)
          Subclass Eutheria: Placental mammals
            Order Edentata: Anteaters, armadillos, sloths
            Order Lagomorpha: Rabbits, hares, pikas
            Order Rodentia: Squirrels, rats, woodchucks
            Order Insectivora: Hedgehogs, tenrecs, moles, shrews
            Order Carnivora: Dogs, wolves, cats, bears, weasels
            Order Pinnipeda: Sea lions, seals, walruses
            Order Cetacea: Whales, dolphins, porpoises
            Order Proboscidea: Elephants
            Order Perissodactyla: Horses, asses, zebras, tapirs, rhinoceroses
            Order Artidactyla: Swine, camels, deer, hippopotamuses, antelopes, cattle, sheep, goats
            Order Tubulidentia: Aardvarks
            Order Scandentia: Tree shrews
            Order Chiroptera: Bats
            Order Primates: Lemurs, monkeys, humans
              Suborder Strepsirhini: Prosimians (lemurs, aye-ayes)
              Suborder Haplorhini: Anthropoids
                Superfamily Tarsioidea: Tarsiers
                Superfamily Ceboidea: New World monkeys
                Superfamily Cercopithecoidea: Old World monkeys
                Superfamily Hominoidea
                    Family Hylobatidae: Gibbons, orangutans, chimpanzees
                    Family Pongidae: Gorillas
                    Family Hominidae: Humans

Glossary

abscisic acid
(ab sis′ik as′əd) (L. ab, away, off + scisso, dividing) A hormonal growth inhibitor that induces and maintains dormancy in plants. p. 541

acetyl CoA
(ə sēt′əl kō′ā) A compound combining the two-carbon acetyl fragment removed during the oxidation of pyruvate with the carrier molecule coenzyme A (CoA). p. 123

acetylcholine
(ə sēt′əl kō′lēn′) The neurotransmitter found at neuromuscular junctions that depolarizes the muscle cell membrane, releasing calcium ions that trigger muscle contraction. p. 257

acetylcholinesterase
(ə sēt′əl kō′lə nes′tə rās′) The enzyme that stops the action of acetylcholine; it is one of the fastest-acting enzymes in the blood. p. 257

acid
(as′əd) Any substance that dissociates to form H⁺ ions when it is dissolves in water. p. 26

acid rain
(as′əd rān) An acid precipitation that falls to the ground as rain, caused primarily by coal-fired plants' emission of sulfur dioxide and nitrogen dioxide, which then combines with water in the atmosphere; results in many devastating environmental effects. p. 705

acrosome
(ak′rə sōm) (Gr. akron, extremity + soma, body) A vesicle located at the tip of a sperm cell. p. 336

ACTH
(ā′sē′tē′āch′) See adrenocorticotropic hormone.

actin
(ak′tən) (Gr. actis, ray) A protein that makes up the thin myofilaments in a muscle fiber; provides support and helps determine cell shape and movement. p. 307

action potential
(ak′shən pəten′chəl) The rapid change in a membrane's electrical potential caused by the depolarization of a neuron to a certain threshold. p. 254

active immunity
(ak′tiv i myoon′ə tē) A type of immunity conferred by vaccination, which causes your body to build up antibodies against a particular disease without getting the disease. p. 227

active site
(ak′tiv sīt) The grooved or furrowed location on the surface of an enzyme where catalysis occurs. p. 105

active transport
(ak′tiv trans′pôrt) The movement of a solute across a membrane against the concentration gradient with the expenditure of chemical energy. This process requires the use of a transport protein specific to the molecule(s) being transported. p. 75

activator
(ak′tə vāt′ər) A chemical that binds to an enzyme and changes its shape so that catalysis can occur. p. 108

adaptation
(ad ap tā′shən) (L. adaptare, to fit) A naturally occurring inherited trait found within a population that makes an individual better suited to a particular environment and produce more offspring than an individual lacking this trait. p. 432

adaptive radiation
(ə dap′tiv rād ē ā′shən) The phenomenon by which a population of a species changes as it is dispersed within a series of different habitats within a region. p. 433

adenine
(ad′ən ēn′) An organic compound that is one of the two purine bases of RNA and DNA. p. 39

adenosine diphosphate (ADP)
(ə den′ə sēn′ dī fos′fāt) The molecule remaining from the breaking off of a phosphate group when adenosine triphosphate is used to drive an endergonic reaction. p. 111

adenosine triphosphate (ATP)
(ə den′ə sēn′ trī fos′fāt) A molecule composed of three subunits: ribose, adenine, and a triphosphate group; ATP captures energy in its high-energy bonds and later releases this energy. ATP is used to fuel a variety of cell processes and is the universal energy currency of all cells. p. 110

ADP
(ā′dē′pē′) See adenosine diphosphate.

adrenal cortex
(ə drē′nəl kôr′teks) (L. near, + ren, kidney; L. rind) The outer, yellowish portion of each adrenal gland that secretes a group of hormones known as corticosteroids in response to ACTH. p. 325

adrenal gland
(ə drē′nəl gland) (L. near, + ren, kidney) Either of two triangular glands, named for their position in the body, having two parts with two different functions, the adrenal cortex and the adrenal medulla. p. 325

adrenal medulla
(ə drē′nəl mə dul′ə) (L. near, + ren, kidney; L. marrow) The inner, reddish portion of each adrenal gland surrounded by the cortex that secretes the hormone adrenaline and noradrenaline. p. 325

adrenaline
(ə dren′əl ən) See epinephrine. p. 326

adrenocorticotropic hormone (ACTH)
(ə drē′nō kôrt′ə kō träp′ik hôr′mōn) (L. near, + ren, kidney + cortex, bark + Gr. tropikos, turning) A tropic hormone secreted by the anterior pituitary that triggers the adrenal cortex to produce certain steroid hormones. p. 320

aerobic
(er rō′bik) (Gr. aer, air + bios, life) Oxygen dependent. p. 116

aerobic respiration
(er rō′bik res′pə rā′shən) The process in which the original glucose molecule has been consumed entirely by the combined actions of glycolysis and the Krebs cycle.

afferent neuron
(af′ər ənt noor′on) See sensory neuron. p. 264

agranulocyte
(ā gran′yə lō sīt′) (Gr. a-, not + L. granulum, granule + Gr. kytos, cell) One of the two major groups of leukocytes; they have neither a cytoplasmic granule nor a lobed nuclei. p. 210

albumin
(al byoo′mən) (L. albumen, white of egg) A protein found in blood that serves to elevate the solute concentration of the blood to match that of the tissues so that the movement of water molecules is regulated. p. 74

G-1

aldosterone
(al dôs′tə rōn) A mineralocorticoid hormone produced by the adrenal cortex that regulates the level of sodium and potassium ions in the blood, thereby promoting the conservation of sodium and water and the excretion of potassium.   p. 244

alga, pl. algae
(al′gə, al′jē) (L. seaweed) A plant-like, photosynthetic eukaryotic organism that contains chlorophyll.   p. 503

allantois
(ə lan′tō əs) (Gr. allas, sausage + eidos, form) An extraembryonic membrane that gives rise to the umbilical arteries and vein as the umbilical cord develops.   p. 360

allele
(ə lēl′) (Gr. allelon, of one another) Each member of a factor pair containing information for an alternative form of a trait that occupy corresponding positions on paired chromosomes.   p. 378

allergen
(al′ər jən) (Gr. allos, other + ergon, work) Any substance that causes manifestations of allergy.   p. 230

all-or-nothing
(ôl′ ôr′ nuth′ing) Refers to the nerve impulse response; the amount of stimulation applied to the receptor or neuron must always be sufficient to open enough sodium channels to generate an action potential; otherwise, the cell membrane will simply return to the resting potential.   p. 255

alternation of generations
(ôl tər nā′shen uv jen ə rā′shənz) A type of life cycle that has both a multicellular haploid phase and a multicellular diploid phase.   p. 529

alveolus, pl. alveoli
(al vē′ə ləs, al vē′ə lī) (L. a small cavity) Microscopic air sacs in the lungs where oxygen enters the blood and carbon dioxide leaves.   p. 185

amino acid
(ə mē′nō as′əd) (Gr. Ammon, referring to the Egyptian sun god, near whose temple ammonium salts were first prepared from camel dung) A molecule containing an amino acid group (-NH$_2$), a carboxyl group (-COOH), a hydrogen atom, a carbon atom, and a functional group that differs among amino acids; an extremely diverse array of proteins is made from the 20 common amino acids.   p. 34

amniocentesis
(am′nē ō sen tē′səs) (Gr. amnion, membrane around the fetus + centes, puncture) A prenatal diagnostic procedure in which a sampling of amniotic fluid is obtained by insertion of a needle into the amniotic cavity and withdrawn into a syringe; the removed fetal cells are then grown in tissue culture and tests are performed to determine if genetic abnormalities are present.   p. 400

amnion
(am′nē ən) (Gr. membrane around the fetus) A thin, protective membrane that grows around the embryo during the third and fourth weeks, fully enclosing the embryo in a membranous sac.   p. 360

amniotic egg
(am′nē ot′ik eg) (Gr. membrane around the fetus) An egg, characteristic of reptiles, birds, and monotremes, which protects the embryo from drying out, nourishes it, and enables it to develop outside of water.   p. 468

amniotic fluid
(am′nē ot′ik floo′əd) (Gr. membrane around the fetus) A fluid in which the fetus floats and moves; it also helps keep the temperature constant for fetal development.   p. 360

amphibian
(am fib′ē ən) (Gr. amphibios, double life) An animal capable of living on land and in the water.   p. 466

amylase
(am′ə lās) (Gr. amylon, starch + asis, colloid enzyme) An enzyme that breaks down starches and glycogen to sugars.   p. 168

anaerobic
(an′er ō′bik) (Gr. an, without + aer, air + bios, life) Literally, without oxygen; any process that can occur without oxygen; includes glycolysis and fermentation.   p. 116

analagous
(ə nal′ə gəs) (Gr. proportionate) Describes a part or organ of an organism that has a similar form and function but possesses different evolutionary origins.   p. 439

anaphase
(an′ə fāz) (Gr. ana, up + phasis, form) The stage of mitosis characterized by the physical separation of sister chromatids and their movement to opposite poles of the cell.   p. 90

angiosperm
(an′jē ə sperm′) (Gr. angeion, vessel + sperma, seed) A vascular, flowering plant with protected seeds.   p. 534

annual ring
(an′yoo əl ring) A growth ring in a tree, the result of cambium cells dividing; one ring equals one year's growth and can thus be used to calculate the age of a tree.   p. 552

antagonists
(an tag′ə nəsts) (Gr. antagonizesthai, to struggle against) An opposing pair of skeletal muscles.   p. 305

anther
(anth′ər) (Gr. anthos, flower) The compartmentalized structure in the male stamen of a flower where haploid pollen grains are produced.   p. 534

antheridium, pl. antheridia
(an′thə rid′ē əm, an′thə rid′ē ə) (Gr. anthos, flower) The structure where sperm is produced in certain algae, the bryophytes, and several divisions of vascular plants.   p. 529

anthropoid
(an′thrə poid) (Gr. anthropos, man + eidos, form) A suborder of mammals that includes monkeys, apes, gorillas, chimpanzees, and humans; they differ from prosimians in the structure of the teeth, brain, skull, and limbs; they are diurnal, live in groups with complex social interactions, and care for their young for prolonged periods.   p. 460

antibody
(ant′i băd′ē) (Gr. anti, against) A protein produced in the blood by a B lymphocyte that specifically binds circulating antigen and marks cells or viruses bearing antigens for destruction.   p. 219

anticodon
(ant′i kō′don) (Gr. anti, against + L. code) A portion of a tRNA molecule with a sequence of three base pairs complementary to a specific mRNA codon.   p. 416

antidiuretic hormone (ADH)
(ant′i dī yoo ret′ik hôr′mōn) (Gr. anti, against + dia, intensive + ouresis, urination) A hormone produced by the hypothalamus but stored and released by the posterior lobe of the pituitary that helps control the volume of the blood by regulating the amount of water reabsorbed by the kidneys.   p. 244

antigen
(ant′i jən) (Gr. anti, against + genos, origin) A foreign substance inducing the formation of antibodies that specifically bind to the foreign substance, marking it for destruction.   p. 219

anus
(ā′nəs) The opening of the rectum for the elimination of feces.   p. 180

aorta
(ā ôrt′ə) (Gr. aeirein, to lift) The largest artery in the body, it carries the blood from the left side of the heart throughout the body except the lungs.   p. 205

aortic semilunar valve
(ā ôrt′ik sem′i loo′nər valv′) A one-way valve that permits blood flow from the left side of the heart and then snaps shut, preventing a backflow from the aorta to the heart.   p. 205

apical meristem
(ap′ə kəl *or* ā′pə kəl mer′ə stem) (L. apex, top + Gr. meristos, divided) A growth tissue at the tips of roots and the tips of shoots in a plant that allows a plant to grow taller and the roots to grow deeper into the ground. p. 548

appendicular skeleton
(ap′ən dik′yə lər skel′ə tən) (L. appendicula, a small appendage) The portion of the human skeleton, containing 126 bones, that consists of the bones of the appendages (arms and legs) and the bones that help attach the appendages to the axial skeleton. p. 299

appendix
(ə pen′diks) (L. appendere, to hang) A pouch that hangs from the beginning of the large intestine and serves no essential purpose in humans. p. 179

aqueous humor
(äk′wē əs hyōō′mər) (L. aqua, water + humor, fluid) A watery fluid filling a chamber behind the cornea that nourishes the cornea and the lens and, together with vitreous, creates a pressure within the eyeball maintaining the eyeball's shape. p. 287

archebacteria
(är′kē bak tir′ē ə) (Gr. arche, beginning + Gr. bakterion, rod) A taxonomic section consisting of bacteria that are chemically different in certain structures and metabolic processes from all other bacteria p. 450

archegonium, pl. archegonia
(är′kə gō′nē əm, är′kə gō′nē ə) (Gr. archegonos, first of a race) A structure in which eggs are formed in certain algae, the bryophytes, and several divisions of vascular plants. p. 529

arteriole
(är tir′ē ōl) A smaller artery that leads from an artery to a capillary. p. 201

artery, pl. arteries
(ärt′ə rē, ärt′ə rēz) (Gr. arteria, artery) A blood vessel that carries blood away from the heart and through the body. p. 201

articulation
(är tik′yə lā′shən) *See* joint.

artificial selection
(ärt′ə fish′əl sə lek′shən) The process of selecting organisms for a desirable characteristic and then breeding that organism with another organism exhibiting the same trait. p. 429

asexual reproduction
(ā′seksh′ə wəl *or* ā′sek′shəl rē′prə duk′shən) A type of reproduction in which a parent organism divides by mitosis and produces two identical organisms; does not introduce variation. p. 85

association area
(ə sō′sē ā′shən er′ē ə) The area in the brain that connects all parts of the cerebral cortex and appears to be the site of higher cognitive activities such as memory, reasoning, intelligence, and personality. p. 266

aster
(as′tər) In animal mitosis, an array of microtubules that radiates outward from the centrioles when the centrioles reach the poles of the cell. p. 89

atherosclerosis
(ath′ə rō sklə rō səs) (Gr. athere, porridge + sklerosis, hardness) A disease in which the inner walls of the arteries accumulate fat deposits, narrowing the passageways and leading to elevated systolic blood pressure. p. 207

atom
(at′əm) (Gr. atomos, indivisible) A core (nucleus) of protons and neutrons surrounded by orbiting electrons. The electrons largely determine the chemical properties of an atom. p. 17

atomic mass
(ə täm′ik mas) The combined mass of all the protons and neutrons of an atom without regard to its electrons. p. 17

atomic number
(ə täm′ik num′bər) The number of protons in an atom; the atomic number is also the same as the number of electrons in that atom. p. 17

ATP
(ā′tē′pē′) *See* adenosine triphosphate.

atrioventricular (AV) node
(ā′trē ō ven trik′yə lər nōd) A group of specialized cardiac muscle cells that receives the impulse initiated by the sinoatrial node and conducts them by way of the bundle of His. p. 207

atrium, pl. atria
(ā′trē əm, ā′trē ə) (L. main room) The upper chamber of each half of the heart; the right atrium receives deoxygenated blood from the body (except the lungs), and the left atrium receives oxygenated blood from the lungs through the pulmonary veins. p. 204

auditory canal
(ôd′ə tōr′ē kə nal′) (L. audire, to hear) A 1-inch long canal that receives sound waves and leads directly to the eardrum. p. 289

Australopithecus afarensis
(ôs′trə lō pith′ə kəs af′ə ren′səs) (L. australis, southern + Gr. pithekos, ape; L. Afar region of Ethiopia) A species of australopithecines considered to be the first hominids, although not human (members of the genus *Homo*); its oldest fossil skeleton is 3.5 million years old. p. 477

autonomic nervous system
(ô′tə nom′ik nurv′əs sis′təm) (Gr. autos, self + nomos, law) A branch of the peripheral nervous system consisting of motor neurons that control the involuntary and automatic responses of the glands and the nonskeletal muscles of the body. p. 264

autosome
(ô′tə sōm′) (Gr. autos, self + soma, body) Any of the 22 pairs of human chromosomes that carry the majority of the genetic information but have no genes that determine gender. p. 81

autotroph
(ô′tə trōf) (Gr. autos, self + trophos, feeder) Self-feeder; an organism that produces its own food by photosynthesis; all organisms live on food produced by autotrophs, including the autotrophs themselves. p. 131

auxin
(ôk′sən) (Gr. auxein, to increase) A hormone that promotes growth through cell elongation and controls phototropism and geotropism in plants. p. 541

AV node
(ā′vē′ nōd′) *See* atrioventricular (AV) node.

axial skeleton
(ak′sē əl skel′ə tən) The central axis of the human skeleton, consisting of 80 bones, that includes the skull, vertebral column, and rib cage. p. 299

axon
(ak′son) (Gr. axle) A single projection extending from a neuron that conducts impulses away from the cell body. p. 250

B lymphocyte (B cell)
(bē′ lim′fə sīt′) A type of lymphocyte that matures in human bone marrow and secretes antibodies during the humoral immune response; named after a digestive organ in birds (the bursa of Fabricius) in which these lymphocytes were discovered. p. 223

bacteriophage (phage)
(bak tir′ē ə fāj) (Gr. bakterion, little rod + phagein, to eat) A virus that infects a bacterial cell. p. 489

bacterium, pl. bacteria
(bak tir′ē əm, bak tir′ē ə) (Gr. bakterion, a little rod) The oldest, simplest, and most abundant organism; it is the only organism with a prokaryotic cellular organization. p. 45

bark
(bärk) The outer protective covering of a woody plant or tree that consists of the

cork, the cork cambium, and the parenchyma cells. p. 552

basal body
(bā′səl bäd′ē) A structure composed of microtubules that serves to anchor a cilium or flagellum to the cell. p. 61

basal ganglion, pl. ganglia or ganglions
(bā′səl gang′glē ən, gang′glē ə *or* gang′glē ənz) Groups of nerve cell bodies located at the base of the cerebrum that control large, subconscious movements of the skeletal muscles, p. 267

base
(bās) Any substance that combines with H^+ ions; a substance that has a pH value higher than water's neutral value of 7. p. 26

basidium, pl. basidia
(bə sid′ē əm, bə sid′ē ə) (Gr. basis, base) In a club fungus, a club-shaped structure from which its sexual spores hang. p. 521

basophil
(bā′sə fil) (Gr. basis, base + philein, to love) A type of granulocyte containing granules that rupture and release chemicals enhancing the body's response to injury or infection; they also play a role in causing allergic responses. p. 210

behavior
(bə hā′ vyər) The range of the pattern of movement, sound (vocalization), and body position (posture) exhibited by an animal; it also embraces competitive, social, submissive, territorial, and reproductive behavior patterns. p. 599

bilateral symmetry
(bī lat′ə rəl sim′ə trē) (L. bi, two + lateris, side; Gr. symmetria, symmetry) Describing an object or an organism whose right side is a mirror image of its left side. p. 562

bile
(bīl) (L. bilis) A collection of molecules secreted by the liver than helps in the digestion of lipids. p. 176

binary fission
(bī′nə rē fish′ən *or* fizh′ən) (L. binarius, consisting of two things or parts + fissus, split) Asexual reproduction in which one cell divides into two with no exchange of genetic material among cells; bacteria use this process. p. 491

binomial nomenclature
(bī nō′mē əl nō′mən klā′chər or nō men′klə chər) (L. bi, twice, + Gr. nomos, law; L. nomen, name + calare, to call) The system of using the last two categories in the hierarchy of classification, genus and species, to provide the scientific name of an organism, usually written in Latin. p. 14

biome
(bī′ōm) (Gr. bios, life + -oma, group) A distinctive, broad, terrestrial ecosystem of plants and animals that occurs over wide areas within specific climatic regions. p. 671

biosphere
(bī′ə sfir) (Gr. bios, life + sphaira, ball) The global ecosystem of life on Earth that extends from the tops of the tallest mountains to the depths of the deepest seas. p. 695

blade
(blād) The flattened portion of a leaf. p. 511

blastocyst
(blas′tə sist) (Gr. blastos, germ + kytos, cell) The stage of development following the morula in which the embryo is a hollow ball of cells. p. 357

blastula
(blas′chə lə) (Gr. a little sprout) The general term used to describe the sac-like blastocyst of mammals and, in other animals, the stage that develops a similar fluid-filled cavity. p. 357

blood pressure
(blud presh′ər) The pressure, determined indirectly, existing in the large arteries at the height of the pulse wave; the systolic intraarterial pressure; normal blood pressures values are 70 to 90 (mm of Hg) diastolic and 110 to 130 systolic. p. 207

bone
(bōn) (A.S. ban, bone) A hard material that forms the vertebrate skeleton; composed of collagen fibers that contribute flexibility and needle-shaped crystals of calcium that impart rigidity. p. 296

bony fish
(bōn′ē fish) A kind of fish that lives in both salt and fresh water and derives its name from its bony internal skeleton; by means of gas exchange between its swim bladder and its blood, a body fish regulates its buoyancy. p. 586

Bowman's capsule
(bō′mənz kap′səl *or* kap′sool) (after Sir William Bowman, British physician) An apparatus at the front end of each nephron tube in a kidney that functions as a filter in the formation of urine. p. 235

brain
(brān) That part of the central nervous system comprised of four main parts: the cerebrum, the cerebellum, the diencephalon, and the brainstem. p. 250

brainstem
(brān′stem) The part of the brain consisting of the midbrain, pons, and medulla that brings messages to and from the spinal cord and controls important body reflexes such as the rhythm of the heartbeat and rate of breathing. p. 265

bronchiole
(brong′kē ōl) (L. bronchiolus, air passage) The last division of bronchi, having thousands of tiny air passageways, whose walls have clusters of tiny pouches, or alveoli. p. 188

bronchus, pl. bronchi
(brong′kəs, brong′kī) (Gr. bronchos, windpipe) One of a pair of airway structures that branches from the lower end of the trachea into each lung, p. 188

brown algae
(broun al′jē) Large, multicellular algae found predominantly on northern, rocky shores, some growing to enormous sizes; kelp is a brown algae. p. 509

bryophyte
(brī′ə fīt) (Gr. bryon, moss + phyton, plant) A single division of small, low-growing plants that are commonly found in moist places, including the mosses, liverworts, and hornworts. p. 529

bulbourethral glands
(bul′bō yoo rē′thrəl glandz) (L. bulbus, bulbous root + Gr. ourethra, urethra) A set of tiny accessory glands lying beneath the prostate that secretes an alkaline fluid into the semen. p. 338

bundle of His
(bun′dəl uv his) (after Wilhelm His, Jr, German physician) A strand of impulse-conducing muscle that conducts heartbeat impulses from the right atrium to the ventricles of the heart. p. 207

C_3 photosynthesis
(sē′thrē′ fō′tō sin′thə səs) A process that uses the normal Calvin-Benson process to fix carbon. p. 141

C_4 photosynthesis
(sē′fôr′ fō′tō sin′thə səs) In hot climates, a process by which plants deal with the problem of CO_2 being released by photorespiration by concentrating CO_2 within the cells where carbon fixation takes place. p. 142

calcitonin (CT)
(kal′sə tō′nən) (L. calcem, lime) A hormone secreted by the thyroid that works with parathyroid hormone to regulate the concentration of calcium in the bloodstream. p. 322

calorie
(kal′ə rē) (L. calor, heat) The measurement of the unit of energy in food. p. 166

Calvin-Benson cycle
(kal′vən ben′sən sī′kəl) (after Calvin and

Benson, American chemists) *See* light-independent reactions.

cambium
(kam′bē əm) (L. combiare, to exchange) Lateral meristem tissue in a dicot with a woody stem or trunk that consists of vascular cambium and cork cambium. p. 552

capillary
(kap′ə ler′ē) (L. capillaris, hair-like) A microscopic blood vessel with a wall only one cell thick that connects the end of an arteriole with the beginning of a venule; the site where gases are exchanged between the blood and tissues, nutrients delivered, and wastes picked up. p. 201

capsid
(kap′sid) (L. capsa, box) A protein covering that covers the nucleic acid core of a virus. p. 487

carbohydrate
(kär′bō hī′drāt) (L. carbo, charcoal + hydro, water) A molecule that contains carbon, hydrogen, and oxygen, with the concentration of hydrogen and oxygen atoms in a 2:1 ratio. p. 30

carbon fixation
(kär′bən fik sā′shən) A process by which organisms use the ATP and NADPH produced by photosynthetic reactions to build organic molecules from atmospheric carbon dioxide. p. 141

carcinogen
(kär sin′ə jən) (Gr. karkinos, cancer + -gen) Any cancer-causing agent. p. 490

cardiac muscle
(kärd′ē ak′ mus′əl) A type of striated muscle fiber arranged in a special way critical for cardiac function; cardiac muscle makes up the heart, acting as a pump for the circulatory system. p. 158

carnivore
(kär′nə vôr) (L. carnivorous, flesh eating) An animal that eats meat. p. 170

carpal
(kär′pəl) (Gr. karpos, wrist) Any of the eight short bones, lined up in two rows of four, that make up the wrist. p. 300

carrying capacity
(ker′ē′ing kə pas′ət ē) The number of individuals within a population that can be supported within a particular environment for an indefinite period. p. 627

cartilage
(kärt′əl ij *or* kärt′lij) (L. cartilago, gristle) A specialized connective tissue that is hard and strong; composed of chondrocytes that secret a matrix consisting of a semisolid gel and fibers. Laid down in long parallel arrays, the result is a firm and flexible tissue that does not stretch. p. 153

cartilaginous fish
(kär′tə laj′ə nəs fish) (L. cartilago, gristle) A marine fish such as a shark, skate, or ray having denticles that cover their skin; many sharks are predators, whereas skates and rays feed on invertebrates that dwell on the ocean floor. p. 584

catabolic
(kat′ə bäl′ik) (Gr. katabole, throwing down) Referring to a process in which complex molecules are broken down into simpler ones. p. 121

catalyst
(kat′əl əst) (Gr. kata, down + lysis, a loosening) A substance that increases the rate of a chemical reaction but is not chemically changed by the reaction; an enzyme is a catalyst. p. 105

cell
(sel′) (L. cella, a chamber or small room) A membrane-bounded unit containing hereditary material, cytoplasm, and organelles; a cell can release energy from fuel and use that energy to grow and reproduce. p. 47

cell body
(sel bäd′ē) The body of a nerve cell or neuron, which contains a nucleus and other cell organelles and two types of cellular projections, axons and dendrites. p. 252

cell cycle
(sel sī′kəl) The life of a cell; the major portion concerned with the activitites of interphase and the minority portion with mitosis. p. 86

cell-mediated immune response
(sel′ mēd′ē āt′əd ri spons′) A chain of events unleashed by T cells during which cytotoxic T cells attack infected body cells. p. 222

cell membrane
(sel′ mem′brān′) A bilayer of phospholipid molecules studded with proteins to which carbohydrates are attached. The membrane proteins allow the cell to interact with the environment and perform various cellular functions. p. 67

cell plate
(sel′ plāt′) During cytokinesis in plants and some algae, the formation of a partition, which begins by the plants manufacturing new membrane sections that then accumulate at the metaphase plate and fuse. p. 92

cell theory
(sel′ thē′ə rē *or* thir′ē) A statement first formulated in the mid-1800s, which states that all living things are made up of cells, that cells are the smallest living units of life, and that all cells arise from preexisting cells. p. 45

cell wall
(sel′ wôl) The structure that surrounds the plasma membrane in cells. In many plants, algae, and fungi the wall is rigid and composed of cellulose and imparts a stiffness to the tissues; in single-celled organisms, cell walls give shape to the organisms and help protect them. p. 61

cellular respiration
(sel′yə lər res′pə rā′shən) The complex series of chemical reactions—glycolysis, the Krebs cycle, and the electron transport chain—by which cells break down fuel molecules to release energy, using oxygen and producing carbon dioxide. p. 116

cellulose
(sel′yə lōs) (L. cellula, little cell) A polysaccharide that is the chief component of plant cell walls; because cellulose cannot readily be broken down, it works well as a biological structural material and occurs widely in plants. p. 32

central nervous system
(sen′trəl nur′vəs sis′təm) The site of information processing within the nervous system, comprising the brain and spinal cord. p. 264

centriole
(sen′trē ōl) (Gr. kentron, center of a circle + L. olus, little one) An organelle surrounded by microtubule-organizing centers that play a key role in the mitotic process; most plants and fungi lack centrioles. p. 86

centromere
(sen′trə mir) (Gr. kentron, center + meros, a part) A constricted region of the chromosome where two identical structures (sister chromatids) are joined. p. 81

cerebellum
(ser′ə bel′əm) (L. little brain) The part of the brain located below the occipital lobes of the cerebrum that coordinates subconscious movements of the skeletal muscles. p. 265

cerebral cortex
(sə rē′brəl kôr′teks) The thin layer of tissue, gray matter, that forms the outer layer of the cerebrum; the major site of higher cognitive processes such as sense perception, thinking, learning, and memory. p. 266

cerebrospinal fluid
(sə rē′brō spī′nəl flōō′əd) A liquid that cushions the brain and the spinal cord, acting like a shock absorber. p. 272

cerebrum
(sə rē′brəm) (L. brain) The largest and most dominant part of the human brain, divided into two hemispheres connected by the corpus callosum. p. 265

cervix
(sur′viks) (L. neck) The narrower part of the uterus that opens into the vagina. p. 343

chemoautotroph
(kē′mō ô′tə trōf) (Gr. chemeia, chemistry + auto, self + trophos feeder) A eubacterium that makes its own food by deriving energy from inorganic molecules. p. 492

chiasma, pl. chiasmata
(kī az′mə, kī az′mə tə) (Gr. a cross) In meiosis, the point of crossing-over where parts of chromosomes have been exchanged during synapsis; under a light microscope, a chiasma appears as an X-shaped structure. p. 93

chitin
(kīt′ən) (Gr. chiton, tunic) A modified form of cellulose, relatively indigestible, that is the structural material in insects and many fungi. p. 29

chlorophyll
(klôr′ə fil) (Gr. chloros, green + phyllon, leaf) The pigment that absorb photons of green, blue, and violet wavelengths; chlorophyll is the primary light gatherer in all plants and algae and in almost all photosynthetic bacteria. p. 133

chloroplast
(klôr′ə plast) (Gr. chloros, green + plastos, molded) An energy-producing organelle found in the cells of plants and algae; the site of photosynthesis in plants. p. 59

cholecystokinin (CCK)
(kol′ə sis′tə kī′nən) (Gr. chole, bile + kystis, bladder + kinein, to move) A hormone that helps control digestion in the small intestine. p. 177

chondrocyte
(kon′drə sīt) (Gr. chondros, cartilage + kytos, cell) A cell that produces cartilage, a specialized connective tissue that is hard and strong. p. 153

chordate
(kôr′dāt) (Gr. chorde, cord) An organism distinguished by three principal features: a nerve cord, a notochord, and pharyngeal slits; includes fishes, amphibians, reptiles, birds, mammals, and humans. p. 454

chorion
(kôr′ə on) An extraembryonic membrane that facilitates the transfer of nutrients, gases, and wastes between the embryo and the mother's body. p. 360

choroid
(kôr′oid′) (Gr. chorioeides, skin-like) A thin, dark-brown membrane that lines the sclera, containing blood-carrying vessels that nourish the retina and a dark pigment that absorbs light rays so that they will not be reflected within the eyeball. p. 288

chromatin
(krō′mə tən) (Gr. chroma, color) The complex of proteins and the hereditary material, DNA, which make up the chromosomes of eukaryotes. p. 81

chromosome
(krō′mə sōm′) (Gr. chroma, color + soma, body) A discrete, thread-like body that forms as the nuclear material condenses during meiosis and that carries genetic information. p. 57

chyme
(kīm) (Gr. chymos, juice) The mixture of partly digested food, gastric juice, and mucus, having the consistency of pea soup, found in the stomach during digestion. p. 175

ciliary muscle
(sil′ē er′ē mus′əl) (L. ciliaris, pert. to eyelid) A tiny circular muscle that slightly changes the shape of the lens by contracting or relaxing. p. 287

ciliate
(sil′ē ət or sil′ē āt′) (L. cilium, eyelash) A protozoan characterized by fine, short, hairline cellular extensions called cilia. p. 507

cilium, pl. cilia
(sil′ē əm, sil′ē ə) (L. eyelash) A whip-like organelle of motility that protrudes from some eukaryotic cells. p. 60

citric acid cycle
(sit′rək as′əd sī′kəl) See Krebs cycle.

cleavage
(klē′vij) The process of cell division that occurs without cell growth. p. 356

cleavage furrow
(klē′vij fur′ō) During cytokinesis in animal cells, the area where a belt of microfilaments pinches the cell at the metaphase plate. p. 91

climax community
(klī′maks′ kə myoo̅′nət ē) A community in which the mix of plants and animals becomes stable; the last stage of succession. p. 652

clitellum, pl. clitella
(klī tel′əm, klī tel′ə) (L. clitellae, a packsaddle) A thickened band on an earthworm's body that secretes a mucus holding the worms together as they exchange sperm. p. 574

clitoris
(klit′ə rəs) (Gr. kleitoris) A small mass of erectile and nervous tissue in the female genitalia that responds to sexual stimulation; it is homologous to the tip of the penis in males. p. 346

cloaca
(klō ā′kə) (L. sewer) In an aquatic animal, the terminal part of the gut into which ducts from the kidney and reproductive systems open. p. 586

clone
(klōn) (Gr. klon, a cutting used for propagation) A group of identical cells that arises by repeated mitotic divisions from one original cell. p. 224

cnidarian
(nī der′ē ən) (Gr. knide, sea nettle) A member of the phylum Cnidaria, which includes jellyfish, hydra, sea anemones, and corals; they have radially symmetrical bodies. p. 566

cochlea
(kok′lē ə) (Gr. kokhlos, land snail) A winding, cone-shaped tube forming a portion of the inner ear that contains the organ of Corti, the organ of hearing. p. 289

cochlear duct
(kok′lē ər dukt) The inner tube of the cochlea containing specialized cells that are the receptors of hearing. p. 289

codominant
(kō′däm′ə nənt) Refers to traits in which the alternative forms of an allele are both dominant. p. 400

codon
(kō′dän) (L. code) A sequence of three nucleotide bases in transcribed mRNA that code for an amino acid, which is the building block of a polypeptide. p. 416

coelom
(sē′ləm) (Gr. koilos, a hollow) A body cavity that is a fluid-filled enclosure within a bilaterally symmetrical organism. p. 153

coenzyme
(kō′en′zīm) (L. co-, together + Gr. en, in + zyme, leaven) A cofactor that is a nonprotein organic (carbon-containing) molecule helping an enzyme catalyze a chemical reaction. p. 110

coevolution
(kō′ev′ə loo̅′shən) (L. co-, together + e-, out + volvere, to fill) Interactions that involve the long-term, mutual evolutionary adjustment of the characteristics of the members of biological communities in relation to one another. p. 457

cofactor
(kō′fak′tər) A special nonprotein molecule that helps an enzyme catalyze a chemical reaction. p. 110

collagen
(käl′ə jən) (Gr. kolla, glue + gennan, to produce) The most abundant protein in the human body, whose fibers are strong and wavy; found in the connective tissue of skin, bone, and cartilage. p. 34

collar cell
(käl′ər sel) A specialized, flagellated cell that lines the inside cavity of the body of a sponge. p. 565

collecting duct
(kə lek′ting dukt) A small duct that receives urine from the kidney nephrons. p. 235

colon
(kō′lən) (Gr. kolon) The large intestine from the small intestine to the rectum; its function is to absorb sodium and water, to eliminate wastes, and to provide a home for friendly bacteria. p. 179

commensalism
(kə men′sə liz′əm) (L. cum, together with + mensa, table) A type of symbiotic relationship in which one species benefits and the other species neither benefits nor is harmed. p. 649

community
(kə myōō′nət ē) (L. communitas, community, fellowship) The various populations of organisms that live together in a particular place. p. 639

compact bone
(käm′pakt′ bōn) A type of bone in the human skeleton that runs the length of long bones and has no spaces within its structure visible to the naked eye; it is hard and dense, which gives the bone the strength to withstand mechanical stress. p. 156

companion cell
(kəm pan′yən sel′) A specialized cell in a vascular plant that actively transports fluid to and from a phloem cell. p. 546

competition
(käm′pə tish′ən) The striving by organisms of different species that live near one another to obtain the same limited resources. p. 640

complement
(käm′plə mənt) A group of proteins that kill foreign cells by creating a hole in their membranes. p. 224

compound
(käm′pound′) A molecule made up of the atoms of two or more elements. p. 18

compound eye
(käm′pound′ ī) A structure that is composed of many independent visual units, each containing a lens; an important structure of many arthropods such as bees, flies, moths, and grasshoppers. p. 576

cone
(kōn) A light receptor located within the retina at the back of the eye that functions in bright light and detects color. p. 286

conjugation
(kän′jə gā′shən) (L. conjugare, to yoke together) A method of genetic recombination in bacteria during which a donor and a recipient cell make contact and the DNA from the donor is injected into the recipient cell. p. 494

connective tissue
(kə nek′tiv tish′ōō) A collection of tissues and its cells that provides a framework for the body, joins its tissues, helps defend it from foreign invaders, and acts as storage sites for specific substances. p. 147

consumer
(kən sōōm′ər) An organism in an ecosystem that feeds on producers and other consumers, passing energy along that was once captured from the sun. p. 657

control
(kən trōl′) In a scientific experiment, a standard against which observations or conclusions may be checked to establish their validity. p. 5

conus arteriosus
(kō′nəs är tir′ē ō′səs) A chamber in the heart of cartilaginous and bony fishes that pumps blood to the gills. p. 586

convergent evolution
(kən vur′jənt ev′ə lōō′shən) The changes over time among different species of organisms having different ancestors that result in similar structures and adaptations. p. 439

coral
(kôr′əl) A general term that refers to a variety of invertebrate animals of the phylum Cnidaria. p. 561

coral reef
(kôr′əl rēf) A reef built by the marine animals corals (phylum Cnidaria) that secrete carbonate, a hard, shell-like substance; coral reefs are complex ecosystems that provide habitat for a variety of invertebrates and fish. p. 688

cork cambium
(kôrk′ kam′bē əm) A lateral meristem tissue whose outermost cells produce cork cells that are waterproof and decay resistant; on maturity, the cork cells lose their cytoplasms and become hardened. p. 552

cornea
(kôr′nē ə) (L. corneus, horny) The transparent portion of the eye's outer layer that permits light to enter the eye. p. 287

corpus callosum
(kôr′pəs kə lō′səm) (N.L. callous body) The single, thick bundle of nerve fibers that connects the two hemispheres of the cerebrum in humans and primates. p. 266

corpus luteum
(kôr′pəs lōō′tē əm) (N.L. yellow body) A structure that emerges from a ruptured follicle in the ovary after ovulation; it secretes increased estrogen and progesterone, preparing the endometrium for the implantation of the fertilized egg. p. 341

cortex
(kôr′teks) (L. rind, bark) (1) The outer layer of an organ as distinguished from the inner, as in the cerebral cortex. (2) The outer superficial portion of the root of a vascular plant. p. 235

cotyledon
(kät′əl ēd′ən) (Gr. kotyledon, a cup-shaped hollow) A seed leaf; in a dicot, the place where most of the food is stored for the developing embryo. p. 538

coupled channel
(kup′əld chan′əl) A type of channel that has binding sites on one membrane transport protein for both sodium (Na^+) and potassium (K^+) ions, which are transported across the cell membrane in opposite directions. p. 75

coupled reaction
(kup′əld rē ak′shən) A reaction in which the energy released in an exergonic reaction is used to drive an endergonic reaction. p. 102

covalent bond
(kō vā′lənt bänd) (L. co-, together + valare, to be strong) A chemical bond created by atoms sharing one or more pairs of electrons. p. 19

cranial nerves
(krā′nē əl nurvz) (Gr. kranion, skull) Any of the twelve pairs of nerves that enter the brain through holes in the skull. p. 275

creatinine
(krē at′ə nēn) (Gr. kreas, flesh) A nitrogenous waste found in urine, derived primarily from creatinine found in muscle cells. p. 239

cri du chat syndrome
(krē dōō shä sin′drōm) A congenital disorder so named because the infant's cry resembles that of a cat; other symptoms include severe mental retardation and a moon face; caused by a deletion located on one of the 5 chromosomes. p. 393

crista, pl. cristae
(kris′tə, kris′tē) (L. crest) The enfoldings of the inner membrane of a mitochondrion. p. 58

Cro-Magnon
(krō mag′nən) (after a cave in southwestern France) An early member of *H. sapiens sapiens* whose anatomical features were more similar to modern humans; they used sophisticated tools, hunted, and made elaborate cave paintings of animals and hunt scenes. p. 481

crossing-over
(krôs′ing ō′vər) An essential element of meiosis, which produces sister chromatids that are not identical with

each other, resulting in new combinations of genes. p. 93

cuboidal
(kyōō boid'əl) (Gr. kubos, cube, + eidos, form) One of the main shapes of epithelial cells. Cuboidal cells have complex shapes but look like cubes when the tissue is cut at right angles to the surface; found lining tubules in the kidney and the ducts of glands. p. 150

cyanobacteria
(sī'ə nō bak tir'ē ə) (Gr. kyanos, dark-blue + bakterion, dim. of baktron, a staff) Formerly blue-green algae; a group of photosynthetic bacteria that played a key role in the evolution of life by gradually oxygenating the atmosphere and the oceans around 2 billion years ago. p. 450

cyclic adenosine monophosphate
(sī'klik *or* sik'lik ə den'ə sēn' mon'ə fos'fāt) A cousin of ATP that triggers enzymes causing a cell to alter its functioning in response to a hormone. p. 59

cyclic electron flow
(sī'klik *or* sik'lik i lek'tron flō) A system that harvests energy in a process involving Photosystem I working independently of Photosystem II. This method of generating energy produces ATP but does not directly produce NADPH as noncyclic electron flow does. Bacteria were the first organisms to use this method. p. 140

cystic fibrosis
(sis'tik fī brō'səs) (Gr. kystis, bladder + L. fibra, fiber + osis, condition) The most common fatal genetic disease of Caucasians in which affected individuals secrete a thick mucus that clogs the airways of the lungs and the passages of the pancreas and liver. p. 399

cytokinesis
(sī'tō kə nē'səs) (Gr. kytos, hollow vessel + kinesis, movement) The physical division of the cytoplasm of a eukaryotic cell into two daughter cells. p. 90

cytoplasm
(sī'tə plaz'əm) (Gr. kytos, hollow vessel + plasma, anything molded) The viscous or gel-like fluid within a cell that contains storage substances, a network of interconnected filaments and fibers, and cell organelles. p. 50

cytoskeleton
(sī'tō skel'ə tən) (Gr. kytos, hollow vessel + skeleton, a dried body) A network of filaments and fibers within the cytoplasm that helps maintain the shape of the cell, move substances within cells, and anchor various structures in place. p. 48

cytotoxic T cell
(sī'tə tôk'sik tē sel) (Gr. kytos, cell + toxikon, poison) A type of T cell that breaks apart cells infected by viruses and foreign cells such as incompatible organ transplants. p. 222

deamination
(dē am'ə nā'shən) The process in which an amino compound loses a (-NH$_2$) group; takes place in the human liver. p. 234

deciduous
(də sij'yōō əs) (L. decidere, to fall off) A tree that drops its leaves and remains dormant throughout the winter. p. 682

decomposer
(dē'kəm pō'zər) An organism in an ecosystem that breaks down organic molecules of dead organisms and contributes to the recycling of nutrients to the environment. p. 657

dehydration synthesis
(dē'hī drā'shən sin'thə səs) The process by which monomers are put together to form polymers. p. 28

deletion
(dē lē'shən) Refers to an abnormally short chromosome that has lost a chromosomal section. p. 393

dendrite
(den'drīt) (Gr. dendron, tree) A projection extending from a neuron that acts as an antenna for the reception of nerve impulses and conducts these impulses toward the cell body. p. 250

density
(den'sə tē) The number of organisms or individuals in a population per unit of area. p. 628

denticle
(den'tə kəl) (L. denticulus, dim. of dentem, tooth) A small, pointed, tooth-like scale that covers the skin of cartilaginous fishes and gives the skin a sandpaper-like texture. p. 585

deoxyribonucleic acid (DNA)
(de ok'sə rī'bō nōō klē'ik as'əd) A nucleic acid present in the chromosomes that is the chemical basis of heredity and the carrier of genetic information; arranged as two long chains that twist around each other to form a double helix. p. 56

depolarization
(dē pō'lə rə zā'shən) The change in electrical potential of a receptor cell or nerve cell membrane. p. 254

dermal tissue
(dur'məl tish'ōō) The outer protective covering of all plants with the exception of woody shrubs and trees, which have bark. p. 546

detritus
(di trīt'əs) (L. worn down) The refuse or waste material of an ecosystem. p. 659

deuterostome
(dōōt'ə rə stōm') (Gr. deutero, second + stoma, mouth) A member of a branch of coelomate animals that during the gastrula stage of development, gives rise to a two-layered embryo; the first indentation becomes the anus and the second indentation becomes the mouth. p. 564

diabetes
(dī'ə bēt'ēz *or* dī'ə bēt'əs) (Gr. passing through) A set of disorders characterized by a high level of glucose in the blood, the underlying cause being the lack or partial lack of insulin; in type I or juvenile onset diabetes, a person has no effective insulin; in type II or maturity onset diabetes, a person has some effective insulin but not enough to meet body needs. p. 328

dialysis
(dī al'ə səs) (Gr. dia, through + lysis, dissolution) The filtering of blood through a selectively permeable membrane to remove toxic wastes, regulate blood pH, and regulate ion concentration; used in renal failure. p. 247

diastolic period
(dī'ə stol'ik pir'ē əd) (Gr. diastole, expansion) The period during the first part of a heartbeat when the atria are filling; at this time the pressure in the arteries leading from the left side of the heart to the tissues of the body decreases slightly as the blood moves out of the arteries, through the vascular system, and into the atria. p. 207

diatom
(dī'ə tom) (Gr. diatemnein, to cut through) A type of microscopic golden algae that possesses a siliceous or calcium-containing cell wall; fossil diatoms are sometimes mined commercially as diatomaceous earth. p. 503

diatomic molecule
(dī'ə tom'ik môl'ə kyōōl') A molecule having two atoms. p. 21

dicot
(dī'kot) An angiosperm in which the food reserves of its seeds is stored in the cotyledons, or seed leaves; a dicot usually has net-like veins in its leaves and its flower parts are in fours or fives. p. 538

diencephalon
(dī'ən sef'ə lôn) (Gr. dia, between + enkephalon, within the head) The part of the brain consisting of the thalamus and hypothalamus. p. 265

diffusion
(di fyōō'zhən) (L. diffundere, to pour

out) The net movement of molecules from a region of higher concentration to a region of lower concentration, eventually resulting in a uniform distribution of the molecules. This movement is the result of random, spontaneous molecular motions. p. 72

dihybrid
(dī hī′brəd) (Gr. dis, twice + L. hybrida, mongrel) The product of two plants that differ from one another in two traits. p. 380

dinoflagellate
(dī′nō flaj′ə lət *or* dī′nō flaj′ə lāt) (Gr. dinos, rotation + N.L. flagellum, whip) A type of flagellated unicellular red algae, characterized by stiff outer coverings; its flagella beat in two grooves, one encircling the cell like a belt and the other perpendicular to it. p. 509

diploid
(dip′loid) (Gr. diploos, double + eidos, form) A cell that contains double the haploid amount; a double set of the genetic information. p. 83

disaccharide
(dī sak′ə rīd) (Gr. dis, twice + sakcharon, sugar) Two monosaccharides linked together; sucrose (table sugar) is a disaccharide formed by linking a molecule of glucose to a molecule of fructose. p. 30

distal convoluted tubule
(dis′təl kôn′və lōō′təd tōō′byōōl) The portion of the kidney tubule whose walls are permeable to water. p. 239

DNA
(dē′ en′ ā) *See* deoxyribonucleic acid.

dominant
(dôm′ə nənt) The form of a trait that will be expressed in a hybrid offspring. p. 375

Down syndrome
(doun sin′drōm) (after J. Langdon Down, British physician) A genetic disorder produced when an individual receives three (instead of two) 21 chromosomes; also called trisomy 21. p. 389

duodenum
(dōō′ə dē′nəm *or* dōō äd′ən əm) (L. duodeni, twelve each; with reference to its length, about twelve finger breadths) The initial, short segment of the small intestine that is actively involved in digestion. p. 175

duplication
(dōō′pli kā′shən) A chromosomal abnormality in which a section of a chromosome has been duplicated; in a karyotype, the duplicated chromosome appears longer than its homologue. p. 393

ecology
(i kôl′ə jē *or* ē kôl′ə jē) (Gr. oikos, house + logos, word) The special field of biology that studies the interactions among organisms and between organisms and their environments in the laboratory and in nature. p. 690

ecosystem
(ē′kō sis′təm) (Gr. oikos, house + systema, that which is put together) A community of plants, animals, and microorganisms that interacts with one another and with their environments and is interdependent on one another for survival. p. 639

ectoderm
(ek′tə durm′) (Gr. ectos, outside + derma, skin) The outer layer of cells formed during the development of embryos; forms the outer layer of skin, the nervous system, and portions of the sense organs. p. 359

ectothermic
(ek′tə thur′mik) (Gr. ectos, outside + therme, heat) Referring to animals like reptiles, amphibians, and fishes that regulate body temperature by taking in heat from the environment. p. 589

effector
(i fek′tər) A muscle or gland that effects (or causes) responses when stimulated by nerves. p. 257

efferent neuron
(ef′ər ənt nōōr′on) *See* motor neuron. p. 264

electrocardiogram (ECG)
(i lek′trō kärd′ē ə gram′) (Gr. elektron, amber + kardia, heart + -gram, written) A recording of the electrical impulses initiated at the SA node as they pass throughout the heart as an electric current and then as a wave throughout the body. p. 208

electron
(i lek′tron) A subatomic particle, having very little mass and carrying a negative charge, that orbits the nucleus of an atom. p. 17

electron-transport chain
(i lek′tron trans′pōrt chān) A term that describes the membrane-associated electron carriers produced by the citric acid cycle; the electrons gleaned from the oxidation of glucose then work as proton pumps. p. 117

element
(el′ə mənt) A pure substance that is made up of a single kind of atom and that cannot be separated into different substances by ordinary chemical methods. p. 18

elimination
(i lim′ə nā′shən) A process that takes place as digestive wastes leave the body during defecation. p. 233

embryo
(em′brē ō′) (Gr. embryon, something that swells in the body) The early stage of development in humans between the second and eighth weeks. p. 361

emigration
(em′ə grā′shən) (L. emigrare, to migrate) The movement of organisms out of a population. p. 625

endergonic
(en′dər gôn′ik) (Gr. endon, within + ergos, work) Used to describe a reaction in which the products of the reaction contain more energy than the reactants, so the extra energy must be supplied for the reaction to proceed. p. 101

endocrine gland
(en′də krən gland) (Gr. endon, within + krinein, to separate) A ductless gland that secretes hormones and spills these chemicals directly into the bloodstream. p. 315

endocrine system
(en′də krən sis′təm) (Gr. endon, within + krinein, to separate) The collective term for the 10 different endocrine glands that secrete 30 different hormones. p. 315

endocytosis
(en′dō sī tō′səs) (Gr. endon, within + kytos, cell) A process in which cells engulf large molecules or particles and bring these substances into the cell packaged within vesicles. p. 77

endoderm
(en′dō durm′) (Gr. endon, within + derma, skin) The inner layer of cells formed during the early development of embryos; it gives rise to the digestive tract lining, the digestive organs, the respiratory tract, the lungs, the urinary bladder, and the urethra. p. 359

endodermis
(en′dō durm′əs) (Gr. endon, within + derma, skin) In the root of a vascular plant, the innermost layer of the cortex, which consists of specialized cells that regulate the flow of water between the vascular tissues and the outer portion of the root. p. 549

endometrium
(en′dō mē′trē əm) (Gr. endon, within + metrios, of the womb) The inner lining of the uterus, which has two layers. p. 341

endoplasmic reticulum
(en′dō plaz′mik ri tik′yə ləm) (Gr. endon, within + plasma, from cytoplasm; L. reticulum, network) An extensive system of membranes that divides the interior of eukaryotic cells into compartments and channels. p. 53

endosperm
(en′dō spurm) (Gr. endon, within + sperma, seed) An extraembryonic tissue in a monocot where most of the food is

stored for the developing embryo. p. 538

endosymbiont
(en'dō sim'bī änt) (Gr. endon, within + bios, life) An organism that is symbiotic within another; the major endosymbionts that occur in eukaryotic cells are mitochondria and chloroplasts. p. 48

endosymbiotic theory
(en'dō sim'bē ôt'ik thir'ē) (Gr. endon, within + bios, life) The idea that mitochondria and chloroplasts originated symbiotically, mitochondria from aerobic bacteria and chloroplasts from anaerobic bacteria. p. 452

endothermic
(en'dō thur'mik) (Gr. endon, within + therme, heat) Referring to organisms such as birds and mammals that regulate body temperature internally. p. 592

end-product inhibition
(end prôd'əkt in'ə bish'ən) The process in which the enzyme catalyzing the first step in a series of chemical reactions has an inhibitor binding site to which the end product of the pathway binds. As the concentration of the end product builds up in the cell, it begins to bind to the first enzyme in the metabolic pathway, shutting off that enzyme. In this way, the end product is feeding information back to the first enzyme in the pathway, shutting the pathway down when additional end product is not needed. p. 108

energy level
(en'ər jē lev'əl) An electron shell that surrounds the nucleus of an atom. p. 18

entropy
(en'trə pē) (Gr. en, in + tropos, change in manner) The energy lost to disorder; it is a measure of the disorder of a system. p. 105

environment
(in vī'rən mənt) (M.E. envirouen, encircle) A general term for the biosphere; the land, air, water, and every living thing on the Earth. p. 695

enzyme
(en'zīm) (Gr. enzymos, leavened, from en, in + zyme, leaven) A protein that lowers the free energy of activation, thus allowing chemical reactions to take place. p. 105

eosinophil
(ē'ə sin'ə fil) (Gr. eos, dawn (rose-colored) + philein, to love) A kind of granulocyte believed to be involved in allergic reactions; they also act against certain parasitic worms. p. 210

epidermis
(ep'ə dur'məs) (Gr. epi, upon + derma, skin) In the root of a vascular plant, the outer layer of the cortex, which absorbs water and minerals. p. 549

epididymis
(ep'ə did'ə məs) (Gr. epi, upon + didymos, testis) A long, coiled tube that sits on the back side of the testes where sperm undergo further development after their formation within the testes. p. 338

epiglottis
(ep'ə glot'əs) (Gr. epi, upon + glotta, tongue) A flap of tissue that folds back over the opening to the larynx, thus preventing food or liquids from entering the airway. p. 172

epinephrine
(ep'ə nef'rən) (Gr. epi, on, over + nephros, kidney) A hormone produced by the adrenal medulla that readies the body to react to stress p. 326

epiphyte
(ep'ə fīt) (Gr. epi, on, over + phyton, plant) A plant of the tropical rain forest that grows on trees or other plants for support but draws its nourishment from rain water. p. 676

epithelial tissue
(ep'ə thē'lē əl tish'oo) A collection of tissues that cover and line internal and external surfaces of the body and compose the glands. p. 147

epithelium
(ep'ə thē'lē əm) (Gr. epi, on + thele, nipple) The collective term for all epithelial cells, which have six different functions in the body: protection, absorption, sensation, secretion, excretion, and surface transport. p. 149

epoch
(ep'ək *or* ē'pôk) (Gr. epechein, to hold on, check) A subdivision of a period of the Cenozoic Era. p. 448

era
(ir'ə *or* er'ə) One of the five major periods of geological time since the formation of the Earth until the present day. p. 448

erythrocyte
(i rith'rō sīt) (Gr. erythros, red + kytos, hollow vessel) A red blood cell, packed with hemoglobin, which acts as a mobile transport unit, picking up and delivering gases. p. 157

esophagus
(i sof'ə gəs) (Gr. oiso, carry + phagein, to eat) The food tube that connects the pharynx to the stomach. p. 172

essential amino acids
(i sen'chəl ə mēn'ō as'ədz) The eight amino acids that humans cannot manufacture and therefore must be obtained from proteins in the food they eat. p. 181

estrogen
(es'trə jən) (Gr. oistros, frenzy + genos, origin) Any of various hormones that develop and maintain the female reproductive structures such as the ovarian follicles, the lining of the uterus, and the breasts. p. 341

estuary
(es'choo er'ē) (L. aestus, tide) A place where the fresh water of rivers and streams meets the salt water of oceans. p. 685

ethology
(ē thôl'ə jē) (Gr. ethos, habit or custom + logos, discourse) The study of animal behavior in the natural environment, which examines the biological basis of the patterns of movement, sounds, and body positions of animals. p. 599

eubacteria
(yoo'bak tir'ē ə) (Gr. eus, good + bakterion, rod) The so-called true bacteria, which include all bacteria except the archaebacteria. p. 492

euglenoid
(yoo glē'noid) (Gr. eu, good + glene, eyeball) A flagellate member of the genus *Euglena*, having chloroplasts and making its own food by photosynthesis. p. 506

eukaryotic cell
(yoo kar'ē ōt'ik sel') (Gr. eu, good + karyon, kernel) A type of cell more complex than a prokaryote that makes up the bodies of plants, animals, protists, and fungi; a plasma membrane encloses the cytoplasm, which contains organelles and the nucleoid. p. 47

eustachian tube
(yoo stā'kē ən *or* yoo stā'shən toob) (after Bartolomeo Eustachio, Italian anatomist) A structure that connects the middle ear with the nasopharynx; it equalizes air pressure on both sides of the eardrum when the outside air pressure is not the same. p. 289

eutrophication
(yoo trof'ə kā'shən) (Gr. eutrophic, thriving) The accumulation of inorganic nutrients in a lake. p. 668

evolution
(ev'ə loo'shən) (L. evolvere, to unfold) Genetic change in a population of organisms over generations; Darwin proposed that natural selection was the mechanism behind evolutionary change. p. 11

excitatory synapse
(ik sīt'ə tōr'ē sin'aps) A type of synapse in which a neurotransmitter depolarizes the postsynaptic membrane, resulting in the continuation of the nerve impulse. p. 259

excretion
(ik skrē'shən) A process whereby metabolic wastes, excess water, and excess salts are removed from the blood and passed out of the body. p. 233

exergonic
(ek′sər gon′ik) (L. ex, out + Gr. ergon, work) Used to describe a reaction in which the products contain less energy than the reactants and the excess energy is released; exergonic reactions take place spontaneously. p. 101

exocrine gland
(ek′sə krən gland) (Gr. exo, outside + krinein, to separate) A term applied to a gland whose secretion reaches its destination by means of ducts. p. 315

exocytosis
(ek′sō sī tō′səs) (Gr. ex, out of + kytos, cell) The reverse of endocytosis; the discharge of material by a cell by packaging it in a vesicle and moving the vesicle to the cell surface. p. 77

exon
(ek′son) (Gr. exo, outside) A nucleotide sequence that encodes the amino-acid sequence of a polypeptide. p. 419

expiration
(ek′spə rā′shən) (Gr. ex, out + L. spirare, to breathe) The expelling of the air from the lungs in breathing; it occurs when the volume of the thoracic cavity is decreased and the resulting positive pressure forces air out of the lungs. p. 192

exponential growth
(ek′spō nen′shəl grōth) A growth rate in which although the rate of increase in population size may stay the same, the actual increase in the number of individuals grows. p. 492

external fertilization
(ek sturn′əl furt′əl ə zā′shən) Fertilization that takes place outside the body of the female. p. 584

external respiration
(ek sturn′əl res′pə rā′shən) The exchange of carbon dioxide and oxygen gases by the red blood cells and alveoli in the lungs. p. 185

extinction
(ik stingk′shən) (L. exstinguere, to quench) Drying out; refers to the serious and ongoing problem of species extinction of organisms taking place on Earth. p. 455

extraembryonic membrane
(ek′strə em′brē on′ik mem′brān) A structure that forms from the trophoblast and provides nourishment and protection; so named because it is not a part of the embryo. p. 360

F plasmid
(ef′ plāz′məd) A certain plasmid with a fertility factor, which are several special genes that promote the transfer of the plasmid to other cells. p. 494

facilitated diffusion
(fə sil′ə tā′təd di fyoo′shən) The movement of selected molecules across the cell membrane by specific transport proteins along the concentration gradient and without an expenditure of energy. p. 74

fatty acid
(fat′ē as′əd) A long hydrocarbon chain ending in a carboxyl (-COOH) group; a fatty acid can be saturated, unsaturated, or polyunsaturated. p. 33

feces
(fē′sēz) (L. faeces) Body waste discharged by way of the anus. p. 178

feedback loop
(fēd′bak′ loop) A mechanism by which information regarding the status of a physiological situation or system is fed back to the system so that appropriate adjustments can be made. p. 162

femur
(fē′mur) The thigh bone in the appendicular skeleton of a human. p. 300

fermentation
(fur′mən tā′shən) (L. fermentum, ferment) An anaerobic process by which certain organisms make ATP. p. 116

fertilization
(furt′ə lə zā′shən) (L. ferre, to bear) The union of a male gamete (sperm) and a female gamete (egg). p. 355

fibroblast
(fī′brə blast′) (L. fibra, fiber + Gr. blastos, sprout) The most numerous of the connective tissue cells, they are flat, irregular, branching cells that secrete fibers into the matrix between them. p. 153

fibrocartilage
(fī′brə kärt′əl ij *or* fī′brə kärt′lij) (L. fibra, fiber + cartilago, gristle) A type of cartilage that has collagen fibers embedded in its matrix; used by the body as a "shock absorber" in the knee joint and as disks between the vertebrae. p. 153

filtrate
(fil′trāt) The water and dissolved substances that are first filtered out of the blood during the formation of urine. p. 234

filtration
(fil′trā′shən) The process by which the blood is passed through nephron membranes that separate blood cells and proteins from the water and small molecules of the blood. p. 235

flagellum, pl. flagella
(flə jel′əm, flə jel′ə) (L. flagellum, whip) A long, whip-like organelle of motility that protrudes from some cells. p. 60

fluid mosaic model
(floo′əd mō zā′ik mod′əl) The model of the cell membrane that describes the fluid nature of a lipid bilayer studded with a mosaic of proteins. p. 69

follicle
(fol′ē kəl) (L. folliculus, little bag) A term for the follicular cells and the oocyte in the ovary. p. 339

follicle-stimulating hormone (FSH)
(fol′ə kəl-stim′yə lāt′ing hôr′mōn) A gonadotropic hormone secreted by the anterior pituitary that triggers the maturation of one egg each month in females; it triggers sperm production in males. p. 320

food chain
(food chān) A series of organisms, from each trophic level, that feed on one another. p. 659

food web
(food web) The interwoven food chain of an ecosystem; a diagram of who eats whom. p. 661

foramen, pl. foramina
(fə rā′mən, fə ram′ən ə) The inner space or tunnel in the vertebral column in which the spinal cord runs down the neck and back; this body casing protects the spinal cord from injury. p. 271

foraminifer, pl. foraminifera
(fôr′ə min′ə fər, fə ram′ə nif′ər ə) (L. forare, to bore) A type of ameba that secretes beautifully sculpted shells made out of calcium carbonate or limestone; also called a foram. p. 505

formed elements
(fôrmd el′ə məntz) The solid portion of blood plasma composed principally of erythrocytes, leukocytes, and platelets. p. 208

fossil
(fos′əl) (L. fodere, to dig) Any record of a dead organism; any trace or impression of an animal or plant that has been preserved in the Earth's crust. p. 434

fovea
(fō′vē ə) (L. a pit) The area of sharpest vision within the retina due to its high concentration of cones. p. 288

free energy of activation
(frē en′ər jē uv ak′tə vā′shən) The energy needed to initiate a chemical reaction. p. 101

free radical
(frē rad′ə kəl) A charged molecule fragment with an unpaired electron that is highly reactive. p. 393

frond
(frond) (L. foliage) The leaf of a fern. p. 531

frontal lobe
(frunt′əl lōb) A section on both hemispheres of the cerebral cortex

dealing with the motor activity movement of the body. p. 266

fruit
(froot) A vessel containing seeds produced by an angiosperm. p. 534

functional group
(fungk′shən əl groop) Special groups of atoms attached to an organic molecule; important because most chemical reactions that occur within organisms involve the transfer of a functional group from one molecule to another. p. 28

fungus, pl. fungi
(fung′gəs, fun′jī) (L. mushroom) A multicellular, eukaryotic organisms that feeds on dead or decaying organic material. p. 516

gall bladder
(gôl blad′ər) In human beings, a sac attached to the underside of the liver, where excess bile is stored and concentrated. p. 175

gamete
(gam′ēt *or* gə mēt′) (Gr. wife) A sex cell; the female gamete is the egg, and the male gamete is the sperm. p. 83

gametophyte
(gə mē′tō fīt) (Gr. gamete, wife + phyton, plant) The haploid phase of a plant life cycle that alternates with the diploid phase. p. 529

ganglion, pl. ganglia
(gang′glē ən, gang′glē ə) A nerve cell body located within the peripheral nervous system. p. 267

gastric glands
(gas′trik glandz) Glands dotting the inner surface of the stomach that secrete a gastric juice of hydrochloric acid and pepsinogen. p. 174

gastrin
(gas′trən) A digestive hormone of the stomach that controls the production of acid. p. 174

gastrulation
(gas′trə lā′shən) (L. little belly) The process by which groups of inner cells mass, migrate, divide, and differentiate into three primary germ layers from which all the organs and tissues of the body develop. p. 361

gene
(jēn) (Gr. genos, birth, race) A piece of hereditary information carried on the X and Y sex chromosomes. p. 81

gene families
(jēn fam′lēz *or* fam′ə lēz) Multiple copies of genes in eukaryotes, which are derived from a common ancestral gene and are a reflection of the evolutionary process. p. 422

gene mutation
(jēn myoo tā′shən) A change in the genetic message of a chromosome due to alterations of molecules within the structure of the chromosomal DNA. p. 394

gene therapy
(jēn ther′ə pē) The technique in which scientists try to cure inherited genetic disorders by inserting genes with the proper genetic message into patients having defective genes. p. 405

general adaptation syndrome
(jen′ə rəl ad ap tā shən sin′drōm) the three stages of reaction by the body to stress: the alarm reaction, resistance, and exhaustion. p. 326

genetic counseling
(jə net′ik coun′səl ing) The process in which geneticists identify couples at risk of having children with genetic defects and help them have healthy children.

genetic engineering (recombinant DNA technology) (jə net′ik en′jə nir′ing) New techniques of molecular biology that involve the purposeful manipulation of genes within organisms. p. 494

genetics
(jə net′iks) (Gr. genos, birth, race) The branch of biology dealing with the principles of heredity and variation in organisms. p. 377

genotype
(jēn′ə tīp) (Gr. genos, offspring + typos, form) The total set of genes that constitutes an organism's genetic makeup. p. 378

geographic isolation
(jē′ə graf′ik ī′sə lā′shən) A term describing organisms that live in sharply discontinuous habitats from one another. p. 432

germination
(jur′mə nā′shən) (L. germinare, to sprout) The sprouting of a seed that begins when it receives water and appropriate environmental cues. p. 538

gestation
(jes tā′shən) (L. gestare, to bear) The time of a developing human from conception until birth, approximately 8½ months. p. 355

gibberellin
(jib′ə rel′ən) (*Gibberella,* a genus of fungi) A hormone that together with auxin, promotes growth in plants through cell elongation. p. 541

gill
(gil) In the body of a mollusk, a filamentous projection of the mantle tissue that is rich in blood vessels and functions as an organ of respiration. p. 571

gizzard
(giz′ərd) The digestive organ of a bird, often filled with grit, that grinds food. p. 591

gland
(gland) (L. glans, acorn) An epithelial cell specialized to produce and discharge substances. p. 149

glial cell
(glē′əl *or* glī′əl sel) (Gr. glia, glue) A supporting nerve cell of the brain and spinal cord. p. 160

glomerulus
(glə mer′ə ləs) (L. a little ball) A tuft of capillaries surrounded by Bowman's capsule that acts as filtration devices in the formation of urine. p. 237

glucagon
(gloo′kə gon) A hormone produced in the islets of Langerhans that raises blood glucose level by converting glycogen into glucose. p. 327

glycogen
(glī′kə jən) (Gr. glykys, sweet + gen, of a kind) Animal starch; a storage form of glucose within animals. p. 30

glycolysis
(glī kol′ə səs) (Gr. glykys, sweet + lyein, to loosen) The first of three series of chemical reactions in cellular respiration, which results in the formation of two ATP molecules and two molecules of pyruvate. p. 117

Golgi complex
(gol′jē käm′pleks) (after Camillo Golgi, Italian physician) The delivery system of the eukaryotic cell; it collects, modifies, packages, and distributes molecules that are made at one location within the cell and used at another. p. 54

gonad
(gō′nad) (Gr. gone, seed) A male or female reproductive sex organ that produces sex cells, or gametes. p. 335

gradient
(grā′dē ənt) Describing the differences in concentration, pressure, or electrical charge in the random motion of molecules, which often results in a net movement of molecules in a particular direction. p. 72

grana, sing. granum
(grā′nə, grā′nəm) (L. grain or seed) Within chloroplasts, stacks of membrane thylakoid sacs. p. 135

granulocyte
(gran′yə lō sīt) (L. granulum, little grain + Gr. kytos, cell) A circulating leukocyte that gets its name from the tiny granules in its cytoplasm; granulocytes are classified into three groups by their staining properties. p. 210

greenhouse effect
(grēn′hous′ ə fekt′) The selective energy

absorption of carbon dioxide (CO_2), which allows heat to enter the Earth's atmosphere but prevents it from leaving; may be a contributing factor to global warming. p. 700

growth hormone (GH)
(grōth hôr′mōn) A hormone secreted by the anterior pituitary that works with the thyroid hormones to control normal growth. p. 320

guard cells
(gärd selz) A pair of cells that brackets a stoma and regulates its opening and closing. p. 554

gymnosperm
(jim′nə spurm′) (Gr. gymnos, naked + sperma, seed) A vascular plant whose seeds are not completely enclosed by the tissues of the parent at the time it is pollinated; the most familiar gymnosperms are the conifers. p. 533

habitat
(hab′ə tat′) (L. habitare, to inhabit) An area or space within an ecosystem, including the factors within it, in which each living thing resides. p. 639

half-life
(haf′ līf′) The length of time it takes for half of the ^{14}C present in a sample to be converted to ^{12}C, which is 5730 years; a useful tool for dating fossils. p. 436

haploid
(hap′loid) (Gr. haploos, single + eidos, form) A sex cell that contains half the amount of hereditary material of the original parent cell. p. 83

haversian canal
(hə vur′shən kə nal′) (after Clopton Havers, British physician and anatomist) A narrow channel that runs parallel to the length of the bone that contains blood vessels and nerve cells. p. 296

heart
(härt) (A.S. heorte) The muscular pump that is the center of the cardiovascular or circulatory system. p. 199

helper T cell
(help′ər tē′ sel) A kind of T lymphocyte that initiates the immune response by identifying foreign invaders and stimulating the production of other cells to fight an infection. p. 222

hemoglobin
(hē′mə glō′bən) (Gr. haima, blood + L. globus, a ball) The iron-containing pigment that imparts the color to red blood cells; hemoglobin is produced within the red bone marrow and carries oxygen in the blood. p. 193

hemophilia
(hē′mə fil′ē ə) (Gr. haima, blood + philein, to love) A hereditary condition in which the blood is slow to clot or does not clot at all. p. 389

herbivore
(hur′bə vôr) (L. herba, grass + vorare, to eat) An animal that feeds on plants. p. 170

hermaphrodite
(hər maf′rə dīt) (Gr. Hermaphroditos, son of Hermes and Aphrodite, who was man and woman combined) An animal or plant having both male and female reproductive organs; a sponge is a hermaphrodite. p. 566

heterotroph
(het′ər ə trōf′) (Gr. heteros, other + trophos, feeder) An organism that cannot produce its own food; a consumer in an ecosystem or food chain. p. 131

hetereozygous
(het′ər ə zī′gəs) (Gr. heteros, other + zygotos, a pair) Refers to a individual having two different alleles for a trait. p. 378

homeostasis
(hō′mē ə stā′səs) (Gr. homeos, similar + stasis, standing) The maintenance of a stable internal environment in spite of a possibly very different external environment. p. 162

hominid
(hom′ə nid) (L. homo, man) The family of hominoids consisting of human beings; the only living hominid is *Homo sapiens sapiens*. p. 475

hominoid
(hom′ə noid) (L. homo, man) A superfamily of the anthropoid suborder that includes apes and humans. p. 474

Homo erectus
(hō′mō i rek′təs) (L. homo, man + erectus, upright) An extinct species of hominids whose fossil record dates back to 1.6 million years ago; *H. erectus* was fully adapted to upright walking, made sophisticated tools, built shelters, used fire, and probably communicated with language. p. 480

Homo habilis
(hō′mō hab′ə ləs) (L. homo, man + habilis, skillful) An extinct species of hominids whose fossil record dates back about 2 million years; considered human because they exhibited a far greater intelligence than their ancestors by making tools and clothing. p. 480

Homo sapiens
(hō′mō sā′pē ənz *or* sap′ē ənz) (L. homo, man + sapiens, wise) A hominid whose fossil record dates back about 200,000 years and most likely evolved from *H. erectus;* their anatomy featured larger brains, flatter heads, more sloping foreheads, and more protruding brow ridges than modern humans. p. 475

Homo sapiens sapiens
(hō′mō sā′pē ənz sā′pē ənz) (L. homo, man + sapiens, wise) Modern man; the subspecies of hominids who made their appearance 10,000 years before the Neanderthal subspecies died out and whose early members are called Cro-Magnons. p. 481

homologous
(hō mol′ə gəs) (Gr. homologia, agreement) Said of the bones of different vertebrates that now differ in structure and function, although having the same evolutionary origin. p. 438

homozygous
(hō′mə zī′gəs) (Gr. homos, same or similar + zygotos, a pair) Refers to an individual having two identical alleles for a trait. p. 378

hormone
(hôr′mōn) (Gr. hormaien, to excite) A chemical messenger secreted and sent by a gland to other cells of the body. p. 315

human chorionic gonadotropic (HCG)
(hyōō′mən kōr′ēən′ik gō′nad ə trō′pik) (Gr. chorion, chorion + gonos, genitals + trope, turning) A hormone secreted by embryonic placental tissue that maintains the corpus luteum so that it will continue to secrete progesterone and estrogen; the detection of this hormone in the urine is the basis for home pregnancy tests. p. 341

human genome project
(hyōō′mən jē′nōm proj′ekt) A monumental, worldwide scientific project, the goal of which is to decipher the DNA code of all 46 human chromosomes. p. 410

human immunodeficiency virus (HIV)
(hyōō′mən im yōō′nō dē fish′ən sē vī′rəs) The virus that causes AIDS, especially deadly because it destroys the ability of the immune system to mount a defense against any infection because it attacks and destroys helper T cells. p. 227

humerus
(hyōō′mər əs) (L. upper arm) The upper arm bone in the appendicular skeleton of a human. p. 300

humoral immune response
(hyōō′mər əl i myōōn ri spons′) A second, longer-range defense than the cell-mediated immune reseponse, in which B cells are converted to plasma cells that secrete the proteins antibodies. p. 222

Huntington's disease
(hunt′ing tənz diz ēz′) (after G. Huntington, U.S. physician) A fatal genetic disorder caused by a mutant dominant allele that causes progressive deterioration of brain cells. p. 398

hyaline cartilage
(hī′ə lən *or* hī′ə līn kärt′əl ij *or* kärt′lij) (Gr. hyalos, glass + L. cartilago, gristle)

A type of cartilage that has very fine collagen fibers in its matrix; it is found on the ends of long bones, ringing the windpipe, and in the ribs and nose. p. 153

hybrid
(hī'brəd) (L. hybrida, mongrel) The offspring of the cross between two different varieties of plants of the same species; also the cross between two different species of organisms. p. 374

hydrogen bond
(hī'drə jən bänd) A molecule formed when the partial negative charge at one end of a polar molecule is attracted to the partial positive charge of another polar molecule. p. 23

hydrolysis
(hī drŏl'ə səs) (Gr. hydro, water + lyse, to break) The process by which a polymer is disassembled by adding a molecule of water. p. 30

hydrolyzing enzyme
(hī'drə līz'ing en'zīm) (Gr. hydro, water + lyse, to break) An enzyme that breaks down substances by hydrolysis. p. 170

hydrophilic
(hī drə fil'ik) (Gr. hydor, water + philos, loving) Referring to the phosphate functional groups in polar molecules, which form hydrogen bonds with water and, consequently, are soluble or dissolve in water. p. 69

hydrophobic
(hī'drə fō'bik) (Gr. hydor, water + phobos, hating) Refers to nonpolar molecules like oil that cannot form hydrogen bonds with water; this is why oil and water do not mix. p. 69

hyperpolarization
(hī'pər pō'lər ə zā'shən) (Gr. hyper, above + polaris, pole) The resulting state in the interior of a rod cell when it becomes even more negatively charged than before because sodium channels have closed because of stimulation. p. 286

hypertonic
(hī'pər ton'ik) (Gr. hyper, above + tonos, tension) Refers to a solution with a solute concentration higher than that of another fluid. p. 73

hypha, pl. hyphae
(hī'fə, hī'fē) (Gr. hyphe, web) A slender filament of a fungus barely visible to the naked eye. p. 516

hypothalamus
(hī'pō thal'ə məs) (Gr. hypo. under + thalamos, inner room) A mass of gray matter lying at the base of the cerebrum that produces two hormones and controls the secretion of hormones by the pituitary gland. p. 269, 318

hypothesis
(hī poth'ə səs) (Gr. hypo, under + tithenai, to put) A plausible answer to a scientific question, based on available knowledge and generalizations made from observations. p. 3

hypotonic
(hī'pə ton'ik) (Gr. hypo, under + tonos, tension) Refers to a solution with a solute concentration lower than that of another fluid. p. 73

ileum
(il'ē əm) (L. groin, flank) The third and final part of the small intestine, following the jejunum. p. 175

immigration
(im'ə grā'shən) (L. immigrare, to migrate) The movement of organisms into a population. p. 625

immune system
(i myoon' sis'təm) The collective term for populations of white blood cells that resist disease. p. 221

immunity
(i myoon'ə tē) (L. immunitas, safe) The specific defense of a human body that consists of cellular and molecular responses to particular foreign invaders. p. 217

implantation
(im'plan tā'shən) (L. in, into + plantare, to plant) The embedding of the developing blastocyst into the posterior wall of the uterus approximately 1 week after fertilization. p. 357

imprinting
(im'print ing) A rapid and irreversible type of learning that takes place during an early developmental stage of some animals. p. 603

incomplete dominance
(in'kəm plēt' däm'ə nəns) A situation in which neither member of a pair of alleles exhibits dominance over the other. p. 399

induction
(in duk'shən) (L. inductio, leading in) The process by which some cells turn "off and on" switches for the genes of neighboring cells. p. 364

inflammation
(in'flə mā'shən) (L. inflammare, to flame) A reaction when cells are damaged by microbes, chemicals, or physical substances. p. 217

inhibitory synapse
(in hib'ə tôr'ē sin'aps) A type of neurotransmitter that reduces the ability of the postsynaptic membrane to depolarize. p. 259

inner ear
(in'ər ir) A complex of fluid-filled canals in the ear where hearing actually takes place; receptor cells of this organ change the mechanical "sound" energy into nerve impulses. p. 289

insertion
(in sur'shən) The attachment of one end of a skeletal muscle to a bone that will move. p. 305

inspiration
(in'spə rā'shən) (L. in, in + spirare, to breathe) Drawing air into the lungs; it occurs when the volume of the thoracic cavity is increased and the resulting negative pressure causes air to be sucked into the lungs; opposite of expiration. p. 190

insulin
(in'sə lən) (L. insula, island) A hormone produced in the islets of Langerhans of the pancreas that promotes the regulation of glucose metabolism by the cells. p. 327

integration
(int'ə grā'shən) A function of the central nervous system in which the brain and spinal cord make sense of incoming sensory information and then produce outgoing motor impulses. p. 265

integument
(in teg'yə mənt) (L. integumentum, covering) The skin, hair, and nails of the human body. p. 296

interneuron
(int'ər noor'on) (L. inter, between + Gr. neuron, nerve) A nerve cell found in the spinal cord and brain situated between other neurons that receives incoming messages and sends outgoing messages in response. p. 251

interphase
(int'ər fāz') The portion of the cell cycle preceding mitosis in which the cell grows and carries out normal life functions. During this time the cell also produces an exact copy of the hereditary material, DNA, as it prepares for cell division. p. 86

intron
(in'tron) (L. intra, within) A segment of DNA transcribed into mRNA but removed before translation. p. 419

inversion
(in vur'zhən) Refers to the process in which a broken piece of chromosome reattaches to the same chromosome but in a reversed direction. p. 393

ion
(ī'ən *or* ī'on) (Gr. going) An electrically charged particle that results from an exchange of electrons in an atom. p. 20

ionic bond
(ī on'ik bänd) An attraction between ions of opposite charge. p. 19

iris
(ī′rəs) (L. rainbow) A diaphragm lying between the cornea and the lens that controls the amount of light entering the eye. p. 288

islets of Langerhans
(ī′ləts uv läng′ər hänz′) (after Paul Langerhans, German anatomist) The separate types of cells within the exocrine cells of the pancreas that produce the hormones insulin and glucagon. p. 327

isomer
(ī′sə mər) (Gr. isos, equal + meros, part) One of two or more molecules that have the same molecular formula but are arranged slightly differently; fructose is an isomer of glucose. p. 447

isotonic
(ī′sə ton′ik) (Gr. isos, equal + tonos, tension) Refers to solutions having equal solute concentrations to one another. p. 73

isotope
(ī′sə tōp) (Gr. isos, equal, + topos, place) An atom of an element that has the same number of protons but different numbers of neutrons in its nuclei. p. 17

jawless fish
(jô′ləs fish) A tubular, scaleless, finless, jawless organism that lives in the sea or in brackish water; a lamprey is a jawless fish that parasitizes bony fish and poses a serious threat to some commercial fisheries. p. 584

jejunum
(ji joo′ nəm) (L. empty) The second portion of the small intestine extending from the duodenum to the ileum. p. 175.

joint
(joint) An articulation; a place where bones or bones and cartilage come together. p. 302

karyotype
(kar′ē ə tīp′) (Gr. karyon, kernel + typos, stamp or print) The particular array of chromosomes that belongs to an individual. p. 81

kidney
(kid′nē) The organ in all vertebrates that carries out the processes of filtration, reabsorption, and excretion. p. 234

kidney stone
(kid′nē stōn) Crystals of certain salts that develop in the kidney and block urine flow. p. 234

kinetic energy
(kə net′ik en′ər jē) The energy of motion. p. 103

Klinefelter syndrome
(klīn′felt ər sin′drōm) (after Harry F. Klinefelter, Jr, U.S. physician) A genetic condition resulting from an XXY zygote that develops into a human male who is sterile, has many female body characteristics, and in some cases, has diminished mental capacity. p. 391

Krebs cycle
(krebz sī′kəl) (after Hans A. Krebs, German-born English biochemist) The series of nine reactions during which pyruvate, the end product of glycolysis, enters the cycle to form citric acid and is finally oxidized to carbon dioxide. Also called citric acid cycle. p. 117

lacteal
(lak′tē əl) (L. lacteus, of milk) A lymphatic vessel within a villus that helps pass nutrients into the lymph and blood. p. 178

lacuna, pl. lacunae
(lə kyoo′nə, lə kyoo′nē) (L. a pit) The tiny chambers where cartilage cells lie. p. 153

laguno
(lə goo′nə) A fine body hair that appears over the body of a fetus toward the end of the third month of pregnancy but that is lost before birth. p. 366

lancelet
(lans′lət) A scaleless, fish-like marine chordate. p. 583

larynx
(lar′ingks) (Gr.) The voice box located at the upper end of the human windpipe. p. 186

lateral line system
(lat′ər əl līn sis′təm) A complex system of mechanoreceptors possessed by all fishes and amphibians that detects mechanical stimuli such as sound, pressure, and movement. p. 585

lens
(lenz) (L. lentil) A body in the eye lying just behind the aqueous humor that plays a major role in focusing the light entering the eye. p. 287

leukocyte
(loo′kə sīt) (Gr. leukos, white + kytos, hollow vessel) Any of several kinds of white blood cells including macrophages and lymphocytes, all functioning to defend the body against invading microorganisms and foreign substances. p. 157

ligament
(lig′ə mənt) (L. ligare, to bind) A bundle or strip of dense connective tissue that holds a bone to a bone. p. 302

light-dependent reactions
(līt di pen′dənt rē ak′shənz) The reactions of photosynthesis that produce ATP and NADPH; so called because they take place only in the presence of light. p. 135

light-independent reactions
(līt in′də pen′dənt rē ak′shənz) The reactions of photosynthesis that use ATP and NADPH to drive the formation of glucose from carbon dioxide; so called because they do not need light to occur as long as ATP is available; also called the Calvin-Benson cycle. p. 136

limb bud
(lim bud) The appearance of arms and legs that appear as microscopic flippers during the fourth week in an embryo. p. 366

limbic system
(lim′bik sis′təm) (L. limbus, border) A network of neurons, which together with the hypothalamus, forms a ring-like border around the top of the brainstem; responsible for many of the most deep-seated drives and emotions of vertebrates. p. 269

lipase
(lī′pās or lip′ās) (Gr. lipos, fat + -ase, enzyme) An enzyme that breaks down the triglycerides in lipids to fatty acids and glycerol. p. 168

lipid
(lip′id) (Gr. lipos, fat) Any of a wide variety of molecules, all of which are soluble in oil but insoluble in water; important categories of lipids are oils, fats, and waxes; phospholipids; and steroids. p. 33

lipid bilayer
(lip′id bī′lā′ər) The basic foundation of biological membranes; it forms a fluid, flexible covering for a cell and keeps the watery contents of the cell on one side of the membrane and the water environment on the other. p. 69

liver
(liv′ər) A large, complex organ weighing over 3 pounds, lying just under the diaphragm, that performs over 500 functions in the body, including aiding in the digestion of lipids. p. 175

loop of Henle
(loop uv hen′lē) (after F.G.J. Henle, German anatomist) The descending and ascending loops of the renal tubule. p. 237

lumen, pl. lumina
(loo′mən, loo′mə nə) (L. light) The hollow core in the three layers of tissue that makes up the walls of the arteries through which blood flows. p. 201

luteinizing hormone (LH)
(loo′tē ən īz′ing hôr′mōn) A gonadotropic hormone secreted by the anterior pituitary that stimulates the release of an egg in females and

lymph
(limf) (L. lympha, clear water) A tissue fluid diffused out of the blood through the capillaries. p. 211

lymph node
(limf nōd) A small, ovoid "spongy" structure located in various places of the body along the route of the lymphatic vessels filtering the lymph as it passes through. p. 211

lymphatic system
(lim fat′ik sis′təm) The system of one-way, blind-ended vessels that collects and returns to the blood the approximately 10% of the fluid that does not return to the blood directly. p. 211

lysogenic cycle
(lī′sə jen′ik sī′kəl) (Gr. lysis, dissolution + gennan, to produce) A pattern of viral replication in which a virus integrates its genetic material with that of a host and is replicated each time the host cell replicates. p. 489

lysosome
(lī′sə sōm) (Gr. lysis, a loosening + soma, a body) A membrane-bounded vesicle containing digestive enzymes that break down old cell parts or materials brought into the cell from the environment and are extremely important to the health of a cell. p. 56

lytic cycle
(lit′ik sī′kəl) (Gr. lysis, dissolution) A pattern of viral replication in which a virus enters a cell, replicates, and then causes the cell to burst, releasing new viruses. p. 489

macromolecule
(mak′rə mol′ə kyool) (Gr. makros, large + L. moliculus, a little mass) A large organic molecule having many functional groups. p. 28

macrophage
(mak′rō fāj) (Gr. makros, long + -phage, eat) A phagocytic cell in the bloodstream that engulfs foreign bacteria and antibody-coated cells or particles in the process of phagocytosis; macrophages act as the body's scavengers. p. 152

malpighian tubule
(mal pig′ē ən too′byool) (after Marcello Malpighi, Italian anatomist) A slender projection from the digestive tract in the body of a terrestrial arthropod that serves as an excretory system. p. 576

mammal
(mam′əl) (L. mamma, breast) A warm-blooded vertebrate that has hair in which the female secretes milk from mammary glands to feed her young. p. 468

mantle
(mant′əl) In the body of a mollusk, a soft epithelium that arises from the dorsal body wall and encloses a cavity between itself and the visceral mass. p. 571

marsupial
(mär soo′pē əl) (L. marsupium, pouch) A subclass of mammals that give birth to immature young that are carried in a pouch; a kangaroo is a marsupial. p. 471

medulla
(mə dul′ə) (L. marrow) The lowest portion of the brainstem, continuous with the spinal cord below; the site of neuron tracts, which cross over one another delivering sensory information from the right side of the body to the left side of the brain and vice versa. p. 235

medusa, pl. medusae
(mə doo′sə *or* mə doo′zə, mə doo′sē *or* mə doo′zē) (Gr. Medousa, a Gorgon slain by Perseus, who gave her head to Athena) A cnidarian that is a free-floating, umbrella-shaped animal. p. 567

megakaryocyte
(meg′ə kar′ē ə sīt) (Gr. megas, large + karyon, nucleus + kytos, cell) A large bone marrow cell that pinches off bits of its cytoplasm, resulting in a cell fragment, or platelet.

meiosis
(mī ō′ səs) (Gr. meioun, to make smaller) The two-staged process of nuclear division in which the number of chromosomes in cells is halved during gamete formation. p. 83

meninges, sing. meninx
(mə nin′jēz, mē′ningks) (Gr. membrane) Any of three layers of membranes covering both the brain and spinal cord. p. 272

menopause
(men′ə pôz) (L. mens, month + pausis, cessation) The period that marks the permanent cessation of menstrual activity in a woman, usually between the ages of 50 and 55; the end of the menses. p. 343

menstruation
(men′stroo ā′shən) (L. mens, month) The monthly sloughing off of the blood-enriched lining of the uterus when pregnancy does not occur; the lining degenerates and causes a flow of blood, tissue, and mucus from the uterus out the vagina. p. 342

meristematic tissue
(mer′ə stə mat′ik tish′oo) (Gr. merizein, to divide) An undifferentiated type of tissue in a vascular plant that produces new plant cells during growth; also called meristem. p. 546

mesoderm
(mez′ə durm′) (Gr. mesos, middle + derma, skin) The layer of cells in the developing embryo that differentiates into the skeleton, muscles, blood, reproductive organs, connective tissue, and the innermost layer of the skin. p. 568

mesophyll
(mez′ə fil′) (Gr. mesos, middle + phyllon, leaf) The parenchymal layer of a leaf, bounded by epidermis, where photosynthesis takes place. p. 554

messenger RNA (mRNA)
(mes′ən jər är′en′ā′) A type of RNA that brings information from the DNA within the nucleus to the ribosomes in the cytoplasm to direct which polypeptide is assembled. p. 413

metabolism
(mə tab′ə liz′əm) (Gr. metabole, change) All the chemical reactions that take place within a living organism. p. 121

metaphase
(met′ə fāz) (Gr. meta, middle + phasis, form) The stage of mitosis characterized by the alignment of the chromosomes in a ring, equidistant from the two poles of the cell. p. 89

microfilament
(mī′krō fil′ə mənt) (Gr. mikros, small + L. filum, a thread) A thin, twisted double-chain fiber of protein within the cytoskeleton that helps support and shape eukaryotic cells. p. 50

microtubule
(mī′krō too′byool) (Gr. mikros, small + tubulus, little pipe) A spiral array composed of protein subunits within the cytoskeleton that provide intracellular support in the nondividing cell. p. 50

microvillus, pl. microvilli
(mī′krō vil′əs, mī krō vil′ī) (Gr. mikros, small + L. villus, tuft of hair) A microscopic, cytoplasmic projection that covers the epithelial cells of the villi on their exposed surfaces. p. 178

midbrain
(mid′brān′) The top part of the brainstem; it contains nerve tracts connecting the upper and lower parts of the brain and nuclei that acts as reflex centers for movement. p. 271

middle ear
(mid′əl ir) The middle portion of the ear containing three bones that act together like an amplifier to increase the force of sound vibrations. p. 289

migration
(mī grā′shən) Long-range, two-way movements by animals, often occurring yearly with the change of seasons. p. 603

mineral
(min′ə rəl) (L. minerale) An inorganic substance transported around as ions dissolved in blood and other body fluids; a variety of minerals perform a variety of functions in the human body. p. 166

mitochondrion, pl. mitochondria
(mīt′ə kon′drē ən, mīt′ə kon′drē ə) (Gr. mitos, thread + chondrion, small grain) An oval, sausage-shaped, or thread-like organelle about the size of a bacterium, bounded by a double membrane whose function is to break down fuel molecules, thus releasing energy for cell work. p. 58

mitosis
(mī tō′səs *or* mi tō′səs) (Gr. mitos, thread) A process of cell division that produces two identical cells from an original parent cell.

molecule
(môl′ə kyōōl′) (L. molecula, little mass) A combination of tightly bound atoms. p. 83

Monera
(mə ner′ə) (Gr. moneres, individual) The kingdom that consists of the bacteria. p. 491

monocot
(mon′ə kot) An angiosperm in which the food reserves within its seeds is stored in extraembryonic tissue called endosperm; a monocot usually has parallel veins in its leaves and its flowers parts are often in threes. p. 538

monohybrid
(mon′ə hī′brəd) (Gr. monos, single + L. hybrida, mongrel) The progeny or product of two plants that differ from one another in a single trait. p. 374

monosaccharide
(mon′ə sak′ə rīd) (Gr. monos, one + sakcharon, sugar) A simple sugar. p. 30

monotreme
(mon′ə trēm′) (Gr. mono. single + treme, hole) A mammal that lays eggs having leathery shells similar to those of a reptile; the only extant monotremes are the duckbilled platypus and two genera of spiny anteaters. p. 471

morphogenesis
(môr′fə jen′ə səs) (Gr. morphe, form + genesis, origin) The early stage of development in a vertebrate when cells begin to move, or migrate, thus shaping the new individual. p. 357

motor area
(mōt′ər er′ē ə) The part of the brain straddling the rearmost portion of the frontal lobe that sends messages to move the skeletal muscles. p. 266

motor neuron
(mōt′ər nōōr′on) A neuron of the peripheral nervous system that transmits commands away from the central nervous system. p. 251

multiple alleles
(mul′tə pəl a lēlz′) A system of alleles in a gene that exhibits either complete dominance or codominance. p. 400

muscle fiber
(mus′əl fī′bər) A long, multinucleated cell packed with organized arrangements of microfilaments capable of contraction. p. 305

muscle tissue
(mus′əl tish′ōō) Any of three different kinds of muscle cells—smooth, skeletal, or cardiac—that are the workhorses of the body; characterized by an abundance of special thick and thin microfilaments. p. 147

mutation
(myōō tā′shən) (L. mutare, to change) A permanent change in the genetic material. p. 649

mutualism
(myōōch′ə wə liz′əm) (L. mutuus, lent, borrowed) A type of symbiosis in which both participating species benefit.

mycelium, pl. mycelia
(mī sē′lē əm, mī sē′lē ə) (Gr. mykes, fungus) In fungi, a mass of hyphae. p. 516

myelin sheath
(mī′ə lən shēth) (Gr. myelinos, full of marrow) The fatty wrapping created by mutliple layers of Schwann cell membranes; the myelin sheath insulates the axon. p. 252

myofibril
(mī′ə fī′brəl) (Gr. myos, muscle + L. fibrilla, little fiber) A cylindrical, organized arrangement of special thick and thin microfilaments capable of shortening a muscle fiber. p. 305

myosin
(mī′ə sin) (Gr. mys, muscle + in, belonging to) One of the two protein components of myofilaments in a muscle fiber; actin is the other. p. 307

natural selection
(nach′ə rəl sə lek′shən) The process in which organisms having adaptive traits survive in greater numbers than organisms without such traits. p. 432

Neanderthal
(nē an′dər thôl′ *or* nē an′dər täl′) (Neander, valley in western Germany) A subspecies of *H. sapiens* that lived from about 125,000 to 35,000 years ago in Europe and the Middle East; Neanderthals were short and powerfully built, with large brains; they made diverse tools, took care of the sick and injured, and buried their dead. p. 480

negative feedback
(neg′ət iv fēd′bak′) The process by which enzyme activity is regulated by inhibitors. p. 108

nephridium, pl. nephridia
(ni frid′ē əm, ni frid′ē ə) (Gr. nephros, kidney) In the body of a mollusk, a tubular structure through which wastes are removed. p. 573

nephron
(nef′ron) (Gr. nephros, kidney) Any of the millions of microscopic tubular units of the kidney where urine is formed. p. 234

nerve
(nurv) A cluster of axons and dendrites surrounded by numerous supporting cells. p. 269

nerve cord
(nurv kôrd) A single, hollow cord along the back that is a principal feature of chordates; in vertebrates, the nerve cord differentiates into a brain and spinal cord. p. 464

nerve impulse
(nurv im′pəls) A rapid electrical signal of a neuron that reports information or initiates a quick response in specific tissues. p. 250

neuromuscular junction
(nyōōr′ō mus′kyə lər jungk′shən) A synapse between a neuron and a skeletal muscle cell. p. 259

neuron
(nōōr′on) (Gr. nerve) A nerve cell specialized to conduct an electric current. p. 159

neurotransmitter
(nōōr′ō trans′mit ər) (Gr. neuron, nerve + L. trans, across + mitere, to send) A chemical released when a nerve impulse reaches the axon tip of a nerve cell. p. 257

neurulation
(nōōr′ə lā′shən) (Gr. neuron, nerve) The development of a hollow nerve cord, which later develops into the central nervous system. p. 363

neutron
(nōō′tron) (L. neuter, neither) A subatomic particle found at the nucleus of atom, similar to a proton in mass but neutral and carrying no charge. p. 17

neutrophil
(nōō′trə fil) (L. neuter, neither + Gr. philein, to love) A type of granulocyte that migrates to the site of an injury and sticks to the interior walls of blood vessels, where it forms projections and phagocytizes microorganisms and other foreign particles. p. 210

niche
(nich) (L. nidus, nest) The special role each organism plays within an ecosystem;

the terms of space, food, temperature, mating conditions, moisture requirements, and behavior of an organism within an ecosystem. p. 639

nitrifying bacteria
(nī′trə fī′ing bak tir′ē ə) (Gr. nitron, salt) Bacteria that live in nodules in the roots of legumes and cycle nitrogen in ecosystems by converting ammonia to nitrates, a form of nitrogen used by plants. p. 493

node of Ranvier
(nōd uv räN vyā′) (after L. A. Ranvier, French histologist) An uninsulated spot between two Schwann cells. p. 252

noncyclic electron flow
(non sik′lik i lek′trŏn flō) A system that harvests energy in a process that involves photosystems I and II working together to produce NAPH and ATP; these two products power the generation of glucose from carbon dioxide in light-independent reactions. p. 137

nondisjunction
(non dis jungk′shən) The failure of homologous chromosomes to separate after synapsis, resulting in gametes with abnormal numbers of chromosomes. p. 389

nonspecific defense
(non spə sif′ik di fens′) A set of defenses that the body uses to act against foreign invaders; they include the skin and mucous membranes, chemicals that kill bacteria, and the inflammatory process. p. 217

notochord
(nō′tə kôrd) (Gr. noto, back + L. chorda, cord) A structure that forms the midline axis along which the vertebral column (backbone) develops in all vertebrate animals. p. 361

nuclear envelope
(nōō′klē ər en′və lōp) The outer, double membrane surrounding the surface of the nucleus of a eukaryotic cell. p. 56

nucleic acid
(nōō klē′ik or nōō klā′ik as′əd) A long polymer of repeating subunits called nucleotides; the two types of nucleic acid within cells are deoxyribonucleic acid (DNA) and ribonucleic acid (RNA). p. 39

nucleolus, pl. nucleoli
(nōō klē′ə ləs, nōō klē′ə lī′) (L. a small nucleus) The site within the nucleus of ribosomal RNA synthesis; consists of ribosomal RNA plus some ribosomal proteins. p. 57

nucleotide
(nōō′klē ə tīd) A single unit of nucleic acid consisting of a five-carbon sugar, a phosphate group, and an organic nitrogen-containing molecule, or base. p. 39

nucleus
(nōō′klē əs) (L. a kernal, dim. fr. nux, nut) (1) The central core of an atom containing protons and neutrons. (2) The double membrane vesicle of a eukaryotic cell that contains the hereditary material, or DNA. p. 267

nutrient
(nōō′trē ənt) (L. nutritio, nourish) A raw material of food; the six classes of nutrients are carbohydrates, fats, proteins, vitamins, minerals, and water. p. 166

occipital lobe
(ok sip′ə təl lōb) The section of the cerebral cortex in each hemisphere of the brain having to do with vision, with different sites corresponding to different positions on the retina. p. 266

omnivore
(om′nə vôr) (L. omnis, all + vorare, to eat) An organism that eats both plant and animal foods; human beings are omnivores p. 170

oogenesis
(ō′ə jen′ə səs)(Gr. oon, egg + genesis, generation, birth) The process of meiosis and development that produces mature female sex cells, or eggs. p. 334

operator
(op′ə rāt′ər) In bacteria and viruses, a gene that acts as an on or off switch for the transcription of the genes. p. 420

operculum
(ō pur′kyə ləm) (L. operire, to shut, cover) A flap in a bony fish that protects the gills and enhances water flow over the gills, thus bringing more oxygen in contact with their gas-exchanging surfaces and allowing a fish to breath while stationary. p. 586

operon
(op′ər on) (L. operis, work) In bacteria and viruses, a cluster of genes having related functions that are regulated as a unit. p. 420

optic nerve
(op′tik nurv) The nerve carrying impulses for the sense of sight. p. 288

organ
(ôr′gən) (L. organon, tool) Grouped tissues that form a structural and functional unit. p. 147

organ of Corti
(ôr′gən uv kôrt′ē) (after Alfonso Corti, Italian anatomist) The organ of hearing; the collective term for the hair cells, the supporting cells of the basilar membrane, and the overhanging tectorial membrane. p. 290

organ system
(ôr′gən sis′təm) A group of organs that function together to carry out the principal activities of the organism. p. 149

organelle
(ôr′gə nel′) (Gr. organella, little tool) Any of a number of highly specialized, membrane-bound, intracellular structures within a cell that perform specific cellular functions; a feature common to most eukaryotes but lacking in bacteria. p. 12

origin
(ôr′ jən) (L. oriri, to arise) The attachment of one end of a skeletal muscle to a stationary bone. p. 305

osmoregulation
(oz′mō reg′yə lā′shən or os′mō reg′yə lā′shən) The control of water movement in and out of an organism's body. p. 585

osmosis
(oz mō′səs or os mō′səs) (Gr. osmos, impulse + osis, condition) A special form of diffusion in which water molecules move from an area of higher concentration to an area of lower concentration across a differentially permeable membrane. p. 72

osmotic pressure
(oz mot′ik or os mot′ik presh′ər) The increase that water pressure exerts on a cell as water molecules continue to diffuse into a cell. p. 73

osteoblast
(os′tē ə blast) (Gr. osteon, bone + blastos, cell) A cell that forms new bone. p. 296

osteocyte
(os′tē ə sīt) (Gr. osteon, bone + kytos, hollow vessel) A cell that produces bone. p. 153

otolith
(ō′tə lith) (Gr. otos, ear + lithos, stone) Small pebbles of calcium carbonate embedded in a layer of jelly-like material that is spread over the surface of ciliated and nonciliated cells within the saccule. p. 291

outer ear
(out′ər ir) The part of the ear that funnels sound waves in toward the eardrum; includes the flaps of skin on the outside of the head called ears. p. 289

oval window
(ō′vəl win′dō) The entrance to the inner ear. p. 289

ovary
(ōv′ə rē) (L. ovum, egg) (1) A female gonad, located in the pelvic cavity of an animal, where egg production occurs. (2) In flowering plants, a chamber at the base of the female pistil that completely encloses and protects the ovules. p. 339

oviparous
(ō vip′ər əs) (L. ovum, egg + parere, to

bring forth) A method of reproduction in which the female lays her eggs and the young hatch outside of the mother; birds and most reptiles are oviparous. p. 586

ovoviviparous
(ō'vō vī vip' ər əs) (L. ovum, egg + vivus, alive + parere, to bring forth) A method of reproduction in which the female retains the fertilized eggs within her oviducts until the young hatch; the young receives its nutrition from the egg yolk and not from the mother. p. 586

ovulation
(ov'yə lā' shən) (L. ovulum, little egg) The monthly process by which an egg is produced and released by the ovary. p. 340

ovule
(ō'vyōōl) (L. ovulum, little egg) A protective structure in which egg cells grow in a naked seed plant. p. 533

ovum, pl. ova
(ō' vəm, ō'və) (L. egg) A mature egg cell. p. 340, 355

oxidation
(ok'sə dā'shən) (Fr. oxider, to oxidize) The loss of an electron by an atom or a molecule. p. 119

oxidation-reduction
(ok'sə dā'shən ri duk'shən) In some chemical reactions, the passing of electrons from one atom or molecule to another; critically important to the flow of energy through living systems and essential to the flow of energy in cellular respiration. p. 119

oxygen debt
(ok'si jən det) A term used to describe the oxygen needed to break down the lactic acid in the liver, delivered by the bloodstream from the muscles during strenuous exercise. p. 128

oxytocin
(ok'si tō'sən) (Gr. oxys, sharp + tokos, birth) A hormone produced by the hypothalamus but stored and released in the posterior lobe of the pituitary that affects the contraction of the uterus during childbirth and stimulates the mammary glands, allowing a new mother to nurse her child. p. 321

ozone
(ō'zōn) (Gr. ozein, to smell) A principal chemical air pollutant, (O_3), formed by photochemical reactions on hydrocarbons and nitrogen oxides in the air, that is extremely irritating to the eyes and upper respiratory tract. p. 707

pancreas
(pang'krē əs *or* pan'krē əs) (Gr. pan, all + kreas, flesh) A long gland that lies beneath the stomach and is surrounded on one side by the curve of the duodenum; it secretes a number of digestive enzymes and the hormones insulin and glucagon. p. 195

parasite
(par'ə sīt') (Gr. para, beside + sitos, food) The organism in a parasitic relationship that is the beneficiary; the host organism is harmed. p. 516

parasitism
(par'ə sə tiz'əm) (Gr. para, beside + sitos, food) A type of symbiosis in which one species benefits but the other (the host) is harmed. p. 649

parasympathetic system
(par'ə sim'pə thet'ik sis'təm) (Gr. para, beside + syn, with + pathos, feeling) A subdivision of the autonomic nervous system that generally stimulates the activities of normal internal body functions and inhibits alarm responses; opposite of sympathetic nervous system. p. 276

parathyroid gland
(par'ə thī'roid gland) (Gr. para, beside + thyreos, shield + eidos, form, shape) One of four small glands embedded in the posterior side of the thyroid that produces parathyroid hormone. p. 325

parathyroid hormone (PTH)
(par'ə thī'roid hôr'mōn) (Gr. para, beside + thyreos, shield + eidos, form, shape) A hormone secreted by the parathyroid glands that works antagonistically to calcitonin to help maintain the proper blood levels of various ions, primarily calcium. p. 322

parenchyma cell
(pə reng'kə mə sel) (Gr. para, beside + en, in + chein, to pour) A cell in the ground tissue of a vascular plant that functions in photosynthesis and storage; the most common of all plant cell types. p. 547

parietal lobe
(pə rī'ə təl lōb) The section of the cerebral cortex of each hemisphere of the brain containing sensory receptors from different parts of the body. p. 266

passive immunity
(pas'iv i myōōn'ə tē) The type of immunity produced by injection of antibodies into the subject to be protected or acquired by the fetus through the placenta. p. 227

passive transport
(pas'iv trans'pôrt) Molecular movement down a gradient but across a cell membrane; the three types of passive transport are diffusion, osmosis, and facilitated diffusion. p. 72

pectoral girdle
(pek'tə rəl gurd'əl) (L. pectus, chest) The part of the appendicular skeleton made up of two pairs of bones: the clavicles, or collarbones, and the scapulae, or shoulder blades. p. 300

pelvic girdle
(pel'vik gurd'əl) (L. pelvis, basin) The part of the appendicular skeleton made up of the two bones called coxal bones, pelvic bones or hip bones. p. 300

penis
(pē'nəs) A cylindrical organ that transfers sperm from the male reproductive tract to the female reproductive tract; the male urinary organ. p. 338

pepsin
(pep'sən) (Gr. pepsis, digestion) An enzyme of the stomach that digests only proteins, breaking them down into short peptides. p. 174

peptide bond
(pep'tīd bänd) A covalent bond that links two amino acids formed during dehydration synthesis when the amino group at one end and the carboxyl group at the other end lose a molecule of water between them. p. 34

pericycle
(per'ə sī' kəl) (Gr. peri, around + kykos, circle) In the roots of vascular plants, a cylinder of cells made up of parenchymal cells that are able to undergo cell division to produce either branch roots or new vascular tissue. p. 549

peripheral nervous system
(pə rif'ə rəl nurv'əs sis'təm) (Gr. peripherein, to carry around) The part of the nervous system made up of the nerves of the body that bring messages to and from the brain and spinal cord. p. 264

peristalsis
(per'ə stôl'səs) (Gr. peri, around + stellein, to wrap) The rhythmic wave of contractions by the muscles of the esophagus that moves food down toward the stomach. p. 172

permafrost
(pur'mə frôst') (Perma[nent] + frost) A layer of permanently frozen subsoil, found about 1 meter down from the surface, that is common throughout most of the Arctic regions. p. 684

petal
(pet'əl) (Gr. petalon, leaf) A part of a flower, frequently prominent and usually colorful, that serves to attract pollinators, usually insects. p. 534

petiole
(pet'ē ōl) (L. petiolus, a little foot) The slender stalk of a leaf. p. 554

pH scale
(pē'āch' skāl) A scale that indicates the relative concentration of H^+ ions in a solution. Low pH values indicate high concentrations of H^+ ions (acids), and

high pH values indicate low concentrations. p. 25

phagocytosis
(fag′ə sī tō′səs) (Gr. phagein, to eat + kytos, hollow vessel) A type of endocytosis in which a cell ingests an organism or some other fragment of organic matter; macrophages and neutrophils are phagocytes. p. 77

pharyngeal (gill) slit
(fə rin′jē əl *or* far′in jē′əl) (Gr. pharynx, gullet) A principal feature of chordates that develops into the gill structure of a fish and into the ear, jaw, and throat structures of a terrestrial vertebrate. p. 464

pharynx
(far′ingks) (Gr. gullet) The upper part of the throat that extends from behind the nasal cavities to the openings of the esophagus and layrnx. p. 172

phenotype
(fē′nə tīp) (Gr. phainein, to show + typos, print) The outward appearance or expression of an organism's genes. p. 378

phloem
(flō′em) (Gr. phloos, bark) A type of vascular tissue that conducts carbohydrates a plant uses as food as well as other needed substances. p. 546

phospholipid
(fos′fō lip′id) (Gr. phosphoros, light-bearer + lipos, fat) A molecule made up of a portion of a fat molecule with a phosphate functional group attached. Because these molecules have polar and nonpolar parts, they form a double layer of molecules (a bilayer) when in a watery environment. A phospholipid bilayer is the foundation of cell membranes. p. 68

photoautotroph
(fōt′ō ôt′ə trōf) (Gr. photos, light + auto, self + trophos, feeder) A eubacterium that makes its own food by photosynthesis using the energy of the sun. p. 492

photocenter
(fōt′ō sint′ər) An array of pigment molecules within the thylakoid membranes of a chloroplast. It acts like a light antenna, capturing and directing photon energy toward a single molecule of a chlorophyll *a* that will participate in photosynthesis. p. 136

photochemical smog
(fōt′ō kem′i kəl smôg) (Gr. photos, light + chemeia, chemistry) Smog that is caused by pollutants reacting in the presence of sunlight. p. 707

photon
(fō′ton) (Gr. photos, light) A discrete packet of energy from sunlight. p. 132

photosynthesis
(fōt′ō sin′thə səs) (Gr. photos, light -syn, together + tithenai, to place) The process whereby energy from the sun is captured by living organisms and used to produce molecules of food; it takes place in the chloroplasts of photosynthetic eukaryotic cells.

Photosystem I
(fōt′ō sis′təm wun) The first photosynthetic pathway to evolve; it is used by most bacteria to produce ATP using energy from the sun. p. 137

Photosystem II
(fōt′ō sis′təm tōō) A photosynthetic pathway in which a molecule of chlorophyll *a* (P680) can absorb more photons of higher energy than in Photosystem I. p. 137

phylum, pl. phyla
(fī′ləm, fī′lə) (Gr. phylon, race, tribe) A major taxonomic group, ranking above a class. p. 13

pineal gland
(pin′ē əl gland) (L. pinus, pine tree) A tiny gland lying deep within the brain whose exact function remains a mystery; it is the possible site of an individual's biological clock. p. 328

pinna
(pin′ə) (L. feather) The projected part of the outer ear, the ear flap. p. 289

pinocytosis
(pī′nō sī tō′səs) (Gr. pinein, to drink + kytos, vessel) A type of endocytosis in which a cell ingests liquid material containing dissolved molecules. p. 77

pistil
(pist′əl) (L. pistillum, pestle) The female sex organ at the center of a flower. p. 534

pith
(pith) A parenchymal (storage) tissue located in the center of the roots of a monocot. p. 549

pituitary
(pə tōō′ə ter′ē) (L. pituita, phlegm) A tiny gland hanging from the underside of the brain, under the control of the hypothalamus, that secretes nine different major hormones. p. 318

placenta
(plə sen′tə) (L. a flat cake) A flat disk of tissue that grows into the uterine wall, through which the mother supplies the offspring with food, water, and oxygen and through which she removes wastes. p. 360

placental mammal
(plə sent′əl mam′əl) (L. a flat cake) A mammal that nourishes its developing embryo with the body of the mother by means of a placenta until development is almost complete. p. 471

plankton
(plangk′tən) (Gr. planktos, wandering) Small organisms that drift in the water, especially at or near the surface. p. 662

plasma
(plaz′mə) (Gr. form) The fluid intercellular matrix within which blood cells float; contains practically every substance used and discarded by cells as well as nutrients, hormones, proteins, salts, ions, and albumin. p. 157

plasma cell
(plaz′mə sel) Any of several different kinds of cells and cell parts suspended within plasma, including erythrocytes, leukocytes, and platelets. p. 224

plasma membrane
(plaz′mə mem′brān) A thin, nonrigid structure that encloses the cell and regulates interactions between the cell and its environment. p. 48

plasmid
(plaz′mid) (Gr. plasma, form) A small fragment of DNA that replicates independently of the main chromosome. p. 63

plasmodium, pl. plasmodia
(plaz mō′dē əm, plaz mo′dē ə) (L. L. plasma, form + Gr. eidos, shape) A nonwalled, multinucleate mass of cytoplasm in a slime mold. p. 515

plate tectonics theory
(plāt tek ton′iks thir′ē *or* thē′ə rē) The theory that the outer shell of the Earth is composed of six large and several smaller plates that are rigid pieces of the Earth's crust; the theory provides insight into the movement of land masses over time. p. 458

platelet
(plāt′lət) (Gr. dim of plattus, flat) A cell fragment present in blood that plays an important role in the clotting of blood. p. 211

predator
(pred′ə tər) (L. praeda, prey) An organism of one species that kills and eats organisms of another. p. 642

prey
(prā) (L. prehendere, to grasp or seize) An organism killed and eaten by a predator. p. 642

primary consumer
(prī′mer ē kən sōōm′ər) A herbivore that feeds directly on green plants. p. 658

primate
(prī′māt or prī′mət) (L. primus, first) A mammal that has characteristics reflecting a tree-dwelling life-style, such as hands and feet able to grasp things, flexible limbs, and a flexible spine. p. 471

producer
(prə dōōs′ər) An organism in an

ecosystem capable of capturing energy from the sun and converting it to chemical energy usable to themselves and consumers. p. 10

prokaryotic cell
(prō kar′ē ot′ik sel) (Gr. pro, before + karyon, kernel) A cell smaller than a eukaryote, it has a simple interior organization with a single, circular strand of hereditary material that is not enclosed with a membrane; bacteria and the blue-green bacteria are prokaryotes. p. 47

prolactin
(prō lak′tən) (L. pro, before + lac, milk) A hormone secreted by the anterior pituitary that in association with estrogen, progesterone, and other hormones, stimulates the mammary glands in the breasts to secrete milk after a woman has given birth to a child. p. 321

prophase
(prō fāz′) (Gr. pro, before + phasis, form) The stage of mitosis characterized by the appearance of visible chromosomes. p. 86

proprioceptor
(prō′prē ə sep′tər) (L. proprius, one's own + ceptor, a receiver) A receptor located within the skeletal muscles, tendons, and inner ear that gives the body information about the position of its parts relative to each other and to the pull of gravity. p. 282

prosimian
(prō sim′ē ən) (Gr. pro, before + L. simia, an ape) A suborder of primates that includes lemurs, indris, aye-ayes, and lorises; they are small animals, mostly nocturnal, with large ears and eyes, elongated snouts, and rear limbs. p. 473

prostaglandin
(pros′tə glan′dən) (from prosta[te] gland + -in) A hormone secreted by cell membranes throughout the body that stimulates smooth muscle contraction and the dilation and constriction of blood vessels. p. 330

pleura, pl. pleurae
(ploŏr′ə, ploŏr′ē) (Gr. side) A thin, delicate, sheet-like membrane that lines the interior walls of the thoracic cavity and folds back on itself to cover each lung; covered by a thin film of liquid, the pleura reduces friction during respiratory movements of the lungs. p. 191

pollen grain
(pol′ən grān) (L. fine dust) A male gametophyte produced by an anther within the male stamen of a flower. p. 534

pollination
(pol′ə nā′shən) The fertilization of the stigmas of flowers by the transfer of pollen from the anthers, usually accomplished by the wind, insects, or birds. p. 534

pollution
(pə loō′shən) (L. polluere, to pollute) Contamination of the water sources and air of the Earth, causing physical and chemical changes that harm living and nonliving things. p. 703

polymer
(pol′ə mər) (Gr. polus, many + meris, part) A giant molecule formed of long chains of similar molecules. p. 28

polyp
(pol′əp) (Gr. polypous, many feet) A cnidarian that exists as an aquatic, cylindrical animal with a mouth at one end ringed with tentacles; it anchors itself to rocks. p. 567

polypeptide
(pol′ē pep′tīd) (Gr. polys, many + peptein, to digest) A long chain of amino acids linked end to end by peptide bonds; proteins are long, complex polypeptides. p. 34

polysaccharide
(pol′ē sak′ə rīd) (Gr. polys, many + sakcharon, sugar) A long polymer composed of insoluble sugar. p. 30

polyunsaturated
(pol′ē un sach′ə rā′təd) Referring to a fat composed of fatty acids that has more than one double bond; polyunsaturated fats have low melting points and are therefore liquid fats, or oils. p. 33

pons
(ponz) (L. bridge) The part of the brainstem consisting of a band of nerve fibers that acts as "bridges" and connects various parts of the brain to one another; it also brings messages to and from the spinal cord. p. 271

population
(pop′yə lā′shən) (L. populus, the people) A group that consists of the individuals of a given species that occur together at one place and at one time. p. 639

positive feedback loop
(poz′ət iv fēd′ bak′ loōp) A feedback loop in which the response of the regulating mechanism is positive with respect to the outcome. p. 163

postsynaptic membrane
(pōst′sə nap′tik mem′brān) The membrane of the target cell of the synaptic cleft. p. 257

potential energy
(pə ten′chəl en′ər jē) Stored energy; energy not actively doing work but having the capacity to do so. p. 103

predation
(pri dā′shən) (L. praeda, prey) The killing and eating of an organism of one species by an organism of another species; an animal that kills and eats members of its own species is a cannibal. p. 629

prostate
(pros′tāt) (Gr. prostates, one standing in front) A gland surrounded by the male urethra that adds a milky alkaline fluid to semen, neutralizing the acidity of the female vagina. p. 338

protein
(prō′tēn *or* prō′tē ən) (Gr. proteios, primary) A linear polymer of amino acids. p. 34

protist
(prō′tist) (Gr. protos, first) A member of the kingdom Protista, which includes unicellular eukaryotes as well as some multicellular forms. p. 503

protocell
(prōt′ō sel′) (Gr. protos, first + L. cella, chamber) An early precursor of a cell. p. 448

proton
(prō′tôn) A subatomic particle in the nucleus of an atom that has mass and carries a positive charge. p. 17

protosome
(prō′tə stōm′) (Gr. proto, first + stoma, mouth) A member of a branch of coelomate animals that during the gastrula stage of development, gives rise to a two-layered embryo; this first indentation becomes the mouth of the organism. p. 564

protozoan
(prō′tə zō′ən) (Gr. protos, first + zoon, animal) An animal-like heterotrophic protist that eats organic matter for energy. p. 503

proximal convoluted tubule
(prôk′sə məl kän′və loō′təd toō′byool) A coiled portion of the nephron closest to Bowman's capsule lying in the cortex of the kidney, the site where reabsorption begins as the filtrate passes through the proximal tubule. p. 237

pseudocoelomate
(soōd′ōsē′lə māt′) (Gr. pseudos, false + koiloma, a hollow) An organism, such as a roundworm, that has a false body cavity. p. 562

pseudopod, pl. pseudopodia
(soō′də pod, soō′də pō′dē ə) (Gr. pseudos, false + pous, foot) A temporary, pushed out cell part of an ameba used for food procurement and locomotion. p. 504

pseudostratified
(soōd′ō strat′ə fīd′) (Gr. pseudos, false + L. stratificare, to arrange in layers) Refers to a type of epithelial

GLOSSARY G-21

tissue that gives the false appearance of being stratified. p. 151

pulmonary circulation
(pul′mə ner′ē sur′kyə lā′shən) The part of the human circulatory system that circulates blood to and from the lungs. p. 204

Punnett square
(pun′ət skwar) (after Reginald C. Punnett, English geneticist) A diagram that visualizes the genotypes of progeny in simple Mendelian crosses and illustrates their expected ratios. p. 378

pupil
(pyōō′pəl) (L. pupilla, little doll) The opening in the center of the iris through which light passes. p. 288

pyruvate
(pī rōō′vāt) The three-carbon molecule left when glycolysis is completed and the beginning material of the citric acid cycle. p. 117

radial symmetry
(rād′ē əl sim′ə trē) (L. radius, a spoke of a wheel; Gr. symmetros, symmetry) Describing the distribution of an animal's body parts that emerge, or radiate, from a central point, much like spokes on a wheel. p. 562

radicle
(rad′ə kəl) (L. radicula, root) The young root that is usually the first portion of the embryo to emerge from the germinating seed. p. 538

radiolarian
(rād′ē ō lar′ē ən) (L. radius, ray) A kind of ameba that secretes a shell made of silica that is glass-like and delicate. p. 505

receptor
(ri sep′tər) A specialized cell component that detects stimuli; it provides the body with information about the internal environment, the position in space, and the external environment. p. 254

recessive
(ri ses′iv) The form of a trait that recedes or disappears entirely in a hybrid offspring. p. 375

rectum
(rek′təm) (L. straight) The lower part of the large intestine, which terminates at the anus. p. 180

red blood cell
(red blud sel) See erythrocyte.

red bone marrow
(red bōn mar′ō) The soft tissue that fills the spaces within the bony latticework of spongy bone; it is the place where most of the body's blood cells are formed. p. 297

reduction
(ri duk′shən) (L. reduction, a bringing back: originally "bringing back" a metal from its oxide) The gain of an electron by an atom or a molecule. p. 119

reflex
(rē′fleks) (L. reflectare, to bend back) An automatic response to a nerve stimulation. p. 264

reflex arc
(rē′fleks ärk) The pathway of nervous activity in a reflex. p. 271

refractory period
(ri frak′tə rē pir′ē əd) The recovery period after membrane depolarization during which the membrane is unable to respond to additional stimulation. p. 255

releasing hormone
(ri lēs′ing hôr′mōn) A hormone produced by the hypothalamus that affects the secretion of specific hormones from the anterior pituitary. p. 318

replication fork
(rep′lə kā′shən fôrk) The split in a DNA molecule where the double-stranded DNA molecule separates during DNA replication. p. 410

respiration
(res′pə rā′shən) (L. respirare, to breathe) The uptake of oxygen and the release of carbon dioxide by the body. Cellular, internal, and external respiration are all part of the general process of respiration. p. 185

respiratory assembly
(res′pə rə tōr′ē ə sem′blē) A special channel that allows protons to pass back from the outer compartment of a mitochondrion into the inner compartment; each passage of a proton back into the inner compartment through a respiratory assembly is coupled to the synthesis of an ATP molecule. p. 125

response
(ri spons′) (L. respondere, to reply) The reply or reaction of an individual organism to a stimulus; it can be an innate or inborn reflex or a learned response such as by operant or classical conditioning. p. 255

resting potential
(rest′ing pə ten′chəl) An electrical potential difference, or electrical charge, along the membrane of the resting neuron. p. 254

restriction enzyme
(ri strik′shən en′zīm) An enzyme that recognizes certain base sequences in a DNA strand and breaks the bonds at that point. p. 496

retina
(ret′ən ə) (L. a small net) The structure in the eye, composed of rod and cone cells, that is sensitive to light. p. 286

rhizoid
(rī′zoid) (Gr. rhiza, root) In a bryophyte, a slender, root-like projection that anchors the plant to its substrate; unlike roots, however, it consists of only a few cells and does not play a major role in the absorption of water or minerals. p. 557

rhizome
(rī′zōm) (Gr. rhizoma, mass of roots) An underground stem of a plant that can produce a new plant by the process of vegetative propagation. p. 540

rhodopsin
(rō dop′sən) (Gr. rhodon, rose + opsis, vision) A complex in the retina formed by the coupling of retinal and opsin. p. 286

ribonucleic acid (RNA)
(rī′bō nōō klē′ik as′əd) One of two types of nucleic acid found in cells; differs from DNA in that its sugar is ribose and uracil is present rather than thymine. p. 406

ribosomal RNA (rRNA)
(rī′bə sō′məl är′en′ ā′) The type of RNA found in ribosomes that plays a role in the manufacture of polypeptides. p. 57

ribosome
(rī′bə sōm′) A minute, round structure found in endoplasmic reticulum; ribosomes are the places where proteins are manufactured. p. 53

ribulose biphosphate (RuBP)
(rib′yə lōs′ bī′fos′fāt) A molecule which joins with carbon dioxide in the first stage of the Calvin-Benson cycle and results in the process of carbon fixation. p. 142

RNA polymerase
(är′en′ā′ pol′ə mə rās′) The special enzyme that transcribes RNA from DNA. p. 413

rod
(räd) A light receptor located within the retina at the back of the eye that functions in dim light and detects white light only. p. 286

root
(rōōt or rŏŏt) The part of a plant usually found below ground; it absorbs water and minerals and anchors the plant. p. 545

root cap
(rōōt or rŏŏt kap) A thimble-like mass of cells that covers and protects the root's apical meristem as it grows through the soil. p. 550

root pressure
(rōōt or rŏŏt presh′ər) The process in which root cells, by means of special cellular pumps, maintain a higher concentration of dissolved minerals than the concentration of minerals in the water of the soil. p. 556

rough endoplasmic reticulum
(ruf en′dō plaz′mik ri tik′yə ləm) (Gr. endon, within + plasma, from cytoplasm; L. reticulum, network) The type of endoplasmic reticulum that makes and transports proteins destined to leave the cell; rough refers to the minute, round structures called ribosomes covering the surface where proteins are made. p. 53

round window
(round win′dō) A membrane-covered hole at the wider end of the cochlea. p. 289

SA node
(es′ā′ nōd) *See* sinoatrial (SA) node.

saccule
(sak′yōōl) (N.L. sacculus, small bag) A sac inside a bulge in the vestibule of the inner ear containing both ciliated and nonciliated cells. p. 291

saliva
(sə lī′və) (L. spittle) The secretion of the salivary glands; a solution consisting primarily of water, mucus, and the digestive enzyme salivary amylase. p. 170

salivary amylase
(sal′ə ver′ē am′ə lās) (L. salivarius, slimy + Gr. amylon, starch -asis, coloid enzyme) A digestive enzyme that breaks down starch into molecules of the disaccharide maltose. p. 170

salivary glands
(sal′ə ver′ē glandz) (L. salivarius, slimy + glans, acorn) The paired glands of the mouth that secrete saliva. p. 170

saltatory conduction
(sal′tə tōr′ē kən duk′shən) (L. saltatio, leaping) A very fast form of nerve impulse conduction in which impulses "jump" along myelinated neurons. p. 256

saprophytic
(sap′rō fit′ik) (Gr. sapros, putrid + phyton, plant) Feeding on dead or decaying organic material; fungi are saprophytic. p. 516

sarcomere
(sär′kə mir′) (Gr. sarx, flesh + meris, part of) The repeating bands of actin and myosin myofilaments that appear between two Z lines in a muscle fiber. p. 307

sarcoplasmic reticulum
(sär′kō plaz′mik ri tik′yə ləm) (Gr. sarx, flesh + plassein, to form, mold; L. reticulum, network) A tubular, branching latticework of endoplasmic reticulum that wraps around each myofibril like a sleeve. p. 310

Schwann cells
(shwän *or* shvän selz) (after Theodor Schwann, German anatomist) The supporting cells associated with nerve fibers of all the other cells that make up the peripheral nervous system. p. 252

scientific method
(sī′ən tif′ik meth′əd) A set of procedures used to answer questions that is common among the various scientific disciplines. p. 3

sclera
(sklir′ə) (Gr. skleros, hard) The tough outer layer of connective tissue that covers and protects the eye. p. 287

sclerenchyma cell
(sklə reng′kə mə sel) (Gr. skleros, hard + en, in + chymein, to pour) A dead, hollow cell with a strong wall found in the ground tissue of a vascular plant that helps support and strengthen the ground tissue. p. 547

scrotum
(skrō′təm) (L. a bag) A sac of skin, located outside the lower pelvic area of the male, which houses the testicles, or testes. p. 335

sea squirt
(sē skwurt) The common name given to most species of tunicates because of its habit of forcefully squirting water out of its excurrent siphons when disturbed. p. 581

secondary consumer
(sek′ən der′ ē kən sōōm′ər) A meat eater, or carnivore, that feeds on a herbivore. p. 658

seed
(sēd) A structure from which a new plant grows. p. 531

selective reabsorption
(sə lek′tiv rē′əb sôrp′shən) The process in which the kidneys select specific substances according to the body's needs from the filtrate. p. 235

semen
(sē′mən) (L. seed) Fluid produced by the accessory glands of the male reproductive system combined with sperm. p. 338

semicircular canal
(sem′i sur′kyə lər kə nal′) Any of three fluid-filled canals in the inner ear that detect the direction of an individual's movement. p. 291

seminal vesicle
(sem′ ən əl ves′ə kəl) One of two accessory glands that secrete a thick, clear fluid forming a part of the semen. p. 338

seminiferous tubule
(sem′ə nif′ər əs tōō′byōōl) (L. semen, seed + ferre, to produce) A tightly coiled tube within a testis where sperm cells develop. p. 336

sensory neuron
(sens′ə rē nōōr′on) A neuron of the peripheral nervous system that transmits information to the central nervous system. p. 250, 264

sepal
(sē′pəl) (L. sepalum, a covering) A member of the outermost whorl, or ring, of a flowering plant. p. 534

septum, pl. septa
(sep′təm, sep′tə) (L. saeptum, a fence) (1) The tissue that separates the two sides of the heart. (2) In a fungus, a cross wall that divides a hypha into cells. (3) In an annelid, an internal partition that divides the segments. p. 516

seta, pl. setae
(sē′tə, sē′tē) (L. bristle) A bristle in the body of an earthworm that helps anchor the worm during locomotion or when it is in its burrow. p. 574

sex chromosome
(seks krō′mə sōm′) The X and Y chromosomes that determine the gender of an individual as well as certain other characteristics. p. 81

sexual reproduction
(seksh′ə wəl *or* sek′shəl rē prə duk′shən) The type of reproduction that involves the fusion of gametes to produce the first cell of a new individual. p. 83

sickle-cell anemia
(sik′əl sel′ ə nē′mē ə) (A.S. sicol, curved + L. cella, chamber + Gr. a, not + haima, blood) A recessive genetic disorder common to African blacks and their descendants in which affected individuals cannot transport oxygen to their tissues properly because the molecules within the red blood cells that carry oxygen, hemoglobin proteins, are defective. p. 398

simple eye
(sim′pəl ī) A structure composed of a single visual unit having one lens; found in some arthropods and some flying insects. p. 576

sinoatrial (SA) node
(sī′nō ā′trē əl nōd) A small cluster of specialized cardiac muscle cells embedded in the upper wall of the right atrium of the heart that automatically and rhythmically sends out impulses initiating each heartbeat. p. 206

sinus venosus
(sī′nəs və nō′səs) A chamber in the hearts of cartilaginous and bony fishes that collects blood from the organs. p. 586

sister chromatid
(sis′tər krō′mə tid) Either of two identical structures held together at the centromere, composed of chromatin material that coils and condenses just

before and during cell division to form chromosomes. p. 81

skeletal muscle
(skel′ət əl mus′əl) A type of muscle that is voluntary because of conscious control over its action; skeletal muscle is connected to bones and allows for body movement. p. 158

slime mold
(slīm mōld) A protist organism that is fungus-like in one phase of its life cycle and ameba-like in another phase of its life cycle. p. 504

small intestine
(smôl in tes′tən) The tube-like portion of the digestive tract that begins at the pyloric sphincter and ends at its T-shaped junction with the large intestine. p. 175

smooth endoplasmic reticulum
(smooth en′dō plaz′mik ri tik′yə ləm) (Gr. endon, within + plasma, from cytoplasm; L. reticulum, network) The type of endoplasmic reticulum that helps build carbohydrates and lipids within the cytoplasm; it does not have ribosomes attached to its surface and does not manufacture proteins. p. 53

smooth muscle
(smooth mus′əl) A type of muscle that contracts involuntarily and is located in the walls of certain internal structures such as blood vessels and the stomach. p. 158

sociobiology
(sō′sē ō bī ol′ə jē) The biology of social behavior that applies the knowledge of evolution to the study of animal behavior. p. 611

sodium-potassium pump
(sōd′ē əm pə tas′ē əm pump) The term given to the coupled channel that uses energy to move sodium (Na^+) and potassium (K^+) ions across the cell membrane. p. 75

soft palate
(sôft pal′ət) (L. palatum, palate) The tissue at the back of the roof of the mouth. p. 172

solute
(sol′yoot) The other kinds of molecules dissolved in water. See solution, solvent. p. 73

solution
(sə loo′shən) A mixture of molecules and ions dissolved in water. p. 73

solvent
(solv′ənt) The most common of the molecules in a solution, usually water. p. 73

somatic nervous system
(sō mat′ik nurv′əs sis′təm) (Gr. soma, body) The branch of the peripheral nervous system consisting of motor neurons that send messages to the skeletal muscles and control voluntary responses. p. 264

somite
(sō′mīt) (Gr. soma, body) A chunk of mesoderm that gives rise to most of the axial skeleton and most of the dermis of the body. p. 364

spawn
(späwn) (M.E. spawnen, to spread out) The depositing of sperm and eggs directly into water by males and females; a method of reproduction used by most fishes, amphibians, and shellfish. p. 584

speciation
(spē′shē ā′shən) The process by which new species are formed during the process of evolution. p. 433

species
(spē′shēz or spē′sēz) A group of related organisms that shares common characteristics and is able to interbreed and produce viable offspring. p. 13

specific defense
(spə sif′ik di fens′) The immune response of the body. p. 217

spermatid
(spur′mə tid) A haploid cell in the testes arising from a diploid cell called a spermatogonium. p. 336

spermatogenesis
(spur′mə tō jen′ə səs) (Gr. sperma, sperm, seed + gignesthai, to be born) The development of sperm cells within the coiled tubules of the testis triggered by follicle-stimulating hormones. p. 336

spinal cord
(spīn′əl kôrd) The part of the central nervous system that runs down the neck and back, receives information from the body, carries this information to the brain, and sends information from the brain to the body. p. 264

spinal nerve
(spīn′əl nurv) A nerve by which the spinal cord receives information from the body; includes sensory and motor nerves. p. 271

spindle fibers
(spin′dəl fī′bərz) Special microtubules that are assembled from a pair of related centrioles during the prophase stage of mitosis. p. 87

spiracle
(spir′ə kəl or spī′rə kəl) (L. spirare, to breathe) A specialized opening in the body of most terrestrial arthropods that allows air to pass into the trachea. p. 576

spleen
(splēn) (Gr. splen) An organ of the lymphatic system that stores an emergency blood supply and also contains white blood cells. p. 211

spongy bone
(spun′jē bōn) A type of bone in the human skeleton that is composed of an open latticework of thin plates of bone; spongy bone is found at the ends of long bones and within short, flat, and irregularly shaped bones. p. 156

spore
(spôr) (Gr. spora, seed) A reproductive body formed by meiosis in a diploid parent and by mitosis in a haploid parent; the spores are always haploid. p. 514

sporophyte
(spôr′ə fīt) (Gr. spora, seed + phyton, plant) In the life cycle of a plant, the diploid (spore-plant) generation that alternates with the haploid phase. p. 529

sporozoan
(spôr′ə zō′ən) (Gr. spora, seed + zoon, animal) A nonmotile protozoan that is a parasite of vertebrates, including humans; they are passed from host to host by various insect species. p. 509

squamous
(skwā′məs) (L. squama, scale) One of the main shapes of epithelial cells. Squamous cells are thin and flat and are found in the air sacs of the lungs, the lining of blood vessels, and the skin. p. 150

stamen
(stā′mən) (L. thread) The male sex organ of a flower. p. 534

starch
(stärch) (A.S. stercan) Stored energy in plants formed by using glucose to form polysaccharides. p. 30

stem
(stem) The framework of a plant that supports and positions the leaves where most photosynthesis occurs. p. 545

stereoscopic vision
(ster′ē ō skôp′ik vizh′ən) (Gr. stereos, solid + skopein, to view) Vision created by two eyes focusing on the same object. p. 472

stoma, pl. stomata
(stō′mə, stō′mə tə) (Gr. mouth) An opening in the epidermis of a leaf that allows the carbon dioxide needed for photosynthesis to enter and the oxygen produced by photosynthesis to escape; it also permits water vapor to escape. p. 554

stomach
(stum′ək) (Gr. stomachos, mouth) A muscular sac in which food is collected and partially digested by hydrochloric acid and proteases. p. 177

stratified
(strat′ə fīd′) (L. stratificare, to arrange in layers) Refers to epithelium that is made up of two or more layers. p. 151

stretch receptor
(strech ri sep′tər) A special nerve that is sensitive to any stretching in its tissue. p. 283

stroma, pl. stromata
(strō′mə, strō′mə tə) (Gr. anything spread out) The fluid that surrounds the thylakoids inside chloroplasts. p. 135

structural gene
(struk′chə rəl jēn) In bacteria and viruses, a gene that directs the synthesis of enzymes having related functions. p. 420

style
(stīl) (L. stilus, stylus) A narrow stalk arising from the top of the ovary that bears the stigma in a flowering plant. p. 534

succession
(sək sesh′ən) The dynamic process of change, during which a sequence of communities replaces one another in an orderly and predictable way. p. 651

suppressor T cell
(sə pres′ər tē sel) A kind of T cell that limits the immune response. p. 222

symbiosis
(sim′bē ō′səs *or* sim′bī ō′səs) (Gr. a living together) The living together of two or more organisms in a close association. p. 48

sympathetic nervous system
(sim′pə thet′ik nurv′əs sis′təm) A subdivision of the autonomic nervous system that generally mobilizes the body for greater activity; produces responses that are the opposite of the parasympathetic nervous system. p. 276

synapse
(sin′aps *or* sə naps′) (Gr. synapsis, a union) A junction between an axon tip and another cell, usually including a narrow gap separating the two cells. p. 257

synapsis
(sə nap′səs) (Gr. union) The lining up of homologous chromosomes during prophase I of meiosis, initiating the process of crossing-over. p. 93

synaptic cleft
(sə nap′tik kleft) The space or gap between two adjacent neurons. p. 257

synovial joint
(sənō′vē əl joint) (L. synovia, joint fluid) A freely movable joint in which a space exists between articulating bones. p. 302

systemic circulation
(sis tem′ik sur′kyə lā′shən) The pathway of blood vessels to the body regions and organs other than the lung. p. 205

systolic period
(sis tol′ik pir′ē əd) (Gr. systole, contraction) The pushing period of heart contraction, which ends with the closing of the aortic valve, during which a pulse of blood is forced into the systemic arterial system, immediately raising the blood pressure within these vessels. p. 207

T cell (T lymphocyte)
(tē sel) A type of lymphocyte that carries out the cell-mediated immune response. p. 222

T lymphocyte
(tē lim′fə sīt) *See* T cell.

taiga
(tī gä′) (Russ.) The northern coniferous forest extending across vast areas of Eurasia and North America; it is a marshy region south of the tundra that has long, cold winters. p. 682

tarsal
(tär′səl) (Gr. tarsos, a broad, flat surface) Any of the seven short bones that make up the ankle in the appendicular skeleton of a human. p. 302

taste bud
(tāst bud) A microscopic chemoreceptor embedded within the papillae of the tongue that works with the olfactory receptors to produce the taste sensation. p. 285

taxonomy
(tak sän′ə mē) (Gr. taxis, arrangement + nomos, law) The classification of the diverse array of species by categorizing organisms based on their common ancestry. p. 12

Tay-Sachs disease
(tā′saks′ diz ēz′) (after Warren Tay, British physician, and Bernard Sachs, U.S. neurologist) An incurable, fatal, recessive, hereditary disorder in which brain deterioration causes death by the age of 5 years; has a high incidence of occurrence among Jews of Eastern and Central Europe and among American Jews. p. 398

tectorial membrane
(tek tôr′ē əl mem′brān) (L. tectum, roof) The membrane that covers the hairs that stick up from the cochlear duct. p. 290

telophase
(tel′ə fāz) (Gr. telos, end + phasis, form) The stage of mitosis during which the mitotic apparatus assembled during prophase is disassembled, the nuclear envelope is reestablished, and the normal use of the genes present in the chromosomes is reinitiated. p. 90

template
(tem′plət) A pattern used as a guide to duplicate a shape or structure; refers to the ability of any polynucleotide, such as a RNA molecule, that can act as a guide for the synthesis of a second polynucleotide based on the complementarity of its bases. p. 447

temporal lobe
(tem′pə rəl lōb) (L. temporalis, the temples) The section of the cerebral cortex in each hemisphere of the brain dealing with hearing; different surface areas correspond to different tones and rhythms. p. 266

tendon
(ten′dən) (Gr. tenon, stretch) A tissue that connects muscles to bones. p. 305

teratogen
(tə rat′ə jən) (Gr. teratos, monster + gennan, to produce) An agent, such as alcohol, that can induce malformations in rapidly developing tissues and organs. p. 361

testcross
(test′krôs′) A cross between a phenotypically dominant test plant with a known homozygous recessive plant; devised by Mendel to further test his conclusions. p. 380

testis, pl. testes
(tes′təs, tes′tēz′) (L. witness) The male gonads where sperm production occurs. p. 335

testosterone
(tes′tos′tə rōn) (Gr. testis, testicle + steiras, barren) A sex hormone secreted by the testes responsible for the development and maintenance of male secondary sexual characteristics. p. 320

tetrad
(te′trad′) (Gr. tetras, four) During prophase I of meiosis I, the name given to the paired homologous chromosomes that together have four chromatids. p. 93

tetrapod
(tet′rə pod) (Gr. tetrapodos, four-footed) A land-living four-limbed vertebrate. p. 584

thalamus
(thal′ə məs) (Gr. thalamos, chamber) A mass of gray matter lying at the base of the cerebrum that receives sensory stimuli, interprets some of these stimuli, and sends the remaining sensory messages to appropriate locations in the cerebrum. p. 269

theory
(thē′ə rē or thir′ē) A synthesis of hypotheses that has withstood the test of time and is therefore a powerful concept that helps scientists made dependable predictions about the world. p. 7

thrombus
(throm′bəs) (Gr. clot) A blood clot that forms in a blood vessel and interferes with the flow of blood. p. 213

thylakoid
(thī′lə koid) (Gr. thylakos, sac + -oides, like) A flat, sac-like membrane in the chloroplast of a eukaryote; stacks of thylakoids are called grana. p. 60

thymus
(thī′məs) (Gr. thymos) A small gland located in the neck that plays an important role in the maturation of certain lymphocytes called T cells, which are an essential part of the immune system. p. 211

thyroid gland
(thi′roid gland) An important gland located in the neck near the voice box that produces hormones regulating the body's metabolism. p. 322

thyroid-stimulating hormone (TSH)
(thi′roid stim′yə lāt ing hôr′mōn) A tropic hormone produced by the anterior pituitary that triggers the thyroid gland to produce the three thyroid hormones. p. 320

thyroxine (T$_4$)
(thī rok′sēn or thī rok′sən) A hormone produced by the thyroid gland that helps regulate the body's metabolism. p. 322

tidal volume
(tīd′əl väl′ yəm or väl′yum) The amount or volume of air inspired or expired with each breath. p. 192

tissue
(tish′$\overline{oo}$) (L. texere, to weave) A member of a group of similar cells that works together to perform a function. p. 8

trachea
(trā′kē ə) (L. windpipe) The windpipe; the air passageway that runs down the neck in front of the esophagus and brings air to the lungs. p. 186

tracheid
(trā′kē əd) (Gr. tracheia, rough) A cell in the xylem that has holes along the length of its walls; it carries water and dissolved minerals through a plant and also provides support. p. 546

trait
(trāt) A distinguishing feature or characteristic of a plant or person due to the transmission of genes. p. 374

transcription
(trans krip′shən) (L. trans, across + scribere, to write) The first step in the process of polypeptide synthesis in which a gene is copied into a strand of messenger RNA. p. 413

transfer RNA (tRNA)
(trans′fur är′en′ā′) A type of RNA that transports amino acids, used to build polypeptides, to the ribosomes; they also align each amino acid at the correct place on the elongating chain. p. 413

translation
(trans lā′shən) (L. trans, across + locare, to put or place) The second step of gene expression in which mRNA, using its copied DNA code, directs the synthesis of a polypeptide. p. 416

translocation
(trans′lō kā′shən) (L. trans, across + locare, to put or place) A situation in which a section of a chromosome breaks off and then reattaches to another chromosome, producing an abnormally long chromosome. p. 393

transpiration
(trans′pə rā′shən) (L. trans, across + spirare, to breathe) The process by which water leaves a leaf through a stoma in the form of water vapor. p. 555

trophic level
(trō′fik lev′əl) (Gr. trophos, feeder) A feeding level in an ecosystem. p. 659

tropomyosin
(trō′pə mī′ə sən or trop′ə mī′ə sən) (Gr. tropos, turn + myos, muscle) A muscle protein that is involved in the formation of cross-bridges during muscle contraction. p. 310

troponin
(trō′pə nən or trop′ə nən) (Gr. tropos, turn) A muscle protein that attaches to both actin and tropomyosin; it is concerned with calcium binding and inhibiting cross-bridge formation. p. 310

true breeding
(tr$\overline{oo}$ brēd′ing) Said of plants that produce offspring consistently identical to the parent with respect to certain defined characteristics after generations of self-fertilization. p. 374

trypsin
(trip′sən) (Gr. tripsis, friction) An enzyme produced by the pancreas, which together with chymotrypsin and carboxypeptidase, completes the digestion of proteins. p. 177

tubular secretion
(t$\overline{oo}$′by$\overline{oo}$ lər sə krē′shən) The process by which the kidneys excrete a variety of potentially harmful substances from the blood. p. 235

tumor
(t$\overline{oo}$′mər) (L. swollen) An uncontrolled growth of a large mass of cells. p. 490

tundra
(tun′drə) (Russ.) A vast, uniform, virtually treeless region encircling the top of the world across North America and Eurasia that covers one-fifth of the Earth's land surface; it has long and cold winters, and the ground below 1 meter is frozen even in summer. p. 684

turgor pressure
(tur′gər presh′ər) (L. turgor, a swelling) Pressure inside a cell that results from water moving into the cell. p. 74

Turner syndrome
(tur′ nər sin′ drōm) (after H.H. Turner, U.S. physician) A genetic condition resulting from an XO zygote that develops into a human female who is sterile, of short stature, with a webbed neck, low-set ears, a broad chest, infantile sex organs, and low-normal mental abilities. p. 392

tympanic membrane
(tim pan′ik mem′brān) (Gr. tympanon, drum) The thin piece of fibrous connective tissue stretched over the opening to the middle ear; the eardrum. p. 289

ultrasound
(ul′trə sound′) A noninvasive procedure that uses sound waves to produce an image of the fetus but that harms neither the mother nor the fetus; allows the fetus to be examined for major abnormalities. p. 401

umbilical artery and vein
(um bil′ə kəl ärt′ə rē and vān) (L. umbilicus, navel) The blood lines of the umbilical cord; the artery provides nourishment and sends wastes from the embryo to the placenta; the fetal blood then travels through the umbilical vein back to the embryo. p. 360

umbilical cord
(um bil′ə kəl körd) (L. umbilicus, navel) The attachment connecting the fetus with the placenta. p. 360

unsaturated
(un sach′ə rā′təd) Referring to a fat composed of fatty acids with double bonds that replace some of the hydrogen atoms. p. 33

urea
(y$\overline{oo}$ rē′ə or y$\overline{oo}$r′ē ə) (Gr. ouron, urine) The primary excretion product from the deamination of amino acids. p. 234

ureter
(y$\overline{oo}$ rē′tər or y$\overline{oo}$r′ə tər) (Gr. oureter) The tube that carries urine from the kidney to the bladder. p. 235

urethra
(y$\overline{oo}$ rē′thrə) (Gr. ourein, to urinate) A muscular tube that brings urine from the urinary bladder to the outside; in men, the urethra also carries semen to the outside of the body during ejaculation. p. 338

uric acid
(y$\overline{oo}$r′ik as′əd) A nitrogenous waste in the urine formed from the breakdown of nucleic acids (DNA and RNA) found in the cells of ingested food and from the metabolic turnover of bodily nucleic acids and ATP. p. 234

urinary bladder
(y$\overline{oo}$r′ə ner′ē blad′ər) A hollow muscular

organ that acts as a storage pouch for urine. p. 244

urine
(yŏŏr′ən) (Gr. ouron, urine) The fluid produced by the kidneys made up of water and dissolved waste products. p. 234

urochordate
(yŏŏr′ə kôrd′āt) (Gr. oura, tail + chorde, cord) A tunicate that has a larval notochord. p. 581

uterus
(yŏŏ′tər əs) (L. womb) The organ in females in which a fertilized ovum can develop; the womb. p. 340

vaccination
(vak′sə nā′shən) (L. vaccinus, pert. to cows) An injection with disease-causing microbes or toxins that have been killed or changed in some way so as to be harmless; it causes the body to build up antibodies against a particular disease. p. 218

vacuole
(vak′yŏŏ ōl′) (L. vacuus, empty) A membrane-bounded storage sac within which such substances as water, food, and wastes can be found; most often found in plant cells, vacuoles play a major role in helping plant tissues stay rigid. p. 56

vagina
(və jīn′ə) (L. sheath) An organ in the body of a female whose muscular, tube-like passageway to the exterior has three functions; it accepts the penis during intercourse, it is the lower portion of the birth canal, and it provides an exit for the menstrual flow. p. 339

valve
(valv) (L. valva, leaf of a folding door) Any of the one-way valves found in the heart and blood vessels, similarly constructed, that prevent the backflow of blood. p. 206

vas deferens
(vas def′ə rənz) A long connecting tube that ascends from the epididymis into the pelvic cavity, looping over the side of the urinary bladder. p. 338

vascular bundle
(vas′kyə lər bund′əl) A cylinder of tissue made up of strands of xylem and phloem tissue positioned next to each other that constitute the transportation system of the stem of a plant. p. 551

vascular cambium
(vas′kyə lər kam′bē əm) A growth tissue in the woody stems of vascular plants that increases the diameter of a stem or root and gives rise to secondary xylem and secondary phloem. p. 552

vascular plant
(vas′kyə lər plant) A plant having a system of specialized vessels within itself that transports water and nutrients. p. 527

vascular tissue
(vas′kyə lər tish′ŏŏ) A differentiated tissue in vascular plants that conducts water, minerals, carbohydrates, and other substances throughout the plant; the principal types are xylem and phloem. p. 546

vector
(vek′tər) The infected genome of a virus having foreign DNA, which then carries it to a bacterial cell it is capable of infecting. p. 405

vein
(vān) (L. vena, a blood vessel) A small vein that collects blood from the capillary beds and brings it to larger veins. p. 201

ventricle
(ven′trə kəl) (L. ventriculus. belly) Either of two lower chambers of the heart; the right ventricle pumps blood into the pulmonary artery and then the lungs, whereas the left ventricle pumps blood through the aorta into the arteries. p. 204

venule
(ven′yŏŏl) (L. vena, vein) A small vein that starts at a capillary and connects it to a larger vein. p. 201

vertebral column
(ver′tə brəl kol′əm) The collection of 26 bones in the middle of the back, stacked one on top of the other, that acts like a strong, flexible rod and supports the head in a human skeleton; the spine or backbone. p. 300

vertebrate
(ver′tə brət or ver′tə brāt) (L. vertebra, vertebra) A member of a subphylum of chordates characterized by having a vertebral column surrounded by a dorsal nerve cord. p. 583

villus, pl. villi
(vil′əs, vil′ī) (L. a tuft of hair) Fine, finger-like projections that increase the surface absorption capability of the small intestine. p. 178

visceral mass
(vis′ə rəl mas) In a mollusk, a group of organs consisting of the digestive, excretory, and reproductive organs. p. 571

vitamin
(vīt′ə mən) (L. vita, life + amine, of chemical origin) An organic molecule that performs functions such as helping the body use the energy of carbohydrates, fats, and proteins; 13 different vitamins play a vital role in the human body. p. 166

viviparous
(vī vip′ər əes) (L. vivus, alive + parere, to bring forth) A method of reproduction in which the young is born alive and the developing embryo derives primary nourishment from the mother and not the egg. p. 586

vocal cord
(vō′kəl kôrd) Two pieces of elastic tissue covered with a mucous membrane stretched across the larynx that are involved in the production of sound. p. 186

vulva
(vul′və) (L. covering) The collective term for the external genitals of a female. p. 343

water mold
(wôt′ər mōld) A mold that thrives in moist places and aquatic environments, parasitizing plants and animals; during sexual reproduction, they produce large egg cells. p. 504

white blood cell
(hwīt or wīt blud sel) See leukocyte.

xylem
(zī′ləm) (Gr. xylon, wood) A principal type of vascular tissue that conducts water and dissolved minerals in a plant. p. 546

yellow marrow
(yel′ō mar′ō) A soft, fatty connective tissue that fills the hollow cylindrical core of long bones. p. 297

yolk
(yōk) (O.E. geolu, yellow) The nutrient material of an ovum that the developing individual can live on until nutrients can be derived from the mother. p. 355

yolk sac
(yōk sak) A membranous sac surrounding the food yolk in the embryo. p. 361

zona pellucida
(zō′nə pə lŏŏ′sə də) (L. zona, girdle + pellucidus, transparent) A jelly-like membranous covering of the ovum. p. 355

zygote
(zī′gōt) (Gr. zygotos, paired together) A cell produced by the haploid nuclei of the sperm and egg; the fertilized ovum. p. 340

CREDITS

CHAPTER 1

Opener (photo) James Aronovsky-(Illu) Frank Crymble/1991 *Discover Magazine;* Laura J. Edwards; Chuck Dressner/St. Louis Zoo; Laura J. Edwards; Mark Moffett/Minden Pictures.
1-1, 1-4, 1-6 Nadine Sokol.
1-2, *A* Stephen Dalton/Natural History Photographic Agency.
1-2, *B* John Trager/Visuals Unlimited.
1-2, *C* Richard Walters/Visuals Unlimited.
1-2, *D* Cabisco/Visuals Unlimited.
1-2, *E* Balkwill-D. Maratea/Visuals Unlimited.
1-3, *A* Triarch/Visuals Unlimited.
1-3, *B* Biophoto Associates/Photo Researchers, Inc.
1-3, *C*, 1-3, *F* E. Rhone Rudder.
1-3, *D*, 1-7 Raychel Ciemma.
1-3, *E* Brian Milne/Animals Animals.
1-3, *G* Jeff Fott/Bruce Coleman Ltd.
1-3, *H* Robert Maier/Animals Animals.
1-3, *I* NASA.
1-5, *A* Patti Murray/Animals Animals.
1-5, *B* Wildtype Products/Bruce Coleman Ltd.
1-5, *C* Adrian Davies/Bruce Coleman Ltd.
1-A John H. Gerard.

CHAPTER 2

Opener Marc Gottfried.
2-1, 2-2, 2-3, *A-B*, 2-7, 2-8, 2-9, 2-17, *B* Nadine Sokol.
2-3, *C* Michael Gadomski/Tom Stack & Associates.
2-4 The Bettmann Archives.
2-5, *A* Jurgen Schmitt/The Image Bank.
2-5, *B*, 2-5, *D* Frank T. Awbrey/Visuals Unlimited.
2-5, *C* Nicholas Foster/The Image Bank.
2-6 Michael & Patricia Fogden.
2-8, *A* Joseph Devenney/The Image Bank.
2-8, *B*, 2-24, *C* George Bernard/Animals Animals.
2-8, *C* Eastcott/Momatiuk/The Image Works.
2-10 Lilli Robbins.
2-15, *B* E.S. Ross.
2-16, *A* Manfred Kage/Peter Arnold, Inc.
2-17, *A* J.D. Litvay/Visuals Unlimited.
2-18 Scott Johnson/Animals Animals.
2-23 Nadine Sokol after Bill Ober.
2-24, *A* Manfred Kage/Peter Arnold, Inc.
2-24, *B* Michael Pasdizor/The Image Bank.
2-24, *D* Oxford Scientific Films/Animals Animals.
2-24, *E* Scott Blakeman/Tom Stack & Associates.
2-A Simon Frasher/Medical Phisics, RVI, Newcastle/Science Photo Library/Custom Medical Stock Photo, Inc.

CHAPTER 3

Opener Allen R.D.: The microtuble; Scientific American, Inc, 1987, p. 43.
3-1, *A* M. Abbey/Visuals Unlimited.
3-1, *B-C*, 3-4, *A*, 3-23, *B* David M. Phillips/Visuals Unlimited.
3-1, *D* John D. Cunningham/Visuals Unlimited.
3-1, *E* Bruce Iverson/Visuals Unlimited.
3-1, *F* Triarch/Visuals Unlimited.
3-1, *G* Carolina Biological Supply Co.
3-2 L.L. Sims/Visuals Unlimited.
3-4, *B*, 3-25, *B* Barbara Cousins.
3-5, 3-6, 3-9, 3-11, *B*, 3-15, *A*, 3-18, *B*, 3-19, *B* Bill Ober.
3-7, *A*, 3-20 Manfred Kage/Peter Arnold, Inc.
3-7, *B* J. David Robertson.
3-8, *A-C* Klaus Weber & Mary Osborn, for **Scientific American,** vol. 153, pp. 110-121, 1985 (unpublished).
3-8 (art) Barbara Cousins after Bill Ober.
3-10, *A* Dr. A. Brody/Science Photo Library/Photo Researchers, Inc.
3-10, *B* Kevein Somerville after Bill Ober.
3-11, *A* Richard Rodewald, University of Virginia.
3-12, *A*, 3-15, *C*, 3-18, *A* Charles J. Flickinger.
3-12, *B*, 3-13 Nadine Sokol.
3-14 K.G. Murti/Visuals Unlimited.
3-15, *B* Dr. Thomas W. Tillack.
3-16, 3-23, *C-D* Ed Reschke.
3-17 C.P. Morgan & R.A. Jersild, *Anatomical Record,* vol. 166, p. 575-586, 1970.
3-19, *A* Kenneth R. Miller.
3-21 Raychel Ciemma.
3-22 Cabisco/Visuals Unlimited.
3-23, *A* J.J. Candamse & B.C. Pugasheti/Tom Stack & Associates.
3-25, *A* L.M. Pope/BPS/Tom Stack & Associates.
3-26 T.D. Pugh and E.H. Newcomb, University of Wisconsin.
3-A Boehringer Ingelham International GmbH/Lennart Nilsson/Bonnier Fakta.

CHAPTER 4

Opener Trevor Wood/Tony Stone Worldwide.
4-1, 4-9 Nadine Sokol.
4-2, 4-3, 4-8 Nadine Sokol after Bill Ober.
4-4 Kevin Somerville after Bill Ober.
4-5, 4-6, 4-7, 4-11, *A* Barbara Cousins.
4-10, *A-B* Christy Krames.
4-10, *C-D* Jim Pearson/Taurus Association.
4-11, *B* Dr. Birgit H. Satir.
4-A William L. Mathews.

CHAPTER 5

Opener CNRI/SPL/Science Source.
5-1 U.K. Laemmli & J.R. Paulson.
5-2, 5-18 Raychel Ciemma.
5-3, *A* Ada L. Olins/Biological Photo Service.
5-3, *B*, 5-4, 5-8 (art), 5-9 (art), 5-10, *B*, 5-13, 5-17 (art) Barbara Cousins.
5-5 Carlyn Iverson.
5-6 John D. Cunningham/Visuals Unlimited.
5-7 Lilli Robins.
5-8 (photos) Carolina Biological Supply Company.
5-9 (photos) Lester V. Bergman, NY.
5-10, *A* Dr. A.S. Bajer.
5-11 David M. Phillips/Visuals Unlimited.
5-12, *A*, 5-14 Nadine Sokol.
5-12, *B* B.A. Palevitz & E.H. Newcomb/BPS/Tom Stack & Associates.
5-15 Kevin Somerville.
5-16 James Kezer, University of Oregon.
5-17 (photo) C.A. Hasenkampf, University of Toronto/Biological Photo Service.

CHAPTER 6

Opener Walter Iooss Jr./The Image Bank.
6-3, 6-5, 6-11, 6-14 Barbara Cousins.
6-6 Gunter Ziesler/Peter Arnold, Inc.
6-7 John Sohlden/Visuals Unlimited.
6-9 Bill Ober.
6-10 Nadine Sokol after Bill Ober.
6-A Tom Tracey/Photographic Resources.

CHAPTER 7

Opener Marc Gottfried.
7-1, 7-6, 7-11, 7-12, *B*, 7-13 Nadine Sokol.
7-2 Robert Barclay/Grant Heilman Photography.
7-3 Barbara Cousins.
7-4 George Klatt.
7-8 Grant Heilman/Grant Heilman Photography.
7-12, *A* Efriam Racker.

C-1

7-15, *A* Custom Medical Stock Photo, Inc.
7-15, *B* Michael Freeman/Bruce Coleman Ltd.

CHAPTER 8
Opener Dennis Di Cicco/Peter Arnold, Inc.
8-1 E.S. Ross.
8-3 Nadine Sokol.
8-5, 8-8, 8-14 Raychel Ciemma.
8-6 Manfred Kage/Peter Arnold, Inc.
8-7, 8-10, 8-12 Barbara Cousins.
8-9 Sherman Thomson/Visuals Unlimited.
8-11 Kevin Somerville after Bill Ober.

CHAPTER 9
Opener, 9-8, 9-11, *A* Lennart Nilsson, *Behold man,* Little, Brown and Co.
9-1, 9-11, *B,* 9-13 Nadine Sokol.
9-2 (art) Christine Oleksyk.
9-2 (photo) Tom Tracy/Photographic Resources.
9-3 Barbara Cousins.
9-4 Emma Shelton.
9-5 J.V. Small & F. Rinnerthaler.
9-6 *St. Louis Globe-Democrat.*
9-7, *A* David J. Hascaro and Associates.
9-7, *B-C* John Hagen.
9-9 David M. Phillips/Visuals Unlimited.
9-10 Bill Ober.
9-12 Cynthia Truner Alexander/Terry Cockerham, Snapse Media Production/Christine Oleksyk.
Table 9-2, Table 9-3 Ed Reschke.
Table 9-4, *A* Triarch/Visuals Unlimited.
Table 9-4, *B-C* John D. Cunningham/Visuals Unlimited.

CHAPTER 10
Opener, 10-13 Chet Hanchet/Photographic Resources.
10-1, 10-2, *A,* 10-5, 10-8, 10-10 Nadine Sokol.
10-2, *B* G. David Brown.
10-3, 10-9, *A* Barbara Cousins.
10-4 Kate Sweeney.
10-6 Bill Ober.
10-7 Christy Krames after Bill Ober.
10-9, *B* David M. Phillips/Visuals Unlimited.
10-12 Lilli Robbins.
10-A Benjamin, M.D./Custom Medical Stock Photos, Inc.

CHAPTER 11
Opener *A* American Cancer Society.
Opener *B* James Stevenson/SPL.
11-1, 11-11 Kate Sweeney.
11-2 Courtesy of AT&T Archives.
11-3 Ellen Dirkson/Visuals Unlimited.
11-4 Lennart Nilsson, *Behold man,* Little, Brown and Co.
11-5 Art Siegel, University of Pennsylvania.
11-6, 11-10 Nadine Sokol.
11-7, 11-8, 11-9 Barbara Cousins.
11-12 M. Moore/Visuals Unlimited.

CHAPTER 12
Opener William Strode/Humana, Inc./Black Star.
12-1, 12-13 Nadine Sokol.
12-2, 12-17 Kate Sweeney.
12-3 Christy Krames.
12-4, 12-7, 12-18 Ed Reschke.
12-5, 12-12 Bill Ober.
12-6 D.W. Fawcett-T. Kuwabara/Visuals Unlimited.
12-8 John D. Cunningham/Visuals Unlimited.
12-9, 12-10, 12-11 Barbara Cousins.
12-14 Makio Murayama/BPS/Tom Stack & Associates.
12-15 Raychel Ciemma.
12-16 Manfred Kage/Peter Arnold, Inc.
12-19 Harry Ransom Humanities Research Center.
12-A Chet Hanchet/Photographic Resources.

CHAPTER 13
Opener Lennart Nilsson.
13-1 Raychel Ciemma.
13-2 Science VU-NLM/Visuals Unlimited.
13-3, 13-12, *A* Bill Ober.
13-5 Kate Sweeney.
13-6, 13-8, 13-9, 13-15 Barbara Cousins.
13-7 Dr. A. Liepins/SPL/Photo Researchers, Inc.
13-10 Secchi, LeCaque, Roussel, Ucalf, CNRI/Science Source/Photo Researchers, Inc.
13-11 Manfred Kage/Peter Arnold, Inc.
13-12, *B* Nadine Sokol after Bill Ober.
13-16 Larry G. Arlain, Wright State University.
13-A M. English/Custom Medical Stock Photo, Inc.

CHAPTER 14
Opener M.P. Kahl/Photo Researchers, Inc.
14-1 Kate Sweeney.
14-2 Dan Gotshall.
14-3 Larry Brock/Tom Stack & Associates.
14-4, 14-5, 14-6, 14-14 Barbara Cousins.
14-13 Raychel Ciemma.
14-16 Organon Teknika/Visuals Unlimited.
14-A William B. Folsom/Uniphoto Picture Agency.

CHAPTER 15
Opener Tom McCarthy 1988 Discover Publications.
15-1 Peter Cohen—Custom Medical Stock Photo, Inc.
15-2, 15-3 (art), 15-7, 15-8, 15-9, 15-10, 15-11, 15-14 Nadine Sokol.
15-3 C.S. Raines/Visuals Unlimited.
15-4 E.S. Ross.
15-6, 15-12, *B,* 15-13 Barbara Cousins.
15-12, *A* Heimer L.: *Human brain and spinal cord,* Springer-Verlag.

CHAPTER 16
Opener James H. Karales/Peter Arnold, Inc.
16-1 Kate Sweeny.
16-3, 16-4, 16-6, 16-9 Nadine Sokol.
16-5 Bill Ober.
16-7, 16-12 Barbara Cousins.
16-8, 16-10 Michael P. Schenk.
16-11 John Daugherty.
16-13 Raychel Ciemma.
16-A Sobel/Klonsky/The Image Bank.

CHAPTER 17
Opener Rod Cardoza.
17-1, *A* Martha Swope, 1990.
17-1, *B-C* Martin/Custom Medical Stock Photo, Inc.
17-2, 17-8, 17-11 Nadine Sokol.
17-3, 17-6, *B,* 17-9 Barbara Cousins.
17-4, *A* Marsha J. Dohrman.
17-4, *B* Christine Oleksyk.
17-5, 17-10, 17-12, 17-13, 17-14 Raychel Ciemma.
17-6, *A* Scott Mittman and Maria T. Maglio.
17-7 Bill Ober.
17-A Fritz Prenzel/Animals Animals.

CHAPTER 18
Opener Vince Rodriguez.
18-1, 18-13, *B* Raychel Ciemma.
18-2 Bill Ober.
18-3, 18-9 Kate Sweeney.
18-4, 18-10 Nadine Sokol.
18-5, 18-6, 18-7, 18-11, 18-15 Barbara Cousins.
18-8 Scott Bodell.
18-12 Richard Rodewald, University of Virginia.
18-13, *A* Raychel Ciemma after Bill Ober.
18-14 John D. Cunningham/Visuals Unlimited.
18-A Jacques Cochin/The Image Bank.
Table 18-1 Nadine Sokol after Rusty Jones.

CHAPTER 19
Opener U.C. Davis Magazine, University of California, Davis.
19-1 Kate Sweeney.
19-2, 19-3, 19-9 Barbara Cousins.
19-4, 19-5, 19-13, 19-14 Nadine Sokol.
19-6 Bettina Cirone/Photo Researchers, Inc.
19-7 *Akhenaten,* The National Museum at Berlin, DDR, Egyptian Museum, no. 14512.
19-8 NMSB/Custom Medical Stock Photo, Inc.
19-10 Custom Medical Stock Photo, Inc.
19-11, 19-12 Raychel Ciemma.
19-15 Ed Reschke.
19-A Susan Lapides/Time Magazine, Inc.

CHAPTER 20
Opener, 20-1 Lennart Nilsson, *Behold man,* Little, Brown, and Co.
20-2, 20-6 Kate Sweeney.

20-3, 20-5 Nadine Sokol after Bill Ober.
20-4 Barbara Cousins.
20-7 Raychel Ciemma.
20-8 Ed Reschke.
20-9 Kevin Somerville.
20-10 David J. Hascuro and Associates.
20-12, 20-13 Joel Gordon.
20-14 Scott Bodell.
20-A Peter Read Miller/Sports Illustrated.

CHAPTER 21

Opener Lauros Giraudon/Art Resource, NY/ from Alfred Roll, The Nurse (detail) Lille, Musee des Beaux-Arts, Paris.
21-1, *A* David M. Phillips/Visuals Unlimited.
21-1, *B*, 21-10, 21-14 Barbara Cousins.
21-2, 21-3, 21-4, 21-5, 21-9, 21-12, 21-13, 21-15, 21-16, 21-17, 21-18 Lennart Nilsson, *A child is born.*
21-4, 21-6, 21-9 Raychel Ciemma.
21-7, 21-20, *A* Scott Bodell.
21-8 Kate Sweeney.
21-11 Kevin Somerville after Bill Ober.
21-19 Nadine Sokol.
21-20, *B* Martin/Custom Medical Stock Photo, Inc.

CHAPTER 22

Opener Bob Krist.
22-1 E. Rhone Rudder.
22-2, 22-3 Nadine Sokol after Bill Ober.
22-4, 22-5, 22-6, 22-8, 22-10 Nadine Sokol.
22-11 Carolina Biological Supply Company.
22-12 Raychel Ciemma.

CHAPTER 23

Opener Science VU-NLM/Visuals Unlimited.
23-1 CNRI/SPL/Science Source/Photo Researchers, Inc.
23-2, *A* The Children's Hospital, Denver, Cytogenetics Laboratory.
23-2, *B* R. Hutchings/Photo Researchers, Inc.
23-3, 23-8 Barbara Cousins.
23-5 Victor A. McKusick/Blackwell Scientific Publications.
23-6 Margaret A. Davee, M.S.; David D. Weaver M.D.; Department of Medical Genetics, Indiana University School of Medicine/Blackwell Scientific Publications.
23-10 Adam Hart-Davis/SPL/Custom Medical Stock Photo, Inc.
23-13 The Bettmann Archive.
23-14 M. Murayama/Biological Photo Service.
23-16 Kevin Somerville after Bill Ober.
23-17 Washington University School of Medicine.
23-A Dr. W. French Anderson/National Institutes of Health.

CHAPTER 24

Opener Richard Feldmann/National Institutes of Health.
24-1 Dennis D. Kunkel/Biological Photo Service.
24-2 Raychel Ciemma.
24-6, 24-7 Cold Springs Harbor Laboratory Archives.
24-8, 24-9 Bill Ober.
24-11 Barbara Cousins.
24-12, 24-17, *A*, 24-*B* Nadine Sokol.
24-13, 24-16 Molly Babich.
24-15, 24-18 Carlyn Iverson.
24-17, *B* 1982 C. Franke, J.E. Edstrom, A.W. McDowall, & O.L. Miller Jr.
24-20 Christian Vioujard/Gamma Liason.
24-A Herb Weitman/Washington University in St. Louis.

CHAPTER 25

Opener Dan Dreyfus & Associates Photography.
25-2 Ron Ervin.
25-3, *A* Cleveland P. Hickman, Jr.
25-3, *B* Peter Gregg/Imagery.
25-4, *A* Frank B. Gill/VIREO.
25-4, *B* John S. Dunning/VIREO.
25-5 Louise Van der Meid.
25-6 E. Rhone Rudder.
25-8, 25-14 Bill Ober.
25-9 The National Portrait Gallery, London.
25-10 Frank S. Balthis.
25-11 Molly Babich.
25-12 John D. Cunningham/Visuals Unlimited.
25-13 S.M. Awramik, University of California/Biological Photo Service.
25-15, 25-18 Nadine Sokol after Bill Ober.
25-16 Don & Pat Valenti/Tom Stack & Associates.
25-17, *A* Johnny Johnson/Animals Animals.
25-17, *B* Dave Watts/Tom Stack & Associates.
25-17, *C* Arthur Gloor/Earth Scenes.
25-17, *D* George H.H. Huey/Earth Scenes.
25-19 Raychel Ciemma.

CHAPTER 26

Opener NASA.
26-1, *A*, 26-6 Barbara Cousins.
26-1, *C* Kevin Walsh, University of California, San Diego.
26-2 Dudley Foster/Woods Hole Oceanographic Institution.
26-4, *A* M.R. Walter, Macquarie University, Australia.
26-4, *B* Nadine Sokol.
26-4, *C-D* Paul F. Hoffman.
26-5 Andrew H. Knoll, Botanical Museum of Harvard University.
26-6 Barbara Cousins.
26-6, *A, D* David Bruton.
26-6, *B, C, E* Simon Conway Morris, University of Cambridge.
26-8 William Boehm.
26-9, *A* William Boehm.
26-9, *B* Frans Lanting/Minden Pictures.
26-10, *A* F.M. Carpenter.
26-10, *B* John Gerlach/Animals Animals.
26-11 Michael W. Nickell.
26-12 Barbara Laing/Black Star.
26-A Mark Mamawal.

CHAPTER 27

Opener, 27-21, *B* John Reader.
27-1, *A* Smithsonian Institution photo no. USNM 57628: Walcott, 1911.
27-1, *B* Heather Angel/Biofotos.
27-2 Raychel Ciemma.
27-3, 27-4 Nadine Sokol after Bill Ober.
27-5 Steve Martin/Tom Stack & Associates.
27-6, *A*, 27-7, *C* Norbert Wu.
27-6, *B* Marty Snyderman/Visuals Unlimited.
27-6, *C*, 27-7, *B* Marty Snyderman.
27-7, *A* Alex Kerstitch.
27-7, *D* John D. Cunningham/Visuals Unlimited.
27-8 Gary Milburn/Tom Stack & Associates.
27-9, 27-11 Bill Ober.
27-10 Stouffer Productions Ltd./Animals Animals.
27-13 Ron Ervin.
27-14, 27-19 Alan E. Mann.
27-15, *A* Luis Castaneda/The Image Bank.
27-15, *B* Alan Nelson/Animals Animals.
27-16, *A* Doug Wechsler/Animals Animals.
27-16, *B* C.C. Lockwood/Animals Animals.
27-17 Russell A. Mittermeier.
27-20 Molly Babich.
27-21, *A* The Living World, St. Louis Zoo, St. Louis, MO.
27-22 Kate Sweeney.
27-23 Douglas Waugh/Peter Arnold, Inc.
27-24 SCODE/Overseas.
27-A Redrawn from Vigilant, L., Stoneking, M., and others: African populations and the evolution of human mitochondrial DNA, *Science*, vol. 253, September 27, 1991, pp. 1205-1207.

CHAPTER 28

Opener Science Source.
28-1, *A* K.G. Murti/Visuals Unlimited.
28-1, *B* K. Namba and D.L.D. Caspar.
28-3, *A* Science VU/Visuals Unlimited.
28-3, *B* Carlyn Iverson.
28-4, 28-5, 28-9, 28-11, 28-14, 28-15 Nadine Sokol.
28-6 Elizabeth Gentt/Visuals Unlimited.
28-7 Runk/Schoenberger/Grant Heilman Photography, Inc.
28-8 Jo Handelsman and Steven A. Vicen.
28-10 Dr. Huntington Potter and Dr. David Dressler/Harvard Medical School.
28-12 Eli Lilly & Co.
28-13 Barbara Cousins.
28-A Monsanto Company.

CHAPTER 29

Opener L.L.T. Rhodes/Earth Scenes.
29-1 Courtesy of Stanley Erlandson.
29-2, *A* Dr. E.W. Daniels, Argonne National Laboratory, Argonne, IL; and University of

Illinois College of Medicine, Dept. of Anatomy & Cell Biology, Chicago.
29-2, *B-C,* 29-18, *A* Cabisco/Visuals Unlimited.
29-3, *A* M. Abbey/Visuals Unlimited.
29-3, *B,* 29-22, 29-23, *C,* 29-24, *B,* 29-27 Carlyn Iverson.
29-4, 29-5, *B* Manfred Kage/Peter Arnold, Inc.
29-5, *A* Richard H. Gross.
29-6 Courtesy of the British Tourist Authority.
29-7, 29-12, 29-A Bill Ober.
29-8, 29-14, *A,* 29-28, *A* David M. Phillips/Visuals Unlimited.
29-9 K.G. Murti/Visuals Unlimited.
29-10 Nadine Sokol after Bill Ober.
29-11 A.M. Siegelman/Visuals Unlimited.
29-13 Sanford Berry/Visuals Unlimited.
29-14, *B* William Patterson/Tom Stack & Associates.
29-15, *A* Philip Sze/Visuals Unlimited.
29-15, *B,* 29-17, *A* William C. Jorgensen/Visuals Unlimited.
29-15, *C* Bob Evans/Peter Arnold, Inc.
29-16, *A* BioPhoto Associates/Photo Researchers, Inc.
29-16, *B-C* John D. Cunningham/Visuals Unlimited.
29-16, *D* London Scientific Films/Oxford Scientific Films.
29-17, *B* Gary K. Robinson/Visuals Unlimited.
29-17, *C* Brian Parker/Tom Stack & Associates.
29-18, *B-F* Higuchi Bioscience Laboratory.
29-19, 29-30, *A,* 29-30, *C,* E.S. Ross.
29-20, *A-C* John Shaw/Tom Stack & Associates.
29-21, *A* Dwight Kuhn.
29-21, *B,* 29-23, *B* Jeremy Burgess/Science Photo Library/Photo Researchers, Inc.
29-23, *A* Sherman Thomson/Visuals Unlimited.
29-24, *A* Bob Evans/Peter Arnold, Inc.
29-25 Gordon Langsbury/Bruce Coleman Ltd.
29-26, *A* Walt Anderson/Tom Stack & Associates.
29-26, *B* Michael and Patricia Fogden.
29-26, *C* Jeff Lepore/Photo Researchers, Inc.
29-28, *B* Biophoto Associates/Photo Researchers, Inc.
29-28, *C* Jack M. Bostrack/Visuals Unlimited.
29-29 Ken Greer/Visuals Unlimited.
29-30, *B* James L. Caster.
29-31 Ed Reschke.

CHAPTER 30

Opener Sneads Ferry/Photo Researchers, Inc.
30-1 Barbara Cousins.
30-2, *A* Kirtley-Perkins/Visuals Unlimited.
30-2, *B* Ken Davis/Tom Stack & Associates.
30-3, *A* John Shaw/Tom Stack & Associates.
30-3, *B,* 30-17 John D. Cunningham/Visuals Unlimited.

30-4, 30-6 Nadine Sokol after Bill Ober.
30-5 Bill Ober.
30-7, 30-9, 30-11 Carlyn Iverson.
30-8, *A,* 30-12, *B, C, F* Whit Bronaugh.
30-8, *B* E.S. Ross.
30-8, *C* Michael & Patricia Fogden.
30-10, *A* Tom J. Ulrich/Visuals Unlimited.
30-10, *B* Ron Spomer/Visuals Unlimited.
30-10, *C-D* James L. Castner.
30-12, *A, D, E* Richard H. Gross.
30-13 Peter C. Hoch.
30-14 Jack M. Bostrack/Visuals Unlimited.
30-15 E. Webber/Visuals Unlimited.
30-16 Sylvan H. Wittwer, Michigan State University.
30-A, *1* George Huey/Earth Scenes.
30-A, *2* Rodale Institute.
30-A, *3* T. Lawrence Mellichamp/Visuals Unlimited.

CHAPTER 31

Opener Ernst van Jaarsueld/Kirstenbosch National Botanic Garden/South Africa.
31-1, 31-4 (art), 31-5 (art), 31-6 (art), 31-10, 31-11 (art), 31-13, 31-17, 31-18 Carlyn Iverson.
31-2 (art), 31-15, 31-16, *A-B* Bill Ober.
31-2, *B* SUNY College of Environmental Science and Forestry Center for Ultrastructure Studies.
31-2, *C* Randy Moore/Visuals Unlimited.
31-3, *A* George J. Wilder/Visuals Unlimited.
31-3, *B,* 31-6 Richard H. Gross.
31-4, *A* Cabisco/Visuals Unlimited.
31-4, *B,* 31-16 (photo) Ed Reschke/Peter Arnold, Inc.
31-4, *C* Fred E. Hossler/Visuals Unlimited.
31-5 Stan Elms/Visuals Unlimited.
31-7 Jack M. Bostrack/Visuals Unlimited.
31-8, *A, C,* 31-9, *B* John D. Cunningham/Visuals Unlimited.
31-8, *B* Salvadore Giordano III.
31-9, *A* Ed Reschke.
31-11 Ray F. Evert.
31-12 USDA Forest Service, Forest Products Laboratory.
31-14, *A-B* Kjell B. Sandved.

CHAPTER 32

Opener, 32-23, *A* Alex Kerstitch.
32-1 Raychel Ciemma.
32-2, 32-6, *B,* 32-15, *E,* 32-20, 32-22, 32-23, *D,* 32-24 Bill Ober.
32-3 Barbara Cousins.
32-4 Scott Bodell after Bill Ober.
32-5, *A,* 32-8, 32-9, 32-11, 32-A Nadine Sokol after Bill Ober.
32-5, *B* D.J. Wrobel, Monterey Bay Aquarium/Biological Photo Service.
32-5, *C* Mickey Gibson/Animals Animals.
32-6, *A* Marty Snyderman.
32-6, *C,* 32-16, *C* Gwen Fidler/Tom Stack & Associates.
32-7, 32-23, *E* Jeff Rotman.
32-10, *A* Jim & Cathy Church.
32-10, *B* E.J. Cable/Tom Stack & Associates.
32-12 V.R. Ferris.

32-13 M.G. Schultz from H. Zaimen (ed.) pictorial presentation of parasites.
32-14, 32-17 Nadine Sokol.
32-15, *A* Milton Rand/Tom Stack & Associates.
32-15, *B* Fred Bavendam.
32-15, *C* A. E. Sprietzer/O.S.U. Museum of Zoology.
32-15, *D,* 32-16, *A,* 32-18, *C,* 32-19, *A,* 32-21 Kjell Sandved.
32-16, *B* David Dennis/Tom Stack & Associates.
32-18, *A* J. MacGregor/Peter Arnold, Inc.
32-18, *B* J. Alcock/Visuals Unlimited.
32-19, *B* Cleveland P. Hickman, Jr.
32-23, *B* Carl Roessler/Tom Stack & Associates.
32-23, *C* David L. Pawson/Harbor Branch Oceanographic Institution, Inc.

CHAPTER 33

Opener D. Griffiths/National Photographic Index of Australian Wildlife.
33-1, 33-2, *C,* 33-5, *B,* 33-12, 33-13 Bill Ober.
33-2, *A* Rick Harbo.
33-2, *B* Jim & Cathy Church.
33-3 Richard Olson.
33-4 Nadine Sokol after Bill Ober.
33-5, *A* Heather Angel.
33-6, *A-B,* 33-18, *E* Marty Snyderman.
33-6, *C* Carl Roessler/Tom Stack & Associates.
33-9, *A,* 33-14, *D* John D. Cunningham/Visuals Unlimited.
33-9, *B* David Dennis/Tom Stack & Associates.
33-9, *C* Cleveland P. Hickman, Jr.
33-9, *D-E* John Shaw/Stack & Associates.
33-10 Michael Fogden.
33-11, *A,* 33-18, *A* Brian Parker/Tom Stack & Associates.
33-11, *B,* 33-14, *E* William J. Weber/Visuals Unlimited.
33-11, *C* John Cancalosi/Tom Stack & Associates.
33-11, *D,* 33-18, *B* Rod Planck/Tom Stack & Associates.
33-14, *A-B* E.S. Ross.
33-14, *C* Alan Nelson/Tom Stack & Associates.
33-15 Raychel Ciemma.
33-16, *A* J.E. Wapstra/NPIAW.
33-16, *B* J. Alcock/Visuals Unlimited.
33-17, *A* Fritz Prenzel/Animals Animals.
33-17, *B* B.G. Murray/Animals Animals.
33-18, *C* Stephen Dalton/Animals Animals.

CHAPTER 34

Opener Paul Lemmons/Frank Lane Picture Agency.
34-1 Jeff Foott/Bruce Coleman, Inc.
34-2 Frans Lanting/Minden Pictures.
34-3 Don & Pat Valenti/Tom Stack & Associates.
34-4, 34-11 Raychel Ciemma.

34-5 Stoufer Productions Ltd./Animals Animals.
34-6 E. Rhone Rudder.
34-7 E.S. Ross.
34-8 Thomas McAvoy, Life Magazine, 1955, Time Warner, Inc.
34-9 James R. Risher.
34-10 FPG International Corp.
34-12 Omikron/Photo Researchers, Inc.
34-13 Lee Boltin.
34-14 Babkin Papers, Osler Library of the History of Medicine, McGill University, Montreal, Canada.
34-15 Bill Eades.
34-A Dale & Marian Zimmerman/Animals Animals.

CHAPTER 35

Opener, 35-1 Jane Burton/Bruce Coleman Ltd.
35-2 Bill Eades.
35-3, 35-9 Nadine Sokol.
35-4 Tui De Roy.
35-5 Dale Sarver/Animals Animals.
35-6 Mark D. Phillips/Photo Researchers, Inc.
35-7 Bob and Clara Calhoun/Bruce Coleman Ltd.
35-8 Oxford Scientific Films/Animals Animals.
35-8 (art) Bill Ober.
35-10 Herman Kacher.
35-11 Fritz Prenzel/Animals Animals.
35-12 Fred Bruemmer.
35-13, *A* John Stern/Animals Animals.
35-13, *B* J.C. Stevenson/Animals Animals.
35-14, 35-15 E.S. Ross.
35-16 E. Rhone Rudder.
35-17 Mark Moffett.
35-18 Tom Stack & Associates.
35-A Mike McKavett/Bruce Coleman Ltd.

CHAPTER 36

Opener Thomas D. Mangelsen/Images of Nature.
36-1 David Scharf/Peter Arnold, Inc.
36-2 Alan Fletcher Research Station, Queensland Department of Lands, Queensland, Australia.
36-4 Nadine Sokol.
36-7 E.S. Ross.
36-9, *A* Bonnie Freer/Peter Arnold, Inc.
36-9, *B* S.L. Nou/Photo Researchers, Inc.
36-11 Fred Griffing/UNICEF.
36-A Stephanie Maze/Woodfin Camp, Inc.

CHAPTER 37

Opener, 37-5 D.P. Wilson/Eric & David Hosking.
37-1 David Swanlung, Save-the-Redwoods League.
37-2, 37-12, 37-13, *D*, 37-15, 37-16, *A-B*, 37-21, 37-23 E.S. Ross.
37-3, *A* Fred Whitehead/Animals Animals-Earth Scenes.
37-3, *B* Stephen J. Krasemann/DRK Photo.
37-3, *C* Jeri Clark.
37-4 Roger del Moral, University of Washington.
37-6 Barbara Cousins.
37-7, *A* George H. Harrison.
37-7, *B* M.P. Kahl/National Geographic Society.
37-7, *C* J.A.L. Cooke/Oxford Scientific Films.
37-8 M. Abbey/Visuals Unlimited.
37-10 Joe McDonald/Animals Animals.
37-11 Calgene, Inc.
37-13, *A* James L. Castner.
37-13, *B* William J. Weber/Visuals Unlimited.
37-13, *C* Alex Kerstitch.
37-14, *A* Paul A. Opler.
37-14, *B* Patty Murray/Animals Animals.
37-14, *C-D*, 37-22 Dwight Kuhn.
37-16, *C-E* James L. Castner.
37-18, *A* Kennan Ward.
37-18, *B* Frank S. Balthis.
37-19 Bill Wood/Bruce Coleman, Inc.
37-20 Cabisco/Visuals Unlimited.
37-24 Gary Braasch.
37-A Scott Camazine/Photo Researchers, Inc.

CHAPTER 38

Opener A. J. Stevens/Bruce Coleman Ltd.
38-1, 38-9, 38-10, 38-11, 38-12, 38-14 Nadine Sokol.
38-2, 38-3 E.S. Ross.
38-4, 38-5 Raychel Ciemma.
38-6 Lilli Robins.
38-13, *A* Gerald A. Peters, Virginia Commonwealth University.
38-13, *B* Tom Stack & Associates.
38-13, *C* Kjell Sandved.
38-15 Norbert Wu.
38-A John D. Cunningham/Visuals Unlimited.

CHAPTER 39

Opener, 39-8, 39-10 James L. Castner.
39-1, 39-4, 39-7 Bill Ober.
39-8, *B*, 39-9, *A, C, D* Michael & Patricia Fogden.
39-9, *B* Doug Wechsler/VIREO.
39-9, *E* E.S. Ross.
39-11 W. Perry Conway/Tom Stack & Associates.
39-12 Patti Murray/Animals Animals.
39-13, *A* Richard Kolar/Earth Scenes.
39-13, *B* Breck P. Kent/Earth Scenes.
39-13, *C* Arthur Gloor/Earth Scenes.
39-14, *A* John Cancalosi/Tom Stack & Associates.
39-14, *B* William J. Hamilton III.
39-15 W.J. Weber/Visuals Unlimited.
39-16 Whit Bronaugh.
39-17 Lee Casebere.
39-18, 39-20, *D* Cleveland P. Hickman, Jr.
39-19 Bill McRae.
39-20, *A* John Gerlach/Tom Stack & Associates.
39-20, *B* Rich McIntyre/Tom Stack & Associates.
39-20, *C* M. & C. Ederegger/Peter Arnold, Inc.
39-21, 39-27 Raychel Ciemma.
39-22, *B-E* Chesapeake Bay Foundation.
39-23 Dan Gotshell.
39-24 Jeff Foott/Tom Stack & Associates.
39-25 Robert Leo Smith.
39-26, *A-B*, 39-28, *D* Norbert Wu.
39-26, *C-E* Dale Glantz.
39-28, *A* Scott D. Taylor, Duke University Marine Laboratory.
39-28, *B* Doc White/Images Unlimited.
39-28, *C* Jeff Rotman/Jeff Rotman Photography.
39-28, *E* Dudley Foster/Woods Hole Oceanographic Institution.
39-A NASA.

CHAPTER 40

Opener Chet Hanchet/Photographic Resources.
40-1 Nadine Sokol.
40-2, *A-B*, 40-14 Thomas A. Schneider.
40-2, *C* Grant Heilman/Grant Heilman Photography.
40-2, *D* Kevin Schafer/Peter Arnold, Inc.
40-2, *E* Stan W. Elms/Visuals Unlimited.
40-4 Frans Lanting/Minden Pictures.
40-6 J.M. Rankin/WorldWide Fund for Nature.
40-7 Agricultural Research Service, USDA.
40-8 Zoological Society of San Diego.
40-9, 40-10 Ray Pfortner/Peter Arnold, Inc.
40-11 Raychel Ciemma.
40-13 Gamma Liaison Network.
40-16, *A* Frank S. Balthis.
40-16, *B* Bruno J. Zehnder/Peter Arnold, Inc.

Index

A

Aberration, chromosomal, 394
Abiotic, 639, 657
Abortion, spontaneous, 391
Abscisic acid, 541
Absorption
　epithelial tissue and, 149
　in small intestine, 175
Absorption spectrum, 133, 286
Abstinence in birth control, 347-350
Abyssal zone, 690
Acacia, 650
Accessory gland, 338
Acetabularia, 46, 405
　in reciprocal graft experiment, 406
Acetaldehyde, 127
Acetyl-CoA, 123
Acetyl group, 123
Acetylcholine, 258, 259
Acetylcholinesterase, 259
Achondroplasia, 398
Acid, 26
　fatty, 33
　folic, 167
　linolenic, 34
　palmitic, 34
　pantothenic, 167
Acid environment, 218
Acid rain, 705-706
Acoelomate, 563
Acromegaly, 320
Acrosome, 336
ACTH; see Adrenocorticotropic hormone
Actin, 51, 307
Action potential, 253, 254-255
Activation energy, free, 101
Activator, 109, 422
Active immunity, 227
Active sites, enzymes and, 105, 106
Active transport, 75, 178
Adam's apple, 172
Adaptation, 432
Adaptation syndrome, general, 326
Adaptive radiation, 433, 455
Adenine, 39
　in adenosine triphosphate, 39, 110
Adenosine, 110
Adenosine diphosphate, 111-112
Adenosine triphosphate, 59, 116-127; *see also* Cellular respiration
　adenine in, 39, 110
　definition of, 110-112
　glycolysis and, 121
　production of, 116
ADH; *see* Antidiuretic hormone
Adhesion, 555
Adipose cell, 147
Adipose connective tissue, 155
ADP; *see* Adenosine diphosphate
Adrenal cortex, 325, 326, 329
Adrenal gland, 325-327
Adrenal medulla, 325, 326-327, 329
Adrenaline; *see* Epinephrine
Adrenocorticotropic hormone, 320
Adventitious root, 551

Aegyptopithecus zeuxis, 477
Aerobic metabolism, 121
Aerobic process, 116
Afferent arteriole, 237
Afferent neuron, 250-251, 267
African grasshopper, 647
African savanna, 679
Afterbirth, 368
Age distribution of population, 630
Agnatha, 465
Agranulocytes, 210
Agriculture in recombinant deoxyribonucleic acid technology, 499
Agrobacterium tumefaciens, 499
Ahnfeltia concinna, 513
Ahnfeltia plicata, 513
AIDS virus, 227-228, 229
Air pollution, 707, 708
Air pressure, 555
Albumin, 74
Aldosterone, 244, 245, 326
Algae, 509-513
　brown, 509, 511
　dinoflagellate, 509-510
　golden, 509, 510
　green, 509, 512
　red, 509, 513
All-or-nothing response, 255
Allantois, 360
Allele, 378
　codominant, 400
　multiple, 400
Allergen, 230
Allergic reaction, 229
Alternation of generation, 529
Alternative energy sources, 697
Aluminum, 27
Alveoli, 185
Ameba, 504-505
Amine, 322
Amino acid, 34, 35, 37, 38, 63
　essential, 181
Amino group, 34
Ammonia, 241
Amniocentesis, 400
Amnion, 360
Amniotic egg, 468, 589, 590
Amniotic fluid, 360
Amphibian, 587-588
　egg protection of, 588
　evolutionary history of, 466-468
Amylase, 168, 169
Amylopectin, 30
Amylose, 30
Anabaena azollae, 667
Anabolic reaction, 101
Anaerobic, defined, 116
Anaerobic metabolism, 121, 127
Anal pore, 507
Anaphase, 87, 88, 90, 91
Anemia, 167
Angina pectoris, 213
Angiosperm, 534, 548
　life cycle of, 535
Animal, 12, 13
　ancestral tree of, 562

Animal—cont'd
　bacterial, compared with plant cell, 48
　bodies of, 563
　cell of, 48
　characteristics of, 561-564
　invertebrate; *see* Invertebrate animal
　vertebrate; *see* Vertebrate
Animal cell, 48
Animalia, 12, 13
Annelid, 573
Annual ring, 552
Ant, bulldog, 575
Ant society, 619
Antagonistic muscle, 305
Anterior pituitary gland, 320-321, 329
Anther, 534
Antheridium, 529
Anthropoid, 473, 474-475
　evolution of, 476-477
Anthropoid primates, 460
Antibody, 219, 224
　heavy or light chain, 226
　molecular structure of, 226
Antidiuretic hormone, 244, 245, 321
Antigen, 219
Antitoxin, 227
Anus, 180
Anvil of ear, 289
Aorta, 204, 205
Aortic semilunar valve, 204, 205
Ape, 474, 475
Apical meristem, 548
Appendages, 300
Appendicitis, 179
Appendicular skeleton, 298, 299, 300-302
Appendix, 179
Aqueous humor, 287
Aquifer, 665
Archaebacteria, 450, 492
Archaeopteryx fossil, 435, 456
Archean Era, 448-452
Archegonium, 529
Arcyria, 515
Arm, 300, 301
Arrowhead spider, 575
Arteriole, efferent, 238, 239, 240, 241
Arteriosclerosis, 213
Artery, 201
　umbilical, 360
Arthropod, 574, 575
Articulation, 302
Artificial breeding, 429-431
Artificial kidney, 247
Artificial selection, 429, 430
Ascaris, anatomy of, 570
Ascending arm of Henle's loop, 237, 238
Asexual reproduction, 85
Aspen, clones of, 540
Aspergillus, 522
Assembly, respiratory, 125
Association area of brain, 266
Aster, 89
Asymmetry, 565-566

Atherosclerosis, 207, 213
Athlete's foot, 522
Atmosphere, 707-708
Atmospheric effect on climate, 672-675
Atom, 17-18
　functional groups of, 28, 29
Atomic energy level, 18
Atomic mass, 17
Atomic number, 17
ATP; *see* Adenosine triphosphate
Atrioventricular node, 207
Atrium
　in bony fish, 586
　left or right, 204, 207
Auditory canal, 289
Australopithecus, 477-479
Autonomic nervous system, 264, 276-277
Autosome, 81, 389
Autotroph, 131, 659
Auxin, 541
AV node; *see* Atrioventricular node
Aves, 591
Axial skeleton, 298, 299-300
Axon, 250
　in giraffe, 251

B

B cell, 209, 223
　memory, 224
B lymphocyte; *see* B cell
Bacteria, 45, 47, 217, 491-494
　binary fission in, 491
　as disease producer, 493-494
　diversity and classification of, 492-493
　eukaryotic cells compared with, 64, 65
　exponential growth in population of, 626
　fossils of, 451
　gene transfer in human-engineered, 496-498
　gene transfer in natural, 494, 495
　nitrogen fixing, 493, 667
　photosynthetic, 59
　reproduction of, 491-492
　size of, 488
Bacteria-like organelle, 58-60
Bacterial cell, 47
Bacteriophage, 489
Balance, 290-291
Balanus, 641, 642
Ball-and-socket joint, 304
Bark of plant, 552, 553
Barnacle, 650
Barrel sponge, 565
Basal body, 61
Basal ganglia, 267, 269
Base, 26
　organic nitrogen-containing, 39
Basidia, 521
Basidiomycota, 520
Basophil, 209, 210
Batesian mimicry, 647

I-1

H.M.S. *Beagle*, 427
Beak, 591
Bee colony, 619
Beetle, 681
Behavior
  innate, 598-609
    ethology in, 599
    fixed action pattern in, 602-603
    genetics and, 600
    kinesis in, 600
    learning and, 603-606, 607
    reflex in, 600
    taxis in, 600
  social, 610-623
    competitive, 611-613, 614
    group, 618-621
    human, 621
    parenting, 618
    reproductive, 615-616, 617
Benign tumor, 490
Benthos, 690
Beriberi, 167
Bicarbonate, pancreas and, 177
Biceps, 305
Bicuspid valve, 204
Bile, 176
Bile pigment, 233
Bile salts, 176
Binary fission, 491
Binomial nomenclature, 14
Bio-gas machine, 696
Bioenergy, 696
Biological concentration, 704
Biological diversity, 701
Biological magnification, 704
Biology, 2-15
  classification in, 12-14
  scientific process in, 3-7
  unifying themes, 7-12
Biomass, 662
Biome, 671-693
  climate of, 672-675
  desert, 680-681
  distribution of, 675
  life zone and
    estuary, 685-687
    fresh water, 684-685
    ocean, 687-690
  savanna, 678, 679
  taiga, 682-683
  temperate
    deciduous forest, 682
    grassland, 681
  tropical rain forest, 676, 677
  tundra, 684
Biosphere in future, 10, 694-710
  atmosphere of, 707-708
  diminishing resources of, 696-700
  land in, 696-703
  solid waste and, 702-703
  species extinction and, 700-701
  water in, 703-706
    acid rain and, 705-706
    pollution of, 703-705
Biotic, defined, 639, 657
Biotin, 167
Bipedal, 475, 478
Bipolar cell, 288
Bird, 591-592
  evolution of, 466-468
  feather of, 11, 592
  song development in, 603, 604
Birth, 368, 369
  labor in, 368

Birth canal, 368
Birth control, 347-351
  abstinence, 347-350
  foam, 347
  implant for, 350
  oral, 347
  pill, 350
  prevention of egg maturation in, 350-351
  rhythm method of, 347
  sperm blockage in, 350
  sperm destruction in, 350
  surgical intervention in, 351
Bivalve, 571
Black bread mold, 518
Bladder, 233, 244
Blade
  of leaf, 554
  of seaweed, 511
Blastocyst, 357, 358
Blastula, 564
Blending inheritance, refutation of, 376
Blood, 155
  heart pathway of, 204
Blood cell, 147, 157
Blood clot, 211
Blood glucose, 180
Blood plasma, 157
Blood sinuses, 571
Blue shark, 584
Blue tunicate, 582
Body cavity, 562-563
Body organization, 146-164
  organs in, 160-162
    interaction of, 162-163
  systems of, 147-149
  tissues in, 149-160
    connective, 152-157
    epithelial, 149-151
    muscle, 158-159
    nervous, 159-160
Bohr, N., 17
Bolus, 172
Bond, high-energy, 110
Bone, 147, 155, 296-304
  compact, 156, 296
  coxal, 300
  organization of, 297
  spongy, 156, 297
Bone cell, 45, 147, 153, 156
Bone marrow, 156, 220
Bone suture, 302
Bony fish, 467, 586
Boron, 27
Bossiella, 513
Bowman's capsule, 235, 237-241
Brachiation, 478
Brachii, 305
Brain, 149
  association area of, 266
  human, 265
  lobes of, 266
  in nervous system, 264, 265-271
  neurons in, 250
  underside view of, 270
  weight of, 265
Brain grafting, 262
Brainstem, 265, 270
Branchial arch, 365
Breastbone, 300
Breathing, 185, 191-192
Bristleworm, shiny, 573
Brittle star, 577

Bronchi, 188
Bronchitis, chronic, 196
Brown-air city, 707, 708
Brown algae, 509, 511
Brownian movement, 72
Bryophyte, 529-530
Bulbourethral gland, 338
Bull elk, 683
Bulldog ant, 575
Bundle of His, 207
Bundle branches, 207
Burgess shale stratification, 453
Bursa of Fabricius, 223
Butterfly, 31

C

Cabbage butterfly, 648
Cactus, prickly pear, 627
Calcitonin, 322, 324
Calcium, 27, 168
Calories, 166
Calvin cycle, 141
Cambium, cork or vascular, 552
Cambrian Period, 452
  fossils of, 453
Camouflage, 648, 649
Cancer, 490
Canine teeth, 171
Capillary, 201, 202-203
Capillary bed, 202
Capillus, 201
Capsid, viral, 487
Capsule, Bowman's, 235, 237-241
Carbohydrate, 30-32, 166
  plasma membrane and, 71
Carbon, 27
  covalent bonds in, 21
Carbon cycle, 665-666
  oxygen cycling and, 665
Carbon fixation, 142
Carboniferous Period, 454
Carboxyl group, 34
Carboxypeptidase, 177
Cardiac muscle, 159-160
Cardiac opening, 172
Cardiac sphincter, 173
Cardio, as prefix, 199
Cardiovascular system, 199
  of embryo, 365
Caribou migration, 604
Carnivore, 170
Carotene, 321
Carotenoid, 133
Carpal, defined, 300
Carrying capacity, 627
Cartilage
  elastic, 153, 155
  hyaline, 153, 154
Cartilage cell, 147, 153
Cartilaginous fish, 584-586
Caste, 619
Catabolic process, 121
Catabolic reaction, 101
Catalysis, 105
Catalyst, 105
Caudal, as term, 562
Cave painting, 481
Cavity
  cranial, 149
  pelvic, 149, 339
Cayuga Lake, 661
CCK; see Cholecystokinin
Cecum, 180

Cell
  animal, 48
  bacterial, 47
  bipolar, 288
  cone, 286
  glial, 160
  Golgi complex and, 47, 48, 54, 55
  of hair, 289
  in human body, 147-149
  in immune system, 219
  in living things, 7-8, 9
  muscle, 307
  plant, 48
  rod, 286
  Schwann, 160, 252
  structure of; see Cell structure
  target, 315
  white blood, 157, 210
Cell body of neuron, 252
Cell culture technique, 540, 541
Cell cycle, 86
Cell division, 80-98
  chromosomes in, 81, 82
  cytokinesis in, 91-92
  eukaryotic versus prokaryotic, 64, 65
  in growth and reproduction, 83-85
  meiosis in, 92-94, 95
  meiotic recombination in, 96, 97
  mitosis in, 86-90
Cell identity, 71
Cell-mediated immune response, 222, 223
Cell membrane, 67-79
  bacteria and, 47
  foundation of, 68-69
  movement across, 72-77
    active transport in, 75
    diffusion in, 72
    endocytosis in, 76, 77
    exocytosis in, 77
    facilitated diffusion in, 74-75
    osmosis in, 72-74
  proteins in, 70-72
Cell plate, 92
Cell structure, 44-66
  bacteria in, 47, 61-63, 64
  characteristics of, 47, 48
  eukaryotic
    cilia of, 60-61
    cytoplasm of, 50-51, 52
    cytoskeleton of, 50-51, 52
    endoplasmic reticulum of, 53
    flagella of, 60-61
    Golgi complex in, 47, 48, 54, 55
    lysosomes of, 56
    nucleus of, 56-58
    organelles of, 53-58, 58-60
    overview of, 49-61
    prokaryotic compared to, 64, 65
    vacuoles of, 56
    wall of, 48, 49, 61
  plasma membrane of, 47, 48, 49-50, 60, 61
  prokaryotic, 61-63, 64
    eukaryotic compared to, 64, 65
    interior organization of, 63, 64
    wall of, 48, 61
  size of, 46-47
Cell theory, 44-66
Cell wall, 48, 49, 61, 62, 63
Cellular respiration, 58, 114-129, 185
  adenosine triphosphate production in, 116

Cellular respiration—cont'd
  adenosine triphosphate release, 116-128
    electron carriers in, 119
    electron transport train in, 125-127
    glycolysis in, 121-123
    Krebs cycle in, 123, 124
    oxidation-reduction in, 119
  chemical energy of, 115
  overview of, 116-117
Cellular slime mold, 514
Cellulose, 29, 32
Cementum, 171
Cenozoic Era, 458-460
Central nervous system, 264, 265-273
  ganglia in, 267
  nuclei in, 267
Centriole, 48
  animal cells and, 86
Centromere, 81
Cephalic, as term, 562
Cephalochodate, 583
Cephalopod, 571
Cerebellum, 265, 270
Cerebral cortex, 265, 266
Cerebral ganglia, 574
Cerebrospinal fluid, 272
Cerebrum, 265, 266-269
Cervical cap, 350
Cervix, 343
Channel, 50
  coupled, 75
  thoroughfare, 202
Chaparral, 658
Chase, M., 414, 415
Chemical bond, 19
Chemical cycling in ecosystems, 664-668
Chemical energy to drive metabolism, 115
Chemical forces, 18-19
Chemical formula, 18
Chemical groups, 28, 29
Chemical reaction
  regulating, 105-110
  starting, 101-102, 103
Chemistry of life, 16-41
  atoms in, 17-18
  carbohydrates in, 30-32
  chemical bonds in, 19
  chemical building blocks in, 28-40
  elements and molecules in, 18-22
  fats and lipids in, 32-34
  nucleic acids in, 39
  proteins in, 34-38
  water in, 22-26
Chemoautotroph, 492
Chemoreceptors, 170
Chesapeake Bay, 686
Chiasmata, 93, 94
Chimpanzee, 475
Chitin, 29, 32
Chiton, 571
Chlamydia, 116
Chlamydomonas, 512
Chloride, 20
Chlorine, 19-20, 168
Chlorophyll, 60, 133
Chlorophyll *a*, 136
Chloroplast, 48, 49, 59
  adenosine triphosphate in, 139
  photosynthesis and, 135
Choking, Heimlich maneuver in, 195

Cholecystokinin, 175, 177
Chondrocytes, 147, 153
Chordate, 454, 464, 465
  embryo, 581
  and vertebrate; see Vertebrate
Chorionic gonadotropin, 341, 358
Chorionic villus sampling, 364
Choroid, 288
Chromatid
  nonsister, 93
  sister, 81, 94
Chromatin, 57, 81, 82
Chromium, 27, 168
Chromosomal aberrations, 394
Chromosomal rearrangement, 394
Chromosomal theory of inheritance, 383
Chromosome, 57, 81, 82
  crossing over, 93
  deoxyribonucleic acid in, 81, 82
  homologous, 93
  human, 392-393, 394
    changes in structure of, 392-393, 394
  sex, 81, 383-384, 389
  X, 384
  Y, 384
Chronic obstructive pulmonary disease, 196
Chthamalus, 641, 642
Chyme, 175
Chymotrypsin, 177
Cilia, 49, 60
  respiratory, 186
Ciliary muscle, 287
Ciliate, 507-509
Circulation, 198-215
  arteries and arterioles in, 201-202
  atmospheric, 672
  blood, 208-211
  capillaries in, 202-203
  heart and blood vessel disease and, 213
  hormone, 201
  lymphatic system in, 211, 212
  nutrient and waste transport in, 199
  ocean, 674
  oxygen and carbon dioxide transport in, 199
  pathways of, 204-208
  temperature maintenance in, 199, 200
  veins and venules in, 203, 204
Circulatory system, 148, 199, 200
  closed, 204, 571
  open, 571
Cisternae, 55
Citric acid cycle, 117
Civilization, beginning of, 481
Clam, 570-573
Classical conditioning, 606
Classification
  of living things, 12
  of mammal, 472
  of squirrel, 13
  of virus, 490
Cleavage, 357
  glycolysis and, 121
Cleft, synaptic, 257, 258
Climate, atmospheric effect on, 672-675
Climax in sexual intercourse, 346
Climax community, 652
Clitellum, 574

Clitoris, 346
Cloaca of cartilaginous fish, 586
Clones of aspen, 540
Closed circulatory system, 204, 571
Closely related, 12
Clot, blood, 211
Clownfish, 650
Club fungi, 517, 520-521
Cnidarian, 566, 567
Cobalt, 27, 168
Cochlea, 289
Cochlear duct, 289
Code of life; see Deoxyribonucleic acid
Codominance, incomplete, 399-400
Codominant allele, 400
Codon, 416
Coelom, 562
  diversity in, 565-578
  human, 149
Coelomate, 562
Coenzyme, 110, 166
Coevolution, 457644-645
Cofactor, 110
Cohesion, 555
Coitus, 346
Cold blooded, defined, 471
Collagen, 34, 38, 153
Collection duct, 236, 238, 239
Colon, 179-180
Colonial organism, 7
Colonial tunicate, 582
Colony of bees, 619
Color blindness, 395-396
Coloration
  protective, 647-648, 649
  warning, 645, 647
Columnar, defined, 147, 150
Comb jelly, 566
Commensalism, 649-650
Community, 10
  climax, 652
  redwood, 639
Community interaction, 638-655
  competition in, 640-642
  in ecosystem, 639
  predation in, 642-648
  succession in, 651-653
  symbiosis in, 649-651
Compact bone, 156, 296
Comparative anatomy, 438-439
Comparative embryology, 440
Competent cell, 494
Competition in nature, 640-642
Competitive behavior, 611-613, 614
Competitive exclusion principle, 640-641
Complement, 224
Compound, 18
Compound eye, 576
Concentration stimulation of endocrine gland, 318
Condensation in cell cycle, 86
Conditioned behavior, 603-606, 607
Condom, 347, 350
Cone cell, 286
Coniferous forest, 682-683
Conjugation, 494, 495, 507
Connective tissue, 152-157, 201
  adipose, 155
  defensive, 152
  dense fibrous, 153, 154
  elastic, 153, 154
  isolating, 152, 156-157
  loose, 153, 154

Connective tissue—cont'd
  matrix, 152
  reticular, 153, 154
  structural, 152
  as tissue type, 147
Connell, J.H., 641
Consumer, 10, 657
  primary or secondary, 658
Continental shelf, 688
Contraceptive; see Birth control
Contraction
  of heart, 206-207
  of muscle, 305-308, 309
Control, 5
  positive and negative, 422
Controlled experiment, 5
Conus arteriosus in bony fish, 586
Convergent evolution, 439, 470, 471
Convolutions, 266
Cookeina tricholoma, 519
COPD; see Chronic obstructive pulmonary disease
Copper, 27, 168
Copulation, 346
  of cartilaginous fish, 586
Coral, 561
Coral reef, 688, 689
  destruction of, 701
Core, viral, 487
Cork cambium, 552
Cornea, 287
Corpus callosum, 265, 266
Corpus luteum, 341
Corpuscle
  Meissner's, 283
  pacinian, 284
Cortex, 236
  adrenal, 325, 326, 329
  of plant root, 549
Corti, organ of, 290
Corticosteroid, 326
Cotyledon, 538
Counseling, genetic, 400-401
Coupled channel, 75
Coupled reaction, 102, 103, 115
Courtship, 615, 616
  hummingbird, 617
  prairie chicken, 616
Covalent bond, 20-22
Cowpox, 218, 220
Coxal bone, 300
Cranial cavity, 149
Cranial nerve, 275
Crayfish, freshwater, 575
Creatinine, 234
Cretaceous Period, 456
Cri du chat syndrome, 393
Crick, F., 409
Cristae, 58
Cro-Magnon, 481
Cross bridge binding site, 310
Cross-fertilization, 374
Crossing-over, 93, 94
Crown, 171
Crown gall disease, 493
Crustacean, 574
CT; see Calcitonin
Cuboidal cell, 147, 150
Culture, 480
Cup fungi, 519
Cuticle, plant, 547, 552
Cyanobacteria, 450, 493
Cyanocobalamin, 167

Cycle
  Calvin, 141
  citric acid, 117
  Krebs, 117, 123, 124
  lysogenic, 489, 490
  menstrual, 341
  ovarian, 341
  reproductive, 341-343
Cyclic electron flow, 140
Cylinder, vascular, 549
Cytokinesis, 90, 91-92
Cytokinin, 541
Cytoplasm, 48, 50
Cytosine, 39
Cytoskeleton, 49, 50, 51, 52
Cytosol, 49
Cytotoxic T cell, 222, 223

D

Dancer, interoception and, 281
Darwin
  observations of, 427-429
  theory developed by, 427-440, 441; see also Evolution
Data collection, 5
Dead air space, 192
Deamination, 234
Deciduous forest, 682
Decomposer, 657, 658
  bacteria as, 493
Deductive reasoning, 5
Deep breathing, 191-192
Defecation, 180
Defense, 645
  disease, 216-231
    allergy in, 229-230
    antigen receptors in, 226-227
    cells of immune system in, 220, 221
    discovery of, 218-219, 220
    immunization in, 227
    mechanisms of, 221-225
    specific and nonspecific resistance in, 217, 218
  genetic engineering and, 645
Defensive connective tissue, 152
Deforestation, 699-700
Dehydration, 28
Dehydration synthesis, 447
Deletion, 393
Delta, 685
Demography, 631-632
Demoiselle dragonfly, 613
Dendrite, 160, 250
Dense fibrous connective tissue, 153, 154
Density of population, 628
Density-dependent limiting factor, 628-630
Density-independent limiting factor, 628
Denticle, 585
Dentin, 171
Deoxyribonucleic acid, 11, 56-57, 63, 406-408
  base pairing in, 409
  in chromosomes, 81, 82
  double helix and size of, 409
  fragment production method for, 497
  genes and, 412
  hereditary information and, 39
  as macromolecule, 29

Deoxyribonucleic acid—cont'd
  nucleotide subunit of, 39, 408
  replication and, 410, 411
  structure of, 39, 408-411
  supercoil of, 81, 82
  Watson-Crick model of, 408
Dependent variable, 5
Depolarization, 253, 254
Depolarization stimulus, 288
Depolarizing, 286
Depth perception, 288
  in evolution, 472, 473
Dermal tissue of plant, 546, 547-548
Descending arm of Henle's loop, 237, 238
Desert, 680-681
Detritus, 659
Deuterostome, 564
Development
  prenatal, 355
  of seed, 538
Devonian Period, 454
Diabetes mellitus, 328
Dialysis, kidney, 246, 247
Diamond sting ray, 466, 584
Diaphragm
  in birth control, 347, 350
  respiratory, 190
    experiment for, 191
    human, 149
Diastolic period, 207
Diatom, 503, 510
Diatomic molecule, 21
Dicot, 538, 539
  leaf of, 554
  root of, 549
Didinium, 642, 643, 644
Diencephalon, 265
Diet, 180-182
Dietary fiber, 180
Differentially permeable membrane, 72
Diffusion, 72
  facilitated, 74-75
Digestion, 164-183; see also Digestive System
  diet and, 180-182
  hormones in, 175
  large intestine in, 179-180
  mouth in, 170-171
  nutrition and, 166-168, 169, 180-182
  organs supporting, 175-176
  small intestine in, 175-179
  stomach in, 174-175
    journey to, 172-173
Digestive enzymes, 169
Digestive system; see also Digestion
  function and components of, 148
  human, 148, 173
  organs of, 148, 173
  upper, 174
Dihybrid, 380
  and probability theory, 381-382
  and Punnett square, 382-383
Diminishing natural resource, 696-700
Dinoflagellate, 509-510
Diploid, 83, 85, 529
Disaccharidase, 176
Disaccharide, 30
Dispersion of seed, 536
Distal convoluted tubule, 237, 238, 239, 240, 241
Distant relative, 12, 428
Distribution pattern in population, 628

Diurnal mammals, 474
Diversity
  and classification of bacteria, 492-493
  of living things, 7, 8
Division, 13
  plant, 527
  reduction, 92
DNA; see Deoxyribonucleic acid
Dodder plant, 651
Dominance, incomplete, 399-400
Dominant genetic disorder, 398, 399
Dominant trait, 375
Dopamine, 263
Dorsal side, 562
Double bond, 21
Double helix, 39
Down syndrome, 389-391
Dragonfly fossil, 455
Drosophila melanogaster, 383-384
Drug testing, 241
Duct, cochlear, 289
Ductus arteriosus, 369
Duodenal ulcer, 175
Duodenum, 175
Duplication, 393
Dura mater, 272
Dutch elm disease, 519

E

Ear, 289, 290
Earthworm, 573-574
ECG; see Electrocardiogram
Echinoderm, 577
Ecosystem, 10, 639, 656-670
  aquatic, 662
  carbon cycling in, 665-666
  community and, 657-658
  food chain and web in, 659-661
  food pyramid in, 661-662
  forest in, 10
  nitrogen cycling in, 666-667
  ocean, 687-690
  phosphorus cycling in, 667-668
  water cycling in, 664-665
Ectoderm, 359, 568
Ectopic pregnancy, 357
Ectotherm, 471
Ectothermic, defined, 589
Effectors, 257
  in peripheral nervous system, 273
Efferent arteriole, 238, 239, 240, 241
Efferent neuron, 251, 264
Egg, 355
Elastic cartilage, 153, 155
Elastic connective tissue, 153, 154
Elastic fiber, 153
Elastin, 153
Electrical potential, 253-254
Electrocardiogram, 208
Electrolyte, 208
Electromagnetic energy, 132
Electromagnetic spectrum, 132
Electron, 17
Electron carrier, 117, 119
Electron flow
  cyclic, 140
  noncyclic, 137-139
Electron gain and reduction, 20
Electron shell, 18
Electron transport chain, 117, 125-127, 139

Elements, 18-19
  periodic table of, 28
Elephantiasis, 570
Elimination, 233
Embolus, 213
Embryo, 361-366
  development patterns of, 564
  in evolutionary history, 440
Emigration, 625
Emperor penguin, 613
Emphysema, 196
Enamel, 171
End-product inhibition, 109
Endergonic reaction, 102, 115
Endocrine gland
  growth and development of, 315
  hormones of, 250, 315-328, 329; see also Hormone
  in reproduction, 315
  response in, 315
  stimulation of, 318
Endocytosis, 77
Endoderm, 359, 568
Endodermis, root, 549
Endometrium, 341
Endoplasmic reticulum, 49, 52
  rough or smooth, 48, 53
Endorphin, 330
Endosperm, 538
Endosymbiont, 49
Endosymbiotic theory, 452
Endothelial cell, 201
Endotherm, 471
Endothermic, as term, 592
Endplate, motor, 310
Energy
  alternative sources of, 697
  atomic, 18
  electromagnetic, 132
  free activation, 101
  geothermal, 696
  kinetic, 103
  of motion, 104
  potential, 103
  in sunlight, 132
Energy extraction, 123
Energy flow, 100-113
  change in living systems and, 103-104
  chemical reactions and enzymes in, 105-110
  start of, 101-102, 103
  in ecosystem, 658
  in Lake Cayuga, 661
  storing and transferring in, 110-112
Energy level, 18
Energy pyramid, 10, 662
Energy transformation in living things, 10
Enkephalin, 330
Entropy, 104
Envelope, viral, 487
Environment, 695
Enzyme, 105-110
  active sites and, 105, 106
  catalysis and, 29
  digestive, 169
  function of, 105-107, 108
  hydrolyzing, 170
  induced fit and, 105, 106
  substrate and, 108
Eocene Epoch, 460
Eosinophil, 209, 210
Epidermal cell, 547

Epidermis of root, 549
Epididymis, 335, 336, 338
Epidural space, 272
Epiglottis, 172
Epinephrine, 326
Epiphyte, 649, 676
Epithelial cell, 147, 149
Epithelial tissue, 147, 149-151
Epithelium, 149
  pseudostratified, 151
  simple, 151
  of stomach, 174
  stratified, 151
  transitional, 151
Epoch, 448
  Eocene, 460
  Oligocene, 460
  Paleocene, 460
Era, 448
  Archean, 448-452
  Cenozoic, 458-460
  Mesozoic, 455-458
  Paleozoic, 452-455
  Proterozoic, 452
Erythrocyte, 157, 209, 210
  oxygen transport and, 156
Esophageal hiatus, 172
Esophageal sphincter, 172
Esophagus, 172
Essential amino acids, 181
Estrogen, 320, 341
Estuary, 685-687
Ethology, 599
Ethylene, 541
Eubacteria, 492-493
Euglena, 61, 506
Euglenoid, 135, 506
Eukaryote, 13
  chromosomes of, 57
  flagellum of, 47, 49, 60, 61
Eukaryotic cell, 47, 49-61
  bacterial cells versus, 64, 65
  nucleus of, 49
  structure of, 48, 52
Eustachian tube, 289
Eutrophic lake or pond, 668
Eutrophication, 703
Evolution, 11-12, 426-443
  convergent, 439, 470, 471
  Darwin's theory of, 427-433
    artificial breeding in, 429-431
    finches in, 432-433
    geology in, 429
    natural selection in, 432
    observations in, 427-429
    populations in, 431
    publication of, 433-434
    testing of, 434-440, 441
  defined, 426-427
  dinosaur extinction in, 455
  drifting continents and, 458, 459, 460
  early history of, 445-446, 447
  embryos in, 440
  evidence of, 427-433
  fossils and, 434-436, 437
  of horse, 437
  human; see Human evolution
  of kingdoms, 444-462
    in Archean Era, 448-452
    and cell origin, 447-448
    in Cenozoic Era, 458-460
    and mass extinction, 455
    in Mesozoic Era, 455-458

Evolution—cont'd
  of kingdoms—cont'd
    and organic molecule origin, 445-446, 447
    in Paleozoic Era, 452-455
    in Proterozoic Era, 452
Excitatory synapse, 259
Excitement phase, 346
Exclusion, competitive, 640-641
Excretion, 232-248
  epithelial tissue and, 149
  kidney in, 234-243
    anatomy of, 235, 236
    filtration of, 235-237
    homeostasis and, 244-245
    nephron of, 235-243
    problems of, 245-247
    renal failure and, 246, 247
    selective reabsorption of, 237-241
    stones of, 245-246
    tubular secretion of, 237-241
  organs of, 233, 234
  substances from body in, 233-234
  urinary system in, 244
Exergonic reaction, 101, 102, 106
Exocrine gland, 315
Exocytosis, 77
Exoskeleton of arthropod, 574, 575
Experiment
  Hershey-Chase, 414, 415
  Miller-Urey, 445, 446
  of Morgan in sex linkage, 383-384
Experimentation, 3, 4, 6
Expiration, 186, 191, 192
Exponential growth, 492, 625-626, 627
External fertilization, 584
External genitals, 338
  in female reproductive system, 343-346
External intercostal muscle, 190
External respiration, 185
Extraembryonic membrane, development of, 360-361
Eye
  compound, 576
  human, 286, 287
  simple, 576
Eyespot, 506

F

F plasmid, 494
Facilitated diffusion, 74-75
Fallopian tube, 340
False rib, 300
Family, 13
Family tree of primate, 476
Fat, 32
  macromolecule of, 29
  nutrition and, 166
  polyunsaturated, 33
  saturated, 33, 34
  triglyceride, 33
  unsaturated, 33, 34
Fat cell, 45
Fatty acids, 33, 68
Feather, 11, 592
Feces, 178
Feedback, 162-163
  negative, 109, 162
  positive, 163
Feedback loop, 162, 318
  negative, 322, 323

Feedback loop—cont'd
  positive, 322
Felis domesticus, 14
Female external genitals, 343-346
Female gamete, 335
Female reproductive system; see Reproductive system, female
Femur, 300
Fermentation, 116, 121, 123
Fern, 532
Fertilization, 83, 85, 355-356, 358
  external, 584
Fetus, human, 366-368
Fever, 222
Fiber
  cellulose, 32
  protein, 38
  skeletal muscle, 159, 305
  spindle, 86
Fibrin, 38
Fibrin thread, 211
Fibroblast, 147, 153
Fibrocartilage, 153, 155
Fibrous protein, 38
Fibrous root, 551
Filial generation
  first, 374
  second, 375
Filtration, kidney, 234, 235
Finch
  Darwin's theory of evolution and, 429, 432-433
  ground, 429, 433
  tree, 433
  warbler, 433
First filial generation, 374
First Law
  of thermodynamics, 104
First law
  of Mendel, 378
First trimester of pregnancy, 366
Fish
  bony, 466, 467, 586
  cartilaginous, 584-586
  evolution of, 465-466, 467
  fleshy-finned, 466
  jawless, 465, 584
  lobe-finned, 466, 586
  lung, 466, 467
  ray-finned, 466, 586
Fission, transverse, 507
Five-carbon sugar, 39
Fixed action pattern, 602-603
Flagellum, 47, 49, 60, 61, 506
Flatworm, 568-569
Fleshy-finned fish, 466
Floating rib, 300
Fluid mosaic model, 69
Fluoride, 168
Fluorine, 27
Foam, birth control, 347
Focusing of human eye, 287
Folic acid, 167
Follicle, 339
Follicle-stimulating hormone, 320, 336
Food chain, 659-661
Food pyramid, 661-662
Food web, 659-661
Foramen, 271
Foramen ovale, 369
Foraminifera shell, 505
Forest
  coniferous, 682-683
  deciduous, 682

Forest—cont'd
  ecosystem of, 10
Formation, reticular, 271
Formed elements, 208
Fossil, 12, 434
  Archaeopteryx, 435, 456
  Burgess shale, 453
  dating of, 436-438
  dragonfly, 455
  evolution of horse based on, 437
  formation of, 435
  Mawsonites spriggi, 436
  mold, 436
  Pikaia gracilens, 464
  process of formation of, 435
  sediment, 434
  trilobite, 454
Fossil fuel, 696-698
Fossil record, 434-436, 437
Fovea, 288
Free activation energy, 101
Free radical, 393
Fresh water life zone, 684-685
Freshwater crayfish, 575
Frog, warning coloration in, 647
Frond, 531
Frontal lobe, 266
Fruit in reproduction, 534, 536
FSH; see Follicle-stimulating hormone
Fuligo, 515
Fundamental niche, 642
Fungus, 516-523
  club, 517, 520-521
  cup, 519
  hyphae of, 516
  imperfect, 517, 522
  as kingdom, 12, 13
  lichen and, 523
  life cycle of, 517
  mycelium of, 516
  sac, 519-520
  spores of, 517
  zygote-forming, 517, 518
Fungus-like protist, 513-514
Fusion of nuclei, 356

G

Galapagos Islands, 428
Galapagos tortoises, 428
Gallbladder, 176
Gamete, 83, 85
  female and male, 335
Gametophyte, 84, 85, 529
Ganglia, 275
  basal, 267, 269
  in central nervous system, 267
  cerebral, 574
Gas
  hydrogen, 17, 21
  transport and exchange of
    hemoglobin and, 192
    respiration and, 192-195
Gastric fluid, 174
Gastric gland, 174
Gastrin, 174, 175
Gastropod, 571
Gastrula, 564
Gastrulation, 361, 363
Gate-control theory of pain, 281
Gause, G.F., 642, 643, 644
Gene, 81
  defined, 413
  deoxyribonucleic acid and, 412

Gene—cont'd
  hereditary information in, 412-413
  structural, 420
Gene expression, 413-419
Gene family, 422
Gene mutation, 394-395
Gene therapy, 422-423
  germ line in, 423
  in recombinant deoxyribonucleic acid technology, 499
  somatic cell, 423
General adaptation syndrome, 326
Generation
  alternation of, 529
  first filial, 374
  parental, 374
  second filial, 375
Generator potential, 275, 282
Genetic code, 416
Genetic counseling, 400-401
Genetic engineering, 494-498
  defense by, 645
Genetics, 377
  behavioral, 600
  human; see Human genetics
Genitals, external, 338, 343-346
Genome, 405
Genotype, 378
Genus, 13
Geographical isolation, 432
Geothermal energy, 696
Germ layer, 359
Germ line gene therapy, 423
Germ plasm bank, 701
Germination, 538, 539
Gestation, 355
GH; see Growth hormone
Giantism, 320
Giardia lamblia, 503
Gibberellin in plant, 541
Gila monster, 647
Gill of mollusk, 571
Gizzard, 591
Gland, 149
  accessory, 338
  adrenal, 325-327, 329
  bulbourethral, 338
  endocrine; see Endocrine gland
  exocrine, 315
  mammary, 346
  parathyroid, 324, 325, 329
  pineal, 328
  pituitary, 318-322
    anterior, 320-321, 329
    posterior, 321-322, 329
  prostate, 244, 338
  salivary, 169
  thymus, 328
  thyroid, 322, 323, 329
Glandular epithelial tissue, 149
Glial cell, 160
Global warming, 700
Globin gene, 441
Globular protein, 38
Glomerulus, 237, 238, 239, 240, 241
Glottis, 172, 186
Glucagon, 327, 328
Glucocorticoid, 326
Glucose, 30, 31
  blood, 180
  mobilization of, 121
  oxidation of, 126
Glyceraldehyde phosphate, 141

Glycerol, 33
  phospholipid, 68
Glycogen, 29, 30, 31
Glycolipid, 55
Glycolysis, 117, 121-123
Glycoprotein, 55
Goiter, 322
Gold tunicate, 582
Golden algae, 509, 510
Golden-brown algae, 510
Golgi, C., 54
Golgi body, 53, 54, 55
Golgi complex, 48, 49, 54, 55
Gonad, 335
Gonadotropin hormone, 320
Gondwana, 458
Goose, 617
Gorilla, 475
Gradient, 72
Grand Canyon, 435
Granulocyte, 210
Granum, 60, 135
Grasshopper
  African, 647
  anatomy of, 576
  copulating, 575
Grassland, 658
  temperate, 681
Gravitaxis, 601
Gray-air city, 707, 708
Gray matter, 266, 271
Gray whale, 650
Great ape, 474
Green algae, 509, 512
Green caterpillar, 648
Greenhouse effect, 700
Grinding teeth, 171
Ground finch, 429, 433
Ground tissue, 546, 547
Ground water, 664, 703
  pollution of, 705
Group behavior, 618-621
Growth
  endocrine gland and, 315
  exponential, 625-626, 627
  plant, 548
  population, 625-627
Growth curve
  of human population, 633
  sigmoid, 627
Growth hormone, 320
Growth rate, 625
Guanine, 39
Guano coast, 668
Guard cell, 554, 555
Gymnosperm, 533

H

Habitat, 639
Habituation, 605
Hair, 38
Hair cell, 289
Hairy-cap moss, 530
Half-life of carbon-14 isotope, 436
Haploid, 83, 85, 529
Hardwood, 552
Haversian canal, 296
Hawkmoth, 649
Head chordate, 583
Hearing, 288-290
Heart, 199
  contraction of, 206-207
Heart attack, 185, 212, 213

Heart disease, prevention of, 206
Heart murmur, 207
Heat regulation, 200
Heavy chain, 226
Heimlich maneuver, 195
Helium, 18
Helix, 39
Helper T cell, 222, 223, 224, 227
Hemoglobin, 210
  and gas transport, 192
Hemophilia, 389, 397
Henle's loop, 237, 238-241, 243
  ascending or descending arm of, 237, 238
Herbivore
  food plants and, 644-645
  teeth of, 170
Heredity of living things, 11; see also Human genetics
Hermaphrodite, 566
Hershey, A., 414
Hershey-Chase experiment, 414, 415
Heterotroph, 131, 659
  bacteria as, 493
Heterozygous, defined, 378
Hiatal hernia, 172
High altitude sickness, 194
High-energy bond, 110
Hinge joint, 304
Hip, 300
His bundle, 207
HIV; see Human immunodeficiency virus
Holdfast, 511
Homeostasis, 162, 245, 276
Hominid, 475
  evolution of, 477-481
Hominoid, 474
Homo erectus, 480
Homo habilis, 480
Homo sapiens, 14, 475, 480-481
Homologous chromosomes, 93
Homology, 438
Homozygous, defined, 378
Homunculus, 373
Hookworm, 651
Hormone, 162, 244, 314-331
  adrenocorticotropic, 320
  aldosterone, 244, 245
  antidiuretic, 244, 245, 321
  in communication, 250
  digestion and, 175
  endocrine gland, 250, 315-328
    adrenal, 325-327
    ovarian and testicular, 328
    pancreatic, 327-328
    parathyroid, 324, 325
    pineal, 328
    pituitary, 318-322
    thymic, 328
    thyroid, 322, 323
  follicle-stimulating, 320, 336
  gonadotropin as, 320
  growth, 320
  local, 315
  luteinizing, 320, 336
  melanocyte-stimulating, 321
  nonendocrine, 328-330
  parathyroid, 322, 324
  peptide, 315, 317
  of plant, 541
  prolactin, 321
  releasing, 318
  steroid, 315, 317

Hormone—cont'd
  thyroid-stimulating, 320
  tropic, 320
  in water and salt balance in kidney, 245
Hornwort, 529, 530
Horseshoe crab, 454
House cat, 14
Human
  body organization of; see Body organization
  brain of, 265
  chromosome of, 392-393, 394
  digestive system of, 148, 173
  eye of, 286, 287
  nervous system of, 148
  tissue of, 149-160
Human-engineered gene transfer, 496-498
Human evolution, 463-483
  amphibian and, 466-468
  bird and, 466-468
  in Cenozoic Era, 458-460
  fish and, 465-466, 467
  hominid in, 477-481
    Australopithecus, 477-479
    Homo erectus, 480
    Homo habilis, 480
    Homo sapiens, 480-481
  mammal in, 468-471
  primate in, 471-475, 476
    anthropoid and, 473, 474-475, 476-477
    prosimian and, 473
  vertebrate in, 464-471
Human fetus, 366-368
Human genetics
  counseling in, 400-401
  dominant and recessive disorders in, 398, 399
  incomplete dominance and codominance in, 399-400
  karyotypes in, 389-395
    autosomes and, 389-391
    chromosome changes and, 392-393, 394
    gene mutations and, 394-395
    Klinefelter syndrome and, 391
    triple X females and, 391
    Turner syndrome and, 392
    XYY males and, 392
  multiple alleles in, 400
  patterns of inheritance in, 389-397
  pedigrees and, 395-397
Human immunodeficiency virus, 228
Human life cycle, 83
Human population, 632-636
Humerus, 300
Hummingbird, 617
Humoral immune response, 223-224, 225
Humulin, 497
Huntington's disease, 398
Hyaline cartilage, 153, 154
Hydra, 566-568
Hydration shell, 24
Hydrocarbon, 33
Hydrochloric acid, 26, 174
Hydrogen, 17, 20-21, 27
  covalent bond of, 20-21
Hydrogen bond, 23
Hydrogen gas, 17, 21
Hydrolysis, 30, 170
Hydrolyzing enzyme, 170

Hydrophilic, defined, 69
Hydrophobic, defined, 69
Hydrophobic bond, 24-25
Hydrophobic compound, 24
Hydrothermal vent, 447
Hydroxyl group, 29
Hypercholesterolemia, 398
Hyperosmotic, defined, 587
Hyperpolarization, rod cell and, 286
Hyperthyroidism, 322, 323
Hypertonic solution, 73
Hyphae of fungi, 516
Hypoosmotic, defined, 587
Hypopituitary dwarfism, 320
Hypothalamus, 163, 265, 329
  in central nervous system, 269, 270
  pituitary gland and, 318
Hypotheses, 3, 6
Hypothyroid goiter, 322
Hypothyroidism, 322
Hypotonic solution, 73

I

Ice age, 480
Ice formation, 24
Ichthyostega, 468
IgM; see Immunoglobulin M
Immigration, 625
Immune response, 219, 221
Immune system, 148, 218-230; see also Defense, disease
Immunity, 217
  active or passive, 227
Immunodeficiency virus, 228
Immunoglobulin, 224
Immunoglobulin M, 227
Imperfect fungi, 517, 522
Implant, contraceptive, 350
Implantation, 357, 358
Imprinting, 603-604
Inborn behavior, 600-602
Inchworm caterpillar, 649
Incisors, 171
Inclusions, 63
Incomplete dominance or codominance, 399-400
Incus of ear, 289
Independent assortment, 381
Independent variable, 3
Induced fit, 105, 106
Inducer T cell, 222
Induction in human development, 364
Inductive reasoning, 3
Infection, 217
Infectious agent, 704
Inferior vena cava, 204, 206
Inflammation, 217
Information
  storage of hereditary, 405, 406
  transmission of, 249-261
    communication system of body and, 250
    nerve impulse in, 252-257
    nerve to target tissue, 257-259
    neuron in, 250-252
Inheritance
  historical view of, 373-374
  karyotypes and, 389-395
  Mendelian, 374-383
    analyzing, 378-379
    conclusions of, 377-378
    crossing-over in, 374
    dihybrid crosses in, 381-383

Inheritance—cont'd
  Mendelian—cont'd
    factors and chromosomes in, 383
    gene interaction in, 377
    questions of, 380-381
    testing of, 379-380
  molecular basis of, 404-424
    deoxyribonucleic acid replication in, 411
    gene in, 412-413
    gene expression in, 413-419
    gene therapy in, 422-423
    location of hereditary material in, 405, 406
    nucleic acid in, 406-411
    prokaryotic and eukaryotic genes in, 419, 420-422
  mysteries of, 385
  pedigrees and, 395-397
  sex linkage, 383-384
  study of, 374
  of traits, 373, 374
  variations in species of, 373-374
Inhibition, end-product, 109
Inhibitor, 109
Inhibitory synapse, 259
Innate behavior; see Behavior, innate
Inner cell mass, 357
Inner ear, 289, 290
Inorganic nutrient, 703
  in plant, 548
Insect, 574-576
  society behavior of, 619-621
Insertion, 305
Insight, 606, 607
Inspiration, 186, 190-191
Instinctive behavior, 600-602
Insulin, 243, 327, 328
  genetic engineering and, 498
Integration, 265, 275
Integrator, 259
Integument, 296
Integumentary system, 148, 296
Interaction of living things, 8-10
Intercostal muscle, 190
Intercourse, 346
Interleukin-1, 222
Intermediate fibers, 50, 51
Internal intercostal muscle, 190
Internal respiration, 186
Interneuron, 251, 275
Interphase, 86, 87, 88, 90
Intertidal zone, 687-688
Intervertebral disk, 300
Intestinal cell, 178
Intestine, 233
  large, 179-180
  small, 175-179
    digestive enzymes and, 169
Intimidation display, 611
Inversion, 393
Invertebrate animal, 560-579
  asymmetry in, 565-566
  bilateral symmetry in, 568-578
    clam, snail, and octopus, 570-573
    earthworm and leech, 573-574
    flatworm, 568-569
    lobster, insect, and spider, 574-576
    roundworm, 569-570
    sea urchin and starfish, 577-578
  characteristics of, 561-564
  radial symmetry in, 566-568
Involuntary response, 264

Iodine, 27, 168
Ion, 19-20
  in vertebrate, 585
Ionic bond, 19-20
Ionization, 25-26
Ionizing radiation, 392
Iris, 288
Iron, 27, 168
Islets of Langerhans, 327, 329
Isolation, geographical, 432
Isotonic solution, 73
Isotopes, 17
  half-life of carbon-14, 436

J

Jackrabbit, 649
Jaw, evolution of, 465
Jawless fish, 584
Jellyfish, 566-568
Joint, 302-304
Jumping spider, 575
Junction, neuromuscular, 258, 259, 310
Jurassic Period, 456

K

Kalanchoe daigremontiana, 540
Kangaroo rat, 235
Karyotype, 81
  and inheritance patterns, 389-395
Kelp, 511
Keratin, 38
Ketone, 328
Kidney
  of fish, 235
  human, 233, 234
    artificial, 247
    dialysis and, 246, 247
    failure of, 246
    filtration of, 234, 235-237
    nephron in, 235-243
    selective reabsorption in, 235, 237-241
    structure of, 236
    tubular secretion in, 235, 241
  of kangaroo rat, 235
  water balance and, 234
Kidney stones, 245-246, 325
Killer cell, 222
Kineses, 600
Kinetic energy, 103, 104
Kinetochore, 89
Kingdoms of life, 12
Klinefelter syndrome, 391
Knee-jerk reflex, 276
Koran angelfish, 467
Krebs cycle, 117, 123, 124

L

Labia majora or minora, 346
Labor in birth, 368
Lac operon, 420, 421
Lac repressor, 422
Lactase, 169
Lacteal, 178
Lactose, 30
Lacunae, 153
Lake Cayuga, 661
Lake zones, 685
Lamprey, 465
Lancelet, 583

Land biome, 675-684, 696-703
Langerhans islets, 327, 329
Lanugo, 366
Large intestine, 179-180
Larval tunicate, 583
Larynx, 172, 186, 187
Lateral bud, 551
Lateral line system, 585
Lateral meristem, 548
Laurasia, 458
Law
  of independent assortment, 381
  of segregation, 378
  of thermodynamics, 103
Leaf, 545, 554-555
Leafcutter ant, 620-621
Learning, 603-606, 607
Leech, 573-574
Left atrium, 204
Left ventricle, 204, 207
Left ventricle assist device, 199
Leg, 300, 301
Lens of eye, 287
Lesser ape, 474
Leukocyte, 157, 210
Levels of human body organization; see Body organization
LH; see Luteinizing hormone
Lichen, 523
Life cycle
  of angiosperm, 535
  of fern, 532
  of fungi, 517
  of mushroom, 521
  of *Obelia*, 567
  of pine, 533
  of plant, 527
  sexual, 528
Ligament, 302
Ligase, 497
Light chain, 226
Light-dependent reaction, 135, 136-140
Light-independent reaction, 135, 141-142
Limb bud, 366
Limbic system, 269, 270
Linolenic acid, 34
Lipase, 168, 177
Lipid, 29, 32-34, 219
Lipid bilayer, 69, 219
Liver, 175, 233
Liverwort, 529, 530
Living things, 7-12
Lobe of brain, 266
Lobe-finned fish, 586
Lobster, 574-576
Local hormone, 315, 328
Loop of Henle, 239, 240, 241, 243
  ascending or descending arm of, 237, 238
Loose connective tissue, 153, 154
Lorenz K., 603, 604
Lower esophageal sphincter, 172
Lucy, *Australopithecus*, 478
Lumen, 201
Lung, 205, 233
Lung volume, 192
Lungfish, 467
Luteinizing hormone, 320, 336
Lymph node, 212, 220
Lymphatic system, 211, 212, 221
Lymphocyte, 210, 221
  B, 209, 223, 224

Lymphocyte—cont'd
  immune system and, 147, 150
  T, 209, 211, 222, 223, 328
Lymphokine, 222, 223, 224
Lyse, 74, 489
Lysogenic cycle, 489, 490
Lysosomes, 48, 49, 56
Lysozyme, 37
Lytic cycle, 489-490

M

Macromolecule, 28, 29
Macrophage, 188, 210
  in body defense, 147, 152, 221, 223, 224
  cell ingestion by, 221
  immune system and, 147, 152
Magnesium, 27, 168
Malaria, 508, 509
Male gamete, 335
Male reproductive system, 335-339
Malleus, 289
Malpighian tubule, 576
Maltase, 169
Mammal, 593, 594
  classification of, 472
  and evolution, 468-471
  placental, 471, 593, 594
Mammalia, 593
Mammary gland, 346
Mandible of arthropod, 574, 575
Manganese, 27, 168
Manipulated variable, 3
Manta ray, 466, 584
Mantle, 571
Marchantia, 529
Marfan's syndrome, 398
Marrow
  bone, 156
  red, 297
  yellow, 297
Marsupial, 471, 593, 594
  with young, 618
Mass extinction, 455
Mass flow, 557
Mast cell, 147, 152
Mathematical progressions, 431
Mating, 615
Matrix in connective tissue, 152
Mawsonites spriggi fossil, 436
Mechanoreceptor, 585
Medulla, 191, 236, 265, 271
  adrenal, 325, 326-327
Medusa, 567
Megakaryocyte, 211
Meiosis, 83, 85, 92-94, 95
  mitosis compared to, 96
Meiotic recombination, 97
Meissner's corpuscle, 283
Melanin, 295, 321
Melanocyte, 321
Melanocyte-stimulating hormone, 321
Melzack, R., 281
Membrane, 48
  nuclear, 56, 57
  photosynthetic, 135
  pleura, 191
  postsynaptic, 257
  presynaptic, 257
  tectorial, 289-290
Membranous organelle, 53
Memory B cell, 224
Memory T cell, 223

Mendel, G.J., 374
  first law of, 378
  second law of, 381
Mendelian genetics; *see* Inheritance, Mendelian
Mendelian trait, 374
Meninges, 272
Menstrual cycle, 341, 342
Menstruation, 342
Meristem, 546, 548
Meristematic tissue, 546, 548
Merkel's disk, 283
Mesentery, 174
Mesoderm, 359, 568
Mesopelagic, defined, 690
Mesophyll, 554
Mesosome, 63
Mesozoic Era, 455-458
Messenger, second, 315
Messenger ribonucleic acid, 413
Metabolic machine, 142
Metabolic pathway, 105, 115
Metabolism, 121, 127
Metamorphosis, 588
Metaphase, 87, 88, 89, 91
Metaphase plate, 89, 93
Microbe, 221
Microfibrils, 32
Microfilament, 50, 51, 52
Microtubule, 50, 52, 60
  tubulin and, 51
Microtubule organizing center, 44, 86
Microvilli, 178
Midbrain, 265
Middle ear, 289
Migration, 603
  caribou, 604
Milk letdown, 346
Miller-Urey experiment, 445, 446
Mimicry, 647-648
Mineral resource, 698-699
Mineralocorticoid, 326
Minerals, 166, 167, 168, 169
Mite, 229
Mitochondria, 48, 49, 52, 58-59
Mitosis, 83, 86-90
  in animal cell, 87, 90, 91
  meiosis compared to, 96
  in plant cell, 88, 90, 91
Mixed nerve, 275
Molars, 171
Mold
  black bread, 518
  slime, 504, 514
  structure of, 516
  water, 504, 515
Mold fossil, 436
Molecular basis of inheritance; *see* Inheritance, molecular basis of
Molecular biology, 440, 441
Molecular motion, kinetic energy of, 104
Molecule, 18-22
  diatomic, 21
  of glucose, 30
  organic, 26, 27, 28
  polar, 23
  stable, 21
Mollusca, 570
Mollusk, 570, 571
Molybdenum, 27
Monarch butterfly, 645
Monera, 12, 13, 491
Monkey, 474

Monocot, 538, 539
  leaf, 554
  root, 549
Monocyte, 152, 209, 221, 222
Monomer, 447
Monosaccharide, 30
Monosynaptic reflex arc, 276
Monotreme, 471, 593
Mons pubis, 343
Moray eel, 467
Morphogenesis, 357, 361
Mortality of population, 630-631
Morula, 357, 564
Mosaic model, fluid, 69
Moss, 530
Motion, energy of, 104
Motor area of brain, 266, 267
Motor endplate, 310
Motor neuron, 251, 264
Motor pathway, 275-277
Mouth in digestion, 170-171
Movement of water in plants, 555-557
MTOC; *see* Microtubule organizing center
Mucin, 355
Mucosa, 174
Mucous membrane, 218
Mucus, 174, 175
Mueller gibbon, 475
Müllerian mimicry, 648
Multicelled, defined, 84
Multiple allele, 400
Muscle, 158-159, 304, 305-311
  ciliary, 287
  intercostal, 190
  microfilament in, 158, 159, 305, 309
  spindle, 283
  as tissue type, 147
Muscle cell, 307
Muscle fiber, 305
  nerve impulses and, 305, 310, 311
Muscle tissue, 158-159
Muscular system, 304, 305-311
  function and components of, 148
Mushroom, life cycle of, 521
Musk-oxen, 618
Mutation, 389
  gene, 394-395
  point, 394
Mutualism, 649, 650-651
Mycelium of fungi, 516
Myelin sheath, 252
Myofibril, 305
Myofilament, 305, 309
Myosin, 307

N

Naked seed, 533
Nasal cavity, 172, 185, 186, 187
Natural gene transfer, 494, 495
Natural history, 3
Natural killer cell, 221, 222
Natural resources, 696-700
Natural selection, 432
Natural variation, 429
Naturalist, 3
Neanderthal man, 480
Nectar, 535
Negative control, 422
Negative feedback, 109
Negative feedback loop, 322, 323
Nekton, 690

Nephridia, 573
Nephron, 235-243
Neritic zone, 688, 689
Nerve, 160
  cranial, 275
  mixed, 275
  optic, 288
  peripheral nervous system and, 269
  spinal, 271
  structure, 274
Nerve cell, 45
Nerve cord, 464
  spinal, 581
Nerve impulse, 250, 252-257
  muscle and, 310, 311
Nerve ring, 578
Nervous system, 148, 250
  autonomic, 264, 276-277
  central, 264, 265-273
    brain in, 265-271
    spinal cord in, 271-272
  organization of, 148, 263-265
  parasympathetic, 265, 276, 277
  peripheral, 264, 273-277
  somatic, 264, 275-276
  sympathetic, 265, 276, 277
Nervous tissue, 8, 159-160
  as tissue type, 147
Net movement, 72
Neural fold, 364
Neural groove, 364
Neuroglia, 252
Neuromuscular junction, 258, 259, 310
Neuron, 159-160, 250-252, 264
Neurotransmitter, 257, 258
Neurulation, 363, 364
Neutral atom, 17
Neutron, 17
Neutrophil, 209
New Jersey toxic chemical dump, 703
New World monkey, 474
New York City dump, 702
Newborn, 368, 369
Niacin, 167
Niche, 639
  fundamental, 642
  realized, 642
Nicholas II of Russia, Czar, 397
Night blindness, 167
Nitrifying bacteria, 493, 667
Nitrogen, 27
  energy levels for, 18
Nitrogen cycle, 666-667
Nitrogen-fixing bacteria, 493, 667
Nitrogenous waste, 234
NK cell; *see* Natural killer cell
Noble gases, 19
Nocturnal, defined, 468
Nodes
  of plant stem, 551
  of Ranvier, 252
Noncyclic electron flow, 137-139
Nondisjunction, 389, 390, 391, 392
Nonendocrine hormone, 328
Nonpoint source, 703
Nonsister chromatid, 93
Nonspecific defense, 217, 218
Nonvascular plant, 527, 557
  patterns of reproduction in, 529-530, 531
Noradrenaline; *see* Norepinephrine
Norepinephrine, 326

Notochord, 361, 454, 464
  in chordates, 581
Nuclear envelope, 48, 56, 57
Nuclear membrane, 56, 57
Nuclear pores, 48, 56, 57
Nuclear power, 696
Nuclei, 47
  in central nervous system, 267
  fusion of, 356
Nucleic acid, 29, 39, 405, 406-411
Nucleoid, 47, 63
Nucleolus, 48, 49, 57, 58
Nucleosome, 82
Nucleotide, 11, 39
  structure of, 406
Nucleus, 47, 48, 56-58
Number pyramid, 662
Nutrition, 180-182

O

Obelia, 567
Obligate parasite, 487
Obstructive pulmonary disease
  chronic, 196
Occipital lobe, 266
Ocean
  ecosystems of, 687-690
  patterns of circulation in, 674
  zones of, 687, 690
Octet rule, 19
Octopus, 570-573
Oil, 33
Old World monkey, 474
Oleic acid, 34
Olfactory receptor, 284
Oligocene Epoch, 460
Omnivore, 170
On the Origin of Species, 434
One gene–one enzyme theory, 412
Onymacris unguicularis beetle, 681
Oocyte, 339, 340
Oogenesis, 340
Oogonia, 340
Open circulatory system, 571
Open dump, 702
Open-sea zone, 684, 687, 690, 691
Operant conditioning, 605-606
Operator, 420
Operculum, 586
Operon, 420
Opsin, 286
Optic nerve, 288
Oral contraceptive, 347
Orangutan, 475
Orb-weaving spider, 575
Orbital, 17
Order, 13
Ordovician Period, 454
Organ, 8
  of Corti, 290
  defined, 147, 160
  of vascular plant, 548-555
  in vertebrate, 160
  vestigial, 438
Organ system, 8
  in human body organization, 148, 149, 160-162, 163
Organelle, 12, 49
  bacteria-like, 58-60
  membranous, 53
Organic molecule, 26, 27, 28
  origin of, 445-446, 447
Organic nitrogen-containing base, 39

Organic nutrient, 703
Organism, 7
Organization
  bone, 297
  of living things, 7-8, 9
Organizing center in microtubule, 86
Organogenesis, 361
Orgasm, 346
Osmoregulation, 585
Osmosis, 72-74
Osmotic pressure, 73
Ossification, 367
Osteoblast, 296
Osteocyte, 147, 153, 156
Osteoporosis, 325
Ostracoderm, 454
Outer ear, 289
Oval window, 289
Ovarian cycle, 341
Ovarian follicle of cat, 341
Ovary, 328, 329, 339-340, 534
Overpopulation, 708
Oviduct, 340
Oviparous, defined, 586
Ovoviviparous, defined, 586
Ovulation, 340
Ovule, 533
Oxidation, 20, 119
  of glucose, 126
  glycolysis and, 121
Oxidation-reduction, 119
Oxidized, defined, 119
Oxygen, 20
Oxygen cycling and carbon cycle, 665
Oxygen debt, 128
Oxygen transport and erythrocyte, 156
Oxytocin, 321, 322
Ozone depletion, 708

P

Pacinian corpuscle, 284
Paleocene Epoch, 460
Paleozoic Era, 452-455
Palisade layer, 554
Palmitic acid, 34
Pancreas, 327-328, 329
  digestion and, 169, 175
Pancreatic amylase, 176
Pancreatic lipase, 177
Pangea, 458
Pantothenic acid, 167
Papillae, 285
Paramecium, 45, 61, 507, 642, 643, 644
  microscopic view of, 47
  pinocytosis and, 76
Parasite, 516
Parasitism, 630, 649, 651
Parasympathetic nervous system, 265, 276
Parathyroid gland, 324, 325, 329
Parathyroid hormone, 322, 324
Parenchyma cell, 547
Parental generation, 374
Parenting behavior, 618
Parietal lobe, 266
Parotid gland, 170
Passive facilitated diffusion, 75
Passive immunity, 227
Passive transport, 72
Pathway
  of air, 186-189
  metabolic, 105, 115

Pathway—cont'd
  motor, 275-277
  sensory, 273-275
Pavlov, I., 606
Pea plant
  anatomy of, 374
  contrasting traits in, 375
Peacock, 615
Peat moss, 530
Pectoral girdle, 300, 301
Pedigree
  and inheritance patterns in humans, 395-397
  red-green color blindness, 395
Pelagic zone, 684, 687, 690, 691
Pelican hunting behavior, 619
Pellicle, 61
Pelvic cavity, 149, 339
Pelvic clasper of cartilaginous fish, 585
Pelvic girdle, 300, 301
Penicillium, 522
Penis, 244, 335, 338-339
Pepsin, 169, 174
Peptidases, 169, 177
Peptide, 34
Peptide bond, 34, 36
Peptide hormone, 315, 317
Peptidoglycan, 62, 63
Perception, depth, 288
Pericycle of plant root, 549
Periodic table of elements, 28
Peripheral nervous system, 264, 273-277
Peristalsis, 172, 173, 244
Permafrost, 684
Permeable membrane
  differentially, 72
  selectively, 73
Permian Period, 455
Pernicious anemia, 167
Petiole, 554
pH scale, 25, 26
Phage, 489
Phagocyte, 221
Phagocytosis, 76, 77, 152, 210, 221
Pharyngeal arch, 464, 581
Pharynx, 172, 186
Phenotype, 378
Pheromone, 620
Phloem, 546
Phosphate, 39, 68
  glyceraldehyde, 141
Phospholipid, 29, 34, 68
Phosphorus, 27, 168
Phosphorus cycle, 667-668
Photic zone, 690
Photoautotroph, 492
Photocenter, 136, 137
Photochemical smog, 707
Photon, 132, 286
Photoreceptor, 506
Photosynthesis, 10, 59, 116, 130-144
Photosynthetic membrane, 65, 135
Photosystem, 136-140
Phototaxis, 601
Phototropism, 541
Phyla, 464
  bilateral symmetry in, 568-578
Phylogenetic tree, 440, 476
Phylum, 13
Phytochrome, 541
Pigment, 133
Pikaia gracilens, 464
Pili, 47
Pill, birth control, 350

Pilus, 494
Pine
  life cycle of, 533
  ponderosa, 32
Pineal gland, 328
Pinna, 289
Pinocytosis, 76, 77
Pipevine swallowtail, 648
Pith, 549
Pituitary gland, 318-322
Pivot joint, 304
Placenta, 360
Placental-fetal circulation, 368
Placental mammal, 471, 593, 594
Plant, 12, 13
  bark of, 552, 553
  competition of, 641
  cuticle of, 547, 552
  dermal tissue of, 546, 547-548
  divisions of, 527
  dodder, 651
  gibberellin in, 541
  growth regulation in, 548
  guard cell of, 554, 555
  herbivore relationship to, 644-655
  hormone in, 541
  hornwort, 530
  life cycle of, 85, 527
  liverwort, 530
  moss, 530
  nonvascular, 527
  reproduction of; *see* Plant reproduction and diversity
  root of, 548-551
  secondary growth of, 548, 553
  seed, 538, 539
    naked, 533
    protected, 534-535
  shoot of, 551-555
  stem of, 551
  structure of, 545
  transpiration in, 548
  vegetative propagation in, 540, 541
  water movement in, 555-557
Plant cell, 45, 48
Plant defense, 645
Plant reproduction and diversity, 544-559
  patterns of, 526-543, 557
  vascular, 545-557
    patterns of, 531-540, 541
    reproduction in, 531-540, 541
Plantae, 12, 13
Plaque, 213
Plasma, blood, 157
Plasma cell, 224
  immune system and, 157
Plasma membrane, 68, 76, 77
  eukaryotic cell, 47, 48, 49-50, 60, 61
  protein, in, 71
Plasmid, 63, 494, 495
Plasmodial slime mold, 515
Plasmodium, 508, 509, 515
Plasmolysis, 74
Plate tectonic theory, 458, 459
Plateau phase, 346
Platelet, 157, 209, 211
Platyhelminthes, 568
Pleura membrane, 191
PMS; *see* Premenstrual tension syndrome
Point mutation, 394
Polar body, 340
Polar molecule, 23

INDEX I-9

Polarized nerve impulse, 254
Pollen grains, 534
Pollination, 534, 535
   mechanisms of, 535
   modes of, 534
Pollinator, 535
Pollution
   air, 707
   water, 703-705
Polydactyly, 398
Polymer, 28, 447
Polyp, 567
Polypeptide, 34, 36
   synthesis of, 420
Polysaccharides, 30
Polyunsaturated fat, 33
Pond food chain, 659
Ponderosa pine, 32
Pons, 191, 265, 271
Population, 8, 431, 625-636
Population ecologist, 625
Population ecology, 624-637
Population mortality, 630-631
Population pyramid, 631
Population size, 627, 628-630
Pore, taste, 285
Porifera, 565
Positive control, 422
Positive feedback, 163
Positive feedback loop, 322
Positive phototaxis, 506
Postcopulatory behavior in goose, 617
Posterior pituitary gland, 321-322, 329
Postsynaptic membrane, 257
Potassium, 27, 168
Potassium pump, sodium, 253, 254
Potential energy, 103
Power
   nuclear, 696
   solar, 696, 697
   tidal, 696, 697
   water, 697
   wave, 696
   wind, 696, 697
Prairie chicken, courtship dance, 616
Pre-embryo, 356-359, 360
Precambrian Period, 453
Precapillary sphincter, 202
Predation, 629, 640, 642-648
Predator, 642
Preformationist, 373
Pregnancy, 355
   ectopic, 357
Premenstrual tension syndrome, 343
Premolars, 171
Prenatal development, 354-371
Preparation reaction, 123
Pressure
   air, 555
   root, 556
   turgor, 74
Pressure gradient, 192
Presynaptic membrane, 257
Prey, 642
Prickly pear cactus, 627
Primary amino acid, 35
Primary bronchi, 188
Primary consumer, 658
Primary germ layer, 359
Primary oocyte, 339
Primary plant growth, 548
Primary structure of protein, 35
Primary succession, 651

Primate
   evolution of, 471-475, 476
   family tree of, 476
Primitive streak, 361
Probability, 381
Probability theory and dihybrid cross, 381-382
Procambarus, 575
Producer, 10, 657
Product, changed substrate, 101
Prokaryote, 12, 63
   cell, 47, 61-63, 64
   and eukaryote gene, 419, 420
   gene expression, 420-422
Prolactin, 321
Promoter, 420
Promoter site, 413
Prophase, 86-89, 90
Proprioceptor, 282
Prosimian, 473
Prostaglandins, 330, 368
Prostate gland, 244, 338
Proteases, 168
Protection
   epithelial tissue and, 149
   support and movement and, 294-313
      bones in, 296-304
      muscles in, 304, 305-311
      skeletal system in, 298-304
      skin in, 295-296
Protective coloration, 647-648, 649
Protein, 50, 53, 55, 166
   chain, 71
   fibrous, 38
   globular, 38
   molecule of, 29
   of plasma membrane, 71
   structure of, 37, 38
   subunits of, 37, 38
   synthesis of, 418
Protein fiber, 38
Proterozoic Era, 452
Prothrombin activator, 211
Protist, 502-515
   fungus and; *see* Fungus
   as kingdom, 12, 13
   phagocytosis and, 76
Protoavis, 456, 457
Protocell, 448
Proton, 17
Protoplasm, 47
Protostome, 564
Protozoan, 503, 504-509
Proximal convoluted tube, 237, 238, 240, 241
Pseudocoelomate, 562
Pseudomonas, 62
Pseudopods, 504
Pseudostratified epithelium, 151
PTH; *see* Parathyroid hormone
Puberty, 336
Puffer, 467
Pulmonary artery, 204, 206
Pulmonary circulation, 204, 205, 206
Pulmonary disease
   chronic obstructive, 196
Pulmonary semilunar valve, 204, 206
Pulmonary vein, 204
Pump, sodium-potassium, 75
Punnett square, 378, 379, 382, 396
   and dihybrid cross, 382-383
Pupil, 288
Purine, 407

Purkinje fiber, 207
Pyloric sphincter, 174, 175
Pyramid of biomass, 662
Pyramid of energy, 662
Pyramid of number, 662
Pyramid of population, 631
Pyramids for aquatic ecosystem, 662
Pyridoxine, 167
Pyrimidine, 407
Pyruvate, 117, 123
Python, vestigial organ of, 438

Q

Qualitative or quantitative data, 5
Quaternary Period, 458
Quaternary structure of protein, 37, 38
Queen Victoria, 396, 397

R

R group, 35
Radial symmetry, 562, 566-568
Radiation, 132
   adaptive, 433, 455
   ionizing, 392
   ultraviolet, 394
Radicle in seed germination, 538
Radioactive isotope, 436
Radiolarian shell, 504, 505
Rain shadow effect, 674
Rank order in group behavior, 621
Ranvier, node of, 252
Rate of growth, 625
Ray in ocean, 466
Ray-finned fish, 586
Reaction
   allergic, 229
   anabolic, 101
   catabolic, 101
   chemical, 101
   coupled, 102, 103, 115
   endergonic, 102
   exergonic, 101, 102, 106
   hydrolysis, 30
   light-dependent, 135, 136-140
   light-independent, 135, 141-142
   preparation for, 123
   redox, 20, 119
Realized niche, 642
Rearrangement, chromosomal, 394
Reasoning, 3, 5
   insight and, 606, 607
Receptor, 50
   in human skin, 283
   olfactory, 284
   in peripheral nervous system, 273
   stretch, 191, 244, 283
Recessive genetic disorder, 398, 399
Recessive trait, 375
Recognition sequence, 496-497
Recombinant deoxyribonucleic acid technology, 498-499
Recombination, meiotic, 97
Rectum, 179, 180
Red algae, 509, 513
Red-and-black African grasshopper, 647
Red blood cell, 157, 202, 210
Red boring sponge, 565
Red-eyed *Drosophila melanogaster*, 383
Red-green color blindness, 395
Red marrow, 297
Red tide, 509, 510

Redox reaction, 20, 119
Reduced, defined, 20, 119
Reduction, 119
   electron gain and, 20
Reduction-division, 92
Redwood community, 639
Reflex, 264, 602
   knee-jerk, 276
   in somatic nervous system, 276
Reflex arc, 271, 276
Refractory period, 255
Regulation, endocrine hormone, 315
Regulatory site, 420
Relation, 12
Releasing hormone, 318
Renal artery, 235, 236
Renal failure, 246, 247
Renal pelvis, 235, 236, 244
Renal pyramid, 235
Renal vein, 236
Repeated trial, 7
Replication
   of deoxyribonucleic acid, 411
   viral, 489-490
Replication fork, 411
Reproduction
   asexual, 85
   of endocrine gland, 315
   sexual, 335
Reproductive behavior, 615-616, 617
Reproductive cycle in human, 341-343
Reproductive system, 148
   female, 339-346
   as major organ system, 148
   male, 335-339
Reptile, 588-591
   evolution and, 466-468
Reptillia, 589
Reservoir, 664
Respiration, 184-197
Respiratory assembly, 125
Respiratory chain, 125
Respiratory cilia, 218
Respiratory diaphragm, 190
Respiratory system, 148
Respiratory tract, 218
Responding variable, 3
Response
   all-or-nothing, 255
   endocrine gland and, 315
   involuntary or voluntary, 264
Resting potential, 253-254
Reticular connective tissue, 153, 154
Reticular formation, 271
Reticulin fiber, 153
Reticulum, 53
   sarcoplasmic, 310
Retina, 286, 288
Retinol, 167
Rhizoid, 557
Rhizome, 540
Rhodopsin, 286
Rhythm method of birth control, 347
Rib, 300
Riboflavin, 167
Ribonucleic acid, 56, 57, 406-408
   messenger, 413
   molecules of, 29
   ribosomal, 413
   structure of, 39
   transfer, 413, 417
Ribonucleic acid polymerase, 413
Ribose, 110

Ribosomal ribonucleic acid, 57, 58, 413
Ribosome, 47, 48, 49, 52, 53, 63, 413
Ribulose bisphosphate, 142
Rickets, 167
Right atrium, 204, 207
RNA; see Ribonucleic acid
Rockweed, 511
Rod cell, 286
Root, 545, 548-551
Root canal, 171
Root pressure, 556
Rotifer, 562
Rough endoplasmic reticulum, 48, 53
Round window, 289
Roundworm, 569-570
rRNA; see Ribosomal ribonucleic acid
RuBP; see Ribulose bisphosphate
Runner, 540
Runner's high, 330

S

SA node; see Sinoatrial node
Sac fungi, 519-520
Saccule, 291
Salinity, 685
Saliva, 170
Salivary gland, 169, 170
Salt
  blood plasma and, 208
  dissolution of, 25
  table, 20, 25
Salt marsh food web, 660
Saltatory conduction, 256-257
Sand dollar, 577
Sanitary landfill, 702
Sap, 557
Saprophytic, defined, 516
Sarcomere, 307
Sarcoplasmic reticulum, 310
Saturated fat, 33, 34
Savanna, 678, 679
Schleiden, M., 45
Schwann, T., 45
Schwann cell, 160, 256
  neuron and, 252
Scientific law, 434
Scientific method, 3
Sclera, 287
Sclerenchymal cell, 547
Scrotum, 335
Scrubber in coal-burning power plant, 706
Scurvy, 167
Sea anemone, 650
Sea cucumber, 577
Sea horse, 467
Sea lilies, 577
Sea otter, 599
Sea perch, 582
Sea squirt, 581
Sea star, 577
Sea urchin, 577-578
Seaweed, 511
Second filial generation, 375
Second law
  of Mendel, 378
  of thermodynamics, 104
Second messenger, 315
Second trimester of pregnancy, 367
Secondary amino acid, 37, 38
Secondary bronchi, 188
Secondary consumer, 658

Secondary oocyte, 340
Secondary phloem, 552
Secondary plant growth, 548, 553
Secondary pollutant, 707
Secondary protein structure, 37, 38
Secondary succession, 653
Secondary xylem, 552
Secretin, 175, 177
Secretion in epithelial tissue, 149
Secretory vesicle, 55
Sediment, 703
Sedimentary rock, 434-435
Seed
  dispersion of, 536
  formation of, 538, 539
  germination, 538, 539
  of vascular plants, 531
Seedless vascular plant, 527, 531-532
Segregation, law of, 378
Selective reabsorption in kidney, 235, 237-241
Selectively permeable membrane, 73
Selenium, 27
Self-fertilization, 374
Self-pollination, 535
Semen, 338
Semicircular canal, 291
Seminal vesicles, 338
Seminiferous tubule, 336
Semiterrestrial, defined, 512
Sensation in epithelial tissue, 149
Senses, 280-293
Sensory area of brain, 266, 267
Sensory neuron, 250-251, 264
Sensory pathway, 273-275
Septum, 207, 573
  of reptile, 588
Serosa, 174
Serum albumin, 208
Sessile, 562
Seta of annelid, 574
Sex chromosome, 81, 383-384, 389
Sex flush, 346
Sex linkage, 383-384
Sex and reproduction, 334-353
  contraception and birth control in; see Birth control
Sexual life cycle, 528
Sexual reproduction, 83, 335
Sexual response, 346-347
Shark, 466
Shell, 18
Shell hydration, 24
Shinbone, 300
Shiny bristleworm, 573
Shockwave lithotripsy, 246
Shoot of plant, 545, 551-555
Shore zone, 684, 687-688
Shoulder, 300
Shrew, 468
Sickle-cell anemia, 398
Sickle-shaped red blood cell, 398
Sight, 286-288
Sigmoid growth curve, 627
Silicon, 27
Silk, 38
Silurian Period, 454
Simple epithelium, 151
Simple eye, 576
Simple sugar, 30
Single bond, 21
Single-celled protist, 503
Sinoatrial node, 206
Sinus venosus in bony fish, 586

Sister chromatid, 81, 94
  anaphase and, 90
Skeletal muscle, 159
  bone attachment of, 305
  contraction of, 305-308, 309
  structure of, 306
  voluntary function and, 158
Skeletal system, 298-304
  function and components of, 148
Skeleton
  appendicular, 298, 299, 300-302
  axial, 299-300
  joint in, 302-304
Skin, 295-296, 296
Skinner, B.F., 605
Skinner box, 605
Skull, 265, 298, 299
Skunk, 647
Slime mold, 504
Small intestine, 175-179
  digestive enzymes and, 169
Smallpox, 218
  immunity to, 220
Smell, 284-285
Smelt, 661
Smog, 707
Smooth endoplasmic reticulum, 48, 53
Smooth muscle, 158-159
Snail, 570-573
Social behavior in animal; see Behavior, social
Society
  ant, 619
  insect, 619-621
Sodium, 19-20, 168
Sodium bicarbonate, 177
Sodium chloride, 19-20
  water and, 24, 25
Sodium hydroxide, 26
Sodium ion, 19-20
Sodium-potassium pump, 75, 253, 254
Soft palate, 172
Softwood, 552
Solar power, 696, 697
Solid waste, 702-703
Soluble water, 24
Solutes, 73
Solution, 73
Solvent, 73
  water as, 24, 25
Somatic cell gene therapy, 423
Somatic nervous system, 264, 275-276
Somite, 364, 581
Soybean root nodule, 493
Spadefoot toad, 681
Spanish moss, 530
Spawn, 584
Speciation, 433
Species, 13
  defined, 427
Species extinction, 700-701
Specific facilitated diffusion, 75
Spectrum
  absorption, 133
  electromagnetic, 132
Sperm, 45, 335, 336, 337
  blockage or destruction of, in birth control, 350
Spermatids, 336
Spermatogenesis, 336, 337
Spermatogonia, 336
Spermatozoon, 335
Spermicide, 347, 350
Sphagnum moss, 530

Sphincter
  lower esophageal, 172
  pyloric, 174, 175
Spider, 574-576
Spinal cord, 149, 264, 271-272
Spinal nerve, 271
Spindle fiber, 86
Spiracle, 576
Spirillum, 62
Spirogyra, 512
Spleen, 211, 220
Sponge, 565-566
  birth control and, 347
Spongy bone, 156, 297
Spongy layer, 554
Spontaneous abortion, 391
Spontaneous generation, 46
Spore, 514, 517
Sporophyte, 84, 85, 529
Sporozoa, 509
Spotted skunk, 647
Squamous cell, 147, 150
Squirrel, 13
Stable molecule, 21
Stable population, 631
Stapes of ear, 289
Staphylococcus, 62
Starch, 29, 30
Starfish, 577-578
Starvation, 635
Stem, 545, 551-552, 553
Stereoscopic vision, 472
Sternum, 300
Steroids, 315, 317
  lipid membrane and, 34
  molecules of, 29
Stickleback courtship, 616
Stigma, 534
Stipe, 511
Stirrup of ear, 289
Stolon, 540
Stoma, 554
Stomach
  digestion in, 174-175
  digestive enzymes and, 169
  epithelium of, 174
Stone Age, 480
Stratification
  of epithelium, 151
  of shale, 453
Stratosphere, 708
Streptococcus, 62
Stress, 326, 327
Stretch receptor, 191, 244, 283
Striated muscle, 159
Stroke, 213
Stromatolite, 450, 451
Style, 534
Subarachnoid space, 272
Submissive behavior, 611
Submucosa, 174
Substrate, 101
  enzymes and, 108
Subunits, 39
Succession, 651-653
Sucrose, 30, 31
Sugar
  five-carbon, 39
  simple, 30
Sugar chain, 30
Sulfur, 27, 168
Sunlight, 132
  climate and, 672, 673
Supercoil, 81, 82

Superior vena cava, 204
Suppressor T cell, 222, 223
Surface tension, 24
Surface transport, 149-150
Surface-to-volume ratio, cell, 46
Surface water pollution, 703-705
Survivorship, population, 630-631
Sustainable society, 696
Suture, bone, 302
Swallowing, 172
Swim bladder, 585
Symbiosis, 49, 640, 649-651
Symmetry, 562, 568-578
Sympathetic nervous system, 265, 276
Synapse, 257, 259
Synapsis, 93
Synaptic cleft, 257, 258
Synaptic transmission, 257, 258
Synovial fluid, 302
Synovial joint, 302
Systemic capillary, 205
Systemic circulation, 205
Systolic period, 207

T

T_3; see Triiodothyronine
T_4; see Thyroxine
T cell, 209, 211, 328
  cytotoxic or suppressor, 222, 223
T tubule, 310
Table salt, 20
Taiga, 682-683
Taproot, 550
Target cell, 315
Tarsal, defined, 302
Taste, 285
Taste bud, 285
Taste pore, 285
Taxis, 600-601, 602
Taxonomy, 12
Tay-Sachs disease, 398
Technicolor fish, 612
Technology as environmental problem, 708
Tectonic plate, 458, 459
Tectorial membrane, 289-290
Teeth, 170-171
Telophase, 87, 88, 90, 91
Temperate deciduous forest, 682
Temperate grassland, 681
Template, 447
Temporal lobe, 266
Tendon, 305
Teratogen, 361
Termite mound, 679
Territorial behavior, 612-613
Tertiary amino acid, 37, 38
Tertiary Period, 458
Testcross, 380
Testis, 328, 329, 335-337
Tetrapod, 466, 584
Thalamus, 265, 269, 270
Theory, 7
  endosymbiotic, 452
  of evolution, 427-434
  plate tectonic, 458, 459
Thermal pollution, 704, 705
Thermodynamics, 103, 104
Third trimester of pregnancy, 368
Thoracic cavity, 149, 190
Thoroughfare channel, 202
Threat display, 611
Threshold potential, 254

Thromboplastin, 211
Thrombus, 213
Thumb, opposable, 473
Thylakoid, 60, 135
Thymine, 39
Thymus gland, 211, 220, 328
Thyroid gland, 322, 323, 329
Thyroid hormone, 322
Thyroid-stimulating hormone, 320
Thyroxine, 322
Tidal power, 696, 697
Tidal volume, 192
Tin, 27
Tissue, 8, 546, 547-548
  in human body, 149-160
Tissue differentiation, 364
Toad, 681
Tobacco mosaic virus, 487
Tocopherol, 167
Tongue, 172
Toxic substance, 704
Toxin, 494
Trachea, 172, 186, 187
  of insect, 576
Tracheid, 546
Tracheole, 576
Trait, 373-375
Transcription, 413, 415
Transduction, 494
Transfer ribonucleic acid, 413, 417
Transformation, 494
Transitional epithelium, 151
Transitional form, 468
Translation, 416, 417
Translocation, 393
Transmission, 250, 255-259
Transpiration in plants, 548, 555, 556
Transport, 72, 75
  surface, 149-150
Transport chain, electron, 117, 125-127, 139
Transverse fission, 507
Transverse tubule, 310
Tree finch, 433
Trial-and-error learning, 605-606
Triassic Period, 454
Tricuspid valve, 204, 206
Triglyceride, 33
Triiodothyronine, 322
Trilobite fossil, 454
Triphosphate group, 110
Triple bond, 21
Triple female, 391
Trophic level, 659
Trophoblast, 357
Tropic hormone, 320
Tropical rain forest, 676, 677
  deforestation of, 699, 700
Tropomyosin, 310
Troponin, 310
True-breeding, 374, 375, 377
True rib, 300
Trypsin, 169, 177
TSH; see Thyroid-stimulating hormone
Tubal ligation, 351
Tubifera, 515
Tubular secretion, 235, 241
Tubule
  convoluted, 237-241
  malpighian, 576
  seminiferous, 336
  transverse, 310
Tubulin, 51
Tundra, 684

Tunicate, 581-582
Turbellaria, 568
Turgor pressure, 74
Turner syndrome, 392
Tympanic canal, 289

U

Ulcer, duodenal, 175
Ultrasound in genetic counseling, 401
Ultraviolet radiation, 394
Umbilical artery, 360
Umbilical cord, 360
Unicorn caterpillar, 649
Unifying themes of biology today, 7-12
Unity of living things, 7, 8
Unsaturated fat, 33, 34
Upper digestive system, 174
Urea, 234
  in cartilaginous fish, 585
Ureter, 235, 236, 244
Urethra, 244, 338
Urey, H.C., 445, 446
Uric acid, 234
Urinary bladder, 244
Urinary system, 148, 244
Urine, 234, 235, 238-241, 243
Uterine tube, 340-341
Uterus, 339, 341, 357
Utricle, 291

V

Vaccinia virus, 218
Vacuole, 48, 56
Vagina, 339, 343
Vanadium, 27
Variable
  dependent, 5
  independent, 3
  manipulated, 3
  responding, 5
Variation in gene interaction, 373-374
Varieties, 374
Vas deferens, 335, 336, 338
Vascular, defined, 199
Vascular cambium, 552
Vascular cylinder, 549
Vascular plant, 545
  life cycle of, 531
  naked seeds in, 533
  organs of, 548-555
  patterns of reproduction in, 531-540
  protected seed in, 534-535
  seedless, 527, 531-532
  tissue of, 545-548
  transpiration in, 548
Vascular tissue, 546-547
Vasectomy, 351
Vasoconstriction, 200
Vegetative propagation in plant, 540, 541
Vein, 201
Ventral side, 562
Ventricle
  in bony fish, 586
  of brain, 272
Venule, 201, 202, 203
Vertebrae, 300
Vertebral column, 149, 300, 465
Vertebrate, 580-596
  digestive tract of, 174
  evolution of, 464-471
  neurons of, 159

Vertebrate—cont'd
  society behavior of, 621
Vesicle, 55
Vestibular canal, 289
Vestibule in inner ear, 290
Vestigial organ, 438
Vibrio cholerae, 64
Viceroy butterfly, 648
Villi
  of chorion, 364
  of small intestine, 178
Vimentin, 51
Viral core, 487
Viral envelope, 487
Virchow, R., 45
Virus, 486-490
  tobacco mosaic, 487
Visceral mass, 571
Vitamins, 166-169, 180
Vitreous humor, 287
Viviparous, defined, 586
Vocal cord, 186
Voluntary muscle, 159
Voluntary response, 264
Volvox, 512
Vulva, 343

W

Waggle dance, 620
Wall, P., 281
Wallace, A. R., 434
Warbler finch, 433
Warm blooded, 471
Warning coloration, 645, 647
Waste, nitrogenous, 234
Water, 22-26, 166
Water balance and kidney, 234
Water cycle, 664-665
Water mold, 504, 515
Water power, 697
Watson-Crick model of deoxyribonucleic acid, 409
Wave power, 696
Whale, 650
White blood cell, 157, 210
White-eyed *Drosophila melanogaster*, 383
White matter, 271
Wind power, 696, 697
Window, oval or round, 289
Wood, 32
World population distribution, 635

X

X Chromosome, 384
Xylem, 546
XYY males, 392

Y

Y Chromosome, 384
Yellow-green algae, 510
Yellow marrow, 297
Yolk sac, 361

Z

Zebra, 645
Zinc, 27, 169
Zona pellucida, 355
Zygospore, 518
Zygote, 340, 356
Zygote-forming fungi, 517, 518